Nanotechnology in the Life Sciences

Series Editor

Ram Prasad, Department of Botany
Mahatma Gandhi Central University
Motihari, India

Nano and biotechnology are two of the 21st century's most promising technologies. Nanotechnology is demarcated as the design, development, and application of materials and devices whose least functional make up is on a nanometer scale (1 to 100 nm). Meanwhile, biotechnology deals with metabolic and other physiological developments of biological subjects including microorganisms. These microbial processes have opened up new opportunities to explore novel applications, for example, the biosynthesis of metal nanomaterials, with the implication that these two technologies (i.e., thus nanobiotechnology) can play a vital role in developing and executing many valuable tools in the study of life. Nanotechnology is very diverse, ranging from extensions of conventional device physics to completely new approaches based upon molecular self-assembly, from developing new materials with dimensions on the nanoscale, to investigating whether we can directly control matters on/in the atomic scale level. This idea entails its application to diverse fields of science such as plant biology, organic chemistry, agriculture, the food industry, and more.

Nanobiotechnology offers a wide range of uses in medicine, agriculture, and the environment. Many diseases that do not have cures today may be cured by nanotechnology in the future. Use of nanotechnology in medical therapeutics needs adequate evaluation of its risk and safety factors. Scientists who are against the use of nanotechnology also agree that advancement in nanotechnology should continue because this field promises great benefits, but testing should be carried out to ensure its safety in people. It is possible that nanomedicine in the future will play a crucial role in the treatment of human and plant diseases, and also in the enhancement of normal human physiology and plant systems, respectively. If everything proceeds as expected, nanobiotechnology will, one day, become an inevitable part of our everyday life and will help save many lives.

Vijayalakshmi Ghosh
Abhishek Kumar Bhardwaj • Naga Raju Maddela
Editors

Applications of Plant Bioactive Compounds-Based Nanoformulations

Sustainable Nanomaterials for Drug Discovery

Springer

Editors
Vijayalakshmi Ghosh
Department of Biotechnology
Rungta International Skills University
Bhilai, Chhattisgarh, India

Naga Raju Maddela
Health Sciences and Research
Universidad Tecnica de Manabi
Portoviejo, Ecuador

Abhishek Kumar Bhardwaj
Department of Environmental Science
Amity School of Life Sciences
Amity University
Gwalior, MP, India

ISSN 2523-8027 ISSN 2523-8035 (electronic)
Nanotechnology in the Life Sciences
ISBN 978-3-032-13085-3 ISBN 978-3-032-13086-0 (eBook)
https://doi.org/10.1007/978-3-032-13086-0

This Springer imprint is published by the registered company Springer Nature Switzerland AG
The registered company address is: Gewerbestrasse 11, 6330 Cham, Switzerland

Contents

Chapter 1
Outlook of the Potential of Medicinal Plant and Scope of Nanoformulation

Swati Rajput, Teeka Ram Sahu, Abhishek Kumar Bhardwaj, and Vijayalakshmi Ghosh

1.1 Introduction

An essential component of human existence on Earth is nature. Human survival is entirely dependent on the natural world. One of the vast components of nature is plants. Phytochemicals obtained from these plants have been used to enhance human health since the distant past. They have been adopted extensively to address multiple ailments. Since they form an integral part of the environment, the demand for medicinal plants is increasing every day. They also improve the general health of the ecosystem. Plants have historically served a variety of purposes, ranging from providing basic necessities to playing a central part in traditional medicine for treating a variety of maladies, laying the groundwork for understanding their therapeutic value (Rabizadeh et al. 2022). The Ayurveda gives reference to the significance of these therapeutic qualities. The Indian medical system has been using therapeutic medicinal plants for the past 5000 years. It includes herbal remedies and diets to prevent and treat mental and physical health issues. Humanity looks to herbal remedies to treat and avoid illnesses. Consumer preference for herbal remedies is frequently motivated by a belief that they are safer than synthetic alternatives, as well as purported benefits in treating various disorders and increasing long-term well-being (Salem et al. 2021). Currently, half of the natural compounds used in pharmaceuticals have some sort of plant origin. Pharmaceutical medications, which may

S. Rajput (✉) · T. R. Sahu · V. Ghosh
Department of Biotechnology, Rungta International Skills University, Bhilai, CG, India

Department of Biotechnology, GD Rungta College of Science and Technology, Bhilai, CG, India

A. K. Bhardwaj
Department of Environmental Science, Amity School of Life Sciences, Amity University, Gwalior, MP, India

V. Ghosh et al. (eds.), *Applications of Plant Bioactive Compounds-Based Nanoformulations*, Nanotechnology in the Life Sciences,
https://doi.org/10.1007/978-3-032-13086-0_1

require complex chemical processes and generate waste, have a greater environmental impact than herbal remedies, which are based on sustainable resources. The internet and social media platforms have considerably helped the broad adoption of herbal medicine by providing simple access to knowledge about various herbs and their applications, as well as personal experiences shared by users and influencers (Sidhu et al. 2022).

Phytochemicals such as carotenoids, tocopherols, ascorbic acids, polyphenols as well as lipoic acids are powerful Herbal antioxidants that bind to free radicals. Endogenous antioxidant enzymes, including, glutathione peroxidase, glutathione reductase, catalase, superoxide dismutase (SOD) as well as minerals including manganese, zinc, copper, selenium and, vitamins A, E, and C, polyphenols, limonoids and carotenoid act synergistically to scavenge free radicals. Free radical scavengers is capable of donating electrons, hydrogen, inhibit enzymes, breakdown peroxide, quench singlet oxygen synergize, and bind metal ions. Both enzymatic and non-enzymatic antioxidants are found inside and outside of cells to detoxify ROS. Antioxidants are essential for shielding the body from harm caused by and reactive oxygen species (ROS) and the free radicals.

They function in four ways: the first prevents free radical formation by neutralizing harmful agents; the second scavenges existing radicals with compounds such as glutathione, vitamins, and flavonoids; the third repairs DNA, proteins, and lipids; and the fourth supports the immune system when other defenses are overwhelmed. Phytochemicals, such as polyphenols, flavonoids and carotenoids are noted for their potent antioxidant activity and associated health advantages, making them essential components in dietary supplements and preventative medicine (Šamec et al. 2021) (Fig. 1.1).

Nanotechnology is already boosting world health and is anticipated to have a substantial impact on the pharmaceutical business as it moves forward into clinical trials. Various research bodies from academia and industry overestimate the

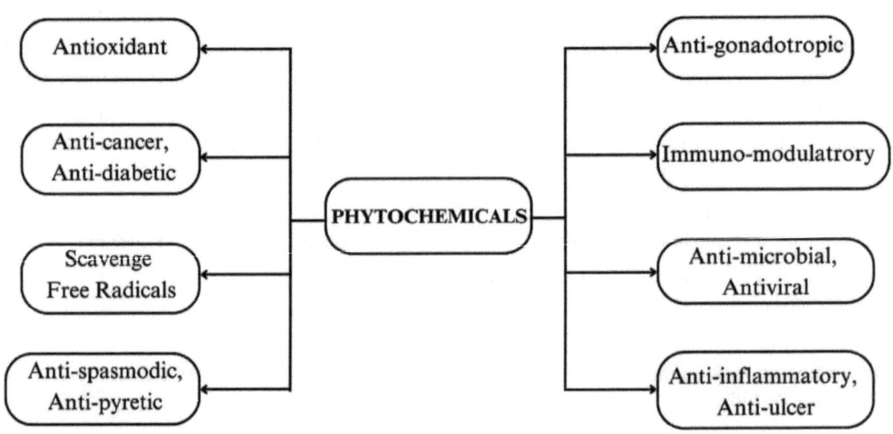

Fig. 1.1 Health benefits of phytochemicals

promising outcomes of proposed technologies, claiming that they are almost ready for market usage despite considerable impediments to acceptance. Nanoparticles do have a long history in medicine, with early treatments such as colloidal silver for eye infections and iron dextran, one of the first nanomedicines (Barhoum et al. 2022). For their possible uses in nanomedicine, nanomaterials are widely used and the subject of much research. On account of their composition, nanomaterials are often divided into three forms: Solid drug nanoparticles (SDNs), which are incredibly tiny drug particles, are the first type. The most varied category of nanomedicine alternatives is nanocarriers, which are nanoscale devices with the API either encased within the carrier's core or coupled to the surface. The second is polymer therapeutics, which are nanoparticles created via the conjugation of an active pharmaceutical ingredient to a hydrophilic polymer, resulting in a polymer based drug or polymer based protein conjugate, commonly referred to as a prodrug (McDonald et al. 2015).

This chapter categorizes and defines various nanomaterial approaches, such as polymer therapies and nanocarriers, in offer a structured insight into the technological environment. The incorporation of medicinal plants into nano formulations is also thought to be a potential field where the bioavailability, stability, and efficacy of plant-derived chemicals could be considerably improved. Such formulations address problems such as poor solubility and quick degradation, which frequently restrict phytochemicals'clinical value. It critically examines possible dangers and moral dilemmas related to the application of nanomedicine, including toxicity, environmental effects, and regulatory conundrums, in order to balance the use.

1.2 Bioactive Compounds in Medicinal Plants

Combinations of the bioactive substances in plants give them their therapeutic qualities, known as phytochemicals. Primary and secondary metabolites are the two categories into which they exist. Whereas the main components of secondary metabolites are alkaloids, flavonoids, and terpenoids, primary metabolites include proteins, sugar, and chlorophyll. Plants produce a class of nitrogenous compounds known as alkaloids in reaction to their biotic or abiotic environment. Alkaloids' unique properties as bioactive compounds are attributed to their variety of structures and biological activity (Mickymaray 2019). Phenolic components are produced by plants using the shikimate, pentose phosphate, and phenylpropanoid pathways. These components feature aromatic rings and hydroxyl groups. Terpenoids are a class of constituents consisting of hydrocarbons and terpenes. This is the most prevalent category of secondary metabolites in plants, and they are generally located in the flowers, vegetative tissues and roots. Their wide spectrum of pharmacological characteristics decreases blood pressure and triglycerides, LDL cholesterol, and total cholesterol (Rabizadeh et al. 2022).

1.2.1 Alkaloids

Numerous creatures, including bacteria, fungi, plants, and mammals, naturally create alkaloids. It has basic nitrogen atoms in it. Alkaloids are separated into the groups listed below. (1) A real alkaloid Atropine, morphine, and nicotine are typical examples. Adrenaline, ephedrine, and mescaline are protoalkaloids. Peptide and cyclopeptide alkaloids, polyamine alkaloids, and pseudoalkaloids are the next three categories (Godswill Awuchi 2020).

1.2.1.1 Structure of Alkaloids

At least one alkaloid normally contains a nitrogen atom in an amine functional group structure, which originates from ammonia by swapping hydrogen atoms by hydrogen-carbon groups. In acid-base interaction, this or an alternative nitrogen atom can operate as a base (Uzor 2020). Figure 1.2 depicts the various alkaloid compounds.

1.2.1.2 Attributes of Alkaloids

Many alkaloids contain oxygen in their structure and frequently appear as colorless crystals under various conditions, whereas oxygen-free alkaloids like nicotine and coniine are characteristically oily, colorless, and volatile liquids; the majority of

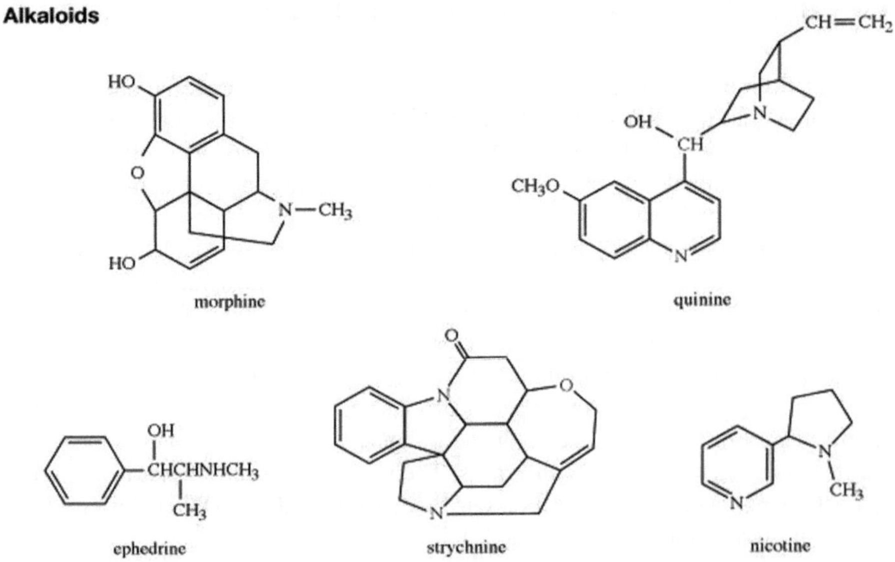

Fig. 1.2 Different types of alkaloids

alkaloids are weak bases, with some exhibiting amphoteric properties, such as theobromine and theophylline (Kumar et al. 2023). Many dissolve easily in organic solvents like diethyl ether, 1,2-dichloroethane, and chloroform, but poorly in water. Salts of various strengths are created by acids and alkaloids. The salts often have an easy solution in ethanol and water but dissolve poorly in a good variety of organic solvents. For example, the salted water-dissolved quinine sulfate and also the salted ethanol-dissolved scopolamine hydrobromide in organic solvents. It is not known what the alkaloids do for the living things that produce them. While early hypotheses suggested that alkaloids were the principal nitrogenous waste products in plants, it is now clear that their content varies dynamically. Furthermore, the existence of alkaloids in ant venom suggests that biosynthetic origins extend beyond plant metabolism; however, particular pathways in insects require further investigation (Ninkuu et al. 2021). The majority of alkaloids' known actions are associated with protection. For instance, the tulip tree's aporphine alkaloid, liriodenine, shields the tree from parasitic fungi. Furthermore, the plant's alkaloids keep insects and chordate mammals from consuming it. Alkaloids are utilized by plants to defend themselves against chordate animals, insects, and other dangers. Nonetheless, certain species may utilize alkaloids in their own metabolism since they are adapted to them. In animals, some alkaloid-related chemicals, including dopamine, histamine, and serotonin, are crucial neurotransmitters. It has also been claimed that many alkaloids control plant growth (Uzor 2020).

1.2.1.3 Application of Alkaloids

Use as a Psychoactive Drug Alkaloids and extracts, as well as pure alkaloids, have been employed as hallucinogenic substances in plant remedies for many decades. Cocaine, cathinone, and caffeine are all stimulants for the central neural system. Mescaline and other indole alkaloids, including ibogaine, psilocybin, and dimethyltryptamine, have hallucinatory effects. Opioid painkillers are morphine and codeine. Certain alkaloids, such as pseudoephedrine and ephedrine in the creation of methamphetamine, are critical precursors in the synthesis of semi-synthetic psychoactive drugs, while others, such as thebaine, are starting materials for medications like oxycodone (Uzor 2020).

1.2.2 Terpenes

Terpenes are hydrocarbons that are found in various plants such as orange peel, lavender, pine, and cannabis and in some insects. They contain aromatic compounds that give them a very strong smell and help them attract a great deal of predators. Terpenoids differ from terpenes in that they contain oxygen and supplementary functional groups. Terpenoids and terpenes are the primary components of many flowers and plants' essential oils. Aromatherapy and traditional medicine both make extensive use of essential oils as fragrances. One type of terpenoid is vitamin A (Boncan et al. 2020).

1.2.2.1 Structure of Terpenes

Terpene biosynthesis includes isopentenyl pyrophosphate units. The numerous categories of terpenes are given in Fig. 1.3. Terpenoid synthesis does not include isoprene, although the terpenoids' structures appear as rearrangements of it (2-methyl-1,3-butadiene). Terpenoid synthesis does not incorporate isoprene, even though the terpenoids' structures appear as rearrangements of it (2-methyl-1,3-butadiene). Terpenoids are synthesized via the 2-C-methyl-D-erythritol 4-phosphate/1-deoxy-D-xylulose 5-phosphate (MEP/DOXP) process and the mevalonic acid pathway (Ninkuu et al. 2021).

1.2.2.2 Types of Terpenes

A head and a tail make up the isoprene unit that characterizes terpene. Terpenoids form bonds through the head-tail connection of isoprene units, claims Ingold. Carotenoids and other tetraterpenes break this rule by establishing a tail–tail (4–4) connection in the middle (Abdallah and Quax 2017). Composed of five-carbon isoprene units that can be assembled in a variety of ways and frequently contain oxygen, terpenes are the biggest and most varied class of secondary metabolites. There are more than 80,000 identified terpenes and terpenoids in the wild, indicating their broad presence in plants (Zhou and Pichersky 2020). Each plant has distinct qualities due to the different concentrations and combinations of these chemicals. Notably, terpenes contribute to the spicy flavor, pleasant scent, and physiological benefits of plants. Plants with particularly obvious terpene concentrations include peppermint, lavender, lemons, and pine trees. Terpenoids are a modified class of terpenes that are distinguished by unique functional groups and the removal or relocation of oxidized methyl groups at various sites. Terpenes are basic hydrocarbons (Perveen 2021). Based on their carbon units, terpenoids are often classified as monoterpenes, sesquiterpenes, diterpenes, sesterpenes, and triterpenes. Different plants have different terpene profiles, and some terpene types are only found in particular species (Arzani et al. 2025). The isoprene rule classifies terpenes and terpenoids according to the quantity and configuration of carbon atoms in isoprene units. Monoterpenes (C10H16), sesquiterpenes (C15H24), diterpenes (C20H32), sesterterpenes (C25H40), triterpenes (C30H48), tetraterpenes (C40H64), hemiterpenes (C5H8), and polyterpenes (>8 isoprene units) are all included in this categorization. Hydrocarbon terpenes like camphene, R-(−)-limonene, terpinene, and (+)-α-pinene have been shown to have weaker antibacterial action than oxygenated terpenes (phenolic terpenoids) such carvacrol, ρ-cymene, and thymol (Guimarães et al. 2019). Many terpenes can be found in a single plant, but some terpenes are frequently present in large amounts in a variety of plant species. Artemisinin,

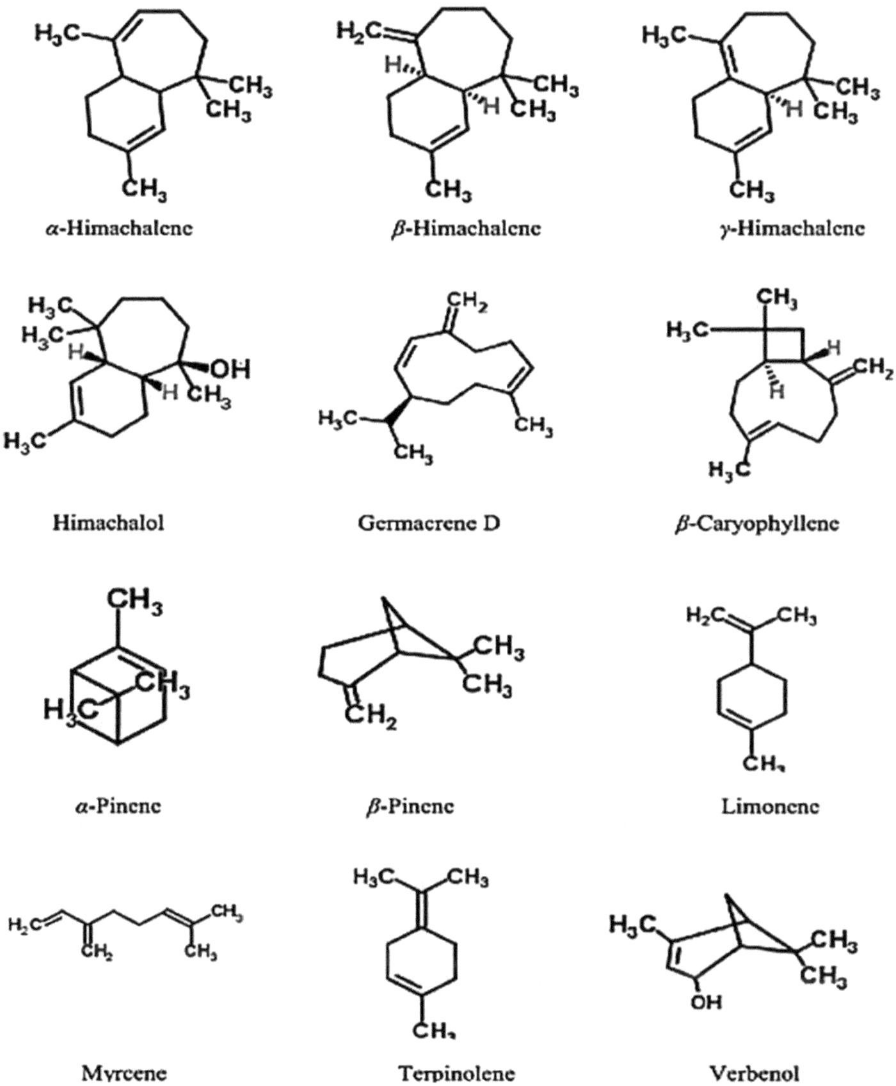

Fig. 1.3 Diverse types of terpenes

β-carotene, β-caryophyllene, ρ-cymene, farnesol, germacrene (germacrene a and d), humulene (α-caryophyllene), limonene, linalool, lupeol, lycopene, menthol, myrcene, ocimene, α-pinene, β-pinene, squalene, taxol, and terpinolene are some of the most significant terpenoid components present in plants (Arzani et al. 2025).

1.2.2.3 Applications of Terpenes

Terpenes have important biological qualities, such as anticonvulsant and analgesic effects. According to research, they can lessen inflammation through the reduction of release of pro-inflammatory cytokines. By blocking certain inflammatory processes, terpenoids can reduce the symptoms of inflammation. They may work better than existing treatments for inflammatory diseases because of their capacity to influence several cell signaling pathways. Because of their varied pharmacological actions, they are also being investigated for their potential to treat a number of different disorders (Del Prado-Audelo et al. 2021). Some isolated compounds have shown photoprotective and regenerative effects on skin surface, highlighting the versatile therapeutic capabilities of fungal terpene compounds. Certain terpenoids can also address skin pigment disorders and cancers related to melanogenesis dysfunction. Other terpenoids and terpenes are characterized by anti-inflammatory and antimicrobial properties that sometimes surpass current treatments, and research into their anticancer and antioxidant qualities suggests potential in preventing and treating skin cancer (Trepa et al. 2024).

1.2.3 Polyphenols

Polyphenols, a broad category of natural and synthetic chemicals distinguished by multiple phenol units, have an extensive range of biological activities, comprising anticancer, anti-inflammatory as well as anti-oxidant effects, which are impacted by structural differences. Examples include flavonoids and phenolic acids (Chen et al. 2023). Also, their bioactivities modify the digestion of macronutrients and regulate oxidative and inflammatory stress. In diverse foods, such as tea, fruits, and vegetables, they enhance their nutritional value and sensory appeal. Research on polyphenols' potential uses in food products is growing since their consumption is associated with a lower risk of conditions including diabetes and cancer (Bertelli et al. 2021).

1.2.3.1 Structure of Polyphenols

The existence of one or more hydroxyl groups and at least one phenyl ring, which are essential to polyphenols' antioxidant qualities, defines them. These substances include complicated polymers like condensed tannins and simple phenolic acids with a single aromatic ring. With two phenyl rings joined by a three-carbon bridge, flavonoids—the most prevalent group—form a number of subclasses with unique structural characteristics. Other important types include lignans, which are made from the dimerization of phenylpropanoid units, and stilbenes, which have two phenyl rings connected by an ethylene bridge. The great diversity and range of biological activity shown among polyphenols are ascribed to the particular number and arrangement of hydroxyl groups as well as other structural differences (Bertelli et al. 2021). Puerarin, the phenol-saccharide conjugate, is a naturally occurring plant compound with a mid-molecular weight that exemplifies the C-glucoside unit in polyphenolic compounds. The carbon-carbon bond (C-C) is how the phenol binds to the saccharide. Polyphenols frequently contain the 10-atom "fused ring" system of isoflavone and benzopyran, which is another structural characteristic of isoflavone (Šamec et al. 2021).

1.2.3.2 Biological Role of Polyphenols

Many plants' ecology depends on both of the bigger polyphenols. The following categories describe their known effects in plant tissues. First, UV screens, which offer color (plant pigments) and protection from ionizing radiation. Furthermore, ripening and other growth stages include signaling molecules. Additionally, the production and downregulation of growth hormones for instance auxin, as well as the avoidance of microbial infections or phytoalexins (Li et al. 2014).

1.2.3.3 Occurrence of Polyphenols

The most prevalent condensed tannins, or polyphenols, are present in almost every plant family. In forest ecology, the breakdown of forest litter and nutrient cycles rely on bigger polyphenols, which are also present in higher concentrations in fruits, bark layers, flowers, epidermis, and leaf tissue. The kind of polyphenols, assay, and literature source all have a significant impact on the total phenolic concentration in plant tissues. They are calculated using dry green leaf mass and lie between 1% and 25% for total polyphenols and natural phenols (Sinha 2019).

1.2.4 Carotenoids

Bright red, yellow, and orange pigments are called carotenoids. They are occurring in bacteria, photosynthetic algae as well as plants. While fruits are particularly abundant in carotenoids, vegetables including spinach, sweet potatoes, carrots, and pumpkins are also rich in them. Carrots, parsley, tomatoes, oranges, daikon radish, cabbage, fenugreek, spinach, green leafy vegetables and round purple turnips are abundant sources of them. Carotenoids are generally those that are regularly seen. The most frequently seen carotenoid is β-carotene. That being said, α-carotene is also usually very common. As the name suggests, oranges, persimmons, and tangerines are exceptionally high in β-cryptoxanthin. Furthermore, the carotenoid landscape of green leafy vegetables is dominated by lutein and zeaxanthin, whereas tomatoes and brown algae contain lycopene (Zakynthinos and Varzakas 2016). Among them, lutein, lycopene, and fucoxanthin are very strong antioxidants, while vitamin A is derived from α and β-carotene as well as β-cryptoxanthin. Besides, lutein is significant for vision. Fine-feature vision is attributed to zeaxanthin, which is an antioxidant and a fat-soluble pigment rich in the macula area of the retina (Kumar et al. 2023).

1.2.5 Flavonoids

One important class of beneficial compounds, flavonoids, has a lower systemic toxicity. Many compounds belonging to the flavonoid family Numerous fruits, vegetables, and other therapeutic plants contain flavonoids, which are common secondary metabolites that include flavanones, flavones, and flavanols. Plant-based foods contain phenolic acids and flavonoids, which are categorized as dietary polyphenols and are noted for their potent antioxidant properties and other advantageous benefits, such as a potential role in treating degenerative and chronic diseases (Del Prado-Audelo et al. 2021). These demonstrate a broad range of additional activities, involving analgesic and antioxidant qualities, as well as anti-diabetic, anticancer, antiproliferative, and immunological stimulatory effects. One plant that is extremely rich in flavonoids is *Warionia saharae*. Because of its ability to prevent diabetes, it is used. Similar to dietary polyphenols, biologically significant amounts of micro- and macronutrients are present. Five to ten percent of the polyphenols found in food are absorbed in the small intestine. Phytochemicals that function as prebiotics and exhibit antibacterial qualities regulate the pathogenic gut microbiota. Additionally, new research indicates that flavonoids have a role in wound healing and may be protective against cutaneous inflammation (Chen et al. 2023).

1.3 Nanoparticles

Structures with dimensions of at least 1 nanometer (nm) and ranging from 1 to 100 nm are known as nanoparticles (NPs). Their physicochemical characteristics differ from those of their derived metals, such as having lower melting points and catalytic qualities. Additional properties include wettability, magnetic, electrical and thermal conductivity, and absorption and scattering of light. Nanotechnology research has expanded in several medical domains, such as gene therapy and drug delivery. Its application in imaging, the diagnosis of various illnesses, and the medical device sector has also advanced significantly. Pharmaceutical medicines are transported by NPs. There are potential clinical uses for various NPs, including solid NPs, polymeric NPs, nano emulsions, and liposomes. Various factors determine how they are used in the clinical setting. Their chemical and physical characteristics, loading efficiency, and drug release all had a role. The ratio of NPs' efficacy to toxicity governs their applications (Omar 2022). NPs are classified into four categories based on the type of substance. Nanomaterials are divided into four types: organic, inorganic, carbon-based, and composite. They include both engineered nanoparticles, which are intentionally synthesized for specific applications, and non-engineered nanoparticles, which occur naturally through processes such as erosion and volcanic activity (Barhoum et al. 2022). Metal and metal oxide-based NPs are used in inorganic-based nanomaterials. Metals such as iron (Fe), cobalt (Co), aluminum (Al), gold (Au), cadmium (Cd), silver (Ag), zinc (Zn), copper (Cu) and lead (Pb) are employed as inorganic nanoparticles in many applications. Zinc oxide (ZnO), aluminum oxide (AlO_3), cerium oxide (CeO_2), iron oxide (FeO_3), magnetite (FeO_4), titanium oxide (TiO_2) and silicon dioxide (SiO_2) are among the metal oxide-based NPs that are frequently utilized. Nanomaterials can have chemical makeup, safety, physicochemical properties, and influence on ecological and human well-being. However, NPs are referred to as organic-based nanomaterials since they are mostly made of organic components and contain no carbon. As illustrated in Fig. 1.4, they include liposome nanoparticles, micelles, dendrimer polymers, gold particles, lipids, solid-lipid nanoparticles, mesoporous nanoparticles, nanospheres, and more. Carbon is a constituent of carbon-based nanomaterials, that can be organized in spheres, ellipses or hollow tubes, among other configurations. Carbon-based nanomaterials include fullerenes (C60), graphene (Gr), carbon nanofibers, carbon nanotubes (CNTs), carbon onions and carbon black. Arc discharge, laser ablation, and chemical vapor deposition (CVD) are some of the most significant procedures to yield carbon-based materials (Omar 2022). NPs can exist in multiple configurations relying on their shape. Compound shapes that are octagonal or polyhedral include simple spherical, rod, triangular, and circular shapes. The shapes of gold nanoparticles (AuNPs) include sub-octahedral, icosahedral tetrahedral, nanocluster, decahedral, octahedral, nanorod, nanosphere, nano cube, nano star, and nano shell. NPs can be grouped into

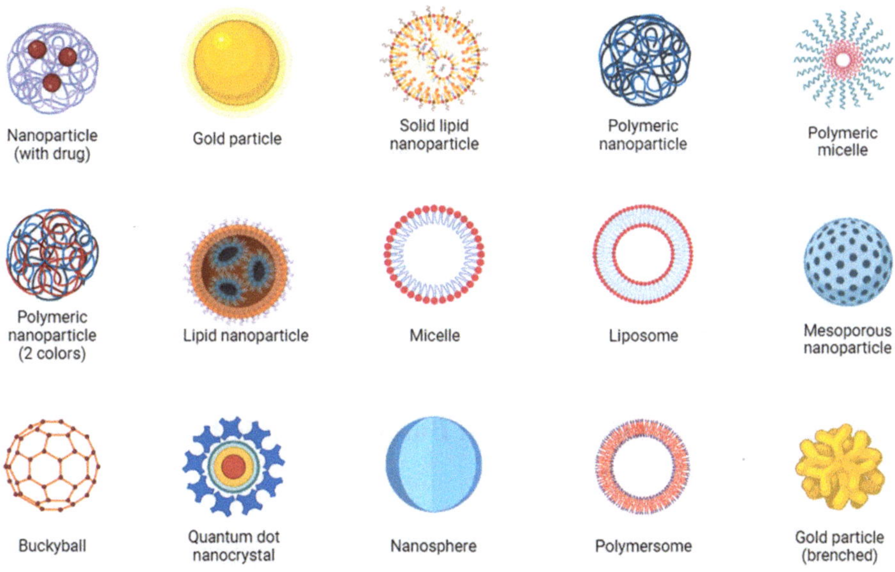

Fig. 1.4 Range of nanoparticles used in medicine

either natural or synthetic based on where they come from. Manmade (designed) NPs are produced via physical, chemical, biological, and hybrid methods, as well as by mechanical grinding, motor exhaust, and smoke. Nowadays, a variety of application-related sources are used to synthesize engineered NPs. The main question with synthetic NPs is whether what is now known about them is sufficient to predict how they behave and how they differ from natural NPs (Barhoum et al. 2022).

1.4 Engineered Nanoparticles and Biomedical Applications

Compared to organic ENPs, inorganic ENPs have a wider variety of uses because of their capability to tolerate unfavorable reaction conditions. Nanoparticles have the potential to increase medication and bioactive component delivery, resulting in higher therapeutic efficacy and bioavailability (Wang and Sun 2021).

The medicines or bioactive chemicals can be delivered via the ENPs in a regulated way. One of the main challenges with the TE and RM approaches is the transient and persistent conveyance of bioactive substances to the designated location or specific region. Targeted delivery, decreased side effects, preservation of drug/bioactive agent concentration, and protection of the therapeutic molecule from degradation are only a few benefits of the ENPs' regulated and prolonged release of the drug/bioactive agent. Therefore, the creation of efficient delivery systems boosts a drug's or bioactive agent's therapeutic efficacy; in this sense, NPs are a promising tool. NPs must pass through the system's physiological barriers, which vary based on the application, target cells, and administration method. For NP-based delivery systems to be successful and efficient, loaded bioactive chemicals must be able to bypass natural obstacles and access the specific region for the best pharmacological impact. ENPs can be eaten by phagocytosis, endocytosis, or pinocytosis, among other processes, relying on the specific kind of nanoparticles and drug conjugate system. The use of ENPs and nanocomposites in medication and bioactive agent delivery may improve the efficacy of treatment for numerous disorders, consisting of cancer-targeted drug delivery. For targeted drug delivery, including cells, tissues, organs, and the specialized compartments within cells can be physically targeted, passively targeted, or actively targeted. Organelle-targeted drug delivery is a novel and fascinating field of research because of its utilization in the treatment of cancer (Rudramurthy and Swamy 2018).

1.5 Types of Nanoformulation Based on Medicinal Plant Bioactives

Figure 1.4 shows various nano formulations based on medicinal plant bioactives such as liposome NPs, micelles, dendrimer polymers, gold particles, lipids, solid-lipid nanoparticles, mesoporous nanoparticles, nanospheres, and other components.

One component of carbon-based nanomaterials is carbon, which can be formed into hollow tubes, ellipses, and spheres, among other shapes. When creating NPs, it is important to take into account their bioavailability, controlled release, enhanced encapsulation and decreased toxicity as naturally degradable nanoparticles. The synthesis technique is controlled by the use for which it was created, such as drug/extract/peptide delivery (dos Santos et al. 2022). Biodegradable nanoparticles are ideal for drug administration due to their regulated release kinetics, capacity to interact at the subcellular level, and biocompatibility with biological systems (Omar 2022). Nanoparticles containing a drug can be manufactured using the solvent displacement approach, which involves dissolving the desired drug in acetone prior to adding it to an aqueous solution, a polymer like PCL/PLGA, resulting in drug entrapment (Heya et al. 2022).

1.5.1 Lipid-Based Nanomaterials

Nanoparticles made from lipids, such as liposomes, solid-lipid nanoparticles, emulsions, and nanostructured lipid carriers, can be made from a range of lipid choices, such as phospholipids, glycerol esters, glycolipids, sphingolipids, free fatty acids, fatty alcohols, and waxes. The most researched nanocarriers for administering various pharmaceutical drugs with different pharmacokinetic and solubility characteristics are lipid-based nanoparticles. Lipid-based nanoparticles are becoming known as bacterial growth inhibitors counter to MDR bacteria in addition to their function as nanocarriers. Lipidic nanocarriers and lipidic nanoparticles are the two categories into which these can be separated. While lipidic nanoparticles are naturally antibacterial, lipidic nanocarriers transport and transmit antibacterial substances like antibiotics and antimicrobial peptides.

1.5.2 Biodegradable Polymeric Nanomaterials

Antimicrobial polymers and polymeric nanoparticles for use as delivery nanocarriers are two categories of biodegradable polymeric nanosystems. The customized structure of polymeric chains gives biodegradable polymeric nanoparticles a variety of benefits, such as improved stability, biocompatibility, extended circulation, antibacterial activity, and bacterial identification (Wang and Sun 2021). By varying the rate, duration, and location of release, biodegradable polymeric nanoparticles offer an attractive delivery system that can improve the efficacy and safety of other chemicals in comparison to lipid-based nanoparticles. Furthermore, the functional groups of the polymer chain provide a suitable matrix for forming polymer-based nanocomposites through interactions with other nanomaterials (Sidhu et al. 2022).

1.5.3 Carbon-Containing Antimicrobial Nanomaterials

Significant characteristics of nanomaterials composed of carbon include intrinsic antibacterial activity, remarkable mechanical qualities, large surface area relative to their volume, exceptional electrical and thermal conductivity, photocatalytic activity photoluminescence, and good stabilities. Owing to these distinctive features, carbon nanoarchitectures exhibit potential across different uses against bacteria, including medication delivery, photothermal therapy, bone and tissue engineering, biosensors, and the potential for new antibacterial substances (Nasir et al. 2018).

1.5.4 Metal Nanomaterials

Metal nanoparticles can be synthesized by using plants as bio-based reductants. Since metal nanoparticles are naturally strong antibacterials, they show the greatest potential choice within this category of substances. While other metal nanoparticles, such as gold, silver, copper, superparamagnetic iron, and zinc have shown encouraging antibacterial abilities, silver nanoparticles (AgNPs) are the most impactful. Silver nanoparticles (AgNPs) have been utilized in a range of biomedical and antibacterial applications in the past, such as topical medications, dental fillings, wound dressings, sanitizing personal care items, solvents, disinfectants, and surface coatings for medical equipment (Bruna et al. 2021).

1.5.5 Metal Oxide Nanoparticles

Strong antibacterial properties are exhibited by metal oxide nanoparticles, such as titanium dioxide and zinc oxide against multidrug-resistant bacteria, making them a possible alternative to conventional antibiotics (Bruna et al. 2021). These numbers of metal oxide nanoparticles—oxides of copper, aluminum, zinc, and magnesium as well as titanium dioxide—are proved to contain antibacterial effects. All these make the following properties place ZnO among the metal oxide nanoparticle candidates well-known. Each year in the world, a billion tons of ZnO is produced; thereby one can conclude that ZnO, in comparison to all those other metal oxides, are very highly valued nanoparticles. The FDA has classified zinc oxide (ZnO), typically considered to be a safe substance because of its ease of biodegradation and absorption in the body. Compared to other metal oxide nanoparticles, ZnO nanoparticles are less toxic and more biocompatible (Jacobsen et al. 2015).

1.6 Biomedical Application of Plant-Based Nanomaterials

Pharmaceutical development and active drug delivery at the nanoscale (1–100 nm) are known as nanosystems, and they can be applied to theragnostics, diagnosis, treatment, and prevention. They can reduce toxicity, change pharmacokinetics and biodistribution, and shield medications against deterioration. Nanocarriers are pharmaceutical systems that have the ability to transport hydrophilic and hydrophobic medications, improving their medicinal effectiveness and enabling more successful drug transport to the intended site. Lipidic systems include liposomes, nanoemulsions, nanostructured lipid systems, and nanostructured lipid carriers (dos Santos et al. 2022).

Because they offer a more environmentally friendly substitute that improves drug administration, releases active components gradually, and permits lower dosages of the active ingredient, NPs show promise as a replacement for synthetic chemical fungicides. Pharmaceutical systems known as nanocarriers can carry medications that can interact with water and/or those that cannot, enhancing their curative potency and facilitating more efficient drug delivery to the intended area. Nanoemulsions, liposomes, nanostructured lipid carriers, and nanostructured lipid systems are examples of lipidic systems. Therefore, plant-derived consumables made of nanoparticles with anti-inflammatory and naturally targeting colonic tissues may be the natural and harmless delivery route.

Intestinal bowel disorder (IBD) is one of the digestive tract conditions that can be treated using this approach. Moreover, research on NPs in leukocytes to identify their precise function in gut inflammation has revealed that they block the cell cycle-regulating protein Cyclin D1. Another study found that metal oxide nanoparticles are either non-toxic to people or exhibit potent antibacterial action at low concentrations, in contrast to their bulk counterparts. Green synthetic techniques for producing nanomaterials from plants, microorganisms, and other natural resources have emerged as a result of the growing demand for green chemistry and nanotechnology. Potential uses for these green plant-based nanoparticles include biotechnology, electronics, imaging, optics, and other biological fields (Trivedi et al. 2022).

1.6.1 Phytochemicals for Disease-Treatment

Because of their ability to reduce inflammation, antioxidant and anti-cancer qualities, phytochemicals are essential for controlling and preventing a wide range of illnesses. They work well against oxidative stress, which has been connected to long-term illnesses like cancer, metabolic problems, and cardiovascular diseases. Phytochemical-rich diets are linked to a decreased risk of several illnesses. Certain phytochemicals have demonstrated promise in controlling obesity, lowering blood sugar, and preventing heart disease and some types of cancer. They accomplish these advantages by a number of methods, such as boosting cellular defenses and modifying gene expression (Luo et al. 2023). Phytochemicals have demonstrated promise in the treatment of inflammatory illnesses including, obesity, cancer, diabetes, arthritis, and cardiovascular disease. Numerous degenerative diseases and health problems are associated with chronic inflammation. Although anti-inflammatory medications are available, prolonged use of them frequently results in unfavorable side effects. By controlling molecular processes and altering inflammatory pathways, phytochemicals, which are present in medicinal plants, have exhibited significant inflammation-reducing properties. This demonstrates how compounds with biological activity extracted from medicinal plants possess anti-inflammatory properties and how they work pharmacologically to treat disorders linked to inflammation (Nisar et al. 2023).

1.6.2 Nanocarriers from Plants: A Potential Treatment for Leishmaniasis

Owing to its elevated detection frequency and potential to induce malformations, especially in the skin-affecting form, leading to disfigurin wounds and significant psychological, social, and economic distress, leishmaniasis is still a major infectious illness around the world, notably in its tegumentary form, which causes deforming wounds with intense psychosocial and economic consequences, while the visceral form, which affects internal organs, can be fatal (dos Santos et al. 2022).

The bite of an infected phlebotomine sandfly can spread leishmania parasites. When the female vector feeds on blood, the parasite grows in her intestine and spreads protozoan's metacyclic forms to the vertebrate host. Following inoculation, the infected organism's phagocytic cells, including macrophages, enclose the infectious forms. These cells then undergo amastigote differentiation, which starts an intracellular binary division process in the lysosome-phagosome fusion. The host cells burst while the protozoan is inside, releasing parasites that infect the vertebrate's new cells. The mosquito consumes cells containing the parasite's amastigote forms when it performs hematophagy on an affected host (Balakrishnan et al. 2021). There are some potential nanocarriers used for the treatment of leishmaniasis mentioned below.

1.6.2.1 Nano-emulsions

Systems of immiscible liquids in which one is distributed in the other as tiny droplets with particle sizes ranging from 20 to 500 nm are called nanoemulsions (NEs). Additionally, more surface contact is made possible by the smaller droplets, which enhances the penetration of active compounds derived from plants. Numerous advantages are provided to pharmaceutical formulations by it, such as improved pharmacokinetics, lipophilic substance solubility, therapeutic agent biodistribution, active ingredient protection from the environment, and the ability to employ multiple delivery routes (Nirale et al. 2020).

1.6.2.2 Liposomes

Spherical nanovesicles known as liposomes (LPs) spontaneously form bilayers. They are frequently made of phospholipids and surfactants. Their polar groups are oriented toward the medium outside the liposome and the watery media as well as their hydrophobic tails pointing inward, the individual lipid molecules are placed in an ordered form within the bilayers. Liposomes are versatile drug delivery systems that can contain medications that are both lipophilic and hydrophilic within their lipid bilayer and aqueous core respectively, and their physical and chemical properties, such as size, surface characteristics, and lamellarity can be tailored to optimize drug transport for particular therapeutic applications (Heya et al. 2022).

1.6.2.3 Lipid Carriers with Nanostructures

The second generation of lipid carriers, known as nanostructured lipid carriers (NLCs), appeared in the 2000s as a substitute for SLNs in order to get past their drawbacks. A complex lipidic matrix is produced by switching out a component of solid lipid with oil, which exists as a liquid oil at ambient temperatures. Higher levels of matrix disorder are encouraged by the liquid lipid, which permits greater drug dissolution and stops drug leakage during storage as a result of recrystallization events (Nisini et al. 2018).

1.6.3 Treatment of Cancer

Nanotechnology developed from plants has attracted attention in the medical and diagnostic domains because it targets a specific location. Peptides, nucleic acids, and small medications are examples of drug molecules that either merge with or confine the naturally-sourced nanomaterials used in therapeutics. Plant-based nanoparticles are a promising strategy for targeted cancer therapy and diagnostics because of their capacity to selectively transport therapeutic chemicals to diseased cells while limiting injury to healthy tissues, as well as their effectiveness in identifying malignant cells (Trivedi et al. 2022). Nanoparticles of many shapes, including nanoshells and nanotubes, are produced here. The antibody recognizes the cancerous cell and gives an indication after attaching to a nanoshell. Particles of metal at the nanoscale are ideal materials for the purpose of delivering therapy to diseases like cancer owing to their distinct physicochemical characteristics. Sol-gel, microemulsion, and hydrothermal precipitation are among the sophisticated methods used to synthesize nanoparticles with tailored chemical and physical properties, including enhanced enzyme activity and unique optical characteristics relevant for targeted drug development (Barhoum et al. 2022). Plant-derived nanoparticles, which have a significantly greater surface area compared to their volume and surface functionalization, hold promise for cancer therapy, improving therapeutic efficacy and decreasing adverse effects (Barathi et al. 2024).

1.6.4 Treatment of Alzheimer's Disease

Dangerous substances are kept out of the brain by the barrier between blood and brain, a intricate network of blood vessels and endothelial cells which implies searching for therapy against neurodegenerative diseases so tough and complicated. This issue may be addressed by the promising prospects offered by the development of nanotechnology. Advances in neurological research are expected to significantly influence the development of novel therapeutic approaches with nanotechnology. Nanoparticles in the protein size range are made for a various therapeutic applications because they can safeguard the desired physiological responses while reducing adverse consequences by stimulating, reacting to, and affecting target cells and tissues. Neurodegenerative disorders (NDs) are a family of chronic, perhaps hereditary diseases characterized by the progressive degradation of neural subtypes (Barathi et al. 2024).

1.7 Antioxidant and Antimicrobial Properties of Plant-Based Nanoparticles

It has been closely related to the numerous chronic diseases that exist, such as neurodegenerative disorders, cancer, respiratory ailments and, thus underlining the significant role ROS may play in the onset, progression, and severity of a disease. At each stage and at different levels, the various activities of ROS have initiation via DNA mutations and oncogenes activated by oxidative stress produced by ROS, with progression marked by the emergence of changed cells. In order to promote cell survival, proliferation, and possibly tissue remodeling or metastasis in specific contexts like cancer, it would also involve increased metalloproteinases. These signaling pathways include Akt/PI3K and ERK, which inhibit pro-apoptotic proteins and facilitate the activation of anti-apoptotic genes (Nasrollahzadeh et al. 2020). In cardiovascular diseases, the ROS modification of LDL has critical involvement in the mechanism of atherosclerosis, leading to the development of, myocyte apoptosis, hypertrophy, and ischemia-reperfusion injury, finally leading to cardiopulmonary arrest. Neurodegenerative conditions like Parkinson's, Huntington's, and Alzheimer's featuring increased oxidative stress caused by neuronal vulnerability, impaired antioxidant defenses, altered cell membrane composition, and increased oxygen demand. Reactive oxygen species contribute significantly to neuronal damage, making ROS regulation a potential therapeutic strategy (Balakrishnan et al. 2021). Overproduction of ROS, typically brought on by cigarette smoke and air pollution, damages alveolar and connective tissues and triggers inflammatory responses in the pulmonary system in respiratory disorders like chronic obstructive pulmonary disease (COPD) and asthma. The synergistic co-function relationship between ROS and such diseases highlighted the challenges that must occur with oxidative stress toward having effective strategies in lowering such overproduction of ROS for the mitigation of deleterious effects on the general population's health (Kumar et al. 2020). Any material or chemical that, even at low concentrations, may prevent the oxidation of a suitable substrate is considered an antioxidant. The standard method for figuring out a compound's antioxidant capacity is to measure how much oxygen is consumed when fat is stored in an oxygen-filled enclosure. The discovery that vitamins C, E and A are antioxidant agents has transformed the field and brought attention to how important antioxidants are to the biochemistry of living things (Parham et al. 2020).

The antioxidant can also be thought of as a stable molecule that neutralizes undesirable free radical species by giving them an electron, so reducing their capacity to do harm. Due to their foraging capabilities, these antioxidants generally avoid or delay cellular damage. These antioxidants' low molecular weight enables them to readily combat free radicals (ROS) and stop their chain reaction before endangering essential components. Among the antioxidant compounds that our body creates during regular metabolic processes are glutathione, ubiquinol, and uric acid. The body contains a number of scavenging enzymes for free radicals, and micronutrients like β-carotene, α-tocopherol (vitamin E) and ascorbic acid (Vitamin C) are among them. These micronutrients are obtained from the food we eat because our bodies are unable to produce them on their own (Salem et al. 2021).

1.8 Conclusion

Medication through a rich phytochemical profile: medicinal plants have traditionally been part of modern medical science. The introduction of nanotechnology has greatly helped their therapeutic performance by enhancing their bioavailability and stability, making targeted drug delivery possible. The possible applications of nanoformulation in plant delivery for antitumor therapies and microbial treatments include reduced side effects and controlled drug release, but challenges including regulatory impediments, environment issues, and scalability problems have arisen. Overcoming those hurdles will open up opportunities in plant-based nanotechnology toward advancing healthcare globally.

1.9 Future Prospects

The future of plant-based nanotechnology shall redefine the therapeutic paradigms by sustainable and environment-friendly innovation. The major role of green synthesis methods using the plant extract and reducing toxic by-products is expected for the synthesis of biocompatible and biodegradable nanomaterials. Advances in nanotechnology are predicted to bolster individualized medicine for complicated diseases such as cancer, neurological disorders, and multidrug-resistant infections. Combining phytochemicals and nanotechnology may lead to novel therapeutic techniques, such as combinatorial medicines that capitalize on the synergistic effects of numerous bioactive molecules. Artificial intelligence and computational modeling may be integrated into the design and prediction of nanoformulations to accelerate their optimization and clinical translation. Moreover, collaborative research efforts supported by governments and industries, along with proper regulatory guidelines, will allow for the safe and efficient use of these technologies. Plant-based nanoformulations are progressing and may be used as a means to address a number of the world's leading healthcare issues, providing unique, efficient, and sustainable solutions for the future.

References

Abdallah II, Quax WJ (2017) A glimpse into the biosynthesis of terpenoids. KnE Life Sci 3(5):81. https://doi.org/10.18502/kls.v3i5.981

Arzani V, Soleimani M, Fritsch T, Jacob UM, Calabrese V, Arzani A (2025) Plant polyphenols, terpenes, and terpenoids in oral health. Open Med 20(1). https://doi.org/10.1515/med-2025-1183

Balakrishnan R, Azam S, Cho D, Su-Kim I, Choi D (2021) Natural phytochemicals as novel therapeutic strategies to prevent and treat Parkinson's disease: current knowledge and future perspectives. Oxid Med Cell Longev 2021:6680935. https://doi.org/10.1155/2021/6680935

Barathi S, Ramalingam S, Krishnasamy G, Lee J (2024) Exploring the biomedical frontiers of plant-derived nanoparticles: synthesis and biological reactions. Pharmaceutics 16(7):923. https://doi.org/10.3390/pharmaceutics16070923

Barhoum A, García-Betancourt ML, Jeevanandam J, Hussien EA, Mekkawy SA, Mostafa M et al (2022) Review on natural, incidental, bioinspired, and engineered nanomaterials: history, definitions, classifications, synthesis, properties, market, toxicities, risks, and regulations. Nanomaterials 12(2):177. https://doi.org/10.3390/nano12020177

Bertelli A, Biagi M, Corsini M, Baini G, Cappellucci G, Miraldi E (2021) Polyphenols: from theory to practice. Foods 10(11):2595. https://doi.org/10.3390/foods10112595

Boncan DAT, Tsang SSK, Li C, Lee IHT, Lam HM, Chan TF et al (2020) Terpenes and terpenoids in plants: interactions with environment and insects. Int J Mol Sci 21(19):7382. https://doi.org/10.3390/ijms21197382

Bruna T, Maldonado-Bravo F, Jara P, Caro N (2021) Silver nanoparticles and their antibacterial applications. Int J Mol Sci 22(13):7202. https://doi.org/10.3390/ijms22137202

Chen S, Wang X, Cheng Y, Gao H, Chen X (2023) A review of classification, biosynthesis, biological activities and potential applications of flavonoids. Molecules 28(13):4982. https://doi.org/10.3390/molecules28134982

Del Prado-Audelo ML, Cortés H, Caballero-Florán IH, González-Torres M, Escutia-Guadarrama L, Bernal-Chávez SA et al (2021) Therapeutic applications of terpenes on inflammatory diseases. Front Pharmacol 12:704197. https://doi.org/10.3389/fphar.2021.704197

dos Santos DB, Lemos JA, Miranda SEM, Di Filippo LD, Duarte JL, Ferreira LAM et al (2022) Current applications of plant-based drug delivery nano systems for leishmaniasis treatment. Pharmaceutics. 14:2339. https://doi.org/10.3390/pharmaceutics14112339

Godswill Awuchi C (2020) The biochemistry, toxicology, and uses of the pharmacologically active phytochemicals: alkaloids, terpenes, polyphenols, and glycosides [Internet]. [cited 2025 Aug 5]. Available from: https://doi.org/10.5281/zenodo.3967809

Guimarães AC, Meireles LM, Lemos MF, Guimarães MCC, Endringer DC, Fronza M et al (2019) Antibacterial activity of terpenes and terpenoids present in essential oils. Molecules 24(13):2471. https://doi.org/10.3390/molecules24132471

Heya MS, García-Coronado P, Cordero-Díaz A, Torres-Hernández AD, Narváez OP (2022) Biomedical application of nanoformulated plant. Int J Pharm Sci Res 13(1):1–12. https://doi.org/10.13040/ijpsr.0975-8232.13(1).1-12

Jacobsen NR, Stoeger T, van den Brule S, Saber AT, Beyerle A, Vietti G et al (2015) Acute and subacute pulmonary toxicity and mortality in mice after intratracheal instillation of ZnO nanoparticles in three laboratories. Food Chem Toxicol 85:84–95. https://doi.org/10.1016/j.fct.2015.08.008

Kumar H, Bhardwaj K, Nepovimova E, Kuča K, Dhanjal DS, Bhardwaj S et al (2020) Antioxidant functionalized nanoparticles: a combat against oxidative stress. Nanomaterials 10(7):1334. https://doi.org/10.3390/nano10071334

Kumar A, Nirmal P, Kumar M, Jose A, Tomer V, Oz E et al (2023) Major phytochemicals: recent advances in health benefits and extraction methods. Molecules 28(2):887. https://doi.org/10.3390/molecules28020887

Li AN, Li S, Zhang YJ, Xu XR, Chen YM, Li HB (2014) Resources and biological activities of natural polyphenols. Nutrients 6(12):6020–6040. https://doi.org/10.3390/nu6126020

Luo Y, Zeng Y, Peng J, Zhang K, Wang L, Feng T et al (2023) Phytochemicals for the treatment of metabolic diseases: evidence from clinical studies. Biomed Pharmacother 165:115274. https://doi.org/10.1016/j.biopha.2023.115274

McDonald TO, Siccardi M, Moss D, Liptrott N, Giardiello M, Rannard S et al (2015) The application of nanotechnology to drug delivery in medicine. In: Nanoengineering: global approaches to health and safety issues. Elsevier, pp 173–223. https://doi.org/10.1016/B978-0-444-62747-6.00007-5

Mickymaray S (2019) Efficacy and mechanism of traditional medicinal plants and bioactive compounds against clinically important pathogens. Antibiotics 8(4):257. https://doi.org/10.3390/antibiotics8040257

Nasir S, Hussein MZ, Zainal Z, Yusof NA (2018) Carbon-based nanomaterials/allotropes: a glimpse of their synthesis, properties and some applications. Materials 11(2):295. https://doi.org/10.3390/ma11020295

Nasrollahzadeh M, Sajjadi M, Dadashi J, Ghafuri H (2020) Pd-based nanoparticles: plant-assisted biosynthesis, characterization, mechanism, stability, catalytic and antimicrobial activities. Adv Colloid Interface Sci 281:102103. https://doi.org/10.1016/j.cis.2020.102103

Ninkuu V, Zhang L, Yan J, Fu Z, Yang T, Zeng H (2021) Biochemistry of terpenes and recent advances in plant protection. Int J Mol Sci 22(11):5710. https://doi.org/10.3390/ijms22115710

Nirale P, Paul A, Yadav KS (2020) Nanoemulsions for targeting the neurodegenerative diseases Alzheimer's, Parkinson's and Prion's. Life Sci 256:117394. https://doi.org/10.1016/j.lfs.2020.117394

Nisar A, Jagtap S, Vyavahare S, Deshpande M, Harsulkar A, Ranjekar P et al (2023) Phytochemicals in the treatment of inflammation-associated diseases: the journey from preclinical trials to clinical practice. Front Pharmacol 14:1177050. https://doi.org/10.3389/fphar.2023.1177050

Nisini R, Poerio N, Mariotti S, De Santis F, Fraziano M (2018) The multirole of liposomes in therapy and prevention of infectious diseases. Front Immunol 9:155. https://doi.org/10.3389/fimmu.2018.00155

Omar AS (2022) Nanoformulation safety versus toxicity; what do recent studies tell us? Int J Pharm Res Allied Sci. 11(4):60–71. https://doi.org/10.51847/spfpldpsvl

Parham S, Kharazi AZ, Bakhsheshi-Rad HR, Nur H, Ismail AF, Sharif S et al (2020) Antioxidant, antimicrobial and antiviral properties of herbal materials. Antioxidants 9(12):1309. https://doi.org/10.3390/antiox9121309

Perveen S (2021) Introductory chapter: terpenes and Terpenoids. In: Biochemistry. IntechOpen. https://doi.org/10.5772/intechopen.98261

Rabizadeh F, Mirian MS, Doosti R, Kiani-Anbouhi R, Eftekhari E (2022) Phytochemical classification of medicinal plants used in the treatment of kidney disease based on traditional Persian medicine. Evid Based Complement Alternat Med 2022:8022599. https://doi.org/10.1155/2022/8022599

Rudramurthy GR, Swamy MK (2018) Potential applications of engineered nanoparticles in medicine and biology: an update. J Biol Inorg Chem 23:1163–1174. https://doi.org/10.1007/s00775-018-1600-6

Salem MZ, El-Hefny M, Ali HM, Abdel-Megeed A, El-Settawy AA, Böhm M et al (2021) Plants-derived bioactives: novel utilization as antimicrobial, antioxidant and phyto reducing agents for the biosynthesis of metallic nanoparticles. Microb Pathog 158:105107. https://doi.org/10.1016/j.micpath.2021.105107

Šamec D, Karalija E, Šola I, Bok VV, Salopek-Sondi B (2021) The role of polyphenols in abiotic stress response: the influence of molecular structure. Plants. 10(1):118. https://doi.org/10.3390/plants10010118

Sidhu AK, Verma N, Kaushal P (2022) Role of biogenic capping agents in the synthesis of metallic nanoparticles and evaluation of their therapeutic potential. Front Nanotechnol 3:801620. https://doi.org/10.3389/fnano.2021.801620

Sinha D (2019) Pharmacological importance of polyphenols: a review. Int Res J Pharm 10:13–23. https://doi.org/10.7897/2230-8407.1009255

Trepa M, Sułkowska-Ziaja K, Kała K, Muszyńska B (2024) Therapeutic potential of fungal terpenes and terpenoids: application in skin diseases. Molecules 29(5):1183. https://doi.org/10.3390/molecules29051183

Trivedi R, Upadhyay TK, Mujahid MH, Khan F, Pandey P, Sharangi AB et al (2022) Recent advancements in plant-derived nanomaterials research for biomedical applications. Processes 10(2):338. https://doi.org/10.3390/pr10020338

Uzor PF (2020) Alkaloids from plants with antimalarial activity: a review of recent studies. Evid Based Complement Alternat Med 2020:8749083. https://doi.org/10.1155/2020/8749083

Wang Y, Sun H (2021) Polymeric nanomaterials for efficient delivery of antimicrobial agents. Pharmaceutics. 13(12):2108. https://doi.org/10.3390/pharmaceutics13122108

Zakynthinos G, Varzakas T (2016) Carotenoids: from plants to the food industry. Curr Res Nutr Food Sci 4:38–51. https://doi.org/10.12944/CRNFSJ.4.Special-Issue1.04

Zhou F, Pichersky E (2020) More is better: the diversity of terpene metabolism in plants. Curr Opin Plant Biol 55:1–10. https://doi.org/10.1016/j.pbi.2020.01.005

Chapter 2
Plant—Based Polysaccharide Nanomaterials: Synthesis and Applications

Niladri Sarkar, Sumit Nath, and Rina Rani Ray

2.1 Introduction

The discovery and subsequent manipulation of nanoparticles have catalyzed a significant shift in scientific inquiry and technological development. Nanoparticles, defined as unique molecules with exclusively small sizes in nanometre ranges (1–100 nm), have appeared as a transformative force in various sectors, including pharma, food, medical, agriculture, and environmental sciences. The distinctive nature of the nanoparticles has made them suitable for several applications (Joudeh and Linke 2022). Extensively smaller nano sizes, higher surface area, stability, and permeability have assisted nanoparticles in entering biological cells easily without any transformation of functional groups and enzymatic degradation. Nanoparticles' action is more site-specific and directed, allowing for the development of drug and gene delivery. Conventionally used nanomaterials are chemically synthesized, which is corrosive to human health and pollutes the environment (Yusuf et al. 2023). Advancement of the formulation of nanoparticles by using natural products such as plant extracts, plant functional molecules, and bioactive materials has been employed to overcome the limitations of chemically synthesized ones.

The plant body is a rich source of bioactive compounds, metabolites, and vitamins. Polysaccharide refers to a biopolymer of monosaccharides, which are structural units of plant cell walls, providing integrity and strengthening to the plant body and protecting it from outer attacks and sensitivity (Díaz-Montes 2022). Several studies have shown that polysaccharide is a very worthy solvent in the formulation of nanoparticles instead of hazardous chemicals. The consisting of functional groups of different polysaccharide units act as a reduction agent, which reduces the

N. Sarkar · S. Nath · R. R. Ray (✉)
Microbiology Research Laboratory, Department of Biotechnology, Maulana Abul Kalam Azad University of Technology, Haringhata, West Bengal, India

V. Ghosh et al. (eds.), *Applications of Plant Bioactive Compounds-Based Nanoformulations*, Nanotechnology in the Life Sciences,
https://doi.org/10.1007/978-3-032-13086-0_2

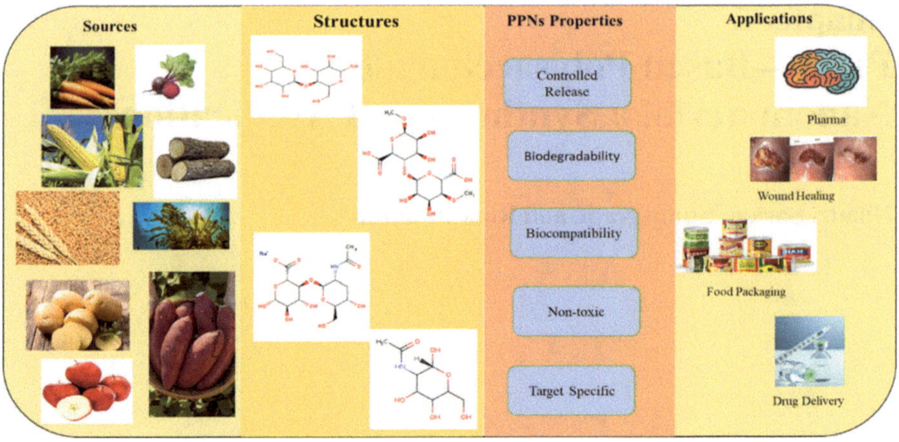

Fig. 2.1 Different plant polysaccharide-nanoparticles properties and applications

bulk or heavy molecular weight particles into nano-sized molecules within particular size ranges. Cellulose, starch, alginate, pectin, lignin, hyaluronic acid, chitin, chitosan, and inulin are commonly used polysaccharides in nanotechnology (Ahmad et al. 2022). Figure 2.1 describes the characteristics of plant polysaccharide nanoparticles.

Biocompatibility of PPNs helps them to easily integrate with cells and tissues without introducing any adverse immune responses. Hence, it is found in many biochemical fields, including gene and drug delivery systems (Raikwar et al. 2023). The target specificity and non-corrosivity of nanoparticles made them acceptable for chemotherapeutics and cancer prevention. Unlike other materials, plant polysaccharides are very versatile. Their various structural and functional configurations support different hydrogels, films, and nano-sized particle formation and promote the fabrication of nanocoating, encapsulation, nanocarriers, and scaffolds. Most importantly, the plant polysaccharide nanoparticles are biodegradable in nature; hence, they are considered preferable in food packaging and environmental remediation to remove heavy metals, pollutants, dyes, and pathogens. Nanoparticles have already shown their antibacterial and antifungal activity in different studies (Thapliyal et al. 2025). Metal, bimetal, and metal oxide nanoparticles and nanocomposites are effectively interfering in the growth and cellular activity of the antimicrobial-resistant microorganism. Integration of polysaccharides in nanoparticles helps twice overcome antibacterial resistance and kill microbes.

This chapter deals with the elucidation of all potential sources of plant polysaccharides for the preparation of nanoparticles and their applications in different fields of science and technology. The modern technologies adopted to ameliorate their efficacies are also discussed.

2.2 Sources of Plant-Based Polysaccharides

Polysaccharides are large, carbohydrate molecules composed of long chains of monosaccharide units, having a crucial role in both plant physiology and human nutrition. Polysaccharides can be broadly divided into two categories: structural and storage polysaccharides. Source, structures of the different plant based polysaccharides are described in the Table 2.1.

2.2.1 *Cellulose*

The most abundant polysaccharide Cellulose consists of large chain glucose units crosslinked with β-1,4-glycosidic bonds, forming microfibrils that constitute the structural scaffoldings of plant cell walls. The applications of cellulose in textiles, food, pharmaceuticals, and biofuel industries become popular due to its biodegradability and biocompatibility. Generally, natural woody plants, such as trees and shrubs, consist of ample amounts of cellulose in their hardroms or xylem transporters. There is a varying distribution of cellulose found in hardwoods (e.g., oak, maple) and softwoods (e.g., pine, spruce) mostly ranging from 37% to 50% of their dry weight. The distinctive chemical composition of cellulose in various plant bodies provides strength and rigidity to the cell wall of the plant, making them vital resources for paper production and construction materials (Mesoppirr et al. 2024). Phytophagous plants, including grasses and annual crops, are biogenic sources of cellulose which has a crucial role in the formulation of biofuels and biocomposites. Bamboo and switchgrass-like green plants have obtained attention for their rich cellulose content (up to 50% by dry weight) and expeditious growth rates, recommending a sustainable alternative for biomass production. Other than that agricultural byproducts, such as straw, corn stover, and sugarcane bagasse, refers to an unexploited source of cellulose. This surplus can be potentially employed for several applications, such as fodder, composting, and the fabrication of cellulose-based products, thereby expanding the quality and making the waste less. A group of algae, especially green and red algae (Dalia et al. 2022), have been observed as significant cellulose sources. Algal cellulose composition is quite lower than compared to higher plants, but their rapid growth rates and ability to thrive in harsh diverse environments make them a promising species for cellulose production.

2.2.2 *Chitin and Chitosan*

Chitin and chitosan are other important polymers procured from natural habitats, primarily crustacean shells and fungal cell walls. They have gathered immense attention in various industries, including pharma, crops, and environmental science.

Table 2.1 Source, structure and applications of the different plant polysaccharides

Sl No.	Plant polysaccharide	Source	Structure	Applications	References
1.	Cellulose	Wood, cotton stalk, wheat straw, bamboo and various plant sources	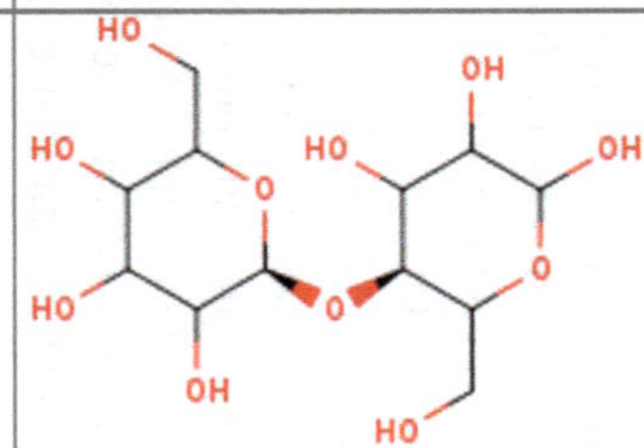	Production of Aerogels for adsorption of agricultural and industrial wastes for bioremediation	Bourakadi et al. (2024)
2.	Starch	Rice, wheat, corn, legumes, tapioca, potato	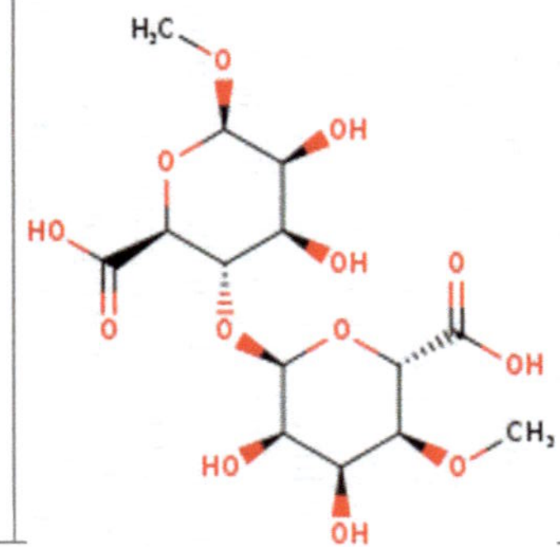	Production of hydrogels for the removal of cationic and anionic dyes, drug coating through injection moulding, arsenic removal from groundwater	Pires et al. (2023)

3.	Chitin/chitosan	<u>Luffa cylindrica</u> plant	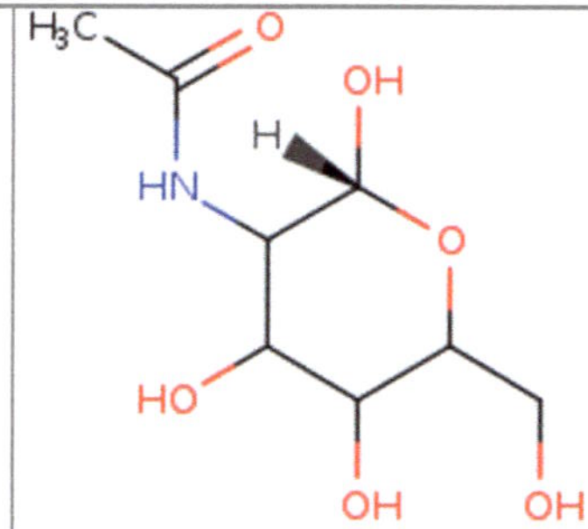	Antibiotic delivery to target site, zeta potential measurement, bacteriostatic, chelating agent, promotes plant growth, adsorption of heavy metals	Saraswat et al. (2023), Soares et al. (2022)
4.	Hyaluronic acid	Potato, Taro, jicama, almonds	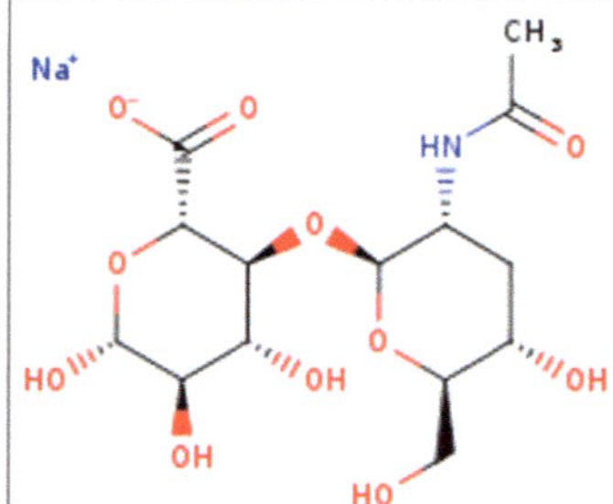	Anticancer drug delivery and food formulation	Kurczewska (2022)
5.	Alginate	Aquatic plants mainly from <u>Laminaria digitata</u>	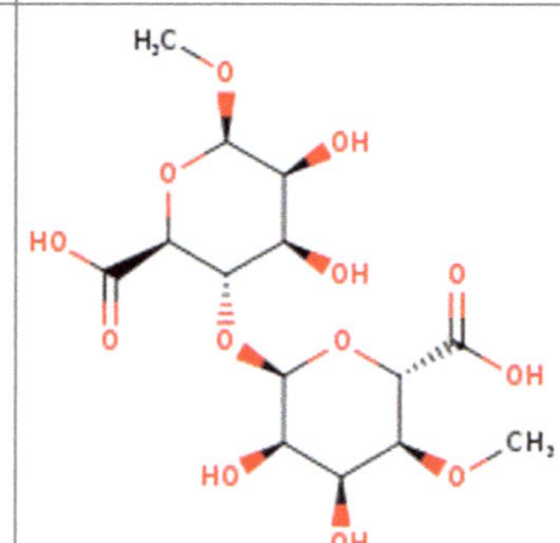	Wastewater treatment, anti-phytopathogenic, antioxidant, photo-catalyter	Sahoo and Biswal (2021)

(continued)

Table 2.1 (continued)

Sl No.	Plant polysaccharide	Source	Structure	Applications	References
6.	Pectin	Apples, wood, citrus fruits	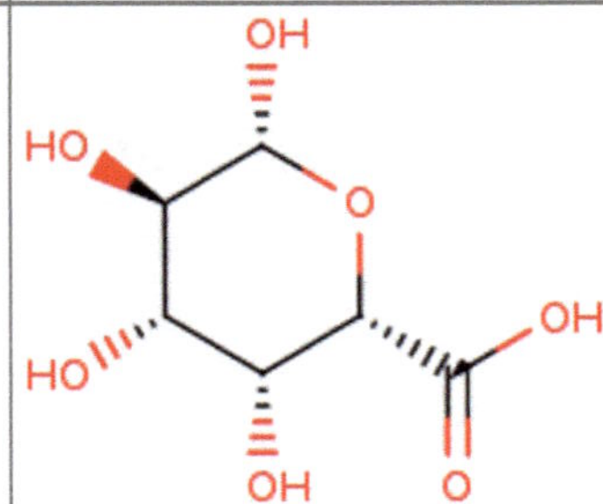	Food, drug industries	Benn et al. (2023)
7.	Lignin	Pine, oak, beech	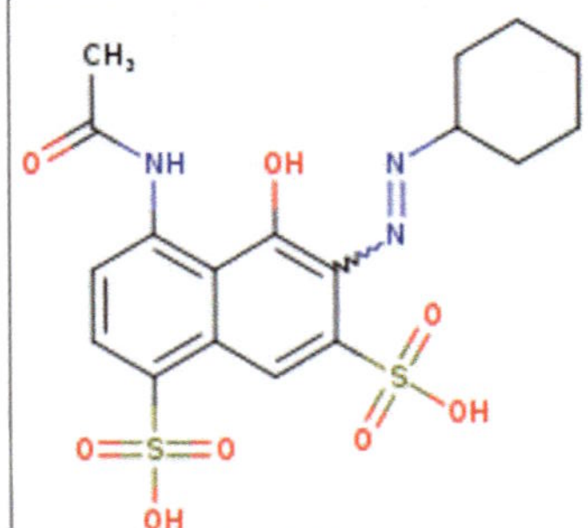	Food packaging	Muddasar et al. (2024)

Chitin is composed of N-acetylglucosamine units associated with β(1→4) glycosidic bonds. Chitosan is acquired from the deacetylation reaction of chitin, which means the removal of acetyl groups. The cationic nature comes from the free amino acid groups derived from the process of chitosan formulation. Grass-land by-products like rice husks, corn husks, and other lignocellulosic materials can be treated such as chemical and enzymatic degradation for extracting chitin-like compounds. Certain algae (primarily red and brown algae) contain higher amounts of polysaccharides that can be transmogrified into chitin-like structures. Similarly, fungi represent a significant source of chitin (Li et al. 2021). The cell walls of filamentous fungi contain chitin, which can be extracted and processed into chitosan. Common fungi include *Chitosan* and *Mycelium* species.

2.2.3 *Starch and Derivatives*

Essential biopolymers such as Starch and their derivatives have become very popular in growing food industries nowadays. It is the most commonly used carbohydrate in the human diet, predominantly sourced from green plants (Chen et al. 2020). Comprising glucose monomers, starch serves as an energy source, which can be broken down into glucose by hydrolysis. Simplest forms like amylose, a dimeric chain of glucose molecules linked by α(1→4) glycosidic bonds, account for 20–30% of starch content in many plants. It plays a distinctive role in the gelling and thickening properties of starch and Amylopectin, a branched polymer composed of glucose units linked by both α(1→4) and α(1→6) glycosidic bonds. It constitutes 70–80% of starch and bestows high viscidity and thickening characteristics.

Generally, cereals such as maize (corn), rice, wheat, and barley are rich in starch content. Maize provides one of the most exorbitant starch compositions, primarily in the form of amylopectin. Rhizomes, bulbs, roots such as potatoes and cassava (manioc), taro, and arrowroots are found as significant sources of these polysaccharides. Cassava, a staple in many developing countries, contains a high starch content, making it a crucial food source. Lentils and peas are also promising sources of starch intake, though their content is somewhat lower than that of cereals and tubers. Some natural fruits, such as bananas and plantains, provide starch, particularly when underripe.

2.2.4 *Alginate*

Alginate exhibits exclusive natural gel-forming capabilities and promotes it as a valuable thickening and gelling agent in various food, drug, and cosmetic industries. Its biocompatibility and non-corrosive nature have assisted in substantial use in the food industry as a stabilizer, emulsifier, and thickener. Beyond food, alginate is employed in pharmaceuticals for drug delivery systems, wound dressings, and tissue engineering scaffolds due to its ability to construct hydrogels (Abka-Khajouei et al. 2022). These biopolymers are composed of generally two monomer units such as β-D-mannuronic acid (M) and α-L-guluronic acid (G). The exact ratio of monomers varies based on the source and quality of the alginate required.

2.2.5 *Hyaluronic Acid*

Hyaluronic acid (HA) is a naturally occurring polysaccharide found in many root vegetables such as potatoes, onions, carrots, sweet potatoes, taro roots, jicama, almonds, edamame, etc. It is derived from the continual linking of disaccharide units—N-acetylglucosamine and glucuronic acid (Valachova et al. 2024). This unique physiology and chemical properties allow HA to attract and retain water, making it a suitable source for hydration and lubrication in the body. These naturally existing potent humectants, help to maintain skin moisture and elasticity, hence it is commonly used in serums, moisturizers, and anti-aging products. Furthermore, Hyaluronic Acid is also used in drugs, artificial synovial fluids, rheumatoid factors, etc.

2.2.6 *Pectin and Lignin*

Pectin is referred to as a polysaccharide, consisting of a linear chain of galacturonic acid units, which is derived by methyl-esterification. Natural sources of pectin include citrus fruits (such as oranges and lemons), apples, pears, and various root vegetables that are easily available. Another type of polymer -lignin, a complex, irregular polysaccharide found in the cell walls of plants, particularly in woody plant species. Both of the polysaccharides are involved for structural integrity, rigidity, and resistance to microbial degradation. The structural configuration of lignin may vary as several phenolic groups have been attached to the molecules in the 3D network system. The food, cosmetics, and packaging industries have shown interest in lignin and pectin for their unique physicochemical properties (Benn et al. 2023, Muddasar et al. 2024).

2.3 Plant—Polysaccharide Based Nanomaterials

Nanotechnology has revolutionized various fields by providing materials with enhanced characteristics at the nanoscale. Nanomaterials, with their unique surface area-to-volume ratios, can exhibit different properties compared to their bulk counterparts. Traditionally, the synthesis of nanomaterials has relied on synthetic polymers and other petroleum-based materials, which raise concerns related to sustainability and environmental impact. Plant-based polysaccharides, such as cellulose, starch, pectin, and alginate, have emerged as green alternatives for producing nanomaterials, allowing researchers to harness their natural abundance, biodegradability, and biocompatibility.

Plant-based polysaccharides are renewable resources extracted from various plants. Their utilization reduces dependency on fossil fuels and decreases the carbon footprint associated with the production of nanomaterials. The biocompatibility and biodegradability of polysaccharides make them appealing for eco-friendly applications, aligning with global movements towards sustainable development (Qiu et al. 2022). Plant-derived polysaccharides are generally recognized as safe (GRAS) and exhibit low toxicity levels, making them suitable for biomedical applications, including drug delivery systems and tissue engineering. Unlike synthetic materials, which often pose health risks, plant-based materials facilitate the development of biocompatible nanomaterials that can safely interact with biological systems. Polysaccharides exhibit various functional properties such as gelation, thickening, and film-forming capabilities, which can be advantageous in creating nanomaterials with specific characteristics. For example, the inherent antioxidant properties of certain polysaccharides can enhance the stability and functionality of nanomaterials. Additionally, the chemical versatility of polysaccharides allows for easy modification and functionalization, enabling tailored properties for specific applications.

2.3.1 Metal Nanoparticles

The green synthesis of metal nanoparticles such as silver (AgNPs), gold (AuNPs), and copper (CuNPs), Zinc (ZnNPs) employing plant polysaccharides is an area of significant concern. The application of green metal nanoparticles has intensified due to their unique physio-chemical properties such as enhanced reactivity, higher optical conductivity, induced electrical capacity, and mechanical strength, making it target specific, biocompatible, permeable, and sustainable. Commonly used plant polysaccharides such as starch, chitin, chitosan, lignin, cellulose, alginate, and pectin act as a reduction agent for metal ions to form their respective nanoparticles. For example, AgNPs are derived from $AgNO_3$ along with using M.*oleifera* plant starch as a reducing agent, while the hydroxyl and aldehyde groups of the polysaccharide promote the redox reaction (Asif et al. 2022). Metal, metal-oxide, and bimetallic nanoparticles have gained extra stability while the polysaccharides provide an outer layer around so that the nanoparticles become more stable from colloidal suspension and enzymatic degradation.

Metal nanoparticles derived from plant polysaccharides have multifaceted attributes on human health, pharma, and food industries, and environments. Most importantly, Metallic nanoparticles act as excellent antibacterial agents, especially silver, zinc, and gold nanoparticles. The metallic nanoparticles are generally effective on a wide range of bacteria, fungi, mycobacteria, and sometimes viruses. The action of nanoparticles affects different stages of microbial life cycles including cell wall damage, alteration of metabolic pathways, Reactive oxygen species (ROS) production, Cell damage, denature of nucleic acid, and interference in protein synthesis (Yin et al. 2020). Other than that, Metal nanoparticles are often used in environmental treatment including wastewater and pollutant treatment. Usually, copper and zinc-based nano polymers and composites are used to degrade and elucidate heavy metals and toxic chemicals (Namasivayam et al. 2020).

2.3.2 Carbon-Based Nanomaterials

Carbon-based nanomaterials, including carbon nanotubes (CNTs) and graphene, carbon quantum dots, and aerogels have been synthesized using plant polysaccharides as precursors. Plant polysaccharides chitin, cellulose, and pectin are commonly used in nanotechnology for green synthesis of carbon nanomaterials. Carbon nanomaterials processing is a complex protocol with two steps such as carbonization and activation. Polysaccharides like cellulose can undergo hydrothermal carbonization to produce carbonaceous materials with porous structures suitable for various applications including supercapacitors. This process involves heating the plant polysaccharide material at high temperatures (usually between 400 °C and 1000 °C) in an inert environment. The thermal putrescence of the polysaccharide leads to the formation of carbonaceous compounds while releasing volatile materials such as water and gases. The structure and composition of the carbon nanomaterials can be influenced by factors such as temperature, time, and the specific type of polysaccharide used.

2.3.3 Quantum Dots

In nanoscale production, quantum dots are semiconductor nanoparticles that possess idiosyncratic Birefringence Properties due to quantum mechanics. The size-dependent fluorescence of quantum dots makes them invaluable in various fields, including bioimaging, photovoltaics, and light-emitting devices. The production of quantum dots by employing plant polysaccharides is a green chemistry approach that minimizes the use of toxic chemicals. For instance, plant polysaccharides can react as the reductant of metal ions to create quantum dots, while the functional groups help to stabilize the nanoparticles and prevent precipitation. Plant polysaccharide-based quantum dots may increase the overall sustainability and bioavailability, hence becoming ideal for industrial applications.

2.4 Fabrication of Plant polysaccharide Nanoparticles (PPNs)

Extraction of plant polysaccharides from plant sources is a stepwise lengthy process including plant material preparation, extraction of plant bioactive compounds by various chemical treatments such as acid, alkali, enzymes and microwave, ultrasound, etc., followed by filtration and purification, nanoform production and characterization.

2.4.1 Extraction of Plant Polysaccharides

Traditionally used methods involve a time-consuming protocol along with uses of high temperature, and prolonged purification steps which sometimes cause degradation of sensitive chemicals and bioactive compounds. Revolutions of technology enhance the quality of extraction procedures. The selection of the polysaccharides-rich plant sources (weeds, fruits, legumes) is a very important phase. The fine plant sources are then dried, and grounded into smooth powder to increase the surface area for extraction. Acid, and alkali, enzymes have been commonly used as a chemical for extracting the polysaccharide by degrading the plant cell walls and membranes (Samrot et al. 2020).

Several organic and inorganic acids such as hydrochloric acid, trichloroacetic acid, citric acid, and chlorosulfonic acid are used for acid extraction whereas alkalis like sodium hydroxide and potassium hydroxide are commonly famous for alkaline extraction of plant polysaccharides. The advantages of acid and alkaline extraction rely on the specificity, capacity to degrade complex biopolymers, and improve yield and time consumption whereas it may react with sensitive polysaccharides. To overcome the limitations of acid and alkaline extraction, another method which become more popular is enzymatic extraction. Specific enzymes e.g. cellulases, amylases, pectinase, lyases, and protease have been widely used for enzymatic degradation of plant cell walls and specific polysaccharide extraction (Tang and Huang 2022). These methods are quite more efficient and very specific than others while it is also more environment friendly and non-toxic. Other than that Citric acid monohydrate-glycerol-based (DES),1-butyl-3-methylimidazolium-bromide ([Bmin] Br), dichloromethane, water-like hydrophilic solvents are popular for microwave and ultrasonic assisted extraction methods.

2.4.2 Plant Polysaccharide-Based Nanoparticle Formulations

Nanoparticles, which show a size range between 1 and 100 nanometers, reveal unique physical and chemical characteristics compared to their bulk counterparts. The incorporation of nanoparticles into various fields, particularly in drug delivery systems, agricultural fields, and industries, is revolutionizing the world. Several synthesis techniques have been used to yield nanoparticles using plant polysaccharides. These techniques can be categorized based on their preparation methods. Green synthesis of nanoparticles has gained interest from researchers and industries as it may be eco-friendly nontoxic and cost-effective as well.

Plant polysaccharides play a vital role as a solvent which offers reduction of chemically bulk molecules so that nanoparticles are being formed. Similarly, the contribution of plant polysaccharides as a stabilizer and capping agent is also a concern. Especially the coating, and encapsulation of nanoparticles by polysaccharides may protect them from enzymatic degradation, heat and salt sensitivity, etc. Several techniques are found in the synthesis of nanoparticles by using plant polysaccharides such as direct solvent evaporation, cross-linking, coacervation, electrospinning, etc. Among these, cross-linking and coacervation are useful for hydrophilic drug encapsulation whereas nanoprecipitation is applicable widely for industrial scalable purposes. Most of the nanofibers are produced by very versatile electrospinning technology which is considered to make nanoparticles with higher surface plasmon. Some synergistic approaches such as nanocomposite formation from biopolymers may increase the overall mechanical strength and specific drug release (Aminu et al. 2023).

2.4.3 Filtration and Purification

Extraction of plant polysaccharide solution needs to be filtered and purified to obtain the specific compounds required as polysaccharide co-existed along with other bioactive metabolites. In general, membrane filtration, centrifugation, and ultrafiltration are used in different laboratories and small production industries. Membrane filtration including micro filter and ultrafilter of different pore sizes has been employed for filtration of impurities to gain the particular required polysaccharide nanoparticles. The membrane filtration process has been used for a long period due to its specificity and high purities.

On the other hand, centrifugation is a more scientific model for large industries with higher production rates. Centrifugation techniques are completely based on the principle of driven centrifugal forces which force the solid and higher molecular weight particles to be stored bottom and lighter particles at the top of the solution. The spontaneous capacity of filtration and high scalability made it suitable for industries and laboratories as well. Upgradation in centrifugation technology, such as the progression of continuous-flow centrifuges and microfluidic systems, can expand the scalability and efficacy of nanoparticle filtration. Moreover, the integration of centrifugation with other separation techniques, such as filtration or chromatography, could yield even purer products. Since purification of filtered products provides more particular and required compounds. In that case, chromatography is appropriate for plant polysaccharide nanoparticle purification such as ion exchange chromatography, gas column chromatography, and liquid chromatography, macroporous resin column chromatography with photodiode array detection. Different purification technologies are being employed as the yield of the purest polysaccharide is very strenuous. Also, precipitation techniques are used for purification. Hexadecyl-trimethylammonium bromide precipitation, salt fractionation, dialysis, and ethanol precipitation are several precipitation techniques available for polysaccharide nanoparticle purification.

2.5 Characterization Techniques

2.5.1 Scanning Electron Microscopy (SEM)

Scanning Electron Microscopy (SEM) furnishes magnificent images of the surface arrangement of Plant polysaccharide Nanoparticles. This technique is used for envisioning size, shape, and appearance, which are considered as fundamental constants for studying their potential interaction with biological and physical environments and analyzing the overall performance.

2.5.2 Transmission Electron Microscopy (TEM)

Internal organization of the nanoparticles has been observed by Transmission Electron Microscopy (TEM) which has been employed as another powerful depicting technique. TEM can provide quantitative data on the crystalline structure and the presence of any core-shell configurations within PPNs.

2.5.3 Dynamic Light Scattering (DLS)

DLS is widely used to analyze the hydromechanical diameter and size administration of nanoparticles. This technique is particularly useful for assessing the stability of PPNs in various environments and can indicate the interaction with biological systems.

2.5.4 Nanoparticle Tracking Analysis (NTA)

NTA is another method that helps to visualize the size distribution data within 50–1000 nm ranges by stalking the Brownian motion. Concentration and uniformity of the nanoparticle suspension are being acquired, which facilitates amelioration of the nanoparticles quality.

2.5.5 Zeta Potential Analysis

Zeta potential measurement is basically an electrical potential which helps to understand the stability and surface interaction of nanoparticles with biological systems. The electrostatic repulsion or attraction has been detected by using Electrophoretic Light Scattering (ELS).

2.5.6 Thermogravimetric Analysis (TGA)

TGA measures the thermal stability of PPNs by recording weight changes as a function of temperature. This information is crucial for determining the thermal properties of nanoparticles and their behavior under processing conditions or in storage.

2.5.7 Differential Scanning Calorimetry (DSC)

DSC provides insights into the thermal transitions of PPNs, such as melting points and glass transition temperatures. This data can be significant for understanding the physical properties of nanoparticles and their interactions with other materials.

2.6 Applications of Plant-Based Polysaccharide Nanomaterials

2.6.1 Biomedical Applications

One of the most promising areas for plant-based polysaccharide nanomaterials is in the biomedical field.

2.6.1.1 Drug Delivery Systems

The application of green synthesized nanoparticles in drug delivery systems has been famous for their biocompatibility and bioavailability. A series of bioactive plant polysaccharides are used for encapsulation of hydrophilic and hydrophobic types of drugs. The advantages of PPNs-based drug delivery facilitate the delivery approach in an advanced way by targeting specific sites and reducing side effects. Simultaneously, the PPNs-based encapsulation of the drug allows the sustained release over time, balancing the need for frequent dosing and enhancing patient compliance. Hence, the approach is useful for chemotherapy. The target tumors or the cancer cells are being particularly targeted instead of damaging the healthy tissues. Several studies reveal that the PPNs based chemotherapeutics potentially able to fight over the cancer cells e.g. gemcitabine, a chemotherapeutics agent generally suggested for pancreatic cancer, had been studied by delivering with sodium alginate nanoparticles, which led to reduced cancer development over conventionally used synthetic based drug delivery systems. Observation of several in vitro studies shows that the pectin-based paclitaxel chemotherapeutics agent is equally potentially able to in tumor suppression by activating the Galectin-3-induced STAT3 and reducing the development of breast and ovarian cancer along with minimizing the cytotoxic effect (Emran et al. 2022).

Other than chemotherapy, oral therapeutics, vaccine antigen delivery, and most popularly, gene delivery is also becoming easy (Fig. 2.2). The better biocompatibility of Plant polysaccharide nanoparticles makes the poorly soluble oral-consuming drugs more soluble. Several studies reveal that plant-based polysaccharide pectin, derived from apple, was combined with the anticancer oral drug doxorubicin may be feasible in treatment where pectin improves the solubility and stability of

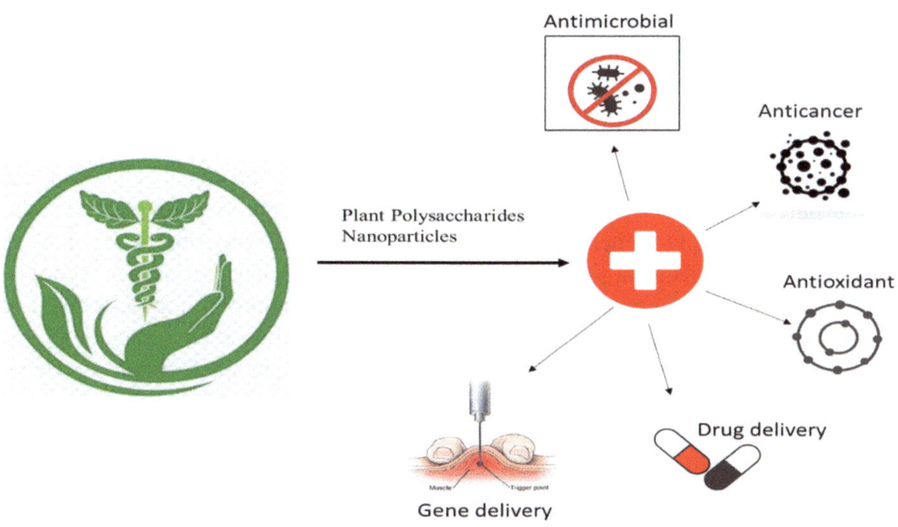

Fig. 2.2 Plant polysaccharide nanoparticles' application in medical field

doxorubicin by defending it from gastric and intestinal enzymatic degradation. A very interesting study in the International Journal of Pharmaceutics divulged that the oral administration of insulin coated with plant-based sodium alginate gives an inflated permeability in the small intestine and saves it from gastro-enzymatic degradation.

Plant polysaccharide nanoparticles (PPNs) are also involved in Vaccines, antigens and gene delivery systems as it provides better stability and enhances bioavailability. Conventionally used vaccines, sensitive rare antigens are often used to store in cool condition whereas the PPNs based vaccines and antigen are somewhere not sensitive to harsh environments.

2.6.1.2 Wound Healing

Plant extracts have been extensively used since the ancient era due to the rich presence of bioactive compounds such as polysaccharides, amino acids, vitamins, several minerals, secondary compounds terpenoids, flavonoids, etc. The usage of plant polysaccharides in various fields of industries including medicine, food, and crops can help to find an alternative to the conventionally used chemicals which are toxic, immunogenic, and side-effective for human health and the environment. The application of plant polysaccharide nanoparticles in wound healing gives a new insight that enhances the regrowth and proliferation of tissues. Prevention of wounds is a complex physiological process having multiple stages including hemostasis, inflammation, cell proliferation, and remodeling of tissue. Plant polysaccharide-based nano molecules can interfere in different stages of wound healing due to their antimicrobial, and anti-inflammatory properties. Otherwise, several plant polysaccharides may act as growth regulators which effectively promotes cell growth and tissue repair. Nanomaterials are more effective than any other form due to their smaller sizes and surface plasmon responses. The higher permeability of nanoparticles can surely help to transport the plant bioactive compounds such as polysaccharides to the cell which facilitates quick healing.

Plant polysaccharides eg. chitin, chitosan, starch, lignin, cellulose, pectin, alginate, and hyaluronic acid play an important role in wound repair. Chitosan and hyaluronic acid provide antibacterial properties whereas chitin, alginate, and cellulose give extensively higher mechanical strength and biocompatibility. Hyaluronic acid, a naturally occurring polysaccharide present in connective tissues, synovial fluids, skin, and membranes mostly regulates the functionality of joints, bones, and lubricates. The wider application of HA nanomaterials can be seen in rheumatology, ophthalmology, and dermatology. Studies by Ahmed et al. reveal that glucocorticoid prednisolone (PD) is encapsulated within solid lipid nanoparticles (SLNs) coated with hyaluronic acid (HA) called HA-SLNs/PD are successfully effective on rheumatoid arthritis (Ahmed et al. 2022). Other than that, Electro spun nanofiber membranes, nanofibril dressings, hydrogels, etc. are commonly used as an application of nano molecules that furnish moist environments and better water retention along with antibacterial efficacies. Nowadays, very commonly used electro-spun nanofibrils for chronic wounds and burns (Tan et al. 2022) become famous due to their higher porosity, extensive surface-to-volume ratio, biodegradability, and ability to adhesion and mimic the extracellular matrix.

2.6.1.3 Bioactive Coatings

Plant polysaccharide nanoparticles are used for coating different types of drugs for delivery to the desired target rather than making the drug hit multiple targets. In this approach, the polysaccharide coating the drug is broken by acid-mediated cleavage which enables the slow degradation of polysaccharide releasing the drug over time (Ahmed et al. 2022). Starch is a polysaccharide that is made up of two types of polymers amylopectin and amylose are non-toxic and are very stable from enzymatic degradation. In recent studies, it has been found that starch capsules produced through the injection molding process serve as an alternate and stable drug delivery system to conventional metallic-oxide-coated drugs and gelatin capsules (Pires et al. 2023). Chitosan on the other hand proved to be a very efficient polysaccharide for antibiotic delivery in the specific infected area. It is also used for biomedical purposes which include transmission electron microscopy and zeta potential measurement (Soares et al. 2022). Dextran is mainly used for the coating of anti-cancer drugs for delivery to its target site. The most advantageous property of this polysaccharide is that it is easily water-soluble. The Dextran-based nanoparticles are highly applicable to anti-cancer therapy but in lower concentrations as higher concentrations of dextran showed hepatotoxicity. Further studies are conducted to determine any kinds of side effects related to these nanoparticles. The future aspects of these plant polysaccharides are very promising and easily available as cost-effective than the conventional drug manufacturing process (Kurczewska et al. 2022).

2.6.2 *Agricultural Application*

Agriculture has always been the primary backbone for the economic development of a country while on the other hand, environmental pollution is becoming a serious global challenge to face for a clean and sustainable environment. With the introduction of nanotechnology in these two fields, more efficient and less complex formulations are consistently generated for better agricultural purposes and environmental bioremediation. These newly formulated materials can alter the physiology of plants, antimicrobial activity on plants, and environmental remediation which includes reductive immobilization of heavy metals in soil and prevention of leaching of groundwater. Various nanoparticles have been used for bioremediation and agricultural purposes (Mittal et al. 2020). Nanomaterials, made of chitosan, showed a promising effect on the growth of tobacco plants by 54–60%. It also showed a significant effect on other crops such as barley, soybean, and corn. It enhanced the process of photosynthesis by interaction with chloroplasts and intensifying seed

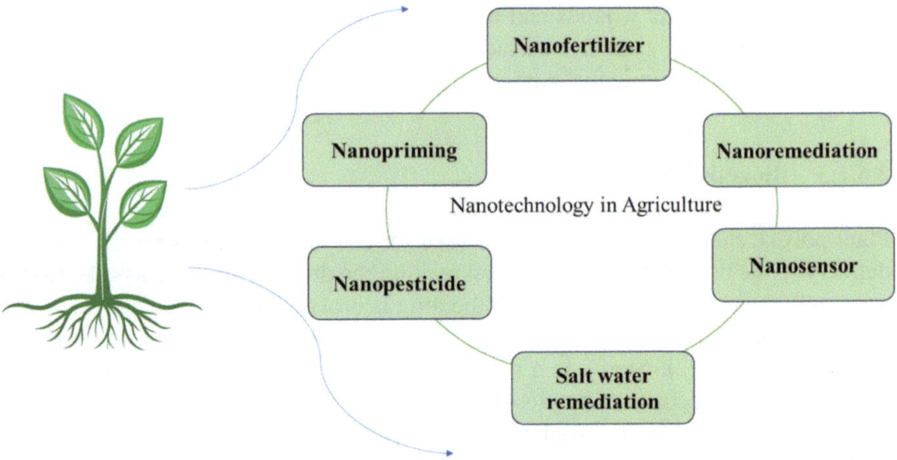

Fig. 2.3 The agricultural applications of plant polysaccharide nanoparticles

germination levels by increasing the protein and chlorophyll content and root elongation in maize plants, to withstand drought conditions. Nanomaterial produced from chitosan (CSNPs) increased seed germination in wheat by upregulating the level of Indole-3-acetic acid (IAA). On the other hand, it is also to protect the agricultural plants from various kinds of pathogens affecting crop production. Alginate-based nanoparticles (ABNs) play a major role in the anti-phytopathogenic reaction against both bacteria as well as fungi. Other important applications include antioxidant actions and photo-catalytic applications. Conventionally, environmental remediation is done by applying various kinds of chemicals and filtering processes.

Due to its smaller size (1–100 nm) and greater surface area to volume ratio, it provides more surface area of reaction leading to the increase of reactivity. In most cases, Starch starch-based nanoparticles are used for environmental remediation purposes and can be easily sourced from any plant source as described in Fig. 2.3. Starch is mainly combined with metallic oxides zero-valent metals and polymer-based compounds to form nanoparticles. Starch-stabilized nanoparticles of Fe_3O_4 are used for the removal of perfluorooctanoic acid (PFOA) in soil and groundwater and are effective in the removal of arsenic the use of nanoparticles in these two fields played a major role in agricultural and environmental remediation. To understand, the study of nanomaterials should be done further for better and successful results (Gamage et al. 2021, Annu et al. 2024).

2.6.3 *Food Packaging*

The planetary concern of food processing and packaging bears very misfortune as our traditionally used food preservatives and packaging materials are derived from petroleum and other chemically toxic products. Packaging that is sustainable non-toxic biodegradable and environmentally friendly is now sought after by the food processing industry. Plant polysaccharides are naturally occurring biopolymers that are easier to modify and have greater strength and durability. Recent studies on plant polysaccharide-based nanoparticles typically reveal new applications such as the food processing and packaging sectors which are directly aided in optimization and modification.

The antimicrobial qualities of plant polysaccharide nanomaterials such as chitosan starch and carrageenan make them useful in food processing for a variety of purposes. Nanomaterials based on pectin cellulose and chitin are typically used to coat fruits and vegetables with edible food. This increases the shelf life of edible foods by adding extra protection against pathogens moisture and gaseous exchange (Fig. 2.4). According to Jurkaninová et al. (2024) naturally occurring *β*-D-glucan from cereals has become popular in bakers industries these days for use in biscuits cakes cereal bars processed flavored milk and yogurts because it adds elasticity and compactness to dough and improves its nutritional value and flavor. (Jurkaninová et al. 2024). Nanotechnology is a rapidly expanding field that is readily linked to environmental and health concerns. Combining nanoparticles with plant polysaccharides offers a number of benefits. G. thermal capacity mechanical and chemical strength and stability which enable food processing and packaging at a new level. New food packaging materials like copper zinc and silver are antimicrobially effective lightweight elastic and resistant to freeze-thaw behavior. Particles in nanoparticle form.

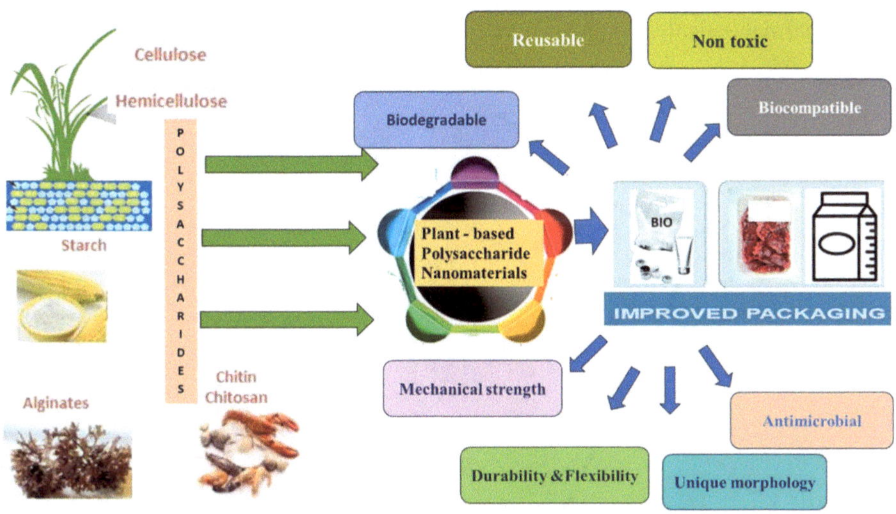

Fig. 2.4 The plant polysaccharide mediated nanoparticles for food packaging

2.6.4 *Environmental Remediation*

Human health and ecosystems are seriously threatened by environmental contamination particularly that caused by organic pollutants heavy metals and pathogens. One of the most important issues of our day is environmental pollution which has an impact on the quality of the air soil and water. Traditional environmental remediation techniques include chemical and physical methods that can also be harmful to the body and cause secondary pollution and contamination. An alternate method for getting rid of contaminants harmful infections and environmental pollutants is to use polysaccharide nanoparticles. Numerous functional groups found in naturally occurring polysaccharides including cellulose hemicellulose pectin and starch offer a range of chemical interference making them suitable options for environmental treatment. As illustrated in Fig. 2.5 the distinct physicochemical properties of these polysaccharides enable the formation of nanoparticles (NPs) which have the ability to encapsulate pollutants or aid in their removal through adsorption or chemical reaction. Heavy metals such as lead cadmium molybdenum arsenic etc. in the body of water is regarded as one of the most important worldwide threads. Aldehyde and keto groups found in polysaccharide-based nanoparticles interact with heavy metals and aid in absorption making them a promising tool for their removal. Electrostatic attraction ion exchange and chelation techniques have been successfully applied to chitosan cellulose and pectin-like polysaccharides (Hazmi et al. 2024). Additionally, plant polysaccharide nanoparticles are thought to be carriers of multiple catalysts and enzymes which promote the degradation of organic chemicals pesticides dyes and pharmaceutical products they can also remove organic pollutants. The

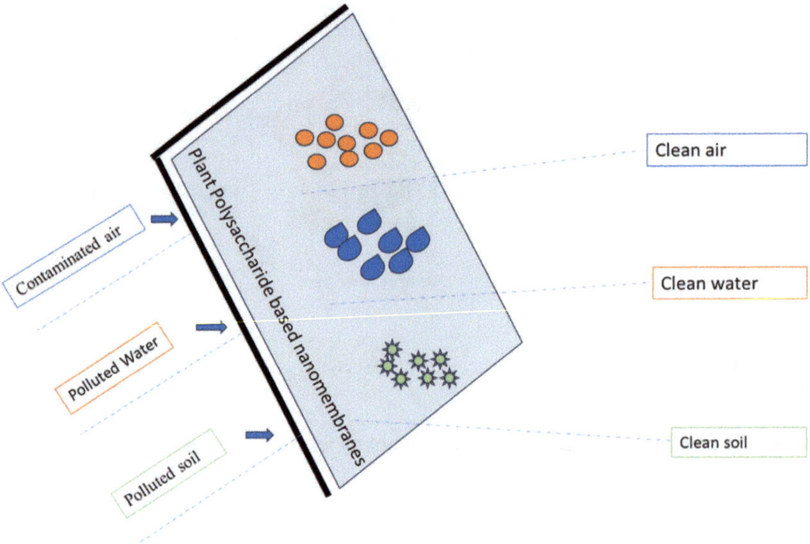

Fig. 2.5 Application of plant polysaccharide nanoparticles on environmental remediation

polysaccharides high reactivity is also caused by its functional groups the majority of which are hydrophilic. Recently chitin cellulose and starch have been employed for remediation and have been found to be effective in achieving this goal.

By altering their functional groups starch-hydrogels can effectively remove a variety of cationic and anionic dyes while also increasing their swelling capacity. Aerogels a class of nanomaterial with low density high flexibility and high porosity and surface area are primarily made from cellulose. These are distributed throughout a three-dimensional structure and work well for adsorption. Chitin-based nanoparticles have the ability to form films have good adsorption capabilities and are bacteriostatic and capable of chelating transition metal ions. Additional research is being done on the application of different plant-based polysaccharide nanoparticles for environmental remediation (Saraswat et al. 2023). Since pathogens are becoming more vulnerable and susceptible due to the development of antimicrobial resistance they are also contributing factors to environmental issues. Because of their antimicrobial properties nanoparticles of silver copper and zinc are widely used. Combining plant polysaccharides with nanoparticles offers far more potent defenses against quorum sensing biofilm formation bacterial or fungal infections etc. In bioremediation procedures plant-based polysaccharides can act as biosorbents assisting in the purification of contaminated environments. Particularly in soil and water bodies their capacity to bind with pollutants aids in the restoration of ecological balance.

2.7 Role of Artificial Intelligence and Machine Learning in Nanotechnological Applications

Artificial Intelligence (AI) and Machine Learning (ML) algorithms are two leading approaches, exhibiting large datasets initiated from several optimization techniques, empowering for swift surveying and characterization of patterns. For instance, machine learning models can be worked on the recognition of correlations between the derivation of conditions of plant polysaccharide nanoparticles and their physicochemical properties, such as size, shape, and stability. This predictive capability can guide researchers in optimizing synthesis protocols and enhancing the quality of the nanoparticles.

AI algorithms can work on vast amounts of related parameters linked to drug-molecule interactions, possessing the detection of minimal formulations. The physicochemical properties, enabling the design of controlled release systems, and targeted drug release can be analyzed by ML. Moreover, reinforcement learning techniques can optimize the drug loading process by adjusting parameters in real time, ensuring maximum encapsulation efficiency. In the contemporary periods, AI and ML can promote the design of drug delivery approaches based on individual patient profiles after observing the genomic and proteomic data which led to revolutionizing the precision medicine strategy.

In the field of other applications of nanoparticles, AL & ML also developed the whole strategy of environmental remediation and food production and packaging. AI and ML can enhance the quality of food products containing polysaccharide nanoparticles as after employing predictive analytics, scientists can analyze the minimal need for nanoparticles usually to gain desired sensory and nutritional imputes. Hence, the conventional trial methods have changed along with the deduction of waste of resources. Again, Polysaccharide nanoparticles have been employed for their potential efficacy in environmental applications, eg. wastewater treatment and soil remediation. AI can enhance the efficiency of environmental remediation processes by examining vast data from various sources, including sensor records and ecosystems' parameters. Machine learning models have been used to predict the effectiveness of polysaccharide nanoparticles in removing contaminants from water or soil based on the composition and structure of the nanoparticles. This approach can lead to the development of more effective and sustainable remediation strategies.

2.8 Conclusion

The approaches of plant polysaccharide nanoparticle research are continuously developing nowadays. The potential applications in various fields of industries made them much more valuable. Scientists are developing new and exclusive methods to evaluate and functionalize these nanoparticles for targeted uses. Their biocompatibility, biodegradability, and versatility offer a huge reply to many of the challenges we face in health, food, and environmental conservation. Moreover, the integration of AI and ML in green nanotechnology will force it to different levels.

References

Abka-Khajouei R, Tounsi L, Shahabi N, Patel AK, Abdelkafi S, Michaud P (2022) Structures, properties and applications of alginates. Mar Drugs 20(6):364. https://doi.org/10.3390/md20060364

Ahmad A, Gulraiz Y, Ilyas S, Bashir S (2022) Polysaccharide based nano materials: health implications. Food Hydrocoll Health 2:100075. https://doi.org/10.1016/j.fhfh.2022.100075

Annu A, Mittal M, Tripathi S, Shin DK (2024) Biopolymeric nanocomposites for wastewater remediation: an overview on recent progress and challenges. Polymers 16(2):294. https://doi.org/10.3390/polym16020294

Bourakadi KE, Semlali FZ, Hammi M, Achaby ME (2024) A review on natural cellulose fiber applications: empowering industry with sustainable solutions. Int J Biol Macromol:135773. https://doi.org/10.1016/j.ijbiomac.2024.135773

Díaz-Montes E (2022) Polysaccharides: sources, characteristics, properties, and their application in biodegradable films. Polysaccharides 3(3):480–501. https://doi.org/10.3390/polysaccharides3030029

Emran TB, Islam F, Mitra S, Paul S, Nath N, Khan Z et al (2022) Pectin: a bioactive food polysaccharide with cancer preventive potential. Molecules 27(21):7405. https://doi.org/10.3390/molecules27217405

Joudeh N, Linke D (2022) Nanoparticle classification, physicochemical properties, characterization, and applications: a comprehensive review for biologists. J Nanobiotechnol 20(1). https://doi.org/10.1186/s12951-022-01477-8

Jurkaninová L, Dvořáček V, Gregusová V, Havrlentová M (2024) Cereal B-D-Glucans in food processing applications and nanotechnology research. Foods 13(3):500. https://doi.org/10.3390/foods13030500

Kurczewska J (2022) Recent reports on polysaccharide-based materials for drug delivery. Polymers 14(19):4189. https://doi.org/10.3390/polym14194189

Li Y, Zheng Y, Zhang Y, Yang Y, Wang P, Imre B et al (2021) Brown algae carbohydrates: structures, pharmaceutical properties, and research challenges. Mar Drugs 19(11):620. https://doi.org/10.3390/md19110620

Mesoppirr LS, Suter EK, Omwoyo WN, Oyaro NM, Nelana SM (2024) Isolative synthesis and characterization of cellulose and cellulose nanocrystals from *Typha angustifolia*. Open J Appl Sci 14(09):2443–2459. https://doi.org/10.4236/ojapps.2024.149161

Mittal D, Kaur G, Singh P, Yadav K, Ali SA (2020) Nanoparticle-based sustainable agriculture and food science: recent advances and future outlook. Front Nanotechnol 2. https://doi.org/10.3389/fnano.2020.579954

Muddasar M, Culebras M, Collins MN (2024) Lignin and its carbon derivatives: synthesis techniques and their energy storage applications. Mater Today Sustain:100990. https://doi.org/10.1016/j.mtsust.2024.100990

Namasivayam SKR, Kumar SN, Kamil TM, Ravi T (2020) Biopolymer-mediated coating influence on wastewater treatment efficacy of silver nanoparticles synthesized from fungal consortium. Natl Acad Sci Lett 43(6):557–561. https://doi.org/10.1007/s40009-020-00914-2

Pires PC, Mascarenhas-Melo F, Pedrosa K, Lopes D, Lopes J, Macário-Soares A et al (2023) Polymer-based biomaterials for pharmaceutical and biomedical applications: a focus on topical drug administration. Eur Polym J 187:111868. https://doi.org/10.1016/j.eurpolymj.2023.111868

Qiu A, Wang Y, Zhang G, Wang H (2022) Natural polysaccharide-based Nanodrug delivery systems for treatment of diabetes. Polymers 14(15):3217. https://doi.org/10.3390/polym14153217

Sahoo DR, Biswal T (2021) Alginate and its application to tissue engineering. SN Appl Sci 3(1). https://doi.org/10.1007/s42452-020-04096-w

Saraswat P, Singh S, Prasad M, Misra R, Rajput VD, Ranjan R (2023) Applications of bio-based nanomaterials in environment and agriculture: a review on recent progresses. Hybrid Adv 4:100097. https://doi.org/10.1016/j.hybadv.2023.100097

Soares Í, Faria J, Marques A, Ribeiro I a C, Baleizão C, Bettencourt A et al (2022) Drug delivery from PCL/chitosan multilayer coatings for metallic implants. ACS Omega 7(27):23096–23106. https://doi.org/10.1021/acsomega.2c00504

Tang Z, Huang G (2022) Extraction, structure, and activity of polysaccharide from Radix astragali. Biomed Pharmacother 150:113015. https://doi.org/10.1016/j.biopha.2022.113015

Yin IX, Zhang J, Zhao IS, Mei ML, Li Q, Chu CH (2020) The antibacterial mechanism of silver nanoparticles and its application in dentistry. Int J Nanomedicine 15:2555–2562. https://doi.org/10.2147/ijn.s246764

Yusuf A, Almotairy ARZ, Henidi H, Alshehri OY, Aldughaim MS (2023) Nanoparticles as drug delivery systems: a review of the implication of nanoparticles' physicochemical properties on responses in biological systems. Polymers 15(7):1596. https://doi.org/10.3390/polym15071596

Chapter 3
Protein-Based Nanoparticles: Advancing the Delivery of Bioactive Compounds for Improved Therapeutic Applications

Vivek Raj, Nishtha Sood, Tushar Singh, Hirak Parikh, Tuhin Subhra Biswas, Harshita Loomba, and Santosh Kumar Karn

3.1 Introduction

3.1.1 Drug Delivery System

Drug delivery refers to the administration of pharmaceutical compounds to achieve therapeutic effects (Tiwari et al. 2012). It is a specific engineered technique to deliver medication in such a way to increase drug concentration in specified body areas. Typically, newly developed drugs are formulated in simple dosage forms that effectively treat the diseased tissue through a protected interaction. However,

V. Raj (✉)
Department of Biotechnology, Patna Women's College (Autonomous), Patna University, Patna, Bihar, India

N. Sood
Centre of Nanotechnology, Indian Institute of Technology, Roorkee, Uttarakhand, India

T. Singh
GreenGrahi Solutions Private Limited, Roorkee, Uttarakhand, India, 247656

H. Parikh
Department of Earth and Environmental Systems, Indiana State University, Terre Haute, IN, USA

T. S. Biswas
Department of Biotechnology, National Institute of Technology, Warangal, Telangana, India

H. Loomba
University School of Biotechnology, Guru Gobind Singh Indraprastha University, Delhi, India

S. K. Karn
Department of Biochemistry and Biotechnology, Sardar Bhagwan Singh University, Balawala, Dehradun, Uttarakhand, India

V. Ghosh et al. (eds.), *Applications of Plant Bioactive Compounds-Based Nanoformulations*, Nanotechnology in the Life Sciences,
https://doi.org/10.1007/978-3-032-13086-0_3

designing and formulation of those drugs are relatively time-intensive and costly. The routes of administration of drug are totally based on the specific target site and the biochemical attributes comprising solubility and permeability. Among the various methods of drug dosage, oral administration is the most common (Alqahtani et al. 2021). Before the 1950s, drugs were primarily formulated as, tablets capsules or pills, syrups or dry powder, releasing their active compounds immediately upon exposure to water, without controlled-release properties (Park et al. 2022). Despite several advantages, these approach face major challenges like poor bioavailability, fluctuation in plasma level, repeated dosing and premature excretion from the body. Smith Kline Beecham introduced Spansule technology in 1952, which contains an increased amount of Active Pharmaceutical Ingredients [API]; enabled the controlled release of drugs, leading to the creation of sustained-release formulations (Park et al. 2022). Several strategies such as purification and structural modification, dose titration, individualized therapy, minimising toxicity and therapeutic drug monitoring have since been employed to improve the safety and efficacy of drugs. Additionally, innovative approaches like controlled, slow, and targeted drug delivery have gained significant attention to increase the sustainability. Now the drug delivery systems (DDS) enhance the pharmacological attributes of conventional drugs by utilizing nanoparticles and microparticles (Afzal et al. 2022). These systems are designed to modify or improve the therapeutic properties of drugs by serving as reservoirs for controlled drug release at specific sites, with altered pharmacokinetics and biodistribution; controlled drug delivery systems maintain a constant drug level over an extended period. Over recent decades, nanoparticle-based DDS has been extensively researched, driven by advances in nanotechnology, which have opened new possibilities in medical sciences, particularly in drug delivery (Lohcharoenkal et al. 2014). Advantages and disadvantages of different drug releasing mechanisms have been listed in (Table 3.1).

3.1.2 Nanoparticles

Nanoparticles are solid colloidal structures that vary in size and shape, typically ranging from 10 to 100 nm. Their physiochemical properties are different from bulk materials and microparticles, that make them a promising candidate for commercial and medical development. Their size, morphology, and surface properties can be tailored, enabling site-specific drug delivery (Hoshyar et al. 2016). The Food and Drug Administration (FDA) in the United States has approved the usage and development of pharmaceuticals based on nanotechnology in a number of pharmaceutical businesses in recent years (Medina et al. 2007). Various factors that influence material selection for nanoparticle synthesis are listed in (Fig. 3.1). There are different classes of nanoparticles like- liposomes, emulsions, polymers, ceramic, metallic and goldshell nanopartciles and quantum dots. In the biopharmaceutical industry,

Table 3.1 Advantages and disadvantages of different drug releasing mechanism

Mechanism	Description	Advantages	Disadvantages	References
Diffusion	Influenced by the concentration gradient between drug loaded hydrogel and medium Drug releases through channel/pores within network	Beneficial for target specific drug delivery; low toxicity	Less stability due to interaction of blood components with micelles	Cao and Ziener (2013), Gupta (2020)
Degradation	Encapsulated drug is released due to breakdown of gel network	Controlled and sustainable drug release Improved stability and bioavailability	Further purification is required Partial degradation may affect drug efficacy	Larsson et al. (2014), Zarekar et al. (2017)
pH shift	In response to change the pH of medium, therapeutics released from nanogel	Targeted drug delivery inside pH sensitive area Intense drug concentration at the intended side helps to achieve desired outcome	High cost of polymer Release of residual materials can be toxic	Sagar et al. (2018), Argentiere et al. (2011)
Temperature	In exposure to specific temperature, bioactive drugs released from temperature sensitive nanogel	Drug directly injected into blood stream, minimising side effects High drug loading capacity (Hydrophobic; hydrophilic)	Poor biocompatibility Temperature variation can affect material properties	Ahmed et al. (2020)
Light sensation	Near-Infra Red light (Wave length 650–600 nm) can trigger the drug release	Mild reaction condition, high biocompatibility Minimise the usage of organic solvents	Limited amount of drug release Time consumed Difficult to scale up	Uthaman et al. (2020), Gonçalves et al. (2010)
Swelling	Nanogels absorb sufficient amount of water, swelling occurs and subsequent release of drugs	Controlled and sustained drug release Improved biodegradability, high drug loading capacity	Premature drug release Breakdown of nanogels	Johansson et al. (2007), Anupama and Priyanka (2018)

these nanoparticles have been proven to enhance drug performance *in vivo* by improving parameters like bioavailability, solubility and targeted delivery (Lohcharoenkal et al. 2014). Beyond drug delivery, nanotechnology has also found applications in therapeutics, diagnostics, energy, and materials sciences.

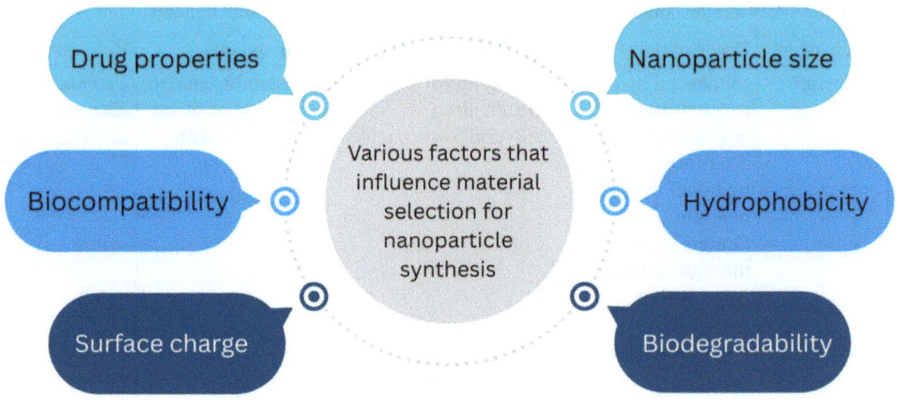

Fig. 3.1 Various factors that influence material selection for nanoparticle synthesis

3.1.3 Nanoparticles as Drug Delivery Systems

Several nano-carrier systems have been developed using liposomes, metals, and synthetic polymers. Despite their advantages, each system has limitations. For instance, liposomes often exhibit low drug encapsulation efficiency, reduced stability, and rapid drug release rates for hydrophilic drugs (Liu et al. 2022b). Metal-based nanoparticles face issues like low clearance rates and limited surface modifications, while synthetic polymers like poly (lactic acid) (PLA) and poly (lactic-co-glycolic acid) (PLGA) face challenges regarding to encapsulation of those therapeutic molecules inside their cores. As a result, the components of engineered nanomaterials must be selected carefully to ensure biosafety and biocompatibility (Rocha et al. 2022). Engineered nanomaterials must exhibit biosafety and biocompatibility, necessitating careful selection of their components. Proteins are amphiphilic in nature, this unique property of proteins allows them to interact with both polar and non-polar moieties, thus making them a perfect candidate for sysnthesis of nanoparticles (Zhang et al. 2020). Protein-based nanomaterials (PNMs) and protein nanocomposites that offer advantages such as limited toxicity, rapid clearance, minimal immune response, and customizable synthesis techniques. These nanoparticles can also be functionalized for targeted drug delivery, with demonstrated applications in various biomedical contexts (Spicer et al. 2018).

Gels an example of collagen-based polymers, above all, stand out as other gels inventive drug carrier systems owing to their elasticity and ability to capture fluids. Through their three-dimensional polymeric frameworks, they can contain water (hydrogels), air (aerogels), or oil (oleogels). (Wang et al. 2025). Historically, wound-healing fillers or hyaluronic acid-based soft tissue repair gels have served as key biomaterials. The focus of more recent work has been on the design of tough cell-compatible gels as well as responsive gels and those intended for other biomedical applications. A new paradigm of drug delivery through blood circulation is enabled by hydrophilic biocompatible microgels which are easily navigated into capillaries and can target tissues due to their small size. The microcapsules are designed for cellular uptake and enable site-specific drug release (Yao et al. 2018). Further developing these systems in regard to biocompatibility and effective therapeutic outcomes, the latest research focuses on advanced systems of drug delivery using nanoparticles and proteins.

3.2 Protein Based Nanoparticles

Bioengineering is a new discipline that combines the unique properties of proteins with new technological advances at the nanoscale level. This has resulted in a newly created class of protein-based nanoparticles (PNPs). Solid nanoparticles are colloidal structures that can be designed into different forms and sizes in the range of 10–100 nm. These nanoparticles are also said to have controllable size, shape, surface properties, and other features that enable targeted delivery of a drug to a specific location. Other than in drug delivery, nanotechnology has sophisticated applications in sensing and therapeutics, energy harvesting, diagnostics, filtration systems, as well as in coating and packaging materials. Advanced nanotechnology, for example, can be applied in biopharmaceuticals to enhance bioavailability in using therapeutic drugs *in vivo* (Zhang et al. 2013). In biomedicine, it has significantly advanced systems for imaging and targeted delivery of drugs.

Due to their small size, protein-based nanoparticles can easily penetrate cells via endocytosis. These nanoparticles exhibit several advantages over other types, including controlled particle size, surface modifications, biodegradability, and stability. They also demonstrate reduced toxicity and immunogenicity. Furthermore, their stability, and half-life can be enhanced significantly while protecting the bioactive drug from renal clearance and enzymatic degradation, making them highly effective for targeted drug delivery applications, such as pulmonary drug delivery, cancer and tumour therapies, vaccine development and so on (Davidov-Pardo et al. 2015). The characteristics of protein based nanoparticles are shown in (Fig. 3.2).

3.2.1 Classification of Proteins Used in PNPs

Protein based nanoparticles can be classified into in multiple ways, based on their origin, source, and their structure- natural or recombinant by their origin, animal or plant -based by their source and globular or fibrous by their structure. This helps to choose the right protein matrix for particular therapeutic application.

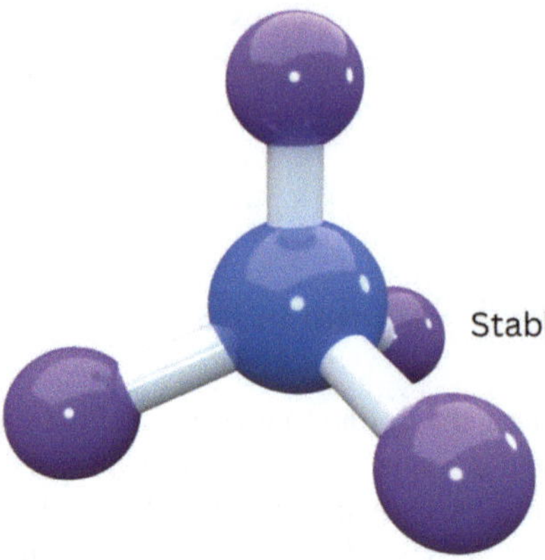

Fig. 3.2 Characteristics of protein based nanoparticles

3.2.1.1 Natural Versus Recombinant Proteins

Natural proteins are derived directly from biological sources with minimal chemical modification. Examples include albumin from serum, gelatin from collagen, and casein from milk. These proteins are inherently biocompatible and biodegradable, making them suitable for clinical applications. In contrast, recombinant proteins are produced through precise genetic engineering in expression systems like bacterial, mammalian, or plant cells, which allows for exact control of amino acid sequence, molecular weight, and overall functional domains to tailor nanoparticle performance. Notably, elastin-like polypeptides (ELPs) have been utilized as thermoresponsive carriers that undergo reversible phase transitions in response to temperature or pH changes, enabling highly targeted, stimuli-controlled release of therapeutic substances.

3.2.1.2 Animal Versus Plant-Based Proteins

Animal-derived proteins, such as albumin, gelatin, silk fibroin, and whey, naturally possess well-defined structures, durable mechanical properties, and a high affinity for a wide range of drugs, making them ideal for engineering nanoparticles that must withstand physiological stresses and afford controlled dosage release. Their established biocompatibility with mammalian tissues facilitates mucosal and oral administration; however, concerns over immunogenicity, potential pathogen transmission, and ethical sourcing have driven interest in non-animal alternatives. Plant-derived counterparts, such as storage proteins like zein, soy globulins, and pea legumin, offer comparable flexibility and self-assembly potential while minimizing potential infectious risks and aligning with vegetarian or vegan formulations. Although plant proteins often require additional processing to achieve the purity and conformational uniformity of their animal-derived analogs, advances in extraction and recombinant expression have begun to narrow this gap, delivering nanoparticle platforms that combine renewable sourcing with manipulative surface chemistry for targeted, low-immunogenicity drug delivery.

3.2.1.3 Globular Versus Fibrous Proteins

It is also possible to classify proteins based on their tertiary and quaternary structures. Globular proteins such as albumin or casein fold into compact, water-soluble spheres rich in surface functional groups, facilitating efficient drug encapsulation, predictable self-assembly into monodisperse nanoparticles, and rapid, enzyme-mediated clearance. On the other hand, Fibrous proteins such as silk fibroin or collagen form extended β-sheet or helix networks that yield high-aspect-ratio fibers or core–shell particles with exceptional mechanical strength and slow, bulk-erosion–driven release. Intermediate systems, such as elastin-like polypeptides, combine globular domains and fibrillar coacervation to achieve a high payload capacity with controllable elasticity and degradation. Finally, by choosing or engineering protein scaffolds along this structural spectrum, scientists can precisely tailor nanoparticle stability, release kinetics, and biodegradation for various therapeutic applications. These overlapping of categories give researchers and formulators better understanding of specific physical, chemical, and therapeutic requirements of their nanoparticle systems.

It can be concluded that classifying proteins used in making protein-based nanoparticles are a very crucial step for strategic formulation and development. Performance in terms of stability, efficacy, biocompatibility and drug release situation of the nanocarrier is purely dependent on a protein's origin, source, and structure.

3.2.2 Criteria for Protein Selection

Selecting appropriate protein for making of protein based nanoparticles is very essential as it determines how successful the functioning of nanoparticle is. the factors critically influence the interaction of protein with body, biodegradability, molecular weight, stability, solubility whether it meets the desired resultant or not.

3.2.2.1 Biocompatibility and Biodegradability

The most fundamental requirement for any biomedical carrier is its compatibility with biological tissues. Proteins used in PNPs should not elicit an adverse immune response and must degrade into non-toxic byproducts that can be readily eliminated from the body. Naturally occurring proteins like albumin and gelatin are inherently biocompatible and biodegradable, which makes them ideal candidates for systemic and localized therapeutic applications.

3.2.2.2 Molecular Weight and Structure

A protein's molecular weight affects its capacity to encapsulate medications, form stable nanoparticles, and regulate release kinetics. For instance, whereas lower molecular weight proteins might promote quicker drug release, higher molecular weight proteins typically form larger, more stable particles. Drug loading capacity and release behavior are also influenced by the protein's tertiary structure, whether it is fibrous or globular.

3.2.2.3 Solubility and Processability

One significant factor influencing nanoparticle fabrication techniques like desolvation, emulsification, or self-assembly is solubility in aqueous or organic solvents. Proteins selected for nanoparticle platforms must dissolve fully and maintain their native fold under the shear forces and moderate heating typical of pharmaceutical manufacturing. Casein and whey proteins meet these requirements, remaining soluble and structurally intact throughout emulsification. Employed as emulsifiers in oil-in-water preparations, they yield uniform droplets that encapsulate hydrophobic compounds such as curcumin or vitamin D_3 with excellent loading efficiency and release profiles that mirror physiological digestion (Chen et al. 2020).

3.2.2.4 Functional Groups and Surface Modifiability

Glutamates, cysteines, and lysines on the surface of proteins act as chemoselective anchors: glutamate carboxylates couple via EDC/NHS to amine-bearing probes; cysteine thiols undergo alkylation with maleimide or haloacetamide to produce stable thioether-linked payloads; and lysine ε-amines react with activated esters (e.g., NHS-PEG) to install PEG or ligands. These simple aqueous bioconjugations improve protein stability, circulation, and therapeutic efficacy by enabling precise polymer attachment, targeting moieties, and imaging labels. For example, albumin's abundant carboxyl and amino groups make it highly amenable to chemical modification.

3.2.2.5 Regulatory and Clinical Acceptance

Proteins that are already approved for medical use by regulatory authorities such as the FDA or EMA provide a clear translational advantage. Albumin (marketed as Abraxane® when loaded with paclitaxel) and gelatin are examples of clinically utilised proteins that streamline the development process for new nanoparticulate formulations.

It can be summarised that the ideal protein choice is the one, that is safe for the body, consist of stable structure, can be easily modified and has regulatory feasibility. These various factors guide researchers in designing efficient, safe, and scalable PNP systems but also practical to produce and tailored to specific therapeutic goals.

3.2.3 Overview of Major Proteins Used in PNPs

This section basically explains the main types of proteins used to formulate protein-based nanoparticles (PNPs), focusing on their physical and chemical characteristics, how they're made, their ability to carry drugs, and how they're used in therapy.

3.2.3.1 Albumin

As α-helical, disulfide-stabilized proteins with numerous hydrophobic nooks and charged residues, serum albumins (such as HSA and BSA) are ideal scaffolding for nanoparticles. Desolvation, thermal gelation, or moderate emulsion-crosslinking techniques are frequently employed to create albumin nanoparticles, which provide homogeneous, biocompatible carriers with high drug loading and prolonged circulation times. Selective tumor accumulation is fueled by their intrinsic affinity for endothelial gp60 receptors as well as the increased permeability and retention (EPR) impact. The translational potential of these nanoparticles is demonstrated by the clinical success of nab-paclitaxel (Abraxane®), an albumin-bound paclitaxel formulation.

3.2.3.2 Gelatin

Gelatin, a low-molecular-weight collagen derivative with an amphoteric surface and RGD-type adhesion motifs, self-assembles into colloidally stable nanoparticles using emulsion-solvent-evaporation, coacervation, or anti-solvent precipitation techniques. To create constructs that combine quick biodegradability with low immunogenicity qualities essential for the localized and systemic delivery of peptide and protein therapies these polymeric cores can be covalently reinforced with biocompatible crosslinkers, such as genipin or aldehydes. The gelatin scaffold's numerous amine and thiol functions enable orthogonal chemical modification, allowing for the precise conjugation of imaging probes, targeting ligands, or responsive moieties to customize cellular internalization pathways, pharmacokinetic profiles, and diagnostic readouts.

3.2.3.3 Casein

Using κ-casein–Ca^{2+} cross-linking, casein, an amphiphilic milk-derived phosphoprotein (19–25 kDa), self-assembles into 100–300 nm micelles. It may also be converted into nanoscale carriers using ionic gelation (e.g., $CaCl_2$) or simple emulsification. The hydrophobic interior of these core–corona complexes sequester weakly soluble medicines, such as paclitaxel or curcumin, while the phosphorylated shell guarantees stability and dispersion in gastrointestinal fluids. Casein nanoparticles protect encapsulated chemicals against enzymatic and photochemical (UV) degradation, allow for regulated release in the stomach, and can increase oral bioavailability by up to twofold by taking advantage of pH-responsive swelling and reversible ionic interactions. Thus, this scalable platform with biological origins provides a flexible method for oral delivery of lipophilic medications. (Chen et al. 2020).

3.2.3.4 Silk Fibroin

Extracted from Bombyx mori cocoons, silk fibroin is a high-molecular-weight (~350 kDa heavy-chain) β-sheet-rich fibrous protein that self-assembles into 100–200 nm nanoparticles through probing sonication or desolvation (LiBr dialysis followed by ethanol precipitation). In addition to providing remarkable mechanical strength and slow proteolytic degradation (ranging from weeks to months), the broad β-sheet crystallinity enables encapsulation efficiencies of over 90% for both proteins and small compounds. For cancer applications or scaffolds used in regenerative medicine, prolonged, controlled release kinetics can be achieved by carefully adjusting the β-sheet composition during formulation. Moreover, silk fibroin nanoparticles offer a flexible and high-performance platform for enhanced drug delivery and tissue engineering, owing to their intrinsic biocompatibility and biodegradability.

3.2.3.5 Whey Protein

Whey proteins represent amphiphilic by-products derived from the cheese manufacturing process, predominantly comprising β-lactoglobulin (18 kDa), α-lactalbumin (14 kDa), and remaining immunoglobulins. Through processes such as heat-induced gelation (≥70 °C) or complex coacervation with polysaccharides, such as chitosan, these proteins can be induced to self-assemble into nanoparticles measuring 50–150 nm. These protein-based carriers leverage inherent antioxidant and antimicrobial peptide sequences, along with enhanced mucosal permeability, to facilitate oral, transdermal, and pulmonary administration, whilst simultaneously safeguarding encapsulated peptides and nutraceuticals from both enzymatic and photochemical degradation. Formulation studies indicate that whey-derived nanoparticles provide a biocompatible and scalable platform conducive to the sophisticated delivery of peptide therapeutics and functional food constituents. Furthermore, they have been shown to enhance peptide bioavailability to exceed 75% in porcine models, thereby suggesting a significant improvement in gastrointestinal stability and absorption in vivo.

3.2.3.6 Zein

Zein, a prolamin with a molecular weight of approximately 22 kDa and possessing around 75% hydrophobicity, undergoes self-assembly to form colloidal nanoparticles within the size range of 150–500 nm through mechanisms such as antisolvent precipitation (transitioning from aqueous to ethanol) or complex coacervation, thereby achieving encapsulation efficiencies exceeding 80% for hydrophobic entities including doxorubicin and curcumin. These nanoparticles exhibit a combination of low immunogenicity and resistance to degradation in gastric environments, along with the capability for sustained release over 48 h in simulated intestinal fluid, rendering them particularly advantageous for the mucosal and oral delivery of poorly soluble bioactive compounds. By meticulously adjusting formulation parameters including zein concentration, pH, ionic strength and incorporating biopolymer or lipid coatings, researchers can effectively regulate particle dimensions, polydispersity, zeta potential, and long-term colloidal stability, thus facilitating the scalable production of zein-based carriers intended for nutraceutical and pharmaceutical applications.

3.2.3.7 Soy Protein

By utilizing their inherent amphiphilicity and colloidal stability, soy glycinin (a 360 kDa hexamer) and β-conglycinin (a 180 kDa trimer), the two primary globulins in soy, can be induced to co-assemble into homogeneous nanoparticles of 100–250 nm in diameter through ethanol-induced desolvation or pH/ionic complexation. The pH-sensitive architecture of the nanoparticles facilitates disassembly under endosomal acidity, thereby enhancing endosomal escape and payload release. Additionally, the abundance of surface lysine and glutamine residues facilitates the easy covalent attachment of targeted ligands or polymers. Beyond gene and vaccine delivery, where these carriers attain transfection efficiencies exceeding 60% in vitro, soy protein nanoparticles are highly versatile platforms for pH-responsive drug, vaccine, and gene delivery due to their excellent emulsifying qualities, favorable safety profile, and renewable, sustainable origin.

3.2.3.8 Elastin-Like Polypeptides (ELPs)

Elastin-like polypeptides (ELPs) are monodisperse polymers with repeating VPGXG motifs (10–60 kDa) that are genetically encoded. They assemble into nanoscale coacervates or block-copolymer micelles by going through an inverted temperature transition (LCST 30–40 °C). One can precisely program transition temperature, particle size, stability, and stimulus-responsive drug release by adjusting guest-residue hydrophobicity, block architecture, and crosslinkable sites (such as disulfides and hydrazones). This allows for in-situ depot formation and targeted payload delivery with little off-target leakage. ELPs are perfect for stimuli-responsive drug delivery because of their precise control over molecular composition, which allows for customizable release and degradation patterns.

3.2.4 Comparative Analysis of Protein Nanoparticles

This comparative analysis highlights the most widely used protein scaffolds for nanoparticle formulation summarizing each platform's molecular features (e.g., size, stability, functionalization sites, biodegradation rate) against critical performance metrics (drug-loading efficiency, release kinetics, circulation time, target accumulation) to reveal their respective advantages, limitations, and optimal therapeutic applications. Plant-derived scaffolds (e.g., zein, soy) are favored for oral delivery and sustainable production, whereas serum-derived albumin and gelatin are preferable for their high-affinity drug-binding and controlled-release capabilities. On the other hand, recombinant proteins stand out for their unique features that offer smart, stimuli-responsive therapies, marking a significant advancement in precision medicine. So, while choosing a protein for PNPs it is important to not only consider biophysical and pharmaceutical characteristics but also factors such as cost, regulatory approval, scalability, and the intended clinical application.

Protein	Source	Biocompatibility	Biodegradability	Drug binding	Release profile	Stability	Typical route	Key advantage	Limitation
Albumin (HSA/ BSA)	Human/ bovine blood serum	Excellent	Excelent	Hydrophobic & electrostatic	Sustained	High	Injectable, Oral	FDA-approved, high drug-binding, long circulation	Expensive (HSA), potential viral contamination (BSA)
Gelatin	Collagen (animal origin)	Good	Good	Hydrophilic & Amphiphilic	pH-responsive, modifiable	Moderate	Injectable, topical	Biodegradable, easy modifiable	Temperature-sensitive, animal-derived
Casein	Milk-derived (bovine)	Moderate	Good	Hydrophobic	Sustained (esp. oral)	Moderate	Oral	Food grade, GRAS status, mucoadhesive	Allergenicity, low pH instability
Silk fibroin	Silkworm cocoons	Excellent	Slow	Hydrophobic & Electrostatic	Sustained	Excellent	Injectable, topical	High mechanical strength, film-forming ability	Complex extraction process, batch variability
Whey protein	Milk- derived	Good	Good	Moderate	Fast	Moderate	Oral, Neutraceuticals	GRAS status, economical, nutritional appeal	Allergenicity, lower drug-loading cpotential
Zein	Maize	Good	Good	Hydrophobic	Sustained (esp. oral)	High	Oral	Plant-based, stable, GRAS-approved	Poor aqueous solubility, ineffective for hydrophilic drugs
Soy protein	Soybean	Moderate to good	Good	Hydrophobic	Moderate	Moderate	Oral, topical	Vegan, sustainable, low cost	Allergy causing potential, variable purity
ELPs	Recombinant (synthetic gene)	Excellent	Tunable	Sequence specific	Thermo/pH responsive	High	Injectable	Customisable, smart delivery, biotechnologically advanced	Expensive, limited clinical use, regulatory barriers

3.3 Role of Protein-Based Nanoparticles in Immunotherapy and Drug Delivery Systems

The immune system plays a pivotal role in maintaining homeostasis, providing protection against pathogens, tissue repair, and clearing dead cells. Immune cells constantly survey the body in order to identify and eliminate invading pathogens and malfunctioning cells. The immune system comprises of two arms: innate and adaptive. Innate immunity responds rapidly to pathogen invasion non-specifically via receptors recognizing molecular motifs, facilitating phagocytosis and chemical signaling. Key innate immune cells include neutrophils and macrophages. Adaptive immunity on the other hand, involves B and T lymphocytes, it activates an antigen-specific response to reinforce innate defences. CD4+ helper T cells release cytokines that regulate the activity of natural killer cells, B cells, and innate cells; while CD8+ killer T cells directly target pathogen infected cells in a more robust way (Qian et al. 2018).

Cancer, a leading cause of global mortality, is characterized by uncontrolled proliferation of abnormal cells. After invasion, cancer cells circulate through blood stream and form new malignant cells, which are more challenging to treat. Traditional treatments like radiotherapy, surgery, and chemotherapy are often limited by inadequate targeting, severe side effects, and drug resistance, leading to disease progression and treatment failure (Anand et al. 2023). Immunotherapy, which leverages the immune system to fight disease, has emerged as a promising approach for cancer treatment. It can be classified into activation immunotherapy (Effector tumour Antigen-specific T cells) or suppression immunotherapy (in autoimmune disease or Organ transplantation) based on its mechanism. Techniques such as specific cytokine administration, monoclonal antibody therapy and adoptive cell transfer (ACT) play pivotal role in modern immunotherapy. Protein based nanoparticles play a significant role in these therapies by acting as nano-carriers or nanovaccines to elicit specific cytotoxic responses against tumours. Nanoparticles can deliver drugs to lymph nodes, isolate and treat immune cells *ex vivo*, and administrate therapeutic agents directly to tumour sites (Qianet al. 2018; Irvine et al. 2015).

Targeted drug delivery using nanoparticles is particularly more desirable in cancer treatment. These carriers can deliver chemotherapeutic agents against tumours with minimal systemic toxicity, thereby sparing healthy tissues. Nanoparticles are widely used as carriers, to enhance the therapeutic index of chemotherapeutic drugs in view of cancer diagnosis and treatment (Larsson et al. 2014). Although initially employed as chemotherapeutic agents, nanoparticles have gained prominence as immunotherapeutic tools following recognition of immunotherapy as a groundbreaking approach. Several nanoparticle formulations that have advanced role in cancer treatment due to progress in nanotechnology are listed in (Fig. 3.3).

These proteins, known for their low toxicity and immunogenicity, are widely utilized in imaging agents and therapeutic preparations. Particularly, serum proteins are being explored as nano-carriers for cancer nanomedicine applications due to

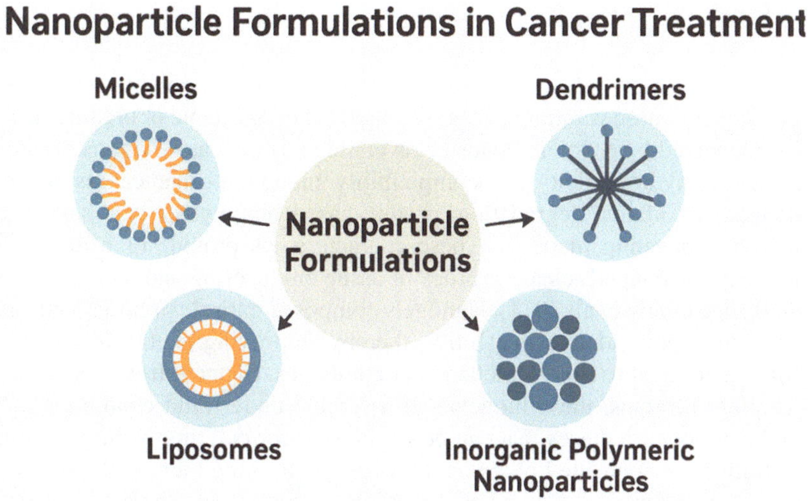

Fig. 3.3 Several nanoparticle formulations that have advanced role in cancer treatment

their biocompatible and biodegradable nature (Jain et al. 2018). Various protein-based nanoparticles have been successfully employed in cancer immunotherapy, including albumin-based nanocarriers like Abraxane, cationic bovine serum albumin (CBSA), gliadin nanoparticles, zein nanoparticles, milk protein nanoparticles, casein nanoparticles, and doxorubicin-loaded lactoferrin nanoparticles (Lohcharoenkal et al. 2014). Albumin-based nanoparticles significantly improve the delivery of chemotherapeutic payloads by preferentially accumulating in solid tumors through caveolae-mediated transcytosis (Murphy et al. 2025). Abraxane®, an albumin-bound paclitaxel formulation that allows for increased dosage without the need for corticosteroid premedication, is based on this idea (Adick et al. 2023). This technology has been further extended to siRNA vectors in lung carcinoma using cysteine-rich bovine serum albumin (CBSA) constructions, which effectively silence genes and induce apoptosis (Ghosh et al. 2021). Concurrently, gliadin-derived nanoparticles designed to treat colon cancer show 48 h of continuous drug release, triggering colorectal cell apoptosis. Despite these developments, several issues remain with nano-immunotherapies, including endothelial interactions, rapid systemic clearance, multidrug resistance pathways, and colloidal instability under physiological stress. A pivotal challenge is the formation of a dynamic "protein corona," in which endogenous proteins (e.g., albumin, apolipoproteins, fibrinogen) adsorb onto nanoparticle surfaces, altering their physicochemical identity and therapeutic performance (Fang et al. 2018; Zhao et al. 2024). Moreover, the immunosuppressive tumour microenvironment can impede nanoparticle–immune cell interactions, critically undermining treatments reliant on molecular recognition at the bio–nano interface.

3.4 Gels

Gels consist of cross-linked, three-dimensional networks of polymers where a continuous solvent phase is immobilized by a scaffold of inorganic or organic structural motifs connected by covalent bonds or reversible physical associations. Because of their inherent hydrophilicity, cytocompatibility, tunable mechanical resilience, and resistance to oxidative degradation, protein-based biopolymer gels have become particularly appealing matrices. These characteristics provide high drug-loading capacity and prolonged release profiles in tissue engineering and therapeutic delivery platforms (Jiang et al. 2014). Nanogels composed entirely of proteinaceous networks reduce peroxidative instability, thereby improving shelf life and in vivo stability, in contrast to lipid-enriched nanoemulsions. Sourced from renewable animal or plant proteins, these biogels self-assemble under mild conditions to yield hierarchical pore structures that can be engineered for cell ingrowth, growth factor sequestration, or controlled pharmacokinetics, positioning them at the forefront of next-generation regenerative and nanomedicine applications (Jiang et al. 2014).

3.4.1 Types of Protein-Based Gels

3.4.1.1 Hydrogels

Hydrogels are three-dimensional networks of hydrophilic polymers that trap a significant amount of water while maintaining their structural integrity. They are assembled through covalent or noncovalent cross-linking (Jia et al. 2021). They split into stimuli-responsive hydrogels, whose polymer mesh dynamically contracts or expands in response to pH shifts, temperature changes, chemical signals, or enzymatic activity, and conventional hydrogels, which achieve a stable equilibrium swelling independent of external stimuli and release payloads through Fickian diffusion (Deng et al. 2022). These "smart" hydrogels improve localized bioavailability and regenerative performance in advanced drug-delivery and tissue-engineering applications by enabling site-specific, on-demand delivery and adaptive tissue scaffolding through in situ modulation of network porosity and polymer–drug interactions (Balan et al. 2022).

3.4.1.2 Nanogels

Constructed from natural or synthetic macromers, nanogels are cross-linked polymeric nanoparticles with dimensions of less than 100 nm and high surface area-to-volume ratios, which promote cellular uptake through direct membrane translocation and endocytic routes. These networks achieve high loading efficiencies, tunable release kinetics, and precise biodistribution by enclosing hydrophilic drugs in their aqueous nanodomains or sequestering hydrophobic actives in hydrophobic junctions. This enables both active targeting through surface-conjugated ligands and passive targeting through the enhanced permeability and retention (EPR) effect. Nanogels are classified based on their cross-linking technology and their responsiveness to environmental stimuli (Soni et al. 2016). These can be further modified by using specific ligand to ensure proper accumulation of therapeutics in target areas. Some prominent Nanogels are Poly (N-Vinyl caprolactum, Xanthate terminated Poly (Ethylene glycol-b-P) and so on. Lipid like components are absent in nanogel because of which, there is a limited risk of oxidation, leading to enhanced stability.

3.4.1.3 Microgels

Microgels are micron-scale 3-D structures formed by cross-linked polymer molecules, typically ranging in size from 0.5 to 5 micrometers. These gels can deform under stress and exhibit surface activity, responding to environmental factors such as temperature and pH. These properties make microgels valuable as colloidal units and stabilizing agents (Farjami and Madadlou 2017). These can be cationic, anionic as well as neutral. Microgels are commonly synthesized using the emulsification method. This involves creating a multiphase mixture by combining an organic phase with a polymeric aqueous phase, which is subsequently homogenized to form an emulsion. The solidification of internal droplets within the aqueous polymeric phase, achieved through physical or chemical cross-linking, produces spherical hydrogels classified as microgels when their size falls within the micron range (Stock and von Klitzing 2022). The preparation method significantly influences the structure, electric charge, and physicochemical properties of microgels. Key factors affecting these properties include the type of cross-linking agent, shear stress, ionic strength, solvent type, temperature, pH, and biopolymer concentration. By coordinating polymer assembly either through noncovalent self-association or covalent cross-linking chemistries and combining it with shaping techniques such as extrusion, atomization, high-shear homogenization, emulsion templating, and micromolding, the architecture and performance of microgels can be precisely adjusted. These combined fabrication approaches offer control over network mesh size, porosity, mechanical stiffness, and surface functionality, adapting microgels for injectable depots, cell-laden scaffolds, or targeted delivery platforms (Thakur et al. 2018).

3.5 Drug Delivery Using Protein-Based Microgels

Advances in human proteomics and high-throughput screening have enhanced the therapeutic potential of biomacromolecules particularly peptides and proteins driving the development of protein-based microgels as versatile drug-delivery depots

(Hu et al. 2019). These hydrophilic, biocompatible networks confer enhanced binding stability and protect labile cargos from enzymatic and chemical degradation. Their tunable mesh architecture enables precise modulation of release kinetics and minimizes immunotoxicity (Malmsten 2006). Upon intravenous administration, submicron microgels significantly smaller than erythrocytes navigate capillary beds and exploit physiological clearance pathways to achieve tissue-specific accumulation; subsequent cellular uptake triggers localized release of the payload. However, rapid clearance by the reticuloendothelial system, mediated by opsonin adsorption (e.g., complement proteins, immunoglobulins), can limit efficacy. Surface grafting with polyethylene oxide chains has emerged as a robust stealth strategy, reducing macrophage recognition, prolonging systemic circulation, and enhancing delivery to non-resident endothelial cells (RES) tissues (Amiri et al. 2019).

3.5.1 Triggering Mechanisms for Microgel-Based Drug Delivery

3.5.1.1 Temperature-Responsive Drug Delivery

Poly(N-isopropylacrylamide) (PNIPAM), polyethylene oxide derivatives, and cellulose ethers are examples of thermoresponsive polymers that undergo a volume phase transition driven by a lower critical solution temperature (LCST), which causes their hydrogel network to collapse and release encapsulated cargo when heated (Das et al. 2006). This reversible deswelling process can be used to protect labile medicines against hydrolytic or enzymatic breakdown at lower temperatures, or it can be used to initiate site-specific drug release. For example, it has been demonstrated that PNIPAM microgels improve pulsatile dosing profiles by releasing insulin more effectively when the temperature rises, similar to a sponge contracting under heat stimuli (Elashnikov et al. 2017). In vivo compatibility and clearance are further enhanced by customizing these networks with biodegradable connections, which also brings thermoresponsive delivery systems into compliance with clinical safety regulations (Das et al. 2006).

3.5.1.2 Electrostatic Triggering

Electrostatic modulation of microgel networks via pH- and ionic-strength–dependent ionization of pendant groups provides a tunable trigger for on-demand drug release. Early work demonstrated that poly (methacrylic acid) microgels swell sharply as carboxyl moieties deprotonate under alkaline conditions, thereby increasing the mesh size and facilitating the diffusion of payloads (Eichenbaum et al. 1999). Subsequent studies extended this principle to copolymeric systems: methacrylic-ethyl acrylate microgels exhibit progressively enhanced swelling and drug liberation at elevated pH due to increased anionic charge density (Tan & Tam), while sodium alginate-acrylic acid networks respond similarly, underscoring the generality of pH-sensitive deswelling in weak-acid microgels (Ramesh Babu et al. 2006). Importantly, integrating chemotherapeutic agents into these electrostatically active matrices has been shown to concentrate drugs within tumour cells leveraging acidic tumour microenvironments to maximize intracellular uptake and cytotoxic efficacy (Kanazawa et al. 2017).

3.5.1.3 External Triggers

Photosensitive, acoustic, and magneto-responsive microgels extend this tunability by incorporating functional moieties or nanofillers that transduce external cues into controlled network remodeling. For instance, azobenzene-decorated dextran microgels undergo reversible cis–trans isomerization under UV/visible light, modulating cross-link density and enabling the pulsatile release of cargos such as rhodamine and aspirin (Patnaik et al. 2007). Embedding plasmonic or magnetic nanoparticles within proteinaceous microgel matrices enables photothermal or magneto-mechanical actuation, where light absorption or alternating magnetic fields generate localized heating or mechanical stresses that transiently enlarge mesh pores and trigger payload egress. Similarly, focused ultrasound can induce cavitation and shear forces that momentarily disrupt microgel networks, creating nanoscopic channels for accelerated drug diffusion. By integrating these stimuli-responsive elements, microgel platforms achieve spatiotemporal precision in release profiles, thereby advancing their applicability in on-demand therapeutics and remotely controlled nanomedicine. As a result gel structure is distorted and bioactive drug diffused into target cells. Such stimuli-responsive systems have potential applications in localized therapies, such as photodynamic treatments (Uthaman et al. 2020).

3.5.1.4 Hydrophilic and Hydrophobic Microgels in Drug Delivery

Microgels can respond to diverse stimuli, including pH, ionic strength, specific metabolites, temperature, and external fields, and can be administered as injectables. However, conventional polymer gels lack hydrophobic domains, limiting their use in delivering poorly soluble drugs. Microgels, with their hydrophilic nature, maintain the biological activity of peptide and protein drugs while minimizing aggregation and conformational changes. The cross-linking density of microgel depends on percentage of hydrophilic residues. More the cross-linking density, lower the percentage of swelling. Although the literature on microgel-based drug delivery is expanding, challenges such as limited production throughput hinder their translation into industrial and clinical applications (Bysell et al. 2011; Kamperman et al. 2018).

3.5.2 Interaction of Microgels with Proteins

Electrostatic interactions are fundamental to the binding of proteins and peptides to microgels due to the gels' charged and hydrophilic structures. Microgel containing hydrophobic domain readily interact with hydrophobic core of protein; especially when the protein is partially folded. A study on oxidized starch microgels demonstrated that increasing starch oxidation enhanced the binding of positively charged lysozyme. However, as ionic strength increased, the electrostatic affinity decreased, resulting in faster drug release, similar to behaviors observed in nonporous substrates (Li et al. 2009). Interestingly, lysozyme binding decreased with increasing pH, emphasizing the role of lysozyme charge over that of starch in determining binding efficiency (Li et al. 2009). The porous structure and cross-linking density of microgels significantly influence protein binding and transport. For instance, lysozyme has been shown to form dense layers within poly (acrylic acid) microgels under specific conditions, influenced by ionic strength, protein concentration, and pH (Li et al. 2010). Unlike lysozyme, smaller proteins like cytochrome C, which lack self-assembly properties, do not form similar layers, highlighting protein-specific interactions (Johansson et al. 2007).

3.6 Limitations of Protein-Based Nanoparticles

Protein-based nanoparticles (PNPs) hold great promise as adaptable carriers of bioactive substances, but a number of interconnected restrictions still prevent their widespread use and clinical translation. The main obstacles are critically examined in this section, with particular attention paid to biological interactions, physicochemical limitations, manufacturing and scalability problems, regulatory obstacles, and long-term stability issues.

3.6.1 Physicochemical Limitations of Protein-Based Nanocarriers

PNPs exhibit severe colloidal instability in physiological environments, primarily because standard fabrication techniques, such as desolvation, emulsion, and coacervation, often produce particles with broad size distributions and heterogeneous surface charges (Khakpour et al. 2024). Under the ionic strengths and shear stresses of the bloodstream, these inconsistencies lead rapidly to aggregation, the formation of a dynamic protein corona, and ensuing recognition and clearance by the mononuclear phagocyte system often leaving only 5–10% of the injected dose in circulation after 8 h (Saif et al. 2023; Lim et al. 2024). Even minor shifts in processing parameters can have outsized effects on particle dimensions and drug exposure: for instance, a slight increase in homogenization pressure during scale-up of Abraxane® bumped mean diameters from 100 nm to 130 nm, correlating with a 15% change in patient area-under-the-curve and exacerbated toxicity (Ifijen et al. 2025). Attempts to stabilize PNPs via chemical crosslinking introduce their complications trace glutaraldehyde residues have been shown to provoke transient pulmonary inflammation in animal models. At the same time, reduction of Schiff bases yields only marginal improvements in structural integrity (Han et al. 2023). The heterogeneity inherent to naturally derived proteins compound these challenges; gelatin and serum albumin preparations often vary in molecular-weight distribution from batch to batch, producing unpredictable payload encapsulation, release kinetics and biodistribution, whereas recombinant analogues, though more uniform, incur higher production costs (Habibi et al. 2022; Li et al. 2024).

Beyond these general limitations, individual protein scaffolds present distinct drawbacks that further impede clinical translation. Silk fibroin nanoparticles may retain immunogenic sericin and degrade too slowly due to their β-sheet–rich structure; human serum albumin carriers, while biocompatible, tend to accumulate nonspecifically in tissues rich in albumin receptors; and gliadin assemblies often exceed 300 nm, aggregate under physiological salt and pH, and require stabilizing polysaccharides (Hassan et al. 2024). Gelatin nanoparticles lack mechanical robustness and degrade enzymatically unless heavily cross-linked, altering drug-release profiles, while legumin coacervates yield only 25–30% recovery with poor reproducibility (Ferraro et al. 2024). Even ferritin nanocages, which boast excellent thermostability, face prohibitive production costs. In contrast, novel platforms such as the silkworm 30Kc19 protein achieve less than 50% nanoparticle yield unless blended with albumin (Gauraha et al. 2024). Finally, the mild processing conditions necessary to preserve protein structure inherently limits drug loading, particularly of hydrophobic molecules, and covalent conjugation strategies frequently disrupt nanoparticle architecture and release behavior (Roszkowski and Durczynska 2025; Aljabali et al. 2022). Together, these interrelated physicochemical constraints colloidal instability, corona formation, size heterogeneity, crosslinker toxicity, scaffold-specific drawbacks, and low encapsulation efficiency constitute formidable obstacles to achieving predictable biodistribution, sustained delivery, and immunocompatibility with PNP-based therapeutics.

3.6.2 Immunogenicity, Rapid Clearance, and Biodistribution Challenges

Proteins, though intrinsically biocompatible, often provoke unintended biological interactions once formulated into nanoparticles. Particulate assembly and chemical cross-linking can expose cryptic epitopes or leave residual macromers (e.g., PEG), eliciting complement activation, neutralizing antibodies, or hypersensitivity reactions that accelerate clearance and undermine safety upon repeat dosing (Saif et al.

2023; Kianfar 2021). In the absence of robust stealth coatings, opsonization by the nascent protein corona drives rapid recognition and phagocytic uptake, resulting in circulation half-lives that lag behind those of lipid- or polymer-based carriers and curtailing tumor accumulation via the EPR effect (Xu et al. 2023). Moreover, surface charge and corona composition lead to uneven tissue deposition PNPs frequently concentrate in the liver, spleen, and inflamed lungs, raising concerns about off-target toxicity and complicating dose optimization for organs or tumor beds (Bashiri et al. 2023). Finally, the inherently hydrophilic protein matrix degrades through proteolysis, pH-mediated denaturation, and mechanical shear, resulting in release profiles characterized by initial bursts or incomplete payload liberation, rather than the sustained delivery required to maintain therapeutic windows in vivo (Hassanzadeh et al. 2021).

3.6.3 Manufacturing and Scalability Issues

Scaling protein-based nanoparticle (PNP) production from bench to industry remains a formidable challenge. Lab-scale techniques, such as emulsification-desolvation and electrospray, can produce high-quality particles; however, they operate within narrow process windows, consume large volumes of organic solvents, and yield only milligram-scale quantities far below the kilogram-scale required for clinical and commercial applications (Fernandes et al. 2023). The specialized equipment and consumables required by these methods increase both capital and operating expenditures, while stringent sterile processing and the need for batch-specific physicochemical and immunotoxicity testing further drive up costs. Compounding these economic hurdles, the intrinsic variability of protein feedstocks whether animal-derived (e.g., silk fibroin, albumin, ferritin) with concerns of zoonotic contamination and prion transmission, or agro-industrial byproducts like zein and gliadin induces batch-to-batch inconsistency in drug loading efficiencies, release kinetics, and pharmacokinetic behaviour (Dissanayake et al. 2025; Elzoghby et al. 2012). Although recombinant and elastin-like polypeptides offer more uniform starting materials, their production infrastructure and scale-up also remain cost-prohibitive (Bhardwaj et al. 2025). Finally, the absence of harmonized regulatory pathways for PNPs each formulation demanding exhaustive characterization of physicochemical attributes, degradation byproducts, and immunological safety has stymied the progression of only a handful of candidates beyond preclinical studies (Khakpour et al. 2024; Mitchell et al. 2021). Together, these factors illustrate why low throughput, high costs, raw material heterogeneity, and regulatory complexity continue to impede the commercial translation of protein-based nanocarriers.

3.6.4 Regulatory and Quality Control

The structural complexity of protein-based nanoparticles (PNPs) with their precise tertiary/quaternary folds, narrow size windows, and sensitive surface chemistries requires exhaustive analytical characterization to meet regulatory mandates. Every batch must demonstrate uniform particle size distribution, intact protein conformation, consistent ligand presentation, and the absence of pyrogens or residual solvents. However, in practice, minor variations in protein source, molecular weight, or trace endotoxin contamination frequently produce batch-to-batch inconsistencies that raise red flags during regulatory review (Creutzenberg 2021). Despite over fifty PNP platforms reaching preclinical testing, only Abraxane® (albumin-bound paclitaxel) and Ontak® (denileukin diftitox) have cleared the FDA an approval rate that underscores both the safety and functional uncertainties of these carriers in vivo and the difficulty of proving clear clinical superiority over established therapies (Namiot et al. 2023). Compounding these translational hurdles, the patent landscape for protein nanoparticles is fragmented, with overlapping claims on protein scaffolds, conjugation chemistries, and manufacturing processes that can stall development and deter investment by smaller companies (dos Santos Fonseca et al. 2024). Even once a candidate advances, its long-term stability poses further barriers: PNPs are prone to aggregation, Ostwald ripening, and irreversible denaturation under ambient storage or during freeze–thaw cycles, with spray-dried albumin formulations losing up to 20% of native structure over 3 months at room temperature and requiring stringent cold-chain logistics (Elzoghby et al. 2012). Mechanical stresses from lyophilization or syringe pumping can exacerbate aggregation, risking not only loss of drug encapsulation but also the formation of immunogenic protein aggregates that provoke dendritic cell maturation, anti-drug antibody formation, and even anaphylaxis (Liu et al. 2022a; Hong et al. 2020). Together, these regulatory, intellectual property, and stability challenges to say nothing of PNPs' poor tolerance of acidic pH and digestive enzymes, which preclude oral delivery illustrate why clinical translation and large-scale deployment of protein nanoparticles remain the exception rather than the rule (Habibi et al. 2022).

3.7 Future Prospects

The escalating costs of healthcare have become a significant concern for both developed and emerging nations, underscoring the need for governments to evaluate the cost-effectiveness of nano-pharmaceuticals to optimize healthcare spending. Conducting standard cost-effectiveness analyses is a critical first step to determine whether the additional health benefits provided by nano-pharmaceuticals justify their higher costs compared to traditional formulations. Further research should concentrate on large scale production techniques in a commercial way. In recent years, intense research is going on to engineer nano- pharmaceuticals in such a way so that these can penetrate the blood brain barrier; leading to create advancement in neurological disease treatment. These analyses will enable governments to establish clear regulations and quantify the financial advantages of fostering this market. The future success of nanomedicine hinges on gaining the trust of patients, the public, and healthcare professionals regarding the safety and efficacy of nano-pharmaceuticals. Raising awareness about the benefits and addressing safety concerns will be essential to overcome these limitations. Despite these challenges, the authors anticipate significant growth in the nano-pharmaceutical market over the next 5 years. This expansion will be driven by advances in bio-nanotechnology and nanoengineering, the implementation of transparent regulations for nanotechnology-based drugs, increased governmental funding, greater consensus on environmental considerations, and strategic partnerships between nanomedicine startups and established pharmaceutical companies.

3.8 Conclusion

Nanotechnology offers significant potential for advancing immunotherapy and developing effective anticancer agents. Proteins, in particular, represent innovative candidates for nanoparticle-based cancer immunotherapy. These biomolecules can be synthesized into nanoparticles through straightforward, cost-efficient, and environmentally friendly processes that minimize the use of hazardous materials. Due to their inherent metabolic pathways, proteins enable the formation of nanoparticles that degrade efficiently within the body, leaving behind non-toxic residues. This property positions them as a natural alternative to existing synthetic and natural sources used for nanoparticle development in targeted biomedical applications. However, further research is critical to addressing limitations associated with microgels and microcapsules as carriers for biomacromolecular drug delivery. Challenges such as uneven drug distribution within microgels (e.g., shell formation), suboptimal drug loading capacity, and slow or incomplete release of therapeutic agents must be resolved to fully harness the potential of these systems. Overcoming these obstacles will enable the optimization of protein-based nanoparticles for advanced medical applications.

References

Adick A et al (2023) Challenges of nanoparticle albumin bound (nab™) technology: comparative study of Abraxane® with a newly developed albumin-stabilized itraconazole nanosuspension. Eur J Pharm Biopharm 193:129–143. https://doi.org/10.1016/j.ejpb.2023.10.022

Afzal O et al (2022) Nanoparticles in drug delivery: from history to therapeutic applications. Nanomaterials 12(24):4494. https://doi.org/10.3390/nano12244494

Ahmed S, Alhareth K, Mignet N (2020) Advancement in nanogel formulations provides controlled drug release. Int J Pharm 584:119435

Aljabali AA, Rezigue M, Alsharedeh RH, Obeid MA, Mishra V, Serrano-Aroca A et al (2022) Protein-based nanomaterials: a new tool for targeted drug delivery. Ther Deliv 13(6):321–338

Alqahtani MS et al (2021) Advances in Oral drug delivery. Front Pharmacol 12:618411. https://doi.org/10.3389/fphar.2021.618411
Amiri M, Salavati-Niasari M, Akbari A (2019) Magnetic nanocarriers: evolution of spinel ferrites for medical applications. Adv Colloid Interface Sci 265:29–44
Anand U et al (2023) Cancer chemotherapy and beyond: current status, drug candidates, associated risks and progress in targeted therapeutics. Genes Dis 10(4):1367–1401. https://doi.org/10.1016/j.gendis.2022.02.007
Anupama S, Priyanka A (2018) Nanohydrogels: emerging trend for drug delivery, organic materials as smart nanocarriers for drug delivery, 8. William Andrew Publishing, Norwich, pp 293–368
Argentiere S, Blasi L, Morello G et al (2011) A novel ph-responsive nanogel for the controlled uptake and release of hydrophobic and cationic solutes. J Phys Chem C 115:16347–16353
Balan KE, Boztepe C, Künkül A (2022) Technology, modeling the effect of physical crosslinking degree of pH and temperature responsive poly (NIPAAm-Co-Vsa)/alginate IPN hydrogels on drug release behavior. J Drug Delivery Sci Technol 75:103671
Bashiri G, Padilla MS, Swingle KL, Shepherd SJ, Mitchell MJ, Wang K (2023) Nanoparticle protein corona: from structure and function to therapeutic targeting. Lab Chip 23(6):1432–1466
Bhardwaj A, Verma S, Agnihotri A, Shukla VK, Chitme HR (2025) Protein based nanoparticles for pulmonary drug delivery: advances, challenges, and future perspectives. J Nanopart Res 27(7):1–20
Boyd BJ, Bergström CA, Vinarov Z, Kuentz M, Brouwers J, Augustijns P, Brandl M, BernkopSchnürch A, Shrestha N, Préat V et al (2019) Successful oral delivery of poorly water-soluble drugs both depends on the intraluminal behavior of drugs and of appropriate advanced drug delivery systems. Eur J Pharm Sci 137:104967
Bysell H, Månsson R, Hansson P, Malmsten M (2011) Microgels and microcapsules in peptide and protein drug delivery. Adv Drug Deliv Rev 63(13):1172–1185
Cao Z, Ziener U (2013) Synthesis of nanostructured materials in inverse miniemulsions and their applications. Nanoscale 5:10093–10107
Chen N, Nicolai T, Chassenieux C, Wang Y (2020) pH and ionic strength responsive Core-Shell protein microgels fabricated via simple coacervation of soy globulins. Food Hydrocolloids 105:105853
Muraleedharan A, Acharya S and Kumar R, (2024) Recent updates on diverse nanoparticles and nanostructures in therapeutic and diagnostic applications with special focus on smart protein nanoparticles: A review. ACS omega, 9(42):pp. 42613–42629
Creutzenberg O (2021) Nanoparticles and their regulation. In: Regulatory toxicology. Springer International Publishing, Cham, pp 1093–1108
Das M, Zhang H, Kumacheva E (2006) Microgels: old materials with new applications. Annu Rev Mater Res 36(1):117–142
Davidov-Pardo G, Joye IJ, McClements DJ (2015) Food-grade protein-based nanoparticles and microparticles for bioactive delivery: fabrication, characterization, and utilization. Adv Protein Chem Struct Biol 98:293–325. https://doi.org/10.1016/bs.apcsb.2014.11.004
De R, Mahata MK, and Kim KT, (2022) Structure-based varieties of polymeric nanocarriers and influences of their physicochemical properties on drug delivery profiles. Advanced Science, 9(10), p. 2105373
Deng Y, Xi J, Meng L, Lou Y, Seidi F, Wu W, Xiao H (2022) Stimuli-responsive Nanocellulose hydrogels: an overview. Eur Polym J 180:111591
Dey S, Hassan S, and Pandey RK, (2024) Nanomedicine in targeted drug delivery: precision therapeutics for personalized medicine. In Nanomedicine: Innovations, Applications, and Breakthroughs in the Quest for Health and Medicine's Future (pp. 179–231). Cham: Springer Nature Switzerland
Dissanayake T, Vidanarachchi JK, Narvaez-Bravo C, Mekonnen T, Bandara N (2025) Advances in protein-based materials for functional food packaging. ACS Food Sci Technol 5:2604–2617
dos Santos Fonseca LM, Machado BAS, Oliveira FO, de Jesus Santos JR, da Silva JW, Hodel KVS et al (2024) An overview on recent patents and technologies on nanoparticles for nucleic acid delivery. Expert Opin Ther Pat 34(3):171–186

Eichenbaum GM, Kiser PF, Dobrynin AV, Simon SA, Needham DJM (1999) Investigation of the swelling response and loading of ionic microgels with drugs and proteins: the dependence on cross-link density. Macromolecules 32(15):4867–4878

Elashnikov R, Slepička P, Rimpelova S, Ulbrich P, Švorčík V, Lyutakov O (2017) Temperature-responsive PLLA/PNIPAM nanofibers for switchable release. Mater Sci Eng C 72:293–300

Elzoghby AO, Samy WM, Elgindy NA (2012) Protein-based nanocarriers as promising drug and gene delivery systems. J Control Release 161(1):38–49

Fang RH, Kroll AV, Gao W, Zhang L (2018) Cell membrane coating nanotechnology. Adv Mater 30(23):1706759

Farjami T, Madadlou A (2017) Fabrication methods of biopolymeric microgels and microgel-based hydrogels. Food Hydrocoll 62:262–272

Fernandes C, Jathar M, Sawant BKS, Warde T (2023) Scale-up of nanoparticle manufacturing process. In: Pharmaceutical process engineering and scale-up principles. Springer Nature Switzerland, Cham, pp 173–203

Ferraro C, Dattilo M, Patitucci F, Prete S, Scopelliti G, Parisi OI, Puoci F (2024) Exploring protein-based carriers in drug delivery: a review. Pharmaceutics 16(9):1172

Gauraha H, Bhadoriya A, Gautam RK, Mishra DK (2024) Protein delivery by nanoparticles. In: Nanocarrier vaccines: biopharmaceutics-based fast track development, pp 305–343

Ghosh S, Javia A, Shetty S, Bardoliwala D, Maiti K, Banerjee S, Khopade A, Misra A, Sawant K, Bhowmick S (2021) Triple negative breast cancer and non-small cell lung cancer: clinical challenges and nano-formulation approaches. J Control Release 337:27–58

Gonçalves C, Pereira P, Gama M (2010) Self-assembled hydrogel nanoparticles for drug delivery applications. Materials 3:1420–1460

Gupta SK (2020) Formulation and evaluation of nanoemulsion based nanomole of Aceclofenac. J Pharm Sci Res 12(4):524–523

Guo H, and Mi P (2024) Polymer–drug and polymer–protein conjugated nanocarriers: design, drug delivery, imaging, therapy, and clinical applications. Wiley Interdisciplinary Reviews: Nanomedicine and Nanobiotechnology, 16(4), p. e1988

Habibi N, Mauser A, Ko Y, Lahann J (2022) Protein nanoparticles: uniting the power of proteins with engineering design approaches. Adv Sci 9(8):2104012

Han M, Liu K, Liu X, Rashid MT, Zhang H, Wang M (2023) Research progress of protein-based bioactive substance nanoparticles. Foods 12(16):2999

Hassan MA, Basha AA, Eraky M, Abbas E, El-Samad LM (2024) Advancements in silk fibroin and silk sericin-based biomaterial applications for cancer therapy and wound dressing formulation: a comprehensive review. Int J Pharm 662:124494

Hassanzadeh S, Nematollahzadeh A, Mirzayi B, Kaboli SF (2021) Protein-based nanoparticles synthesized at a high shear rate and optimized for drug delivery applications. J Mol Liq 335:116133

Hong S, Choi DW, Kim HN, Park CG, Lee W, Park HH (2020) Protein-based nanoparticles as drug delivery systems. Pharmaceutics 12(7):604

Hoshyar N et al (2016) The effect of nanoparticle size on in vivo pharmacokinetics and cellular interaction. Nanomedicine 11(6):673–692. https://doi.org/10.2217/nnm.16.5

Hu J-P, Wu Z-X, Xie T, Liu X-Y, Yan X, Sun X, Liu W, Liang L, He G, Gan YJP et al (2019) Applications of molecular simulation in the discovery of Antituberculosis drugs: a review. Protein Pept Lett 26(9):648–663

Ifijen IH, Awoyemi RF, Faderin E, Akobundu UU, Ajayi AS, Chukwu JU et al (2025) Protein-based nanoparticles for antimicrobial and cancer therapy: implications for public health. RSC Adv 15(19):14966–15016

Irvine DJ, Hanson MC, Rakhra K, Tokatlian T (2015) Synthetic nanoparticles for vaccines and immunotherapy. Chem Rev 115(19):11109–11146

Jain A, Singh SK, Arya SK, Kundu SC, Kapoor S (2018) Engineering, protein nanoparticles: promising platforms for drug delivery applications. ACS Biomater Sci Eng 4(12):3939–3961

Jia F, Song J, Kubiak JM, Onoda M, Santos PJ, Sano K, Holten-Andersen N, Zhang K, Macfarlane R (2021) Brush polymers as nanoscale building blocks for hydrogel synthesis. Chem Mater 33(14):5748–5756

Jia X, Fan X, Chen C, Lu Q, Zhou H, Zhao Y, Wang X, Han S, Ouyang L, Yan H, and Dai H (2024) Chemical and structural engineering of gelatin-based delivery systems for therapeutic applications: a review. Biomacromolecules, 25(2), pp. 564–589
Jiang Y, Chen J, Deng C, Suuronen EJ, Zhong Z (2014) Click hydrogels, microgels and Nanogels: emerging platforms for drug delivery and tissue engineering. Biomaterials 35(18):4969–4985
Johansson C, Hansson P, Malmsten M (2007) Interaction between lysozyme and poly (acrylic acid) microgels. Science 316(2):350–359
Kamperman T, Karperien M, Le Gac S, Leijten J (2018) Single-cell microgels: technology, challenges, and applications. Trends Biotechnol 36(8):850–865
Kanazawa T, Tamano K, Sogabe K, Endo T, Ibaraki H, Takashima Y et al (2017) Intra-articular retention and anti-arthritic effects in collagen-induced arthritis model mice by injectable small interfering RNA containing hydrogel. Biol Pharm Bull 40:1929–1933
Khakpour S, Hosano N, Moosavi-Nejad Z, Farajian AA, Hosano H (2024) Advancing tumor therapy: development and utilization of protein-based nanoparticles. Pharmaceutics 16(7):887
Kianfar E (2021) Protein nanoparticles in drug delivery: animal protein, plant proteins and protein cages, albumin nanoparticles. J Nanobiotechnol 19(1):159
Larsson M, Bergstrand A, Mesiah L, Van Vooren C, Larsson A (2014) Nanocomposites of polyacrylic acid nanogels and biodegradable polyhydroxybutyrate for bone regeneration and drug delivery. J Nanomater 2014:1–9
Segneanu AE, Bejenaru LE, Bejenaru C, Blendea A, Mogoşanu GD, Biţă A, and Boia ER (2025) Advancements in hydrogels: a comprehensive review of natural and synthetic innovations for biomedical applications. Polymers, 17(15), p. 2026
Li Y, de Vries R, Slaghek T, Timmermans J, Cohen Stuart MA, Norde W (2009) Preparation and characterization of oxidized starch polymer microgels for encapsulation and controlled release of functional ingredients. Biomacromolecules 10(7):1931–1938
Li Y, Vries RD, Kleijn M, Slaghek T, Timmermans J, Stuart MC, Norde W (2010) Lysozyme uptake by oxidized starch polymer microgels. Biomacromolecules 11(7):1754–1762
Li C, Chen L, McClements DJ, Peng X, Xu Z, Meng M et al (2024) Encapsulation of polyphenols in protein-based nanoparticles: preparation, properties, and applications. Crit Rev Food Sci Nutr 64(31):11341–11355
Lim SH, Wong TW, Tay WX (2024) Overcoming colloidal nanoparticle aggregation in biological milieu for cancer therapeutic delivery: perspectives of materials and particle design. Adv Colloid Interf Sci 325:103094
Liu H, Zhang Y, Zhang J, Xiong Y, Peng S, McClements DJ et al (2022a) Utilization of protein nanoparticles to improve the dispersibility, stability, and functionality of a natural pigment: Norbixin. Food Hydrocolloids 124:107329
Liu P, Chen G, Zhang J (2022b) A review of liposomes as a drug delivery system: current status of approved products, regulatory environments, and future perspectives. Molecules 27(4):1372. https://doi.org/10.3390/molecules27041372
Lohcharoenkal W, Wang L, Chen YC, Rojanasakul Y (2014) Protein nanoparticles as drug delivery carriers for cancer therapy. Biomed Res Int 2014:1–12. https://doi.org/10.1155/2014/180549
Malmsten M (2006) Soft Drug Delivery Systems. Soft Matter 2(9):760–769. https://doi.org/10.1039/b608348j
Medina C et al (2007) Nanoparticles: pharmacological and toxicological significance. Br J Pharmacol 150(5):552–558. https://doi.org/10.1038/sj.bjp.0707130
Mitchell MJ, Billingsley MM, Haley RM, Wechsler ME, Peppas NA, Langer R (2021) Engineering precision nanoparticles for drug delivery. Nat Rev Drug Discov 20(2):101–124
Murphy G et al (2025) Albumin-based delivery systems: recent advances, challenges, and opportunities. J Control Release 380:375–395. https://doi.org/10.1016/j.jconrel.2025.01.035
Namiot ED, Sokolov AV, Chubarev VN, Tarasov VV, Schiöth HB (2023) Nanoparticles in clinical trials: analysis of clinical trials, FDA approvals and use for COVID-19 vaccines. Int J Mol Sci 24(1):787
Park H, Otte A, Park K (2022) Evolution of drug delivery systems: from 1950 to 2020 and beyond. J Control Release 342:53–65. https://doi.org/10.1016/j.jconrel.2021.12.030

Patnaik S, Sharma AK, Garg B, Gandhi R, Gupta K (2007) Photoregulation of drug release in Azodextran Nanogels. Int J Pharm 342(1–2):184–193

Qian H, Liu B, Jiang X (2018) Application of nanomaterials in cancer immunotherapy. Mater Today Chem 7:53–64

Ramesh Babu V, Krishna Rao K, Sairam M, Naidu BVK, Hosamani KM, Aminabhavi TM (2006) pH sensitive interpenetrating network microgels of sodium alginate-acrylic acid for the controlled release of ibuprofen. J Appl Polym Sci 99(5):2671–2678

Rocha CV et al (2022) PLGA-based composites for various biomedical applications. Int J Mol Sci 23(4):2034. https://doi.org/10.3390/ijms23042034

Roszkowski S, Durczynska Z (2025) Advantages and limitations of nanostructures for biomedical applications. Adv Clin Exp Med 34(3):447–456

Sagar V, Arti V, Rashi G, Madhavan N (2018) Nanogels for biomedical applications: challenges and prospects. In: Nanogels for biomedical applications. The Royal Society of Chemistry, London, pp 290–300

Saif A, Anjum L, Faisal Z, Akram N, Shah YA, Irfan R et al (2023) Recent advances in protein-based nanoparticles and their applications in the delivery of bioactive compounds. Int J Food Prop 26(2):2866–2880

Spada A, Emami J, Tuszynski JA, and Lavasanifar A (2021) The uniqueness of albumin as a carrier in nanodrug delivery. Molecular pharmaceutics, 18(5), pp.1862–1894.

Soni KS, Desale SS, Bronich TK (2016) Nanogels: an overview of properties, biomedical applications and obstacles to clinical translation. J Control Release 240:109–126

Spicer CD, Jumeaux C, Gupta B, Stevens MM (2018) Peptide and protein nanoparticle conjugates: versatile platforms for biomedical applications. Chem Soc Rev J 47(10):3574–3620

Stock S, von Klitzing R (2022) Microgels at droplet interfaces of water-in-oil emulsions—challenges and Progress. Curr Opin Colloid Interface Sci 58:101561

Thakur G, Rodrigues FC, Dathathri E, Nayak SK, Pal K (2018) Protein-based gels: preparation, characterizations, applications in drug delivery, and tissue engineering. In: Polymeric gels, pp 31–54

Tiwari G et al (2012) Drug delivery systems: An updated review. Int J Pharm Investig 2(1):2–11. https://doi.org/10.4103/2230-973X.96920

Uthaman S, Pillarisetti S, Mathew AP, Kim Y, Bae WK, Huh KM, Park I-K (2020) Long circulating Photoactivable Nanomicelles with tumor localized activation and ROS triggered self-accelerating drug release for enhanced Locoregional chemo-photodynamic therapy. Biomaterials 232:119702

Wang E et al (2025) Gels as promising delivery systems: physicochemical property characterization and recent applications. Pharmaceutics 17(2):249. https://doi.org/10.3390/pharmaceutics17020249

Xu M, Qi Y, Liu G, Song Y, Jiang X, Du B (2023) Size-dependent in vivo transport of nanoparticles: implications for delivery, targeting, and clearance. ACS Nano 17(21):20825–20849

Yao S et al (2018) Development and evaluation of novel tumor-targeting paclitaxel-loaded nanocarriers for ovarian cancer treatment: in vitro and in vivo. J Exp Clin Cancer Res 37(1):29. https://doi.org/10.1186/s13046-018-0700-z

Zarekar N, Lingayat V, Pande V (2017) Nanogel as a novel platform for smart drug delivery system. Nanosci Nanotechnol 4:25–31

Zhang Y, Chan HF, Leong KW (2013) Advanced materials and processing for drug delivery: the past and the future. Adv Drug Deliv Rev 65(1):104–120

Zhang N et al (2020) Protein-based artificial Nanosystems in cancer therapy. Small 16(23):1907256. https://doi.org/10.1002/smll.201907256

Zhao T et al (2024) Engineering the protein corona: strategies, effects, and future directions in nanoparticle therapeutics. Biomed Pharmacother 175:116627. https://doi.org/10.1016/j.biopha.2024.116627

Chapter 4
Recent Advances in Medicinal Plants Mediated Nanomaterials Synthesis for Biomedical Applications

Ajay Kumar Tiwari, Abhishek Kumar Bhardwaj, Mohini Bhardwaj, Ram Raseele Awasthi, Sharad Kumar Tripathi, Vijayalakshmi Ghosh, and Shalini Pathak

4.1 Introduction

From last decades, researchers are paying attention to producing eco-friendly and sustainable nanomaterials (NMs) due to their advanced physicochemical properties. Currently, NMs are attracting global attention in the era of medicinal sciences, due to their unique characteristics, such as antioxidants, antifungal, antimicrobial, and anti-cancerous activities. These characteristics have made them promising

A. K. Tiwari
School of Sciences, Department of Physics, G. H. Raisoni International Skill Tech University, Pune, MH, India

A. K. Bhardwaj (✉)
Department of Environmental Science, Amity School of Life Sciences, Amity University, Gwalior, Madhya Pradesh, India

M. Bhardwaj
Department of Botany, Rajiv Gandhi Government PG College, Ambikapur, Chhattisgarh, India

R. R. Awasthi
Faculty of Engineering and Technology, Khwaja Moinuddin Chishti Language University, Lucknow, UP, India

S. K. Tripathi
Department of Life Science, Arni University, Kangra, HP, India

V. Ghosh
Department of Biotechnology, G.D. Rungta College of Sciences & Technology, Bhilai, Chhattisgarh, India

S. Pathak
Amity Institute of Biotechnology, Amity School of Life Sciences, Amity University Madhya Pradesh, Gwalior, MP, India

V. Ghosh et al. (eds.), *Applications of Plant Bioactive Compounds-Based Nanoformulations*, Nanotechnology in the Life Sciences,
https://doi.org/10.1007/978-3-032-13086-0_4

candidates in emerging medical technologies. Medicinal plant-mediated NPs have a great potential application in biomedical fields, along with shortening limitations. In recent trends, researchers are using several synthesis methods to produce NPs. However, it is broadly categorized, and the most accepted methods are (1) Physical, (2) Chemical, and (3) Biological methods (Tiwari et al. 2024c). Over the last few decades, researchers have been adopting eco-friendly and environmentally friendly methods, which are safer, cost-effective, biocompatible, and stable for the synthesis of NMs over chemical and physical methods (Tiwari et al. 2025). However, chemical methods have potential for industrial-scale production of NMs, but they are associated with several environmental concerns. Furthermore, physical synthesis methods are time-consuming, costlier, tedious, and difficult in the synthesis process. However, the biological synthesis method is not new; it was used from the ancient period of Ayurvedic medicines as Bhasma including, Tambra (copper), Lauh (iron), etc., where it was prepared by the decoction of several medicinal plants (Valikarimwala et al. 2021).

In modern biological synthesis of NMs, production of NMs is taking place via extracts of almost all parts of the plants. The extract of medicinal plants contains various bioactive compounds, including alkaloids, flavonoids, coumarin, phenolics, saponins, terpenoids, vitamins, etc. These bioactive compounds and biomolecules can play a crucial role in producing a various types of NMs by reducing metal ions. The important bioactive molecules can be characterized as functional capping agents over NMs for curing specific health issues. Medical plant extracts and plant-based biopolymers are often preferred for the preparation of NMs due to their capability in diminishing the particle size and providing more stability (Jafarzadeh et al. 2024). This chapter different types of extract isolation, synthesis of NMs via greener route using medicinal plants and highlights their biomedical applications. This chapter also covers various types of extraction methods, factors affecting during synthesis process, characterization techniques and their potential application in clinical and diagnostic prospects.

4.2 Extraction Methods from Medicinal Plants

The extraction method is a systematic procedure used to isolate active compounds present in medicinal plants from their inactive or inert components by employing selective solvents under standardized procedures. This method is fundamental in the preparation of various medicinal plant-based products used in traditional and modern medicine. The substances obtained through this process typically take the form of relatively impure liquids, semisolids, or powders. The intermediate products of extract were primarily designed for oral or external applications, depending on their intended therapeutic purpose, which can be directly utilized as medicinal agents, such as in the form of tinctures or fluid extracts. These formulations are prepared, making them ready for immediate use. However, all the extracts contain desired bioactive compounds generally not suitable for direct application. Before it requires

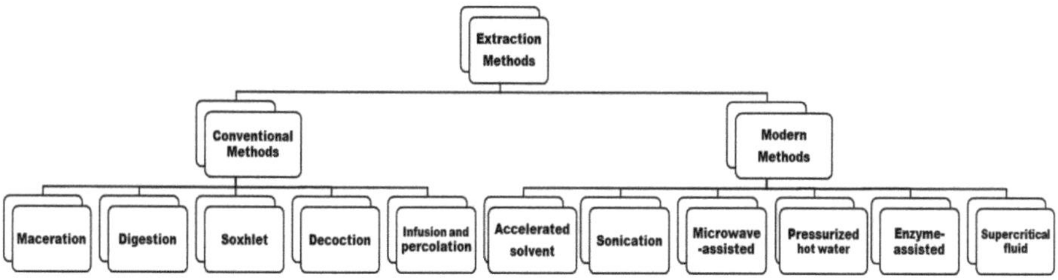

Fig. 4.1 Classification of different extraction processes

additional processing steps to remove impurities, purify, and refine them for specific medicinal applications (Lončarić et al. 2020). The process of extracting active compounds from medicinal plants and their natural sources can generally be categorized into two main methodologies. These approaches differ in their techniques, solvents, and specific conditions but share the common goal of maximizing the yield and quality of the desired bioactive compounds while minimizing the presence of undesirable substances. In the green synthesis of nanoparticles and nano formulation, with the required appropriate extraction process, can be categorised into two types and further subdivided, as shown in Fig. 4.1.

4.2.1 Conventional Methods

Traditional plant extraction methods, including maceration, percolation, and Soxhlet extraction are widely used for extracting bioactive compounds using organic solvents. However, these methods are often associated with the disadvantages of requiring significant amounts of solvent and extended durations to achieve satisfactory yields. On the other hand, techniques like decoction and hydro distillation utilize water as a solvent, offering simpler and more sustainable approaches. Some of the conventional extraction methods are as follows-

4.2.1.1 Maceration

Maceration is a widely employed and straightforward extraction technique utilized in the preparation of bioactive compounds from plant materials. The process begins by immersing coarsely ground or finely powdered plant material in a solvent of choice under ambient conditions, ensuring that the solvent thoroughly interacts with the plant matrix. In order to ensure optimal extraction, the resulting mixture is left unaffected for at least 3 days. Throughout this span, stirring or agitation might be necessary to help the target phytochemical integrate into the solvent. The mixture is separated to separate the liquid extract from the solid plant residue, subsequently referred to as the marc, when the extraction time is judged adequate. By using sieves or fine-mesh nets to eliminate enormously particles, mechanical straining is used to accomplish this separation. To recover any solvent that may have been trapped in its matrix, the remaining marc is further compressed or squeezed. Afterwards making the ingredients settle for a predetermined amount of time, the resultant liquid extract is next eliminated to get rid of any tiny particles or contaminants. This is usually done by decantation or filtration using the proper filter substance. Maceration is often carried out in a tightly sealed or stopper container to reduce solvent loss through evaporation during the extraction process. However, this process is frequently denounced for its inherent limitations, such as the long time required for complete extraction and its relatively low efficiency in extracting phytochemical as compared to modern or advanced extraction techniques, despite its simplicity and inexpensive equipment needed. These drawbacks highlight the necessity of process optimization or the use of different approaches in order to increase yield and efficiency (Chuo et al. 2022).

4.2.1.2 Digestion

In applying moderate temperatures to the extraction process, the fact closely related approach improves on the doctrines of maceration. This moderate heating, usually between 35 °C and 50 °C, improves the recovery of bioactive chemicals by increasing the solvent's effectiveness in dissolving plant cell walls (Doellinger et al. 2020). In digestion, the extraction procedure entails combining plant material with a heated solvent, keeping the mixture at the appropriate temperature for 30–24 hours, and stirring frequently to promote solubilisation.

4.2.1.3 Soxhlet

This superior and versatile technique is frequently utilized to the continuous extraction of chemical constituents using a heated solvent. This method involves storing the ground phyto-parts in a porous glove composed of long-lasting materials like filter paper or cellulose. The arrangement makes it possible to repeatedly wash the plant material with new solvent, guaranteeing the highest possible recovery of bioactive chemicals. This technique is particularly useful for removing chemicals from plant materials that need to be exposed to solvents over an extended period for complete extraction (Lόpez-Bascón and De Castro 2020).

4.2.1.4 Decoction

This technique has been developed especially for boiling plant materials to extract water-soluble chemicals. Depending on the kind of tissue and the desired phytochemicals and phyto-parts are soaked in hot water and then boiled for 15 to 60 minutes. A steady period usually 15 minutes is adequate for soft plant parts. Although given with significant temperatures, like 100 °C, decoction has been shown to be an effective method for removing substances like phenols and flavonoids from a variety of plant sources, such as fruits, rhizomes, and leaves (Shikov et al. 2022).

4.2.1.5 Infusion

A simple yet effective extraction technique, infusion produces a diluted solution with readily soluble plant material components. To extract the plant material's bioactive components, this method involves immersing it in a boiling solvent, usually water. To maximize the release of the targeted phytochemicals into the solvent, the mixture remains quiescent in a sealed or capped container for around 15 minutes. After this steeping time, the liquid extract, often known as "tea," is carefully poured off or decanted, and the solid residue (marc) is filtered off to provide a clear solution. A typical instance of an infusion that illustrates the broad range of applications for this technique is tea. In addition to being used as a cool drink, infusions are often utilized in herbal remedies and traditional medicine to treat several illnesses, such as asthma, bronchitis, and diarrhoea. The antioxidant qualities of substances like flavonoids and phenols, for example, have been used to infuse ginger rhizomes. This involves boiling the plant material in water for 10 minutes to effectively extract these bioactive molecules (Bitwell et al. 2023). The result demonstrates the procedure by which infusions can be used therapeutically to give useful bioactive chemicals for overall health and wellbeing.

4.2.1.6 Percolation

In the domains of phytochemistry and natural product extraction, percolation is a prominent and often used technique for creating fluid extracts, such as tinctures. "Percolation" is the process of permitting a liquid gradually, usually in a methodical and regulated way, move through a solid medium. This method operates especially well for effectively removing bioactive substances from plant matrices. The plant material is first soaked and placed within a cylindrical percolator apparatus to help the solvent flow downward during the percolation process. At the top of the device, a solvent is added usually ethyl alcohol or a combination of alcohol and water. The solvent dissolves and removes the phytochemicals as it seeps through the plant material, getting more saturated as it descends. The freshly prepared solvent is gradually introduced from the top to maintain a continuous extraction process. This ensures that the flow is constant and that the phytochemicals are effectively leached from the plant matrix. Since it can yield concentrated extracts with small solvent waste, the percolation process is well respected. Furthermore, the process's regulated and progressive nature minimizes the co-extraction of undesirable components while enabling the selective extraction of desired chemicals. This technique's dependability, scalability, and capacity to produce high-quality extracts have led to its widespread use in both conventional and contemporary phyto-pharmaceutical preparations (Senapati and Behera 2023).

4.2.2 *Modern Methods*

4.2.2.1 Accelerated Solvent Extraction

The stainless-steel extraction cell is packed with the plant sample and layered with inert silica and cellulosic filter papers to prevent clogging from plant material aggregation. Once filled with solvents, the cell is exposed to high temperature and pressure for a set period, conditions that enhance extraction by increasing the diffusion rate and lowering solvent viscosity. This promotes deeper solvent penetration into the plant matrix. After extraction, the solution is collected in vials, and the cell is cleaned by flushing with fresh solvent followed by nitrogen gas. The inert packing materials help ensure even distribution of the sample and prevent system blockages (Rahmalia et al. 2015).

4.2.2.2 Sonication Extraction

Sonication extraction is a widely utilized technique for the recovery of bioactive compounds. This method employs ultrasonic waves to generate cavitation in the solvent, which disrupts plant cell walls and enhances the release of phytochemicals into the extraction medium. It has proven particularly effective for extracting phenolic compounds from fruit matrices. Comparative studies with conventional methods, such as maceration, highlight the advantages of sonication, including significantly higher yields and reduced extraction time. For example, sonication achieved superior extraction efficiency within 30 minutes, whereas maceration required 120 minutes and yielded lower concentrations of target compounds (Ranjha et al. 2021).

4.2.2.3 Microwave-Assisted Extraction

Microwave-assisted extraction (MAE) utilizes electromagnetic radiation in the frequency range of 300 MHz to 300 GHz to generate heat through direct interaction with polar molecules in the plant matrix. This process induces dipolar rotation, wherein polar molecules attempt to align with the rapidly oscillating electromagnetic field, resulting in frictional heating. The efficiency of microwave heating is influenced by the dielectric properties of the solvent. By simultaneously heating both the solvent and plant tissue, MAE enhances mass transfer and significantly accelerates the extraction process. This technique has proven effective for the rapid recovery of bioactive compounds, including phenolic compounds and anthocyanins, as demonstrated in extractions from grape peels (Nour et al. 2021).

4.2.2.4 Pressurized Hot Water Extraction

This technique effectively facilitates the release of bound compounds by employing hot water as the extraction solvent under elevated pressure. Operating at pressures around 5 MPa and temperatures ranging from 100 °C to 250 °C, the method maintains water in its liquid state, enabling efficient solubilization of target compounds. As a green and sustainable alternative to conventional extraction methods, it offers both environmental and operational advantages (Plaza and Marina 2023).

4.2.2.5 Enzyme-Assisted Extraction

Enzyme-assisted extraction is frequently integrated with complementary extraction techniques to improve the release and solubility of non-extractable phytochemicals. Enzymes function by degrading complex cellular matrices, thereby liberating bioactive compounds that are typically inaccessible through conventional extraction methods. A synergistic approach combining enzymatic treatment with microwave-assisted extraction enhances phytochemical recovery by utilizing elevated temperatures and rapid heating, leading to significantly higher extractability compared to traditional water-based solvent extraction (Das et al. 2021). Similarly, a sequential strategy incorporating enzymatic pre-treatment followed by ultrasound-assisted extraction has proven highly effective in increasing pectin yields from sisal waste. The enzymatic pre-treatment disrupts the cell wall architecture, enabling ultrasonic waves to further facilitate the release of pectin. This combined approach yields substantially greater pectin recovery than conventional extraction methods lacking enzymatic intervention (Shen et al. 2023).

4.2.2.6 Supercritical Fluid Extraction

Supercritical fluid extraction (SFE) has gained prominence as an efficient and sustainable technique for the large-scale extraction of high-value compounds from food products and plant-based materials. Utilizing supercritical fluids, most commonly carbon dioxide—under high pressure and moderate temperature, SFE offers a cleaner and more selective alternative to conventional solvent-based methods. This approach is particularly advantageous in addressing the limitations posed by complex plant matrices, where phytochemicals are often sequestered within the cell cytoplasm and remain inaccessible through traditional extraction techniques (Uwineza and Waśkiewicz 2020).

4.3 Synthesis of Medicinal Plant-Assisted Nanoparticles

Over the past decades green chemistry or sustainable chemistry has become a key member of sustainable development in various industries. Sustainable development generally includes, growing in a way that meets today's needs without harming future generations. This is especially important for chemical industries which often face challenges like pollution and excessive uses of resources. An example of green chemistry in recent context is plant-based synthesis of nanoparticles (NPs). Instead of using harmful chemicals this method is totally relies on natural extracts of medicinal plants along with safe solvents like water or ethanol. It also uses non-toxic reducing and stabilizing agents, making the process more sustainable. Nanoparticles can be synthesized through physical, chemical, and biological methods. Among these, the biological or "green" synthesis route has gained popularity due to its eco-friendly nature and reduced environmental footprint. Traditional chemical methods, while effective, are often costly and involve hazardous chemicals that pose risks to both human health and the environment (Salem and Fouda 2021). In contrast, biological synthesis using plant extracts offers a safer, biocompatible, and environmentally sustainable alternative. This green approach has shown immense potential for applications in diverse fields, including nanomedicine, agriculture, food technology, bioengineering, and cosmetics (Bhardwaj et al. 2022). Biological synthesis is not limited to plants; fungi, algae, and bacteria have also been employed for NP production. However, plant-based synthesis is particularly attractive due to the availability and diversity of plant materials. Various parts of plants, including leaves, fruits, roots, stems, and seeds, have been effectively utilized to synthesize the NPs of different sizes, shapes, and compositions (Bhardwaj et al. 2019). Plant extracts contain a wide range of phytochemicals, such as flavonoids, terpenoids, alkaloids, and phenolic compounds, which act as natural reducing and stabilizing agents during NP synthesis. These compounds not only facilitate the formation of NPs but also contribute to their biocompatibility and functionality. Furthermore, medicinal plant-based NPs are associated with minimal side effects in humans compared to their chemically synthesized counterparts. This makes them highly suitable for biomedical applications such as drug delivery, nanomedicine, and human health protection. Additionally, their use in agriculture for pest control and soil enhancement, as well as in food science for packaging and preservation, underscores their versatility. The ability to control NP size, shape, and composition through careful selection of plant extracts and synthesis conditions further enhances their potential in a wide range of technological and scientific applications (Fig. 4.2).

4.4 Factors Affecting the Synthesis of NPs

Numerous factors influence the preparation of NPs from medicinal plants and their sources, directly as well as indirectly. Among them, the extract and salt concentration, temperature, pH, Pressure, reaction time, and environment are the most important factors that affect the physical and chemical properties of NPs (Fig. 4.3).

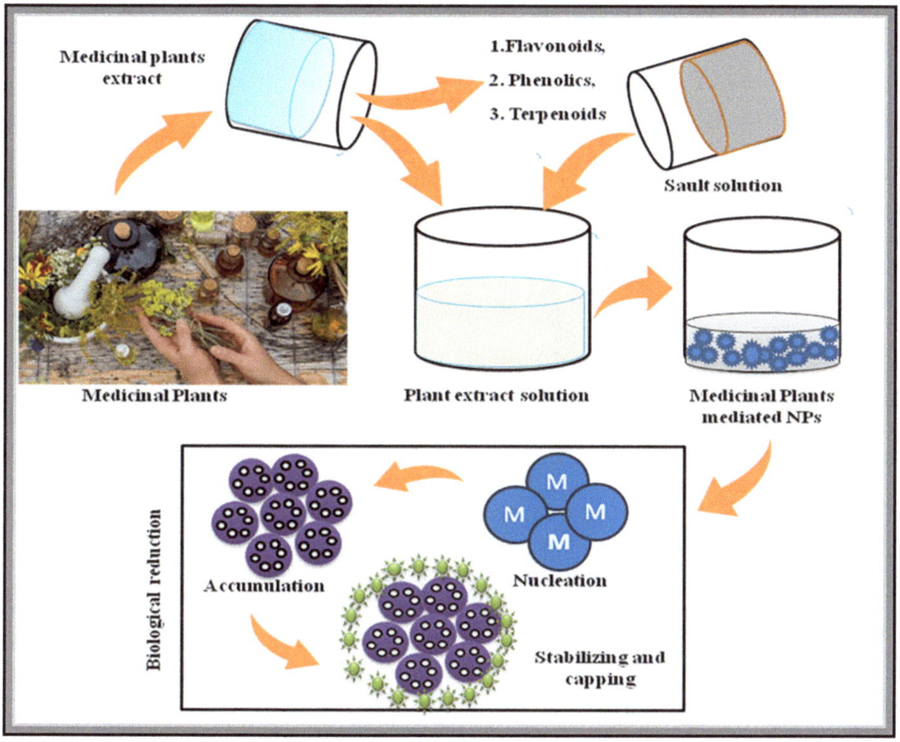

Fig. 4.2 Schematic representation for medicinal plant-based synthesis of nanoparticles

4.4.1 Concentration of Extract or Precursor

The size, shape, and rate of building up of the NPs are significantly influenced by the concentration of plant extract employed in their synthesis. It has been observed that the size of NPs grows relative to the concentration of extracts and salt. This is because the extract's reducing agent concentration improves as well, leading to a modification of particle size. However, if the extract concentration is increased too much, the process may continue, and the NPs' size could be increased (Dogiparthi et al. 2021).

4.4.2 Temperature

The temperature is one of the most significant physical parameters that affect the crystallinity of NPs during the synthesis of NPs using medicinal plant sources. While secondary reduction ceases on the surface of constructed nuclei, the faster rate of reduction happens at higher temperatures because metal ions are consumed during nuclei production. The narrow peak at high temperature implies the synthesized NPs are bigger in size, whereas the broadening peak at low temperature

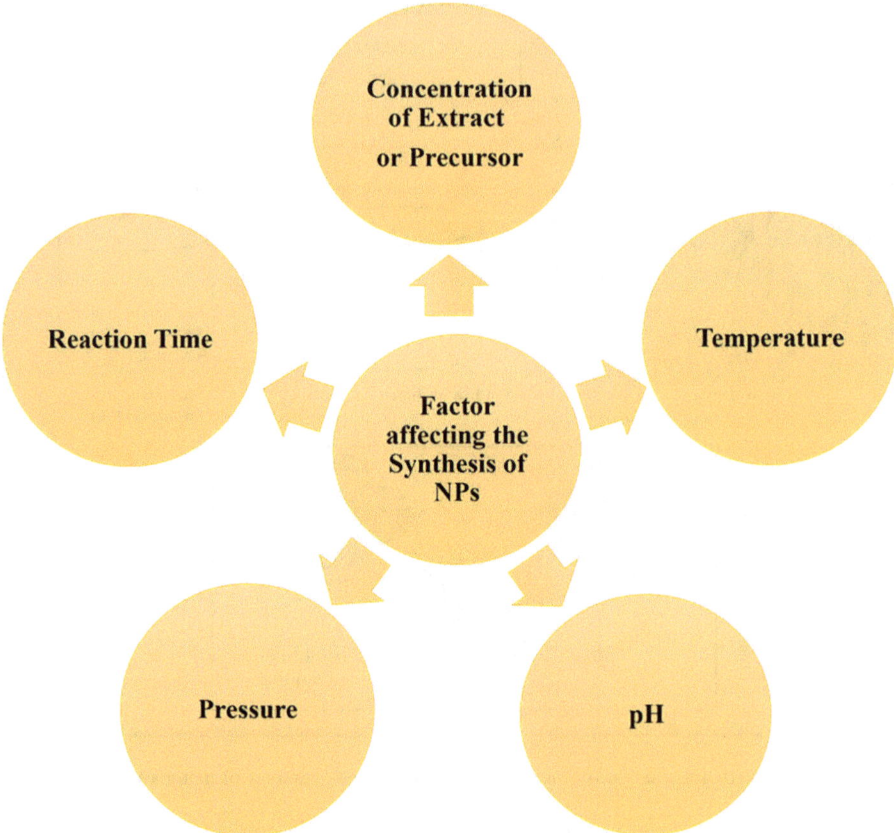

Fig. 4.3 Factor affecting the synthesis of NPs using medicinal plants sources

suggests the production of smaller NPs. Therefore, it can be concluded that a greater temperature is ideal for manufacturing massive NPs synthesis. Therefore, temperature is a crucial factor that influences the physical and chemical behavior of NPs. Green technology often requires temperatures below 100 °C or room temperature to synthesize NPs, even physical and chemical procedures require a very high temperature up to 350 °C (Chan et al. 2022)

4.4.3 pH

The pH has an essential parameter effect on the NPs during the green synthesis method. The size and texture of the synthesized NPs are influenced by the pH of the solution medium, similar to scientific reports (Taha et al. 2024). Consequently, by adjusting the pH of the solution media, NPs size of NPs can be modified.

4.4.4 Pressure

During the synthesis of NPs, pressure is a crucial parameter. The size, shape, and surface area of the created NPs are influenced by the pressure imposed on the reaction media. It has been identified that at ambient pressure instances, the rate at which biological agents reduce metal ions is significantly better (Salem 2023).

4.4.5 Reaction Time

The length of time the reaction media is incubated has a significant impact on the type and quality of NPs produced utilizing green technology. Likewise, the properties of the synthesized NPs changed with time and were significantly impacted by the fabrication procedure, light exposure, storage conditions, and other factors. Variations in time can happen in a variety of ways, including particle aggregation from prolonged storage, particle growth or shrinkage during storage, shelf life, and other factors that impact potential (Hosseingholian et al. 2023).

4.5 Spectroscopic Investigation of Synthesized NPs

The shape, texture, size and surface area of the synthesized NPs are characterized using a variety of characterizations techniques, such as Powder X-ray diffraction (XRD), field emission scanning electron microscopy (FE-SEM), transmission electron microscopy (TEM), energy dispersive X-ray spectroscopy (EDX), UV-visible absorption spectroscopy (UV-Vis), Fourier transform infrared spectroscopy (FTIR), and dynamic light scattering (DLS) (Fig. 4.4).

4.5.1 X-Ray Diffraction (XRD)

In addition to assisting with molecular modeling to ascertain the material's structure, XRD data provide details on crystallinity, crystallite size, orientation, and phase composition. This method may describe the NPs and compare them to known structural databases by looking at the crystalline structure (orientation, phases, and lattice characteristics) of powdered materials. A diffraction pattern is created when X-rays strike the crystal surface and interact with the atoms because the atoms are spaced appropriately apart in a crystal. Two peaks are produced by the XRD experiment; a broad peak indicates an amorphous powder, and a sharp peak reveals the crystalline structure of the synthesized NPs. In numerous instances, specimens exhibited a significant diffraction pattern that was directly linked to the structure of

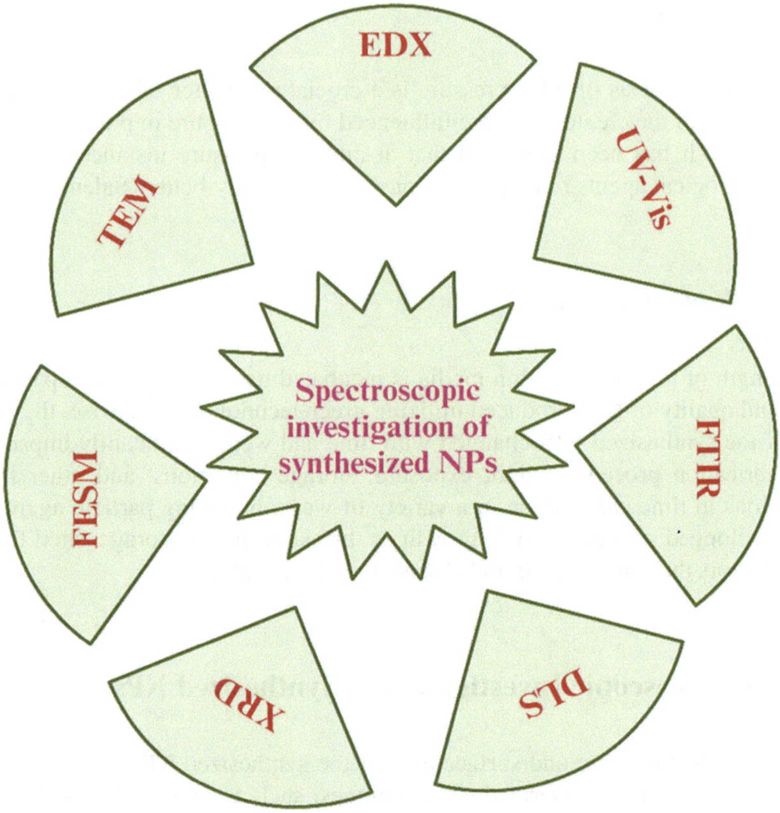

Fig. 4.4 Spectroscopic investigation of medicinal plant-mediated NPs

the NPs. The XRD method has been utilized in many studies ~~to~~ describe the phase purity of NPs and determine the crystalline characteristics of samples (Kumar et al. 2022). Simple preparation, quick measurement, mixed phase analysis, and sample purity determination are all benefits of XRD. Its drawbacks include the need for uniforms, powdered material and the ambiguous data caused by exceeding extensions.

4.5.2 Field Emission Scanning Electron Microscope (FE-SEM)

The sample surface ~~is~~ imaged using a field emission scanning electron microscope using a high-energy electron beam. Signals in the form of secondary electrons, backscattered electrons, and distinctive X-rays are produced when the electron beam hits the specimen's surface and interacts with the sample's atoms. These signals provide details on the sample's composition, surface topography, and other characteristics. SEM gives two-dimensional imaging, simplicity in sample preparation, and availability of digital data forms as its advantages (Joudeh and Linke 2022a). One of its drawbacks is that poor sample preparation might cause artifacts and real data to be confused. Size, expense, and upkeep are other evident restrictions.

4.5.3 Transmission Electron Microscopy (TEM)

The crystalline sample interacts with the electron beam in a TEM mostly by diffraction as opposed to absorption. The orientation of the planes within a crystal determines the diffraction intensity. This results in a change in electron intensity that provides crystal structural information. A TEM image can also be used to visualize the exfoliation, intercalation, orientation, and grain distribution and dispersion of NPs. High-quality, precise, and effective magnification of element and compound structures is provided by the TEM's analysis. Its drawbacks include time-consuming sample preparation, sample preparation artifacts, and unquestionably big and expensive samples (Joudeh and Linke 2022b).

4.5.4 Energy Dispersive X-Ray (EDX)

The Energy dispersive X-ray (EDX) is a combination with the SEM micrograph employed for elemental analysis of the sample. The EDX technique uses distinctive X-rays to determine the composition of a sample, providing an overall mapping of the sample. On the other hand, EDX is a special and versatile instrument for completing an elemental analysis specific to a given location or creating a map of the composition of a sample. This technique entails using electrons or high-energy photons to stimulate samples, followed by the detection of the photons' spectrum (Filippov et al. 2023). The benefits of EDX include enhanced quality control, process optimization, contaminant identification, and increased production yield. The precision is limited by the fluorescence of emitted X-rays, and quantitative analysis needs standards of known composition.

4.5.5 Ultraviolet-Visible Spectroscopy (UV-Vis)

The size, shape, and material composition of NPs affect their peak magnitude, wavelength, and spectral bandwidth. The rapid analysis, extremely high precision and accuracy, applicability to a large range of compounds, and the ability to be employed both quantitatively and qualitatively are some of the benefits of UV-VIS spectroscopy (Khrennikov et al. 2024). The failure to distinguish between substances that absorb at the same wavelength is a shortcoming of the technique.

4.5.6 *Fourier Transforms Infrared Spectroscopy (FT-IR)*

Chemical information, especially functional groups (amide, hydroxyl, and carboxyl groups) and bond parameters, is displayed via Fourier-transform infrared spectroscopy (FTIR). FTIR analysis is utilized to obtain useful information about the surface chemistry of fabricated NPs by identifying the bioactive compounds found in plant extracts that may act as coating/capping and stabilizing agents. To show that a functional group can transform from a raw plant extract to a manufactured NPs, FTIR not only detects functional groups but also stretching and vibrational frequencies like C=O, C=C, C–H, and C–O (Eid 2022).

4.5.7 *Dynamic Light Scattering (DLS)*

It is the most common technique to measure a molecule's or particle's size. Particle size can be determined due to the size-dependent variation in the intensity of the scattered light from laser-illuminated particles. DLS can, therefore, be used to study particle size and size distribution (Rasmussen et al. 2020). Benefits include consistent analysis, no sample preparation, a liquid sample, precision of ±1%, and determination of particle sizes of 1 nm. Multiple light scattering and limited resolution of polydisperse samples are its drawbacks.

4.6 Recent Advances of Medicinal Plant-Mediated NPs

The synthesis of medicinal plant-based nanoparticles (NPs) is gaining increasing attention due to their simplicity, cost-effectiveness, and environmental friendliness. Among these, zinc oxide nanoparticles (ZnO NPs) are particularly easy to prepare using various parts of medicinal plants. The fabrication of ZnO NPs via the green route typically involves the use of a zinc metal ion solution in combination with a reducing biological agent derived from plant biomolecules. This method employs a range of naturally occurring compounds to reduce and stabilize zinc ions, offering an inexpensive and straightforward approach for ZnO NPs fabrication (Rodríguez-Barajas et al. 2024; Osman et al. 2024). Remarkably, almost all medicinal plants possess the potential for utilization in the fabrication of ZnO NPs due to their rich biochemical diversity. Gold nanoparticles (AuNPs) have garnered significant interest in recent years, primarily due to their ease of fabrication, versatile surface fictionalization capabilities, and unique properties. Gold nanoparticles (AuNPs) are widely used in medicines due to their high biocompatibility, low toxicity, and ease of function (Amina and Guo 2020; Venkatas and Singh 2024). Natural biomolecules like flavonoids, phenols and protein help reduce metal ions and stabilizes AuNP, enhancing their performance (Perveen et al. 2021). Copper nanoparticles (CuNPs) are another promising option offering a cost-effective alternative to gold and silver nanoparticles. They are synthesized by plant extracts, which contain bioactive compounds that reduce copper ions and keep the nanoparticle stable (Antonio-Pérez et al. 2023). The green synthesized ZnO, Au, TiO2, FeO, CaO, Mg and Cu NPs highlights the potential of plant-based nanotechnology which is an ecofriendly approach that uses natural compounds to produce advanced nanomaterials shown Table 4.1.

4.7 Applications of Medicinal Plant-Mediated NPs

The goal of nanotechnology is to comprehend and work with matter at the nanoscale. Because of their unique properties and smaller size, they are more advantageous in the biomedical field. Medicinal plants mediated by nanomaterials are advantageous due to their wide range of biomedical applications, which include anticancer, antimicrobial, antifungal, antioxidant, and other clinical applications.

4.7.1 Anti-Cancer Potential

Nanomedicine offers comprehensive strategies for combating cancer through predictive diagnostics, prevention, therapeutic interventions, and the potential for personalized medicine (Anjum et al. 2021a). Many herbal-based nanoparticles (NPs) have demonstrated anticancer potential. For example, zinc oxide nanoparticles (ZnO NPs) synthesized from Cassia auriculata leaf extract exhibited selective tumoricidal activity against MCF-7 breast cancer cells without adversely affecting normal MCF-12A human breast cells (Prasad et al. 2020). Similarly, gold nanoparticles (AuNPs) derived from *Trachyspermum ammi* seed extract showed concentration-dependent inhibition of HepG2 cancer cell growth, primarily through a mechanism involving reactive oxygen species (ROS)-mediated apoptosis (Perveen et al. 2021). This apoptotic process has been associated with mitochondrial dysfunction, driven by ROS-induced disruption of mitochondrial membrane potential and the subsequent activation of Caspase-3 gene expression and enzymatic activity. Despite these promising findings, a deeper understanding of the molecular mechanisms underlying NPs' anticancer effects is essential. Additionally, questions about the fate of medicinal plant-based NPs within the body remain unanswered. To advance nanoparticles (NPs) applications in medicines, researchers must first study their degradation rates and identify the factors that affect how long they remain active in the body. It is also important to develop methods for controlling their retention time and assessing their short-term and long-term effects at both cellular and systematic levels. Standardizing NPs is essential to ensure consistency in experiments and establish safety profile before clinical use. Further research should focus on evaluating potential risks to prevent unexpected side effects. Scientists must

Table 4.1 Medicinal plant-mediated NPs and their antimicrobial/antibacterial applications

S. No.	NPs	Plants Used	Parts of the plants	Shape/Size	Application	Efficiency (MIC/ ZE or ARP values (mm))	Reference
1.	ZnO NPs	*Cymbopogon olivieri*	Leaves	Spherical 28 nm	Antimicrobial Anticancer	*Streptococcus pyogenes* (12.2 ± 0.8), *Klebsiella pentomonium* (nill) *F. oxyspo-rum* (11.6 ± 0.6) *C. albinx* (nill)	Golzarnezhad et al. (2025)
2.	ZnO NPs	*Calendula* sp.	Leaf	Hexagonal 28.23 nm	Antibacterial and Antioxidant	*S. aureus* (35.2 mm), *K. pneumoniae* (23.6) *E. coli* (13.5)	Tiwari et al. (2024a)
3.	ZnO NPs	*Calendula* sp.	Petal	Spherical 16 nm	Antioxidant	ZnO (0.07) Ascorbic acid (0.12)	Tiwari et al. (2024b)
4.	ZnO NPs	*Murraya* sp.	Leaf	Hexagonal 55 nm	Antibacterial	*E. coli* (36.775 ± 1.27) *S. typhimurium* (32.375 ± 1.96) *S. aureus* (24.35 ± 0.43)	Tiwari et al. (2022)
5.	ZnO NPs	*Platycladus orientalis*	Leave	High degree agglomeration 26.9 nm	Antimicrobial Antifungal	*L.monocytogenes* (13.50) *E. coli* (13.65), *E. faecalis* (14.05) *Mucor mucedo* (0.05 mg/mL), *Penicillium chrysogenum* (0.05 mg/mL), *Aspergillus niger* (0.03 mg/mL).	Ogwuegbu et al. (2025)
6.	ZnO NPs	*Monoonlongifolium*	Leave	Hexagonal 49.25 nm	Antibacterial, Antioxidant and Antifungal	*S. aureus* (22 ± 0.57 mm) *E. coli* (19 ± 1 mm), as well as antifungal activity against *C. albicans* (21 ± 0.16)	Bekele et al. (2024)

7.	TiO2	*Spinach*	Leaf	Spherical 10–40 nm	Antibacterial	*E. coli* (0.5 mg/ml) *S. aureus* (2 mg/ml)	Ansari and Daneshjou (2024)
8.	TiO2	*Mangifera indica and Azadirachta indica*	Leaf	Agglomerated 11.9 nm 11.2 nm	Antimicrobial, Antioxidant	*A. indica* (18.9 ± 0.4 mm) *M. indica* (16.3 ± 0.65 mm) against *A. alternata* fungus	Rana et al. (2024)
9.	TiO2	*Piper nigrum*, *Coriandrum sativum* and *Syzygium aromaticum*	Leaf	Spherical 5–20 nm	Drug delivery, biosensing and as antimicrobial agents	*Aspergillus Niger* (18.76 ± 2.45) *Candida Albicans* (19.12 ± 1.43) *Aspergillus Flavus* (18.45 ± 1.30)	Ali et al. (2024)
10.	TiO2 NPs	*Tinospora cordifolia*	Stem	Triangular 15.02 nm	Antibacterial	*E. coli* (13 ± 0.39) at 150 μg/ml and (26.5 ± 0.79) at μg/ml	Saini and Kumar (2023)
11.	TiO2 NPs	*Caricaceae (Papaya)*	Shell	Agglomeration 15 nm	Antifungal	*S. sclerotiorums* (60.5%)	Saka et al. (2022)
12.	CaO NPs	*Citrullus colocynthis ethanolic*	Fruit	Cubic 35.93 ± 2.54 nm	Cytotoxicity, antimicrobial, antioxidant, hemolytic activity	*M. luteus* (25.6 ± 1.1), *V. cholera* (28.5 ± 0.3) *V. parahaemolyticus* (27.2 ± 1.3)	Mazher et al. (2023)
13.	CaO NPs	*Piper betel*	Leaf	Spherical 24.9 nm	Antibacterial and anticancer	*P. aeruginosa* (10), *S. aureus* (10). *E. coli* (18) at 500 mg/mL concentration	Yoonus et al. (2021)
14.	CaO NPs	*Annona squamosa*	Seeds	Spherical 3.72 nm	Antimicrobial	*P. aeruginosa* (12.5 ± 0.7) *P. acnes* (15.5 ± 0.7) at 500 μg μl^{-1}	Sarathamani et al. (2025)
15.	CaO NPs	*Moringa oleifera*	Leaf	Spherical 32.08 nm	Antibacterial	*E. coli* (24 mm) *S. aureus* (11 mm) at 100 mg/ml	Jadhav et al. (2022)

(continued)

Table 4.1 (continued)

S. No.	NPs	Plants Used	Parts of the plants	Shape/Size	Application	Efficiency (MIC/ ZE or ARP values (mm))	Reference
16.	CaO NPs	*Papaya*	*Leaf*	Spherical or quasi-spherical 24.89 nm	Antimicrobial	*E. faecalis* (26.67 ± 0.5)	Kazi et al. (2024)
17.	CuO NPs	*Aloe vera*	Leaf	Sago-shaped 15.8 nm	Antibacterial	*S. typhi* (13 ± 0.02) *L. monocytogenes* (15 ± 0.04)	Jabeen et al. (2024)
18.	CuO NPs	*okra*	Fruit	Spherical 57 nm	Antibacterial	*S. aureus-1112* (12 ± 0.53), *E. faecalis-29212* (11 ± 0.48), *K. pneumonia- 10031* (11 ± 0.56)	Javid-Naderi et al. (2025)
19.	CuO NPs	*Balanites aegyptiaca*	Stem bark	Spherical 22–35 nm	Antibacterial	*B. subtilis* (13)	Teklu et al. (2023)
20.	CuO NPs	*Morinda citrifolia*	Leaf	Spherical 20 to 50 nm	Antibacterial	*Aspergillus flavus* (13.0 ± 1.1), *A. niger* (14.3 ± 0.7), *Penicillium frequentans* (16.8 ± 1.4)	Priya et al. (2023)
21.	Fe_2O_3 NPs	*Carica papaya*	Leaf	Agglomeration form 4.58 nm	Antibacterial	*S. aureus* (12.5 ± 0.5) *Klebsiella* spp. (9 ± 1) *E. coli* (9 ± 1) *Pseudomonas* spp. (10 ± 0.5) at 30 mg/ml	Bhuiyan et al. (2020)
22.	Fe_2O_3 NPs	*S. officinali*	Leaf	Spherical 42.327 nm	Antibacterial	*E. coli* (21.14 ± 0.16), *P. aeruginosa* (17.26 ± 0.26), *MRSA strains* (20.56 ± 0.62)	Yassin et al. (2024)
23.	Fe_2O_3 NPs	*Hibiscus rosa sinensis*	Flower	Spherical 12.99–22.72	Antibacterial	*S. aureus 25 mm* *E. coli 24 mm*	Elkhateeb et al. (2024)
24.	Fe_2O_3 NPs	*Thymus migricus*	Leaf	Spherical 14.74 ± 0.23	Antimicrobial	*S. aureus* (6.15 ± 0.19) *S. enterica* (13.31 ± 0.46)	Ashrafi-Saiedlou et al. (2025)

25.	Fe_2O_3 NPs	*Clitoria ternatea*	Flower	Partially rhombic-shaped 68.41 nm	Antibacterial	*E. coli* 18–26 mm	Kachhawaha et al. (2025)
26.	MgO NPs	*P. guajuava*	Leaf	Spherical 44.7 nm	Antibacterial	*S. aureus* 5 mm, *E. faecalis* 7 mm *K. pneumonia* 6 mm	Abinaya and Kavitha (2023)
27.	MgO NPs	*Neem*	Leaves	Rod like with needle edges 60 nm	Antibacterial	*E. faecalis* 8 mm *S. typhimurium* 6 mm, *Enterococcus* 15 mm	Rabi (2025)
28.	MgO NPs	*Abrusprecatorius*	Leaf, Bark	Agglomerate 100–200 nm	Photocatalytic, Antioxidant, Antibacterial and Cytotoxicity Activities	*S. epidermidis* (20 ± 0.26 mm) *B. subtilis* (31 ± 0.25 mm) *P. aeruginosa* (22 ± 0.50 mm) *E. coli* (30 ± 0.10 mm)	Ali et al. (2023)
29.	MgO NPs	*Mariposa christia vespertilionis*	*Leaves*	Spherical 17 nm (600 °C) 90 nm (400 °C) 158 nm (900 °C)	Antibacterial	*Better result for S. aureus than E. coli*	Farizan et al. (2023)
30.	MgO NPs	*Rosa floribunda charisma*	Petals	Polyhedral 35.25–55.14	Antioxidant, antiaging, and antibiofilm activities	*S. epidermidis,* (15.63) *S. pyogenes* (7.81) *Pseudomonas* (31.81)	Younis et al. (2021)

carefully design and optimize nano formulations to actively target cancer cells while maximizing safety and effectiveness. For future treatments rigorous experimental testing determined the best therapeutic approach (Anjum et al. 2021b).

4.7.2 Antimicrobial Potential

Antibiotic resistance has become a serious issue worldwide. Bacteria are changing quickly and becoming resistant to common medicines. To fight this scientist are looking at neutral nanoparticles as a new way to treat these microbes (Jassal et al. 2022). Silver nanoparticles show special promise because it can kill many types of bacteria. Researchers made these nanoparticles from the leaves of *Carissa carandas* plant, and they worked well against these infectious bacteria, especially a type called *Shigella flexneri* that causes serious stomach illness (Singh et al. 2021). Similarly, bimetallic nanostructures, such as Pd-Ag nanostructures coated with reduced graphene oxide from *Stevia* leaf extract, effectively inhibited the growth of Gram-negative bacteria like *Escherichia coli* (Mallikarjuna et al. 2021). Zinc oxide nanoparticles (ZnO NPs) have also exhibited notable antimicrobial properties. For instance, ZnO NPs synthesized from *Cinnamomum verum* bark extract showed activity against *E. coli* and *Staphylococcus aureus* (Hheidari et al. 2024), while ZnO NPs derived from *Cassia auriculata* leaf extract disrupted bacterial cell integrity through direct contact (Padalia et al. 2018). Copper oxide nanoparticles (CuO NPs) derived from *Cymbopogon citratus* further demonstrated potent antimicrobial activity, including antibiofilm properties, underscoring their potential in managing resistant bacterial infections (Talebian et al. 2023). Cuprous oxide hollow microspheres (Cu_2O-HMs) were successfully synthesized using *D. carota* pulp waste extract, and bactericidal activity was observed significantly against *S. aureus* (Bhardwaj et al. 2017).

4.7.3 Antifungal Potential

The antimicrobial properties discussed earlier may hold significant potential for agricultural pathogen management, particularly in crop protection. Zinc oxide nanoparticles (ZnO NPs) have garnered considerable attention for their anti-phytopathogenic effects, demonstrating activity against bacteria, such as *Dickeya dadantii*, the soft rot pathogen, using ZnO NPs synthesized from lemon fruit (Hossain et al. 2019), and against fungi, as shown by the fungicidal activity of ZnO NPs derived from Eucalyptus globulus extract against key apple orchard pathogens (Ahmad et al. 2020). Notably, titanium dioxide nanoparticles (TiO_2 NPs) produced from lemon fruit exhibited antibacterial activity comparable to ZnO NPs against *D. dadantii* (Khan et al. 2023a). Silver nanoparticles (Ag NPs) synthesized from wheat extract have shown significant efficacy in mitigating the adverse effects of salinity stress in wheat by modulating abscisic acid levels, maintaining ion homeostasis, and enhancing defense mechanisms through both enzymatic and non-enzymatic antioxidants (Khan et al. 2023b). Additionally, ZnO NPs have been noted for their low toxicity and their ability to boost the antioxidant response in flax seedlings (Zhor et al. 2023).

4.7.4 Antioxidant Potential

Elevated oxidative stress, originating from mitochondrial activity or other endogenous and exogenous sources, can induce oxidative modifications in critical cellular macromolecules such as membrane lipids, proteins, and DNA. These alterations contribute to functional impairments, the progression of degenerative pathologies, and the mechanisms underlying aging (Rajput et al. 2020). Antioxidants have the potential to mitigate or reverse these harmful processes and may serve as therapeutic agents for aging and age-associated diseases. Certain nanoparticles (NPs) derived from green medicinal plants have demonstrated antioxidant properties, as exemplified by silver nanoparticles (Ag NPs) synthesized using *Carissa carandas* leaf extract (Dua et al. 2023). Gold nanoparticles (AuNPs) and silver nanoparticles (Ag NPs) synthesized from *Clerodendrum inerme* leaf extract, as well as nickel oxide nanoparticles (NiO NPs) derived from stevia leaf extract, have been studied for their antioxidant properties (Moghadam et al. 2022). The phytochemicals coating the nanoparticle surfaces play a critical role in the observed antioxidant effects. Typically, antioxidant activity is assessed using a single in vitro assay, such as the DPPH (2, 2-diphenyl-1-picrylhydrazyl) assay. However, given the complex nature of phytochemicals and the reaction mechanism-dependent variability in antioxidant activity, relying on a single assay is insufficient (Tungmunnithum et al. 2020). Consequently, the results from in vitro cell-free antioxidant tests should be interpreted cautiously, emphasizing chemical reactivity, while comprehensive *in vivo* (cellular) validation is essential for a more accurate assessment.

4.7.5 Other Clinical Potential

Medicinal plants-mediated NMs such as silver, platinum, titanium, and silica are utilized in various cosmetic and personal care products because these NMs serve as active ingredients in products including toothpaste, mouthwash, sunscreens, hair care products, perfumes, and anti-aging products. Additionally, medicinal plant medicated silica NPs act as excellent pesticide control in the agricultural industry. In the agriculture sector, a major challenge with current livestock additives and nutrient supplements is to boost livestock production while maintaining product quality, safeguarding the environment from hazardous substances, and ensuring food security. Nanotechnology offers novel vehicles for nutrient supplementation, enhancing the functionality of feed molecules. Moreover, minerals in the form of NPs can bypass the intestinal wall and cells more easily due to their size, thus improving their bioavailability (Thatyana et al. 2023).

4.8 Future Prospective

Medicinal plants have immense biochemical components which are still not explored in systematic manner because of high complexity and diversity of secondary metabolites. Some of the bioactive compounds explore by researchers in synthesis of nanomaterials and applications in the field of biomedical sectors. However, medicinal plant assisted synthesis of various NMs carry the potential for next generation sustainable drug designing. Because it can engineered/encapsulate/conjugate for targeted drug delivery and improve for control and sustainable drug release. Moreover, inherent nature of biocompatible NMs based drug still point of attraction for researchers, in future research scientists could isolate purest form of functional compound from extract and design more advanced highly efficient, sustainable drugs and diagnostic systems. It includes antibiotics, biosensing, analyzing, imaging, tissue engineering, and gene therapy. In interdisciplinary work we able to access very few new research of medicinal plant assisted NMs synthesis using microfluidic devices using AI, machine learning approaches providing the fin to unveil new discoveries.

4.9 Conclusions

The existing chapter highlights the conventional and nonconventional extraction process to determine higher selectivity and purity to obtain desired extract. After studying more than one hundred fifty recent research paper we have compiled this chapter which shows the immense opportunities are available in medicinal plant assisted nanomaterials due to the presence of huge number of secondary metabolites including flavonoids, alkaloids, phenolics terpenes and saponins. These compounds are playing a significant role in synthesis of NMs (Ag, ZnO, FeO, MgO, Au, Cu NPs etc.). This chapter gives opportunity to understand systematic extraction and synthesis processes. In addition to discusses different characterization techniques (XRD, SEM, TEM, UV-Vis, FTIR, and DLS) to explore physicochemical attributes. Furthermore, it can be easily explored for biomedical applications including development of antibiotics, antioxidant, diagnostics fields. This emerging field has significant potential for advancing nanotechnology, with diverse applications in medicine and implications for environmental sustainability.

Conflict of Interest No conflict of interest was reported by the authors.

Funding Source No funding sources are available.

References

Abinaya S, Kavitha HP (2023) Magnesium oxide nanoparticles: effective antilarvicidal and antibacterial agents. ACS Omega 8(6):5225. https://doi.org/10.1021/acsomega.2c01450

Ahmad H, Venugopal K, Rajagopal K, De Britto S, Nandini B, Pushpalatha HG, Konappa N, Udayashankar AC, Geetha N, Jogaiah S (2020) Green synthesis and characterization of zinc oxide nanoparticles using eucalyptus globules and their fungicidal ability against pathogenic fungi of apple orchards. Biomolecules 10(3):425. https://doi.org/10.3390/biom10030425

Ali S, Sudha KG, Thirumalaivasan N, Ahamed M, Pandiaraj S, Rajeswari VD, Vinayagam Y, Thiruvengadam M, Govindasamy R (2023) Green synthesis of magnesium oxide nanoparticles by using Abrus precatorius bark extract and their photocatalytic, antioxidant, antibacterial, and cytotoxicity activities. Bioengineering 10(3):302. https://doi.org/10.3390/bioengineering10030302

Ali H, Dixit S, Alarifi S (2024) Green synthesis of titanium dioxide nanoparticle using combination of plants and its antimicrobial study with their characterization. Mater Res Express 11(9):095403. https://doi.org/10.1088/2053-1591/ad7dd3

Amina SJ, Guo B (2020) A review on the synthesis and functionalization of gold nanoparticles as a drug delivery vehicle. Int J Nanomed:9823–9857

Anjum S, Ishaque S, Fatima H, Farooq W, Hano C, Abbasi BH, Anjum I (2021a) Emerging applications of nanotechnology in healthcare systems: grand challenges and perspectives. Pharmaceuticals 14(8):707. https://doi.org/10.3390/ph14080707

Anjum S, Khan AK, Qamar A, Fatima N, Drouet S, Renouard S, Blondeau JP, Abbasi BH, Hano C (2021b) Light tailoring: impact of UV-C irradiation on biosynthesis, physiognomies, and clinical activities of Morus macroura-mediated monometallic (Ag and ZnO) and bimetallic (Ag–ZnO) nanoparticles. Int J Mol Sci 22(20):11294. https://doi.org/10.3390/ijms222011294

Ansari FS, Daneshjou S (2024) Optimizing the green synthesis of antibacterial TiO2-anatase phase nanoparticles derived from spinach leaf extract. Sci Rep 14(1):22440. https://doi.org/10.1038/s41598-024-73344-5

Antonio-Pérez A, Durán-Armenta LF, Pérez-Loredo MG, Torres-Huerta AL (2023) Biosynthesis of copper nanoparticles with medicinal plants extracts: from extraction methods to applications. Micromachines 14(10):1882. https://doi.org/10.3390/mi14101882

Ashrafi-Saiedlou S, Rasouli-Sadaghiani M, Fattahi M (2025) Green synthesis of iron oxide nanoparticles using thymus migricus for multifunctional applications in antioxidant, antimicrobial, photocatalytic, and seed priming processes. Heliyon 11(5). https://doi.org/10.1016/j.heliyon.2025.e42933

Bekele SG, Ganta DD, Endashaw M (2024) Green synthesis and characterization of zinc oxide nanoparticles using Monoon longifolium leave extract for biological applications. Discover Chem 1(1):5. https://doi.org/10.1007/s44371-024-00007-9

Bhardwaj AK, Shukla A, Singh S, Uttam KN, Nath G, Gopal R (2017) Green synthesis of Cu2O hollow microspheres. Adv Mater Proc 2(2):132–138. https://doi.org/10.5185/amp.2017/215

Bhardwaj AK, Manish KG, Naraian R (2019) Myco-nanotechnological approach for improved degradation of lignocellulosic waste: its future aspect. In: Mycodegradation of lignocelluloses. Springer International Publishing, Cham, pp 227–245

Bhardwaj AK et al (2022) Biogenic and non-biogenic waste for the synthesis of nanoparticles and their applications. In: Bioremediation. CRC Press, pp 207–218

Bhuiyan MS, Miah MY, Paul SC, Aka TD, Saha O, Rahaman MM, Sharif MJ, Habiba O, Ashaduzzaman M (2020) Green synthesis of iron oxide nanoparticle using Carica papaya leaf extract: application for photocatalytic degradation of remazol yellow RR dye and antibacterial activity. Heliyon 6(8). https://doi.org/10.1016/j.heliyon.2020.e04603

Bitwell C, Indra SS, Luke C, Kakoma MK (2023) A review of modern and conventional extraction techniques and their applications for extracting phytochemicals from plants. Sci Afr 19:e01585. https://doi.org/10.1016/j.sciaf.2023.e01585

Chan YB, Selvanathan V, Tey LH, Akhtaruzzaman M, Anur FH, Djearamane S, Watanabe A, Aminuzzaman M (2022) Effect of calcination temperature on structural, morphological and optical properties of copper oxide nanostructures derived from Garcinia mangostana L. leaf extract. Nanomaterials 12(20):3589. https://doi.org/10.3390/nano12203589

Chuo SC, Nasir HM, Mohd-Setapar SH, Mohamed SF, Ahmad A, Wani WA, Muddassir M, Alarifi A (2022) A glimpse into the extraction methods of active compounds from plants. Crit Rev Anal Chem 52(4):667–696. https://doi.org/10.1080/10408347.2020.1820851

Das S, Nadar SS, Rathod VK (2021) Integrated strategies for enzyme assisted extraction of bioactive molecules: a review. Int J Biol Macromol 191:899–917. https://doi.org/10.1016/j.ijbiomac.2021.09.060

Doellinger J, Schneider A, Hoeller M, Lasch P (2020) Sample preparation by easy extraction and digestion (SPEED)-a universal, rapid, and detergent-free protocol for proteomics based on acid extraction. Mol Cell Proteomics 19(1):209–222. https://doi.org/10.1074/mcp.TIR119.001616

Dogiparthi LK, Sana SS, Shaik SZ, Kalvapalli MR, Kurupati G, Kumar GS, Gangadhar L (2021) Phytochemical mediated synthesis of silver nanoparticles and their antibacterial activity. SN Appl Sci 3(6):631. https://doi.org/10.1007/s42452-021-04641-1

Dua TK, Giri S, Nandi G, Sahu R, Shaw TK, Paul P (2023) Green synthesis of silver nanoparticles using Eupatorium adenophorum leaf extract: characterizations, antioxidant, antibacterial and photocatalytic activities. Chem Pap 77(6):2947–2956. https://doi.org/10.1007/s11696-023-02676-9

Eid MM (2022) Characterization of nanoparticles by FTIR and FTIR-microscopy. In: Handbook of consumer nanoproducts. Springer Singapore, Singapore, pp 1–30. https://doi.org/10.1007/978-981-15-6453-6_89-1

Elkhateeb O, Atta MB, Mahmoud E (2024) Biosynthesis of iron oxide nanoparticles using plant extracts and evaluation of their antibacterial activity. AMB Express 14(1):92. https://doi.org/10.1186/s13568-024-01746-9

Farizan AF, Yusoff HM, Badar N, Bhat IU, Anwar SJ, Wai CP, Asari A, Kasim MF, Elong K (2023) Green synthesis of magnesium oxide nanoparticles using mariposa christia vespertilionis leaves extract and its antimicrobial study toward S. aureus and E. coli. Arab J Sci Eng 48(6):7373–7386. https://doi.org/10.1007/s13369-022-07282-7

Filippov SK, Khusnutdinov R, Murmiliuk A, Inam W, Zakharova LY, Zhang H, Khutoryanskiy VV (2023) Dynamic light scattering and transmission electron microscopy in drug delivery: a roadmap for correct characterization of nanoparticles and interpretation of results. Mater Horiz 10(12):5354–5370. https://doi.org/10.1039/D3MH00717K

Golzarnezhad F, Allahdou M, Mehravaran L, Naderi S (2025) Green synthesis of ZnO nanoparticles from the extract of Cymbopogon olivieri and investigation of their antimicrobial and anticancer effects. Discover Appl Sci 7(3):196. https://doi.org/10.1007/s42452-025-06623-z

Hheidari A, Mohammadi J, Ghodousi M, Mahmoodi M, Ebrahimi S, Pishbin E, Rahdar A (2024) Metal-based nanoparticle in cancer treatment: lessons learned and challenges. Front Bioeng Biotechnol 12:1436297. https://doi.org/10.3389/fbioe.2024.1436297

Hossain A, Abdallah Y, Ali MA, Masum MM, Li B, Sun G, Meng Y, Wang Y, An Q (2019) Lemon-fruit-based green synthesis of zinc oxide nanoparticles and titanium dioxide nanoparticles against soft rot bacterial pathogen Dickeya dadantii. Biomolecules 9(12):863. https://doi.org/10.3390/biom9120863

Hosseingholian A, Gohari SD, Feirahi F, Moammeri F, Mesbahian G, Moghaddam ZS, Ren Q (2023) Recent advances in green synthesized nanoparticles: from production to application. Mater Today Sustain 24:100500. https://doi.org/10.1016/j.mtsust.2023.100500

Jabeen S, Siddiqui VU, Bala S, Mishra N, Mishra A, Lawrence R, Bansal P, Khan AR, Khan T (2024) Biogenic synthesis of copper oxide nanoparticles from aloe vera: antibacterial activity, molecular docking, and photocatalytic dye degradation. ACS Omega 9(28):30190–30204. https://doi.org/10.1021/acsomega.3c10179

Jadhav V, Bhagare A, Wahab S, Lokhande D, Vaidya C, Dhayagude A, Khalid M, Aher J, Mezni A, Dutta M (2022) Green synthesized calcium oxide nanoparticles (CaO NPs) using leaves aqueous extract of moringa oleifera and evaluation of their antibacterial activities. J Nanomater 2022(1):9047507. https://doi.org/10.1155/2022/9047507

Jafarzadeh S, Nooshkam M, Zargar M, Garavand F, Ghosh S, Hadidi M, Forough M (2024) Green synthesis of nanomaterials for smart biopolymer packaging: challenges and outlooks. J Nanostructure Chem 14(2):113–136. https://doi.org/10.1007/s40097-023-00527-3

Jassal PS, Kaur D, Prasad R, Singh J (2022) Green synthesis of titanium dioxide nanoparticles: development and applications. J Agric Food Res 10:100361. https://doi.org/10.1016/j.jafr.2022.100361

Javid-Naderi MJ, Sabouri Z, Jalili A, Zarrinfar H, Sammak S, Darroudi M (2025) Green synthesis and characterization of Ag/CuO nanoparticles: exploring their antifungal, antimicrobial, and cytotoxic properties. Environ Technol Innov 38:104147. https://doi.org/10.1016/j.eti.2025.104147

Joudeh N, Linke D (2022a) Nanoparticle classification, physicochemical properties, characterization, and applications: a comprehensive review for biologists. J Nanobiotechnol 20(1):262. https://doi.org/10.1186/s12951-022-01477-8

Joudeh N, Linke D (2022b) Nanoparticle classification, physicochemical properties, characterization, and applications: a comprehensive review for biologists. J Nanobiotechnol 20(1):262. https://doi.org/10.1007/s11696-023-02676-9

Kachhawaha AS, Elizabeth M, Hari P, Rajan A, Verma N (2025) Green synthesis and antibacterial evaluation of iron oxide nanoparticles using Clitoria Ternatea flowers. Biomed Pharmacol J 18(1). https://doi.org/10.13005/bpj/3125

Kazi FM, Parvez K, Asghar A, Akbar S, Jawaed NU, Kumar N, Palma PJ (2024) In vitro evaluation of the influence of biosynthesized calcium oxide nanoparticles on the antibacterial activity, pH, microleakage and cytotoxicity of conventional intracanal medicaments. Int J Mol Sci 25(22):11991. https://doi.org/10.3390/ijms252211991

Khan I, Awan SA, Rizwan M, Akram MA, Zia-ur-Rehman M, Wang X, Zhang X, Huang L (2023a) Physiological and transcriptome analyses demonstrate the silver nanoparticles mediated alleviation of salt stress in pearl millet (Pennisetum glaucum L). Environ Pollut 318:120863. https://doi.org/10.1016/j.envpol.2022.120863

Khan AR, Azhar W, Fan X, Ulhassan Z, Salam A, Ashraf M, Liu Y, Gan Y (2023b) Efficacy of zinc-based nanoparticles in alleviating the abiotic stress in plants: current knowledge and future perspectives. Environ Sci Pollut Res 30(51):110047–110068. https://doi.org/10.1007/s11356-023-29993-6

Khrennikov D, Labuntsov V, Ladutenko K, Terekhov I, Bogdanov A, Ågren H, Karpov S (2024) Unique features of plasmonic absorption in ultrafine metal nanoparticles: unity and rivalry of volumetric compression and spill-out effect. Nanophotonics 13(26):4795–4804. https://doi.org/10.1515/nanoph-2024-0475

Kumar M, Ranjan R, Dandapat S, Srivastava R, Sinha MP (2022) XRD analysis for characterization of green nanoparticles: a mini review. Glob J Pharm Pharma Sci 10(1):555779. https://doi.org/10.19080/GJPPS.2022.10.555779

Lončarić A, Celeiro M, Jozinović A, Jelinić J, Kovač T, Jokić S, Babić J, Moslavac T, Zavadlav S, Lores M (2020) Green extraction methods for extraction of polyphenolic compounds from blueberry pomace. Foods 9(11):1521. https://doi.org/10.3390/foods9111521

López-Bascón MA, De Castro ML (2020) Soxhlet extraction. In: Liquid-phase extraction. Elsevier, pp 327–354. https://doi.org/10.1016/B978-0-12-816911-7.00011-6

Mallikarjuna K, Nasif O, Ali Alharbi S, Chinni SV, Reddy LV, Reddy MR, Sreeramanan S (2021) Phytogenic synthesis of Pd-Ag/rGO nanostructures using stevia leaf extract for photocatalytic H2 production and antibacterial studies. Biomolecules 11(2):190. https://doi.org/10.3390/biom11020190

Mazher M, Ishtiaq M, Hamid B, Haq SM, Mazhar A, Bashir F, Mazhar M, Mahmoud EA, Casini R, Alataway A, Dewidar AZ (2023) Biosynthesis and characterization of calcium oxide nanoparticles from Citrullus colocynthis fruit extracts; their biocompatibility and bioactivities. Materials 16(7):2768. https://doi.org/10.3390/ma16072768

Moghadam NC, Jasim SA, Ameen F, Alotaibi DH, Nobre MA, Sellami H, Khatami M (2022) Retracted article: nickel oxide nanoparticles synthesis using plant extract and evaluation of their antibacterial effects on Streptococcus mutans. Bioprocess Biosyst Eng 45(7):1201–1210. https://doi.org/10.1007/s00449-022-02736-6

Nour AH, Oluwaseun AR, Nour AH, Omer MS, Ahmed N (2021) Microwave-assisted extraction of bioactive compounds. In: Microwave heating-electromagnetic fields causing thermal and non-thermal effects. IntechOpen. https://doi.org/10.5772/intechopen.96092

Ogwuegbu MC, Olatunde OC, Pfukwa TM, Mthiyane D, Fawole OA, Onwudiwe DC (2025) Green synthesis, characterization and antimicrobial activity of ZnO and Co-doped ZnO nanoparticles obtained using aqueous extracts of Platycladus orientalis leaves. Discov Mater 5(1):1–2. https://doi.org/10.1007/s43939-025-00230-w

Osman AI, Zhang Y, Farghali M, Rashwan AK, Eltaweil AS, Abd El-Monaem EM, Mohamed IM, Badr MM, Ihara I, Rooney DW, Yap PS (2024) Synthesis of green nanoparticles for energy, biomedical, environmental, agricultural, and food applications: a review. Environ Chem Lett 22(2):841–887. https://doi.org/10.1007/s10311-023-01682-3

Padalia H, Moteriya P, Chanda S (2018) Synergistic antimicrobial and cytotoxic potential of zinc oxide nanoparticles synthesized using Cassia auriculata leaf extract. BioNanoScience 8(1):196–206. https://doi.org/10.1007/s12668-017-0463-6

Perveen K, Husain FM, Qais FA, Khan A, Razak S, Afsar T, Alam P, Almajwal AM, Abulmeaty MM (2021) Microwave-assisted rapid green synthesis of gold nanoparticles using seed extract of Trachyspermum ammi: ROS mediated biofilm inhibition and anticancer activity. Biomolecules 11(2):197. https://doi.org/10.3390/biom11020197

Plaza M, Marina ML (2023) Pressurized hot water extraction of bioactives. TrAC Trends Anal Chem 166:117201. https://doi.org/10.1016/j.trac.2023.117201

Prasad KS, Prasad SK, Ansari MA, Alzohairy MA, Alomary MN, AlYahya S, Srinivasa C, Murali M, Ankegowda VM, Shivamallu C (2020) Tumoricidal and bactericidal properties of ZnONPs synthesized using Cassia auriculata leaf extract. Biomolecules 10(7):982. https://doi.org/10.3390/biom10070982

Priya M, Venkatesan R, Deepa S, Sana SS, Arumugam S, Karami AM, Vetcher AA, Kim SC (2023) Green synthesis, characterization, antibacterial, and antifungal activity of copper oxide nanoparticles derived from Morinda citrifolia leaf extract. Sci Rep 13(1):18838. https://doi.org/10.1038/s41598-023-46002-5

Rabi SA (2025) Green synthesis of magnesium oxide nanoparticles using neem leaves and their antimicrobial activity. https://www.researchgate.net/publication/390695563_Green_Synthesis_of_Magnesium_Oxide_Nanoparticles_using_Neem_Leaves_and_their_Antimicrobial_Activity

Rahmalia W, Fabre JF, Mouloungui Z (2015) Effects of cyclohexane/acetone ratio on bixin extraction yield by accelerated solvent extraction method. Procedia Chem 14:455–464. https://doi.org/10.1016/j.proche.2015.03.061

Rajput S, Kumar D, Agrawal V (2020) Green synthesis of silver nanoparticles using Indian Belladonna extract and their potential antioxidant, anti-inflammatory, anticancer and larvicidal activities. Plant Cell Rep 39:921–939. https://doi.org/10.1007/s00299-020-02539-7

Rana A, Pathak S, Kumar K, Kumari A, Chopra S, Kumar M, Kamil D, Srivastava R, Kim SK, Verma R, Sharma SN (2024) Multifaceted properties of TiO 2 nanoparticles synthesized using Mangifera indica and Azadirachta indica plant extracts: antimicrobial, antioxidant, and non-linear optical activity investigation for sustainable agricultural applications. Mater Adv 5(7):2767–2784. https://doi.org/10.1039/D3MA00414G

Ranjha MM, Irfan S, Lorenzo JM, Shafique B, Kanwal R, Pateiro M, Arshad RN, Wang L, Nayik GA, Roobab U, Aadil RM (2021) Sonication, a potential technique for extraction of phytoconstituents: a systematic review. Processes 9(8):1406. https://doi.org/10.3390/pr9081406

Rasmussen MK, Pedersen JN, Marie R (2020) Size and surface charge characterization of nanoparticles with a salt gradient. Nat Commun 11(1):2337. https://doi.org/10.1038/s41467-020-15889-3

Rodríguez-Barajas N, Ponce-Regalado MD, Segura-Almendárez MS, Rodríguez-Razon CM, Ghotekar S, Fellah M, Pérez-Larios A (2024) Plant-mediated synthesis and interaction of ZnO against breast and prostate cancer. Results Chem:101654. https://doi.org/10.1016/j.rechem.2024.101654

Saini R, Kumar P (2023) Green synthesis of TiO2 nanoparticles using Tinospora cordifolia plant extract & its potential application for photocatalysis and antibacterial activity. Inorg Chem Commun 156:111221. https://doi.org/10.1016/j.inoche.2023.111221

Saka A, Shifera Y, Jule LT, Badassa B, Nagaprasad N, Shanmugam R, Priyanka Dwarampudi L, Seenivasan V, Ramaswamy K (2022) Biosynthesis of TiO2 nanoparticles by Caricaceae (Papaya) shell extracts for antifungal application. Sci Rep 12(1):15960. https://doi.org/10.1038/s41598-022-19440-w

Salem SS (2023) A mini review on green nanotechnology and its development in biological effects. Arch Microbiol 205(4):128. https://doi.org/10.1007/s00203-023-03467-2

Salem SS, Fouda A (2021) Green synthesis of metallic nanoparticles and their prospective biotechnological applications: an overview. Biol Trace Elem Res 199(1):344–370. https://doi.org/10.1007/s12011-020-02138-3

Sarathamani T, Kalaiselvi V, Blessymol B, Gopi S, Khuwaja G, Palanismy S, Ismail KS, Ali SK, Sillanpaa M, Ayrilmis N (2025) Annona squamosa seeds capped calcium oxide nano particles–anti-microbial, antioxidant, anti-ulcer analysis. RSC Adv 15(7):4904–4914. https://doi.org/10.1039/D5RA00375J

Senapati MR, Behera PC (2023) Novel extraction conditions for phytochemicals. In: Recent frontiers of phytochemicals. Elsevier, pp 27–61. https://doi.org/10.1016/B978-0-443-19143-5.00019-0

Shen L, Pang S, Zhong M, Sun Y, Qayum A, Liu Y, Rashid A, Xu B, Liang Q, Ma H, Ren X (2023) A comprehensive review of ultrasonic assisted extraction (UAE) for bioactive components: principles, advantages, equipment, and combined technologies. Ultrason Sonochem 101:106646. https://doi.org/10.1016/j.ultsonch.2023.106646

Shikov AN, Mikhailovskaya IY, Narkevich IA, Flisyuk EV, Pozharitskaya ON (2022) Methods of extraction of medicinal plants. In: Evidence-based validation of herbal medicine. Elsevier, pp 771–796. https://doi.org/10.1016/B978-0-323-85542-6.00029-9

Singh R, Hano C, Nath G, Sharma B (2021) Green biosynthesis of silver nanoparticles using leaf extract of Carissa carandas L. and their antioxidant and antimicrobial activity against human pathogenic bacteria. Biomolecules 11(2):299. https://doi.org/10.3390/biom11020299

Taha A, Farooq N, Hashmi AA (2024) Medicinal plant-mediated nanomaterials. In: Green synthesis of nanomaterials: biological and environmental applications, pp 22–45. https://doi.org/10.1002/9781119900931.ch2

Talebian S, Shahnavaz B, Nejabat M, Abolhassani Y, Rassouli FB (2023) Bacterial-mediated synthesis and characterization of copper oxide nanoparticles with antibacterial, antioxidant, and anticancer potentials. Front Bioeng Biotechnol 11:1140010. https://doi.org/10.1016/j.heliyon.2020.e04508

Teklu B, Kadiri SK, Vidavalur S (2023) Green synthesis of copper oxide nanoparticles using Balanites aegyptiaca stem bark extract and investigation of antibacterial activity. Results Chem 6:101152. https://doi.org/10.1016/j.rechem.2023.101152

Thatyana M, Dube NP, Kemboi D, Manicum AL, Mokgalaka-Fleischmann NS, Tembu JV (2023) Advances in phytonanotechnology: a plant-mediated green synthesis of metal nanoparticles using phyllanthus plant extracts and their antimicrobial and anticancer applications. Nanomaterials 13(19):2616. https://doi.org/10.3390/nano13192616

Tiwari AK, Jha S, Singh AK, Mishra SK, Pathak AK, Ojha RP, Yadav RS, Dikshit A (2022) Innovative investigation of zinc oxide nanoparticles used in dentistry. Crystals 12(8):1063. https://doi.org/10.3390/cryst12081063

Tiwari AK, Jha S, Tripathi SK, Shukla R, Awasthi RR, Bhardwaj AK, Singh AK, Dikshit A (2024a) Spectroscopic investigations of green synthesized zinc oxide nanoparticles (ZnO NPs):

antioxidant and antibacterial activity. Discover Appl Sci 6(8):399. https://doi.org/10.1007/s42452-024-06049-z

Tiwari AK, Jha S, Tripathi SK, Awasthi RR, Bhardwaj AK, Singh AK, Dikshit A (2024b) Antioxidant response of Calendula officinalis L. assisted synthesized zinc oxide nanoparticles. Mater Res Express 11(8):085005. https://doi.org/10.1088/2053-1591/ad6ca9

Tiwari AK, Jha S, Awasthi RR, Singh AK, Bhardwaj AK, Tripathi SK (2024c) Recent trends of zinc oxide nanoparticles for emerging biomedical application. IntechOpen. https://doi.org/10.5772/intechopen.1007653

Tiwari AK, Jha S, Awasthi RR, Singh AK, Ojha RP, Dikshit A, Rai S, Bhardwaj AK (2025) Nanomaterial production using agro-waste: a current status and future outline. In: Sustainable management of agro-food waste. Academic Press, pp 249–261. https://doi.org/10.1016/B978-0-443-23679-2.00017-3

Tungmunnithum D, Drouet S, Kabra A, Hano C (2020) Enrichment in antioxidant flavonoids of stamen extracts from Nymphaea lotus L. using ultrasonic-assisted extraction and macroporous resin adsorption. Antioxidants 9(7):576. https://doi.org/10.3390/antiox9070576

Uwineza PA, Waśkiewicz A (2020) Recent advances in supercritical fluid extraction of natural bioactive compounds from natural plant materials. Molecules 25(17):3847. https://doi.org/10.3390/molecules25173847

Valikarimwala M, Gadale S, Waghmode S (2021) Applications of nanotechnology in ayurveda: a review. Res Rev Biotechnol Biosci 8(1):163–169. https://doi.org/10.5281/zenodo.5118448

Venkatas J, Singh M (2024) Chemical and green synthesis of gold nanoparticles for mRNA delivery in vitro. Adv Nat Sci Nanosci Nanotechnol 15(1):015009. https://doi.org/10.1088/2043-6262/ad2c7b

Yassin MT, Al-Otibi FO, Al–Askar AA (2024) Green synthesis, characterization and antimicrobial activity of iron oxide nanoparticles with tigecycline against multidrug resistant bacterial strains. J King Saud Univ-Sci 36(4):103131. https://doi.org/10.1016/j.jksus.2024.103131

Yoonus J, Resmi R, Beena B (2021) Greener nanoscience: piper betel leaf extract mediated synthesis of CaO nanoparticles and evaluation of its antibacterial and anticancer activity. Mater Today Proc 41:535–540. https://doi.org/10.1016/j.matpr.2020.05.246

Younis IY, El-Hawary SS, Eldahshan OA, Abdel-Aziz MM, Ali ZY (2021) Green synthesis of magnesium nanoparticles mediated from Rosa floribunda charisma extract and its antioxidant, antiaging and antibiofilm activities. Sci Rep 11(1):16868. https://doi.org/10.1038/s41598-021-96377-6

Zhor C, Wafaa L, Ghzaiel I, Kessas K, Zarrouk A, Ksila M, Ghrairi T, Latruffe N, Masmoudi-Kouki O, El Midaoui A, Vervandier-Fasseur D (2023) Effects of polyphenols and their metabolites on age-related diseases. Biochem Pharmacol 214:115674. https://doi.org/10.1016/j.bcp.2023.115674

Chapter 5
Extending Shelf-Life Through Eco-friendly Nanotechnology in Food Systems

Deepika, Monisha Soni, Arati Yadav, Manish Kumar Jha, Kishor Kumar Paul, Neha Upadhyay, Nawal Kishore Dubey, and Abhishek Kumar Dwivedy

5.1 Introduction

Improving the shelf-life of fresh products relies on preserving freshness by deferring ripening, controlling transpiration, and preventing microbial growth (Kumar et al. 2021). An estimated 20–40% of fresh product is wasted owing to inefficient handling procedures, suboptimal or inadequate preservation technology and insufficient postharvest facilities, globally (Fig. 5.1) (Sagar et al. 2018). Therefore, addressing these challenges through advanced technological interventions is imperative to mitigate post-harvest losses in fresh product and enhance the efficiency of food system.

For that, various preservation technologies, including cold storage, irradiation, chemical treatments and biopreservation, have been developed to enhance the shelf-life of fresh product by preventing microbial growth, regulating respiration, and

Deepika
Laboratory of Herbal Pesticides, Department of Botany, Institute of Science, Banaras Hindu University, Varanasi, Uttar Pradesh, India

Department of Botany, Mahila Vidyalaya Degree College, Aminabad, Lucknow, Uttar Pradesh, India

M. Soni · A. Yadav · M. K. Jha · K. K. Paul · N. K. Dubey · A. K. Dwivedy (✉)
Laboratory of Herbal Pesticides, Department of Botany, Institute of Science, Banaras Hindu University, Varanasi, Uttar Pradesh, India
e-mail: akdwivedy.bot@bhu.ac.in

N. Upadhyay
Laboratory of Herbal Pesticides, Department of Botany, Institute of Science, Banaras Hindu University, Varanasi, Uttar Pradesh, India

Department of Botany, Shaheed Hira Singh Government P.G. College, Dhanapur, Chandauli, Uttar Pradesh, India

V. Ghosh et al. (eds.), *Applications of Plant Bioactive Compounds-Based Nanoformulations*, Nanotechnology in the Life Sciences,
https://doi.org/10.1007/978-3-032-13086-0_5

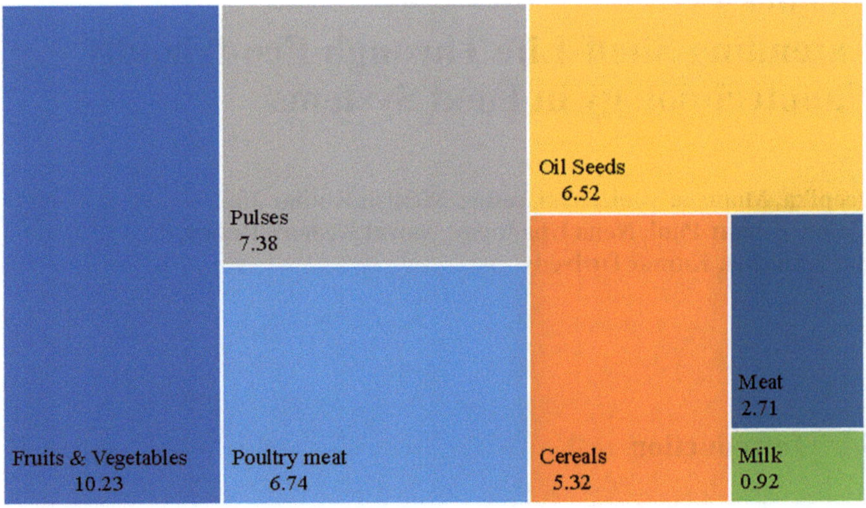

Fig. 5.1 Agro-economic products and their cumulative loss by improper practices (in %)

minimizing water loss. However, these approaches have a number of limitations. Cold storage, while effective, is energy-intensive and prone to inducing chilling injuries if temperatures are poorly managed (Makule et al. 2022). Chemical preservation, though cost-effective and simple, is increasingly scrutinized due to safety and health concerns (Jurić et al. 2024). Similarly, while irradiation preserves structural integrity, it requires stringent radiation control, incurs high operational costs, and demands extended processing times (Lisboa et al. 2024). As a result, there is a critical need for advanced, consumer-friendly, and sustainable methods to maintain produce quality and safety.

In the contemporary world, nanoparticles have gained tremendous interest due to its unique characteristics, encompassing their exceedingly nanoscale dimensions, remarkable surface-area-to-volume ratio, surface adjustability, favorable magnetic characteristics properties, and exceptional biocompatibility that can bridge the gap between molecular structures and bulk materials, offering novel functionalities (Thakkar et al. 2010; Bogunia-Kubik and Sugisaka 2002; Daniel and Astruc 2004). This approach presents advantages characterized by cost-effectiveness, environmental sustainability, security, durability, and suitability for large-scale production. Specifically, plants and microbes (bacteria, algae, fungi, etc.) exhibit the capacity to biosynthesize a substantial array of secondary (2°) metabolites, including phenolics, terpenoids and alkaloids. These extracts, owing to their inherent reducing and stabilizing capabilities, have found to be beneficial in synthesizing nanoparticles (Fig. 5.2).

So, nanotechnology in food packaging extends the shelf-life of food products by creating a suitable environment that reduces microbial load, loss of moisture, limits transmission of gases, and changes in organoleptic property by regulating metabolic

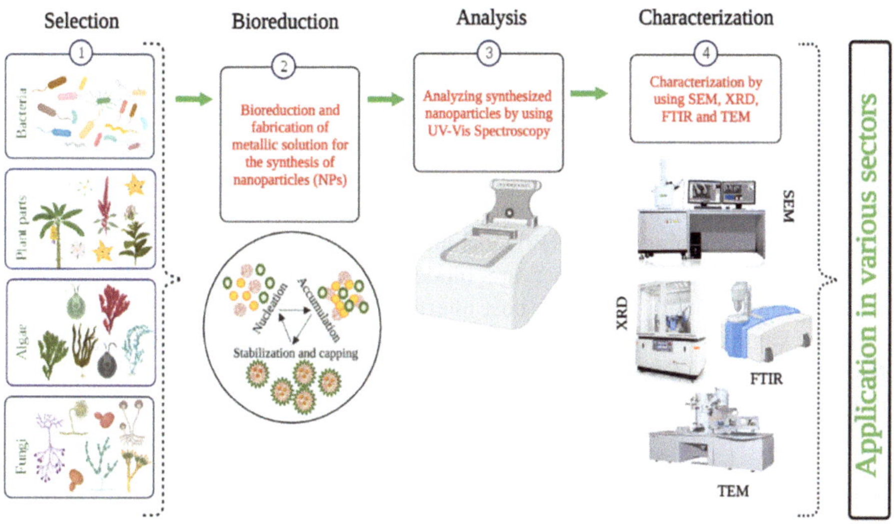

Fig. 5.2 Schematic illustration of biogenic nanoparticles synthesis

and enzymatic activities (Rai et al. 2019). Despite these promising applications, the research on direct effects of nanoparticles on the shelf-life of fresh product remains limited. This chapter explores eco-friendly nanomaterials and the biogenic processes involved in their development, focussing on their potential to improve the shelf-life of vegetables and fruits sustainably.

5.2 Eco-friendly Nanomaterials for Prolonged Viability of Food

The application of eco-friendly nanomaterials for shelf-life extension is revolutionizing the way we can preserve perishable items for quite some time. These nanomaterials are extracted from natural sources viz. plants, microbes, and others offer sustainable resolutions for increasing the shelf-life of food products. These materials also function as active and safe packaging substances, antimicrobial agents, and oxygen scavengers, effectively reducing spoilage and maintaining nutritional quality. There are some examples of eco-friendly nanomaterials for increasing durability of food and food items.

5.2.1 Plant-Based Biopolymers

Numerous studies have proven that nanomaterial can improve the shelf-life and nutritional value of food while retaining its flavour and nature. These materials also offer functional properties, such as improved barrier properties, a large specific surface area, and antimicrobial activity, which can assist overcome the limitations of traditional methods for extending shelf-life (Manikandan and Min 2023). Plant-based nanobiopolymers are those derived from plants or produced using plant-based technologies and recently have achieved economic significance in active packaging for providing safe, high-quality foods that may applicable for both fresh and processed foods. These nanobiopolymer are attracting interest because of their sustainability, renewable characteristics, and environmentally friendly attributes in contrast to traditional nanobiopolymer (Priyadarshi and Rhim 2020).

Plant based nanobiopolymer are biodegradable polymers such as cellulose, starch, protein which serve as a best alternative to synthetic nanomaterials in the food packaging. These biopolymers can decompose naturally without producing harmful residues. Several significant plant-based nanobiopolymer derived from plants are mentioned below.

5.2.1.1 Polysaccharides

(a) **Starch**

Starch is abundant in nature, and the primary energy source for animals and humans. It consists of two types of linkages: amylose is made up of D-glucose units connected by α-(**1–4**) bonds, whereas amylopectin contains α-(**1–6**) linkage. When mixed with plasticizers, exposed to high temperature, and subjected to mechanical pressing, starch exhibits thermoplastic properties due to its water insoluble units. Among various polysaccharides, starch is significant because it is relatively inexpensive and available in diverse forms. Starchy films, however, tend to have limitations such as deprived thickness, elasticity, and transparency, along with poor mechanical properties and high-water vapour permeability, restricting their use in film production. Despite these drawbacks, the main challenge lies in the fact that raw starch is weak, highly hygroscopic, and has minimal mechanical strength (Dutta and Sit **2023**). Despite these obstacles, starch acts an appealing biopolymer due to its affordability, ease of availability, and capacity to perform diverse roles chiefly in eco-friendly farming and food packaging.

(b) **Cellulose**

Cellulose is a glucan polymer known for its hydrophilic properties and it is thermally more stable compared to hemicellulose. Cellulose interacts with pectin to form a stronger cell wall, which helps to preserve structural integrity of cell in water. The glucose units connect to create linear chains through β (**1–4**) glycosidic linkages, providing strength, stability, and rigidity, subsequently forming the cellulose structure with a high degree of polymerization. Several investigators have investigated the role of cellulose and its derived nanocomposites in food packaging. It creates a cellulose nanocomposite film with high performance and a tensile strength of **140** MPa from agricultural low value waste, specifically bagasse of sugarcane, and is deemed appropriate for food packaging (Pandey et al. **2013**). Xio et al. (**2020**) developed a film of nanocomposite based on soy protein which included chitosan nanofiber, which exhibited improvements in tensile strength, water vapour and oxygen barrier characteristics as well as resistance to water and heat. Similarly, a nanocomposite film prepared by using whey protein and polydextrose containing chitosan nanofiber showed improved mechanical and physical barrier capabilities (Karimi et al. **2020**).

(c) **Alginate**
Alginate is naturally occurring polymer derived from seaweed, as well as produced by various bacteria. Alginate structurally consist of a group of linear copolymers contain blocks of L- guluronate and (**1–4**)- linked D-mannuronate residues. It is the second most plentiful polymer on earth following cellulose and is known for its distinct features, including biodegradable, non-toxic, biocompatibility, and gelling properties. These attributes make it highly suitable for applications in biomedicals, pharmaceutics, and tissue engineering. Among the different types of alginates, sodium alginate is highly stable and soluble in the water and hydrophilic natured organic solvents like ethanol and isopropanol at ideal doses (**50–70%**). As a result, sodium alginate continues to be the most preferably used alginate (Xie et al. **2024**). Frank et al. (**2018**) formulated antibacterial films mainly based on alginate by integrating nanoemulsion of cinnamon essential oil. The most efficient antibacterial films had **20%** cinnamon loaded alginate nanoemulsion and demonstrated considerable inhibition against several bacteria for prolonging the shelf-life of food products.

5.2.1.2 Proteins

Proteins are standard polymers made up of amino acids derived from various sources. It is mined from plant sources such as corn, wheat, soy, sorghum, pea, and rice bran for use in packaging. Due to their abundance and exceptional film forming characteristics, proteins are extensively utilized in packaging applications. Protein based films exhibit remarkable barrier properties, effectively blocking CO_2 and O_2 gases and volatile compounds. Naturally derived proteins are gaining recognition as effective active agents that are both abundant and cost-effective. Protein isolated from *Phaseolus vulgaris*, when combined with chitosan showed enhanced flexibility and reduced surface free energy, which resulted in increased hydrophobic traits. This makes them ideal for dehydrated packaging essential for moisture-sensitive food items. The film forming characteristics of proteins are influenced by factors such as heat, solvent, and pH (Bora et al. 2020).

(a) **Zein**
Zein is naturally occurring protein, derived from corn kernels and soluble in alcohol. It contains a significant number of residues of hydrophobic amino acids like leucine, proline, and alanine which are generally recognized as safe (GRAS) excipient. It can be formed into fibres, films, and nanoparticles. Zein is abundant, hydrophobic, biocompatible, biodegradable in nature, and possesses film-forming characteristics, making it an attractive material for designing nanocomposite edible films and coatings. However, zein-based films face challenges like poor

processability, brittleness, limited elongation, and inadequate thermal properties. To address these limitations, plasticizers, including glycerol, ethers, long chain fatty acids and aldehydes are commonly used by chemical, physical or enzymatic process to enhance properties such as solubility in water, foaming, and emulsifying capabilities (Vimala Bharathi et al. 2020). Kashiri et al. (2017) synthesized *Zataria multiflora* essential oil (5% and 10%) based nanocomposite thin films and coatings for food packaging systems against microbes. They applied these nanocomposite coating to the interior of poly-propylene bags and evaluated their effects on preserving pasteurized milk kept for 6 days at 4 °C. The findings indicate a notable decrease in *E. coli* levels in comparison to control bags, with reductions of 0.51, 0.92, and 0.99 log CFU/mL on days 1, 3, and 6, respectively during storage.

(b) **Wheat gluten**

Wheat gluten is a byproduct of wheat processing industry abundantly utilized in plant-based food packaging due to its biocompatibility. Wheat gluten protein (WGP) consists of equal portions of glutenin and gliadin. Glutenin are formed through the polymerization of polypeptide chains via intermolecular disulfide linkages, making them water and solvent resistant, and are also responsible for providing strength and durability to gluten. Conversely, gliadins enhance the elasticity of gluten and are soluble in aqueous ethanol. Protein films with high opacity are most effective for light sensitive food packaging while barrier films are more suitable for hygroscopic products such as fruits and candies. However, to increase the mechanical strength, optical characteristics, and barrier efficacy of these films, it is necessary to incorporate modifiers like plasticizers and antioxidants. (Morel et al. 2022). Chavoshizadeh et al. (2020) created composite conductive films in various colors utilizing chlorophyll, pyrrole, and wheat gluten nanocomposite. They observed that chlorophyll and polypyrrole can increase the antibacterial and physico-mechanical characteristics of wheat gluten film, making it an effective antimicrobial packaging option.

5.2.1.3 Lipids

Lipids such as saturated, mono-unsaturated, and poly-unsaturated fatty acids, as well as resins and waxes, are crucial components of films and coating, serving as either fundamental component or plasticisers. The addition of hydrophobic materials (oils, waxes, and fats) into films can also improve their water-resistance (Avramescu et al. 2020). Furthermore, lipid-based edible coatings can act as an efficient barrier to oxygen and moisture, preserving food quality for an extended period. These lipids-based edible films and coatings provide nutritional and functional benefits.

(a) **Carnauba wax**

Waxes are a diverse collection of organic or chemical compounds known for their hydrophobic properties, often derived from alcohol and long chain fatty acids.

Because of their unique physical and chemical qualities, they are employed in a variety of industries. Wax coatings are commonly used on vegetables and fruits to assist prevent shrinkage, dehydration and shrivelling. Carnauba, a natural wax obtained from *Copernica cerifera* leaves, is usually considered safe (GRAS). Since 1930, carnauba has been widely used to coat vegetables and fruits to reduce weight loss and give them a shiny appearance (Islam et al. 2024). Jo et al. (2014) found that a mixture of carnauba-shellac wax combined with a nanoemulsion of lemongrass oil decreased the respiration rate, weight loss, ethylene production as well as preserved the firmness of stored postharvest apples.

Recent research explored the role of emulsion coatings with carnauba wax along with nanoclay to enhance the quality of Valencia orange after harvest. Coating formulations with varying concentrations of nanoclay (0.0, 0.5, 1.0%) were implemented to orange, and their effectiveness was assessed and compared to both fruit (in uncoated form) and two type of commercially accessible waxes. They stated that adding 1% nanoclay to carnauba wax significantly increased the textural, sensory profile, and nutritional attributes, extended longevity, and successfully reduced dehydration of fruits during storage (Motamedi et al. 2018).

5.2.2 *Biodegradable and Edible Nanocoatings*

Vegetables and fruits generally have a short life span due to their water logged (75–95%) condition, which causes them to deteriorate because of unwanted microbial proliferation and rapid oxidation reactions imparting an unappealing appearance to them (Otoni et al. 2017; Valencia et al. 2019). To limit this deterioration, a number of processing and preservation methods along with different packaging materials are employed to keep food commodities fresh for an extended amount of time by controlling oxygen, light, moisture, and providing physical protection.

However, a significant portion of packaging materials are made of imperishable petroleum-based polymers, which presents a substantial threat to the environment (Ncube et al. 2020). Using biopolymer-based edible packaging materials is best alternative to non-biodegradable polymers and preservatives which can readily extend the shelf-life and durability of fresh produce without altering its organoleptic attributes (Álvarez et al. 2017). The edible nano-coatings are composed of biodegradable polymers that serves as an efficient barrier to regulate the movement of gases (CO_2 and O_2) and moisture from the fruit and its surroundings, reducing respiration, delaying ripening, and minimizing the loss/evaporation of naturally occurring volatile compounds (Pratiwi et al. 2015). Some examples of plant-based biodegradable and edible nanocoatings used to preserve vegetables and fruits has been provided in Table 5.1.

Table 5.1 Different types of biodegradable nano-coatings used for preservation of vegetables and fruits

Biopolymers used for nano-coatings		Source	Applied on food systems	Application	References
Protein	Soy protein	Soybean (*Glycine max*)	Banana	Post harvest preservation of Banana involves efficiently delaying ripening and weight loss	Akrami et al. (2023)
			Apple	Delays the climacteric peak in apple and increased shelf life	Mikus and Galus (2025metabolites, including phenolics,)
			Fish	Preservation of fish and seafood products	Cetinkaya and Wijaya (2024)
	Whey	Cow's milk	Pineapple	Long-term preservation by inhibiting microbial growth	Uthairatanakij et al. (2021)
	Corn Zein	Maize (*Zea mays*)	Common carp fillet (*Cyprinus carpio*)	Zein edible film along with β-cyclodextrin improved its thermal stability and appearance during cold storage	Zolfaghari et al. (2023)
Lipids	Wax (Carnauba wax)	Carnauba palm (*Copernicia prunifera*)	Orange	Carnauba wax containing nanoparticles lowered weight loss and respiration rates in oranges while improving nutritional quality	Babarabie et al. (2024)
	Essential oil	Thyme, Clove, Cardamom	Kiwi fruit (*Actinidia deliciosa*)	Nano emulsion coating from essential oil with carboxymethyl cellulose significantly reduced weight loss, ethylene generation, and bacterial population	Iqbal et al. (2024)
	Shellac resin	Female lac bug (*Laccifer lacca*)	Fresh eggs	Shellac bio coating application enhanced the egg shell strength and storage stability by sealing the pores, to maintain the desired functional characteristics	Şahansoy et al. (2024)

Polysaccharides	Pectin	Guava	Papaya	Pectin-based protective coating increased the shelf-life and retained the freshness in papaya	Sharma et al. (2024a)
	Starch	Potato	Strawberry and Apple	Starch based nano film along with silver nano particles shown an elevated anti-microbial property and extended the shelf life during storage	Díaz-Silvestre and Ramirez (2024))
	Alginate	Marine brown algae	Grapes	Alginate with sodium, inhibited the food borne microbial growth and extended the shelf-life of grapes	Jamali et al. (2024)
	Cellulose	Sugarcane bagasse	Mango	Lignin containing nano-cellulose from sugarcane bagasse significantly influenced the ripening process in post harvested mangoes	Quan et al. (2024)
Nano-composites	Chitosan+Silver (Ag-NPs)	Crustaceans' exoskeleton	Melon	Using Ag-chitosan nanocomposites as edible coatings can increase the shelf-life of melon by lowering respiration and ethylene generation	Lieu et al. (2024)
	Carrageenan+ Se-AgNPs	Red seaweed	Longan fruits (*Dimocarpus longan*)	Longan fruits were kept for an extended time without any depletion in weight loss, color change and sensorial quality	Chen et al. (2025)

5.2.3 *Nanocomposites for Active and Intelligent Packaging*

Intelligent packaging is anticipated to account for approximately 5% of the whole value of the packaging business, and demand is expected to climb even faster in the future years (Wang et al. 2023). It enables the monitoring and documentation of essential parameters for food quality during packaging and storage. This quality is typically achieved by the addition of equipment or materials either inside or outside the package (Fig. 5.3). In active packaging, specific compounds are deliberately released or absorbed within the food matrix or packaging headspace to prolong product's shelf life by mitigating degradative procedures such as lipid oxidation, moisture fluctuations and microbial proliferation. Oxygen scavengers, ethylene scavengers, and moisture regulators constitute the predominant categories of active packaging; however, the utilization of active packaging emitters has seen a notable increase in recent years (Sharma et al. 2022).

Conversely, intelligent packaging integrates advanced functionalities that facilitate interaction with the external environment beyond conventional containment. This system incorporates diagnostic and indicator mechanisms to enhance real-time monitoring, automation, consumer engagement, and marketing. Typically, intelligent packaging employs sensor and indicator technologies to provide critical information regarding product integrity, including parameters such as package seal integrity, temperature fluctuations, and freshness indicators. One approach to create intelligent packaging films is to include indicators in the packaging system, such as pH or gas sensors, biosensors and temperature detectors (Kerry et al. 2006). Natural pigments (anthocyanins) present in vegetables and fruits are ideal pH indicators in food packaging because of their pH-dependent colour variations (Grotewold 2006). Furthermore, packaging with oxygen absorbers and antioxidants reduces food oxidation by removing free radicals (Pereira Jr et al. 2015). As a result, by reducing reliance on synthetic preservatives and food waste, these technologies help to promote environmental sustainability, market value improvement and consumer health, solidifying their position as a cornerstone in contemporary food preservation and packaging industries.

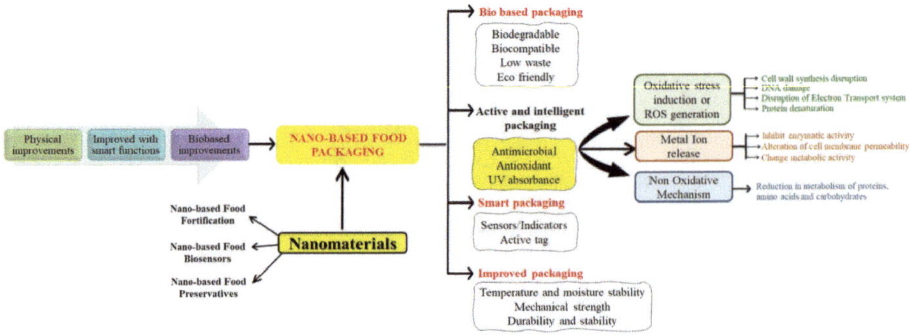

Fig. 5.3 Nanotechnology in food packaging

Ongoing research into eco-friendly nanoparticles derived from renewable sources aims to improve environmental compatibility and provide optimal packaging solutions for certain food categories. Notably, biodegradable and edible nanocoatings hold significant potential for perishable foods, extending shelf-life directly at the product level. Furthermore, putting green nanoparticles into active and intelligent packaging systems opens up new options to enhance food safety, decrease waste and revolutionise food preservation, all of which are consistent with global sustainability goals.

Eco-friendly nanotechnology offers revolutionary potential in food systems, particularly in terms of increasing shelf-life through the application of antioxidant, antimicrobial and gas regulating nanoparticles. The use of nanocomposites in active and intelligent packaging allows for real-time regulating the food quality using spoilage detection sensors. Furthermore, edible and biodegradable nanocoatings protect fresh product by inhibiting microbial growth and decreasing respiration rates, whilst nanocarriers allow for regulated release of preservatives and antioxidants, thereby increasing food safety (Fig. 5.4).

5.3 Mechanisms of Extending the Shelf-life of Vegetables and fruits Using Nanomaterials

Nanomaterials have revolutionised the food packaging industry by offering novel ways for increasing food shelf-life. They provide various benefits such as improved packaging options, effective antibacterial activity, higher barrier properties and regulated bioactive release behaviour (Gupta et al. 2024). These characteristics are primarily attributable to the nano-range size of nanoparticles, which results in large surface area, toughness, reactivity and interaction with food components (Khan et al. 2019). Their mode of action includes a number of physical and biological mechanisms that lowers the decay rate of fruits and vegetables while preserving the qualitative characteristics, texture, and overall shelf-life. The key mechanism of action of nanomaterials for prolonging the shelf-life of fruits and vegetables is discussed below:

5.3.1 Moisture Retention and Barrier Properties

Nanoemulsion-based edible coatings increase the shelf-life of vegetables and fruits by forming a physical barrier surrounding the surface, reducing water loss, which is the principal cause of shrivelling and drying. It acts as a shield that protects the vegetables and fruits from internal damage and reduces the rate of transpiration (Sharma et al. 2024b). Bashir et al. (2023) found that water-chestnut starch infused with 0.5% rosemary essential oil (REO) based coating was able to prevent water

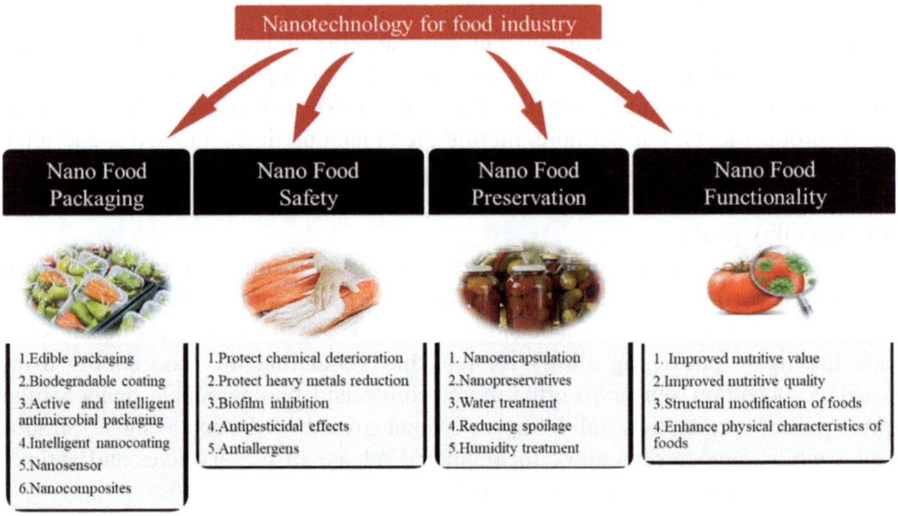

Fig. 5.4 Various applications of nanotechnology in food industry

loss from apples and keep them firm for 120 days of storage, thus, extending their shelf life. The study revealed that weight loss percentage decreased considerably by increasing the concentration of REO in the formulation, presumably as a result of covalent and H-bond interactions between REO and starch, that restrict the -OH groups to form hydrophilic interaction with water, thereby, retaining the moisture content in apples. Through moisture retention, the edible coatings help to maintain the turgidity and freshness of vegetables and fruits with high water content, such as tomatoes, cucumbers, leafy vegetables, strawberries, apples, oranges, etc. Another important characteristic of nanoemulsion is the formation of gaseous barrier around vegetables and fruits. Generally, vegetables and fruits continue their respiration process even after harvest, which is crucial for ripening as well as senescence. Nanoemulsion based coatings creates a mechanical barrier, which selectively allows the gaseous exchange, reducing O_2 intake and CO_2 emission, which reduces the overall metabolic activity of the produce, decline the ripening process and extends the product's shelf-life (Kondle et al. 2022). However, the thickness of the coating and the degree of permeability has a negative correlation. Thick coatings often provide an unwanted barrier that prevents proper exchange of CO_2 and O_2. This condition often leads to accumulation of ethanol, acetaldehyde, and CO_2 which causes fermentation and imparts unpleasant taste to vegetables and fruits (Nor and Ding 2020). Hence, thickness of edible coating is crucial parameter to consider before applying it on vegetables and fruits. Boyacı et al. (2019) developed flexible antimicrobial zein films incorporated with thymol, carvacrol, and eugenol effective against *Listeria innocua* and *Escherichia coli*, major pathogens infecting peel surface of whole melons. Eugenol-loaded zein (3%) films showed the best oxygen barrier performance having oxygen transmission rate (OTR) 7700 $cm^3\ m^{-2}$ per 24 h to 2200 $cm^3\ m^{-2}$ per 24 h. These OTR values were appropriate for the intended use since, coating the surface of melons would reduce the respiration rate, leading to delayed senescence without completely blocking the transport of oxygen between fruits and its surrounding environment.

5.3.2 *Antimicrobial and Antioxidant Properties*

Many bioactive substances like antimicrobials and antioxidants can be incorporated into nanoemulsion coatings which safeguard vegetables and fruits against oxidative stress caused due to generation of free radicals. Oxidation is a major contributor to nutritional loss in vegetables and fruits. It leads to enzymatic activity and develops off-flavors (rancidity) rendering the product unfit for human consumption (Gómez-Estaca et al. 2014; Khan et al. 2021). The integration of bioactive substances (α-tocopherol, catechin, curcumin, caffeic acid, plant extracts, essential oils, etc.) into biopolymers might prevent oxidation, defend against microbial degradation and improve the shelf-life of vegetables and fruits (Sanches-Silva et al. 2014). Some bioactive such as essential oils (EOs) possess antimicrobial activity against bacteria, fungi, and other pathogens responsible for biodeterioration of vegetables and fruits. The edible coatings infused with EOs offers a sustainable approach to inhibit growth of spoilage pathogens, thereby, preventing rotting and extending shelf-life of products. A study has been done by Nkede and co-workers to develop edible films enriched with *Helichrysum italicum* EO (HIEO) using sodium alginate as polymer for enhancing the shelf-life of variety of cherry tomatoes. It has been found that the edible coating of alginate with 0.5% HIEO decreased the rapid growth of B*otrytis cineria* mycelia (*in-vitro*) and reduced the diameter of lesions caused by *Rhizopus stolonifer* up to 61.36% on cherry tomatoes. Additionally, improved water vapor permeability, antioxidant capacity, thermal stability, and UV light barrier qualities were demonstrated by 0.5% HIEO infused alginate films (Nkede et al. 2024). Similarly, coatings made of sodium caseinate, Arabic gum, cinnamon, and lemon grass EO increased the oxidative stability of guava. After 7 days of storage, the polyphenol activity for uncoated guava samples was found to be double of the coated samples. The coating was able to change the environment around guavas, lowering the O_2 availability and preventing the polyphenol oxidase from rising rapidly (Murmu and Mishra 2018).

5.3.3 *Controlled release of Bioactive Components*

The movement of bioactive components from their primary zone within the delivery system is referred to as release. In such systems, the diffusion process is typically the mass transfer route of bioactive substances. Controlled release can be defined as the process that ensures targeted delivery of bioactive components at specific concentration, while sustained release refers to the release of bioactives at a specific or calculated rate in order to maintain the concentration of bioactive compounds at the target site for a certain amount of time (Boostani and Jafari 2021). To control the release rate from biopolymers, a number of strategies have been employed. Common technique includes encapsulation, more specifically nanoencapsulation that is thought to be an effective strategy to control the pace at which bioactive compounds are released from biopolymeric structures. These bioploymeric matrices protect the bioactive compounds by enclosing them inside tiny compartments and helps in ensuring their targeted and sustained release rate, hence improving food quality and safety. A large number of encapsulation techniques, including nanoemulsions, nanoliposomes, electrospun nanofibers, lipid nanoparticles, nanotubes, etc. are being used to maintain the bioactive compounds and control their release behaviour (Sabaghi et al. 2022). In a study, ginger EO (GEO) having antimicrobial and antioxidant properties was encapsulated into sodium alginate composite films stabilized using cellulose nanocrystals (SA/CNCs) and their effect was seen on shelf life of mangoes. The SA/CNCs film had a homogenous morphology and showed sustained release behaviour of GEO which helped in preserving the post-harvest status of mangoes by allowing prolonged bioavailability of GEO for mangoes (Zhang et al. 2024).

5.3.4 Reduction in Ethylene Sensitivity

Ripening is a complex, irreversible, genetically programmed process that alters the biochemical, physiological, and organoleptic attributes of vegetables and fruits (Alonso-Salinas et al. 2024). Ethylene is a key metabolite that speeds up the ripening process in climacteric fruits. As the fruit ripens, it starts producing more ethylene causing ethylene accumulation which eventually leads to over ripening and spoilage of many fruits (Bapat et al. 2010). The biopolymeric coatings and nanoemulsions containing certain plant-derived bioactive compounds have been found to lower ethylene sensitivity in fruits. They work by either trapping ethylene or reduce its production, slowing the ripening process and allowing the fruits to stay fresh over longer periods without impairment in quality attributes. For example, Arabic gum with unique film-forming and emulsification properties can lower the respiration and ethylene generation in fruits. Huang et al. (2021) found the effect of 9%, 12%, and 15% Arabic gum (GA) coating on Ponkan fruits and found that 15% GA coating was highly effective in delaying the ripening index (RI) in coated fruits (7.52 ± 0.11) as compared to uncoated fruits (8.66 ± 0.21). Moreover, the coatings reduced the fruit decay, maintained the total soluble solids, weight loss and acidity of Ponkan fruits during storage. Another study showed the effect of nanoemulsion prepared using orange EO and *Opuntia oligacantha* extract on quality attributes of avocados during post-harvest storage. A 25% nanoemulsion (N25) coating significantly increased polyphenol oxidase (PPO) activity (0.26 ± 0.01) in avocados after 60 days of storage compared to the control set (0.33 ± 0.01). PPO enzyme is generally responsible for darkening of mesocarp in fruits, hence, PPO activity rises in case of accelerated ripening which is caused due to high ethylene production. In this study, it has been found that nanoemulsion coating was able to reduce the browning of avocados by limiting the activity of PPO enzyme, thus, conserving the nutritional and quality parameters of avocados (Cenobio-Galindo et al. 2019).

5.3.5 *Improvement in Handling and Transportation*

Vegetables and fruits face challenges of mechanical damage during handling and transport which often leads to economic losses (Al-Dairi et al. 2022). The coating of nanomaterials on the surface of vegetables and fruits forms a semi-permeable protective barrier that helps to reduce the impact of stress by providing cushioning effect and increases the durability of fresh produce. Since, the coating is extremely thin and nearly undetectable, it retains the textural properties of vegetables and fruits, especially those that are more prone to mechanical injuries. The coating act as shock absorber, provides flexibility, and diminishes the irregularities on surface of fresh produce by forming a smooth layer. A smooth surface is less prone to bruising and scratches when in contact with other products, providing better protection and perhaps increasing the longevity of vegetables and fruits (Kondle et al. 2022). Deng and colleagues, for example, developed a cellulose nanoparticle emulsion coating to regulate post-harvest physiology and improve banana storage under ambient conditions. The nanoemulsion coating increased the surface wettability of fruit, delayed the formation of ethylene and CO_2, and altered the surface morphology of bananas by forming a homogenous layer on the surface, protecting it from mechanical damage and delaying the ripening process (Deng et al. 2017).

5.4 Regulatory and Safety Issues Around the Use of Nanomaterials in Food System

The unique properties of nanomaterials make them beneficial in food preservation but they also pose safety risks to human health and the environment. International agencies such as United State Food and Drug Administration (USFDA), World Health Organization (WHO), European Chemicals Agency (ECHA), European Food Safety Authority (EFSA), European Medicines Agency (EMA), Joint Research Centre (JRC) and others pre-established agenda for regulating the action of nanomaterials in food products (EFSA Scientific Committee et al. 2021). As a result, novel regulatory solutions are required to meet the specific challenges raised by the use of nanomaterials in food preservation. Traditional food safety regulations are insufficient for nanomaterials because of their unique features. Standard toxicity studies may fail to reveal the risks associated with nanoparticle use, necessitating the development of standardized approaches for assessing nanoparticle toxicity, migration and possible long-term effects in food (Singh and Kumar 2023). Regulatory agencies require producers to provide detailed information on nanomaterial size, shape, surface properties, and their composition. However, characterization can be challenging and varies according to the form and application. Clear guidelines are however essential for consistency and safety (Isibor 2024). Transparency is vital for increasing consumer awareness of threats. The main safety issues with the use of nanomaterials in food systems include:

(a) Toxicity and health risks
Nanoparticles, due to their tiny size and large surface area, have the ability to potentially penetrate cell membranes and enter the bloodstream, reaching numerous organs and tissues. This raises concerns regarding bioaccumulation, which can lead to unanticipated toxicological effects (Dolai et al. 2021). Some nanomaterials, such as silver and gold nanoparticles can release harmful ions that disrupt biological functions, potentially resulting in oxidative stress, inflammation, or DNA damage (Panzarini et al. 2018). Furthermore, certain nanomaterials can interact with proteins and other biological components, altering their function and perhaps causing detrimental health effects (Mu et al. 2014).

(b) Unwanted biological interactions
Nanomaterials can easily interact with other dietary components such as fats/lipids, proteins and carbohydrates, influencing their nutritional value and safety. These interactions may affect food stability, resulting in changes in flavor, texture, or nutrient content. For example, the inclusion of nanoparticles in food may disrupt the digestion of particular nutrients, changing their bio-availability as well as function in the body (Moradi et al. 2022).

(c) Environmental impact and sustainability
The environmental safety of nanomaterials is a major concern, particularly considering their potential release into ecosystems. Nanomaterials can enter the environment through packaging disposal, food waste or decay. Since nanoparticles are very small in size and highly reactive, they may be more damaging to the environment than larger particles (Souza and Fernando 2016). Unlike large materials which dissolve or degrade over time, certain nanomaterials can stay in the environment for extended periods of time, potentially accumulating in soils or water sources. This might have unintended ecological consequences, such as changing microbial populations or contaminating the food chain (Lead et al. 2018). Rajput et al. (2018) investigated zinc-oxide nanoparticles in the environment, which might be absorbed by plants, animals, and soil organisms. If these nanoparticles accumulate in living species, they may pose a significant danger to biodiversity, resulting in long-term environmental imbalances. This is especially concerning for aquatic environments, as nanoparticle may easily penetrate water bodies and be consumed by fish or other aquatic species (Singh et al. 2023).

(d) Uncertainty in long-term effects
The long-term behavior of nanomaterials in food systems is currently poorly understood. While short-term research has shed some light on their toxicity, the long-term

impacts of nanomaterials on human health and the environment remain largely unclear. Nanoparticles can accumulate in tissues or disrupt biological systems in ways that alter over time, raising risk assessment difficulties. Thus, ongoing research should focus on thoughtful ideas to understand the cumulative impact of these nanomaterials (Warheit 2018).

5.5 Conclusion and Future Prospects

Nanotechnology has enormous potential to transform food science by increasing food quality, safety and sustainability. Its applications include enhancing shelf-life and storage efficiency, detecting contaminants, pollutants and augmenting food with nutritional ingredients. The combination of nanotechnology with traditional germ control methods significantly reduces foodborne risks, while breakthroughs in nanoscience open the way to more efficient, sustainable food processing and packaging. Green synthesized nanoparticles, in particular, are a considerable improvement, with antioxidant, antibacterial and sensory stabilizing properties. Their environmental friendliness, cost-effectiveness and durability in a range of environments make them an excellent choice for modern food packaging and preservation. These nanoparticles not only prevent spoilage and microbial contamination, but also aid to preserve freshness and reduce food waste by replacing synthetic, non-biodegradable components.

Despite these advantages, maximising the potential of nanotechnology in the food industry requires a detailed understanding of safety and regulatory challenges. Comprehensive study is required to understand nanoparticle mobility, quantify cytotoxicity and assess long-term environmental consequences. With further innovation and caution implementation, nanotechnology has the potential to play a significant role in establishing sustainable, secure and resilient food systems for future generations.

References

Akrami S, Saki M, Marashi Hossaeini SM, Sabahi S, Noori SMA (2023) Application of soy protein-based films and coatings on the shelf life of food products: a mini-review of recent publications with emphasis on nanotechnology. J Food Meas Charact 17(2):1393–1401

Al-Dairi M, Pathare PB, Al-Yahyai R, Opara UL (2022) Mechanical damage of fresh produce in postharvest transportation: current status and future prospects. Trends Food Sci Technol 124:195–207

Alonso-Salinas R, López-Miranda S, Pérez-López AJ, Acosta-Motos JR (2024) Strategies to delay ethylene-mediated ripening in climacteric fruits: implications for shelf life extension and postharvest quality. Horticulturae 10(8):840

Álvarez K, Famá L, Gutiérrez TJ (2017) Physicochemical, antimicrobial and mechanical properties of thermoplastic materials based on biopolymers with application in the food industry. In: Advances in physicochemical properties of biopolymers: Part, 1, pp 358–400

Avramescu SM, Butean C, Popa CV, Ortan A, Moraru I, Temocico G (2020) Edible and functionalized films/coatings—performances and perspectives. Coatings 10(7):687

Babarabie M, Sardoei AS, Jamali B, Hatami M (2024) Carnauba wax-based edible coatings retain quality enhancement of orange (Citrus sinensis cv. Moro) fruits during storage. Sci Rep 14(1):4133

Bapat VA, Trivedi PK, Ghosh A, Sane VA, Ganapathi TR, Nath P (2010) Ripening of fleshy fruit: molecular insight and the role of ethylene. Biotechnol Adv 28(1):94–107

Bashir O, Amin T, Hussain SZ, Naik HR, Goksen G, Wani AW, Manzoor S, Malik AR, Wani FJ, Proestos C (2023) Development, characterization and use of rosemary essential oil loaded water-chestnut starch based nanoemulsion coatings for enhancing post-harvest quality of apples var. Golden delicious. Curr Res Food Sci 7:100570

Bogunia-Kubik K, Sugisaka M (2002) From molecular biology to nanotechnology and nanomedicine. Biosystems 65(2–3):123–138

Boostani S, Jafari SM (2021) A comprehensive review on the controlled release of encapsulated food ingredients; fundamental concepts to design and applications. Trends Food Sci Technol 109:303–321

Bora H, Kamle M, Mahato DK, Tiwari P, Kumar P (2020) Citrus essential oils (CEOs) and their applications in food: an overview. Plants 9(3):357

Boyacı D, Iorio G, Sozbilen GS, Alkan D, Trabattoni S, Pucillo F, Farris S, Yemenicioğlu A (2019) Development of flexible antimicrobial zein coatings with essential oils for the inhibition of critical pathogens on the surface of whole fruits: test of coatings on inoculated melons. Food Packag Shelf Life 20:100316

Cenobio-Galindo ADJ, Ocampo-López J, Reyes-Munguía A, Carrillo-Inungaray ML, Cawood M, Medina-Pérez G, Fernández-Luqueño F, Campos-Montiel RG (2019) Influence of bioactive compounds incorporated in a nanoemulsion as coating on avocado fruits (Persea americana) during postharvest storage: antioxidant activity, physicochemical changes and structural evaluation. Antioxidants 8(10):500

Cetinkaya T, Wijaya W (2024) Advanced nanomaterials for enhancing the shelf life and quality of seafood products. Food Biosci 59:104018

Chavoshizadeh S, Pirsa S, Mohtarami F (2020) Conducting/smart color film based on wheat gluten/chlorophyll/polypyrrole nanocomposite. Food Packag Shelf Life 24:100501

Chen X, Ding X, Huang Y, Zhao Y, Chen G, Xu X, Xu D, Jiao B, Zhao X, Liu G (2025) Recent advances in polysaccharide-based nanocomposite films for fruit preservation: construction, applications, and challenges. Foods 14(6):1012

Daniel MC, Astruc D (2004) Gold nanoparticles: assembly, supramolecular chemistry, quantum-size-related properties, and applications toward biology, catalysis, and nanotechnology. Chem Rev 104(1):293–346

Deng Z, Jung J, Simonsen J, Zhao Y (2017) Cellulose nanomaterials emulsion coatings for controlling physiological activity, modifying surface morphology, and enhancing storability of postharvest bananas (Musa acuminate). Food Chem 232:359–368

Díaz-Silvestre SE, Ramirez L (2024) Biofilm of potato starch and silver nanoparticles. J Exp Syst 11(30):1–5

Dolai J, Mandal K, Jana NR (2021) Nanoparticle size effects in biomedical applications. ACS Appl Nano Mater 4(7):6471–6496

Dutta D, Sit N (2023) Application of natural extracts as active ingredient in biopolymer-based packaging systems. J Food Sci Technol 60(7):1888–1902

EFSA Scientific Committee, More S, Bampidis V, Benford D, Bragard C, Halldorsson T, Hernández-Jerez A, Hougaard Bennekou S, Koutsoumanis K, Lambré C, Machera K (2021) Guidance on risk assessment of nanomaterials to be applied in the food and feed chain: human and animal health. EFSA J 19(8):e06768

Frank K, Garcia CV, Shin GH, Kim JT (2018) Alginate biocomposite films incorporated with cinnamon essential oil nanoemulsions: physical, mechanical, and antibacterial properties. Int J Polym Sci 2018(1):1519407

Gómez-Estaca J, López-de-Dicastillo C, Hernández-Muñoz P, Catalá R, Gavara R (2014) Advances in antioxidant active food packaging. Trends Food Sci Technol 35(1):42–51
Grotewold E (2006) The genetics and biochemistry of floral pigments. Annu Rev Plant Biol 57(1):761–780
Gupta RK, Abd El Gawad F, Ali EA, Karunanithi S, Yugiani P, Srivastav PP (2024) Nanotechnology: current applications and future scope in food packaging systems. Measurement: Food 13:100131
Huang Q, Wan C, Zhang Y, Chen C, Chen J (2021) Gum arabic edible coating reduces postharvest decay and alleviates nutritional quality deterioration of ponkan fruit during cold storage. Front Nutr 8:717596
Iqbal SZ, Haider A, ur Rehman F, Cui G, Waseem M, Iqbal M, Khaneghah AM (2024) Application of carboxymethylcellulose in combination with essential oils nano-emulsions edible coating for the preservation of kiwifruit. Int J Biol Macromol 261:129947
Isibor PO (2024) Regulations and policy considerations for nanoparticle safety. In: Environmental Nanotoxicology: combatting the minute contaminants. Springer Nature Switzerland, Cham, pp 295–316
Islam RU, Khan T, Junaid PM, Naseem S (2024) Recent developments in sustainable bio-based wax Nanoemulsions as edible coatings for food commodities: A comprehensive review. Packag Technol Sci 37(12):1175–1194
Jamali V, Emamifar A, Beiginejad H, Moradi M, Rasouli M (2024) Enhancing grape freshness and quality with nano zinc oxide coating: a study on post-harvest preservation and grape molasses. J Food Meas Charact 18:9959–9970
Jo WS, Song HY, Song NB, Lee JH, Min SC, Song KB (2014) Quality and microbial safety of 'Fuji' apples coated with carnauba-shellac wax containing lemongrass oil. LWT-Food Science and Technology 55(2):490–497
Jurić M, Bandić LM, Carullo D, Jurić S (2024) Technological advancements in edible coatings: emerging trends and applications in sustainable food preservation. Food Biosci 58:103835
Karimi N, Alizadeh A, Almasi H, Hanifian S (2020) Preparation and characterization of whey protein isolate/polydextrose-based nanocomposite film incorporated with cellulose nanofiber and L. plantarum: A new probiotic active packaging system. LWT 121:108978
Kashiri M, Cerisuelo JP, Domínguez I, López-Carballo G, Muriel-Gallet V, Gavara R, Hernández-Muñoz P (2017) Zein films and coatings as carriers and release systems of Zataria multiflora Boiss. essential oil for antimicrobial food packaging. Food Hydrocoll 70:260–268
Kerry JP, O'grady MN, Hogan SA (2006) Past, current and potential utilisation of active and intelligent packaging systems for meat and muscle-based products: a review. Meat Sci 74(1):113–130
Khan I, Saeed K, Khan I (2019) Nanoparticles: properties, applications and toxicities. Arab J Chem 12(7):908–931
Khan MR, Di Giuseppe FA, Torrieri E, Sadiq MB (2021) Recent advances in biopolymeric antioxidant films and coatings for preservation of nutritional quality of minimally processed fruits and vegetables. Food Packag Shelf Life 30:100752
Kondle R, Sharma K, Singh G, Kotiyal A (2022) Using nanotechnology for enhancing the shelf life of fruits. In: Food processing and packaging technologies-recent advances. IntechOpen
Kumar N, Petkoska AT, Al-Hilifi SA, Fawole OA (2021) Effect of chitosan–pullulan composite edible coating functionalized with pomegranate peel extract on the shelf life of mango (Mangifera indica). Coatings 11(7):764
Lead JR, Batley GE, Alvarez PJ, Croteau MN, Handy RD, McLaughlin MJ, Judy JD, Schirmer K (2018) Nanomaterials in the environment: behavior, fate, bioavailability, and effects—an updated review. Environ Toxicol Chem 37(8):2029–2063
Lieu DM, Dang TKT, Nguyen TH (2024) Green synthesized silver nanoparticles, a sustainable approach for fruit and vegetable preservation: an overview. Food Chem: X 23:101664
Lisboa HM, Pasquali MB, dos Anjos AI, Sarinho AM, de Melo ED, Andrade R, Batista L, Lima J, Diniz Y, Barros A (2024) Innovative and sustainable food preservation techniques: enhancing food quality, safety, and environmental sustainability. Sustainability 16(18):8223

Makule E, Dimoso N, Tassou SA (2022) Precooling and cold storage methods for fruits and vegetables in sub-Saharan Africa—A review. Horticulturae 8(9):776

Manikandan V, Min SC (2023) Roles of polysaccharides-based nanomaterials in food preservation and extension of shelf-life of food products: a review. Int J Biol Macromol 252:126381

Mikus M, Galus S (2025) Extending the shelf life of apples after harvest using edible coatings as active packaging—a review. Appl Sci 15(2):767

Moradi M, Razavi R, Omer AK, Farhangfar A, McClements DJ (2022) Interactions between nanoparticle-based food additives and other food ingredients: a review of current knowledge. Trends Food Sci Technol 120:75–87

Morel MH, Pincemaille J, Lecacheux L, Menut P, Ramos L, Banc A (2022) Thermodynamic insights on the liquid-liquid fractionation of gluten proteins in aqueous ethanol. Food Hydrocoll 123:107142

Motamedi E, Nasiri J, Malidarreh TR, Kalantari S, Naghavi MR, Safari M (2018) Performance of carnauba wax-nanoclay emulsion coatings on postharvest quality of 'Valencia'orange fruit. Sci Hortic 240:170–178

Mu Q, Jiang G, Chen L, Zhou H, Fourches D, Tropsha A, Yan B (2014) Chemical basis of interactions between engineered nanoparticles and biological systems. Chem Rev 114(15):7740–7781

Murmu SB, Mishra HN (2018) The effect of edible coating based on Arabic gum, sodium caseinate and essential oil of cinnamon and lemon grass on guava. Food Chem 245:820–828

Ncube LK, Ude AU, Ogunmuyiwa EN, Zulkifli R, Beas IN (2020) Environmental impact of food packaging materials: A review of contemporary development from conventional plastics to polylactic acid based materials. Materials 13(21):4994

Nkede FN, Wardak MH, Fanze M, Kondo N, Wardana AA, Jothi JS et al (2024) The potential of helichrysum italicum essential oil-infused alginate coatings and film for prolonging the shelf-life of cherry tomatoes. Food Packag Shelf Life 46:101381

Nor SM, Ding P (2020) Trends and advances in edible biopolymer coating for tropical fruit: a review. Food Res Int 134:109208

Otoni CG, Avena-Bustillos RJ, Azeredo HM, Lorevice MV, Moura MR, Mattoso LH, McHugh TH (2017) Recent advances on edible films based on fruits and vegetables—a review. Compr Rev Food Sci Food Saf 16(5):1151–1169

Pandey JK, Nakagaito AN, Takagi H (2013) Fabrication and applications of cellulose nanoparticle-based polymer composites. Polym Eng Sci 53(1):1–8

Panzarini E, Mariano S, Carata E, Mura F, Rossi M, Dini L (2018) Intracellular transport of silver and gold nanoparticles and biological responses: an update. Int J Mol Sci 19(5):1305

Pereira VA Jr, de Arruda INQ, Stefani R (2015) Active chitosan/PVA films with anthocyanins from Brassica oleraceae (Red Cabbage) as Time–Temperature Indicators for application in intelligent food packaging. Food Hydrocoll 43:180–188

Pratiwi AS, Dwivany FM, Larasati D, Islamia HC, Martien R (2015) Effect of chitosan coating and bamboo FSC (fruit storage chamber) to expand banana shelf life. In: AIP conference proceedings, vol 1677, No. 1. AIP Publishing

Priyadarshi R, Rhim JW (2020) Chitosan-based biodegradable functional films for food packaging applications. Innovative Food Sci Emerg Technol 62:102346

Quan VM, Somboonsub P, Kongsin K, Li B, Sukyai P (2024) Isolation and utilization of lignin-containing nanocellulose as a bio-based coating for preserving perishable fruit. Ind Crop Prod 222:119597

Rai M, Ingle AP, Gupta I, Pandit R, Paralikar P, Gade A, Chaud MV, dos Santos CA (2019) Smart nanopackaging for the enhancement of food shelf life. Environ Chem Lett 17:277–290

Rajput VD, Minkina TM, Behal A, Sushkova SN, Mandzhieva S, Singh R, Gorovtsov A, Tsitsuashvili VS, Purvis WO, Ghazaryan KA, Movsesyan HS (2018) Effects of zinc-oxide nanoparticles on soil, plants, animals and soil organisms: A review. Environ Nanotechnol Monitor Manag 9:76–84

Sabaghi M, Tavasoli S, Hoseyni SZ, Mozafari MR, Degraeve P, Katouzian I (2022) A critical review on approaches to regulate the release rate of bioactive compounds from biopolymeric matrices. Food Chem 382:132411
Sagar NA, Pareek S, Sharma S, Yahia EM, Lobo MG (2018) Fruit and vegetable waste: bioactive compounds, their extraction, and possible utilization. Compr Rev Food Sci Food Saf 17(3):512–531
Şahansoy H, Caner C, Yüceer M (2024) The shellac and shellac nanocomposite coatings on enhanced the storage stability of fresh eggs for sustainable packaging. Int J Biol Macromol 261:129817
Sanches-Silva A, Costa D, Albuquerque TG, Buonocore GG, Ramos F, Castilho MC, Machado AV, Costa HS (2014) Trends in the use of natural antioxidants in active food packaging: a review. Food Addit Contamin Part A 31(3):374–395
Sharma S, Jaiswal S, Duffy B, Jaiswal AK (2022) Advances in emerging technologies for the decontamination of the food contact surfaces. Food Res Int 151:110865
Sharma B, Nigam S, Verma A, Garg M, Mittal A, Sadhu SD (2024a) A biogenic approach to develop guava derived edible copper and zinc oxide Nanocoating to extend shelf life and efficiency for food preservation. J Polym Environ 32(1):331–344
Sharma S, Nakano K, Kumar S, Katiyar V (2024b) Edible packaging to prolong postharvest shelf-life of fruits and vegetables: a review. Food Chem Adv 4:100711
Singh R, Kumar S (2023) Regulatory and safety concerns regarding the use of active nanomaterials in food industry. In: Nanotechnology advancement in agro-food industry. Springer Nature Singapore, Singapore, pp 269–306
Singh S, Prasad SM, Bashri G (2023) Fate and toxicity of nanoparticles in aquatic systems. Acta Geochimica 42(1):63–76
Souza VGL, Fernando AL (2016) Nanoparticles in food packaging: biodegradability and potential migration to food—a review. Food Packag Shelf Life 8:63–70
Thakkar KN, Mhatre SS, Parikh RY (2010) Biological synthesis of metallic nanoparticles. Nanomedicine 6(2):257–262
Uthairatanakij A, Wote YH, Laohakunjit N, Jitareerat P, Kaisangsri N (2021) Whey protein incorporated with olive oil as novel edible coating for fresh cut pineapples. In: V Asia symposium on quality management in postharvest systems 1336, pp 351–356
Valencia GA, Luciano CG, Monteiro Fritz AR (2019) Smart and active edible coatings based on biopolymers. In: Polymers for agri-food applications, pp 391–416
Vimala Bharathi SK, Maria Leena M, Moses JA, Anandharamakrishnan C (2020) Zein-based anti-browning cling wraps for fresh-cut apple slices. Int J Food Sci Technol 55(3):1238–1245
Wang Y, Liu K, Zhang M, Xu T, Du H, Pang B, Si C (2023) Sustainable polysaccharide-based materials for intelligent packaging. Carbohydr Polym 313:120851
Warheit DB (2018) Hazard and risk assessment strategies for nanoparticle exposures: how far have we come in the past 10 years? F1000Res 7:376
Xie F, Gao C, Avérous L (2024) Alginate-based materials: enhancing properties through multiphase formulation design and processing innovation. Mater Sci Eng R: Rep 159:100799
Zhang Y, Pu Y, Jiang H, Chen L, Shen C, Zhang W, Cao J, Jiang W (2024) Improved sustained-release properties of ginger essential oil in a Pickering emulsion system incorporated in sodium alginate film and delayed postharvest senescence of mango fruits. Food Chem 435:137534
Zolfaghari A, Bazargani-Gilani B, Aghajani N (2023) Edible film based on corn zein containing dill extract and essential oil/β-cyclodextrin inclusion complex: shelf life enhancement of common carp fillet. Food Sci Nutr 11(7):4275–4288

Chapter 6
Techniques for Developing Plant Bioactive-Based Nanomaterials

Kushagra Nagori, Anjali Patel, Amrita Thakur, Krishna Yadav, Madhulika Pradhan, Kartik T. Nakhate, Ajazuddin, Mukesh Kumar Sharma, Vinay Sagar Verma, Parag Jain, and Kulvinder Kaur

6.1 Introduction

Nanomaterial synthesis is gaining popularity because of its numerous uses in fields such as the delivery of drugs, environmental remediation, and electronics. In particular, the utilization of plant based biologically active compounds in nanomaterial manufacturing has lately emerged as a key trend, with recognition as a sustainable and ecologically friendly alternative to traditional chemical-based approaches (Deka et al. 2025; Nagori et al. 2011). This green synthesis is the process of creating plant-based bioderived nanomaterials that use the naturally existing features of

K. Nagori (✉) · K. Kaur
Department of Pharmaceutical Chemistry, Rungta College of Pharmaceutical Sciences and Research, Bhilai, Chhattisgarh, India

A. Patel · K. Yadav · Ajazuddin
Department of Pharmaceutics, Rungta College of Pharmaceutical Sciences and Research, Kohka, Bhilai, Chhattisgarh, India

A. Thakur
Department of Pharmaceutics, School of Pharmacy, Vishwakarma University, Pune, Maharashtra, India

M. Pradhan
Department of Pharmaceutical Technology, Rungta College of Pharmaceutical Sciences and Research, Bhilai, Chhattisgarh, India

K. T. Nakhate
Department of Pharmacology, School of Pharmacy & Technology Management, SVKM NMIMS Global University, Dhule, Maharashtra, India

M. K. Sharma
Department of Pharmacognosy, Rungta College of Pharmaceutical Sciences and Research, Bhilai, Chhattisgarh, India

V. Ghosh et al. (eds.), *Applications of Plant Bioactive Compounds-Based Nanoformulations*, Nanotechnology in the Life Sciences,
https://doi.org/10.1007/978-3-032-13086-0_6

these chemicals found in plants. They are the most effective reducing, stabilizing, and capping agents used in nanoparticle manufacturing.

As natural reducing agents, plant bioactives, by their polyphenols, flavonoids, alkaloids, and terpenoids, are excellent candidates for forming nanoparticles. These compounds can carry out metal ions reduction to get their respective nanoparticles and stabilize the nanoparticles to prevent aggregation and provide a uniform particle size distribution (Dauthal and Mukhopadhyay 2016). For example, it was shown that polyphenolic compounds contained in the leaf extract of *Camellia Sinensis* (green tea) could be used for the silver and gold ions reduction to the nanoparticles (Nadaf et al. 2022). In addition to reducing metal ions, plant bioactives serve as capping agents to prevent nanoparticles from aggregating. Metal ions can form stable complexes with bioactive compounds like flavonoids, resulting in the stability and dispersion of nanoparticles in solution (Dhaka et al. 2023). The inherent feature of plant bioactives to stabilize nanoparticles makes them highly attractive for green synthesis of nanomaterials as an eco-friendly replacement to the conventional method involving toxic compounds use. The nanomaterials synthesized with plant bioactives using this process have unique properties like large surface area, biocompatibility and enhanced reactivity. Due to these properties, they are very suitable for applications in biomedicine, e.g. in drug delivery systems and anticancer therapy. For instance, nanoparticles procured from extracts of plant have been utilized to encapsulate bioactive molecules and enhance their stability, solubility, and controlled release properties (Nair et al. 2019). Additionally, nanoparticles extracted from plants have been employed in environmental applications like the removal of heavy metals from wastewater. Plant based nanoparticles are proving to be a promising tool for environmental cleanup due to their high surface area and presence of functional groups on the surface, which makes them extremely effective for adsorbing pollutants (Verma et al. 2021). Compared to conventional chemical methods, plant bioactives have advantages for nanomaterial synthesis. This green synthesis approach is cost effective since plant extracts are available and inexpensive, and it does not involve use of toxic chemicals, henceforth, this is safer and environmentally friendly (Fawcett et al. 2017). Moreover, this method requires mild conditions including atmospheric temperature and pressure for the synthesis process, which adds to the sustainability of this method. Further, plant bioactives offer multifunctional nature that can assist to impart the additional benefits of antioxidant, antimicrobial and anticancer activities to the synthesized nanoparticles (Salem et al. 2021). This versatility introduces additional uses of plant-based nanomaterials, specifically on the therapeutic and environmental field.

V. S. Verma
Department of Pharmaceutical Chemistry, Kamla Institute of Pharmaceutical Sciences, SSTC, Junwani, Bhilai, Chhattisgarh, India

P. Jain
Department of Pharmacology, Rungta College of Pharmaceutical Sciences and Research, Bhilai, Chhattisgarh, India

6.2 Role of Phytochemicals in Nanoparticle Formation

Phytochemicals, bioactive compounds found in plants, play a critical role in the formation, reduction, stabilization, and capping of nanoparticles. These compounds, including flavonoids, alkaloids, polyphenols, terpenoids, and proteins, possess distinct chemical properties that allow them to interact with metal ions, facilitating nanoparticle formation in green synthesis methods. The diversity of phytochemicals and their mechanisms of action in nanoparticle synthesis have made them an attractive alternative to conventional chemical reducing agents. Phytochemicals exhibit excellent reducing abilities owing to functional groups presence like hydroxyl, carbonyl, and amine groups, which can donate electrons to reduce metal ions to their elemental form. Polyphenols, for example, are rich in hydroxyl groups, which enable them to reduce metal salts into nanoparticles efficiently. The phenolic compounds present in plant extracts, such as those from *Tamarindus indica* and *Eucalyptus camaldulensis*, have been shown to reduce silver and gold ions, respectively, outcoming in nanoparticles formation with controlled size and morphology (Jayaprakash et al. 2017). Flavonoids, another important group of phytoconstituents, are also engaged in the metal ions reduction. Studies have demonstrated that flavonoids, like quercetin and catechins, reduce metal ions as well as prevent the aggregation of nanoparticles by stabilizing them through electrostatic and steric interactions (Karuvantevida et al. 2022). This reduction process is essential for the synthesis of well-defined nanoparticles with preferred properties, making phytochemicals a valuable tool in nanotechnology. In addition to their reducing properties, phytochemicals serve as stabilizing and capping agents for nanoparticles. The stabilization process is critical in preventing nanoparticle agglomeration, which can affect the performance and functionality of nanomaterials. Phytochemicals such as polyphenols, flavonoids, and proteins interact with the nanoparticles surface, creating a protective or shielding layer that maintains nanoparticle stability. For instance, the stabilization of gold nanoparticles by *Syzygium cumini* leaf extract has been attributed to the interaction between polyphenolic compounds and the nanoparticle surface (Kumar and Yadav 2012). Similarly, terpenoids, which are widely distributed in plant species, have been shown to interact with metal ions and form stable metal-nanoparticle complexes. These complexes are resistant to aggregation due to the shielding effect provided by the terpenoid molecules, which gets adsorbed on the nanoparticles surface, thereby preventing agglomeration and improving their dispersibility (Dhaka et al. 2023). The ability of phytochemicals to functionalize nanoparticles further enhances their utility in various applications. Phytochemicals can provide nanoparticles with additional properties, like antioxidant, antimicrobial, and anticancer action (Sahu et al. 2024). For example, the surface functionalization of nanoparticles with phytochemicals like curcumin, a polyphenolic compound from *Curcuma longa*, can enhance the nanoparticles' biological activity and therapeutic potential (Fuloria et al. 2022). These functionalized nanoparticles exhibit improved bioavailability and target specificity, making them potential candidates for delivery of drug and cancer therapy. Furthermore, the bioactivity of phytochemicals can be transferred to the synthesized nanoparticles, imparting therapeutic benefits. For example, nanoparticles of silver synthesized utilizing *Cinnamomum verum* (cinnamon) extract exhibited significant antimicrobial activity, which was attributed to both the phytochemicals in the extract and the properties of the nanoparticles themselves (Nayar et al. 2023). This dual-action mechanism makes phytochemical-functionalized nanoparticles highly flexible in environmental and biomedical implementations.

6.3 Key Bioactive Compounds: Flavonoids, Terpenoids, Alkaloids, and Polyphenols

The synthesis of nanomaterials utilizing extracts of plants often involves bioactive compounds that play crucial part in metal ions reducing and stabilizing nanoparticles. Among the most important bioactive compounds are flavonoids, terpenoids, alkaloids, and polyphenols. These compounds, abundant in various plant species, are responsible for the unique properties of the synthesized nanomaterials, including their size, morphology, and bioactivity. Table 6.1 shows various examples of the plants and the bioactive compounds that are involved in nanoparticle synthesis. Figure 6.1 describes the different biological activities of the phytochemicals obtained from plants, that can be helpful in the formation of green nanoparticles.

6.3.1 Flavonoids

Flavonoids are polyphenolic substances broadly distributed in the plant species, designated for their anti-inflammatory, antioxidant and antimicrobial properties. They are particularly effective in the green synthesis of metal nanoparticles because of their electron-donating ability, which assists in reducing metal ions and metal salts into nanoparticles. Flavonoids such as quercetin, kaempferol, and catechins have been shown to facilitate the reduction and stabilization of nanoparticles like silver, gold, and copper oxide (Edo et al. 2025; Gupta et al. 2024). Their chelating ability also allows them to bind with metal ions, preventing nanoparticle aggregation and ensuring a stable dispersion (Shen et al. 2024).

6.3.2 Terpenoids

Terpenoids, or isoprenoids, are a large and diverse class of organic compounds produced by plants, known for their aromatic properties and therapeutic benefits. Terpenoids such as limonene, eucalyptol, and cineole have been utilized in the

Table 6.1 List of plants and their key role bioactive compounds involved in nanoparticle synthesis

Sr. No.	Plant (Botanical)	Part Used	Nanoparticle Type	Key Bioactive Compounds	Typical Size (nm)	Shape	Applications	Reference
1	*Azadirachta indica* (Neem)	Leaf	AgNP	Flavonoids, terpenoids	10–50	Spherical	Antimicrobial, anticancer	Asimuddin et al. (2020)
2	*Ocimum sanctum* (Tulsi)	Leaf	AgNP, PtNP	Eugenol, ursolic, gallic acids	20–80	Irregular	Biomedicine, catalysis	Sahoo et al. (2020)
3	*Curcuma longa* (Turmeric)	Flower	AgNP	Curcuminoids, flavonoids	~5	Spherical	Antibacterial	Rajak et al. (2023)
4	*Ficus carica* (Fig)	Leaf	Fe_3O_4 NP	Polyphenols	43–57	–	Antioxidant	Aksu Demirezen et al. (2019)
5	*Peltophorum pterocarpum*	Leaf	Fe_3O_4 NP	Tannins, flavonoids	85	–	Dye degradation	Anchan et al. (2019)
6	*Terminalia chebula*	Peel/ Seed	ZnONPs	Tannins, gallic, chebulagic acid	NA	–	Antibacterial	Akhter et al. (2024)
7	*Punica granatum* (Pomegranate)	Peel	ZrNPs	Ellagic acid, flavonoids	20–60	Spherical	Antibacterial, Antioxidant	Chau et al. (2022)
8	*Dysphania ambrosioides*	Extract	ZnO NP	Phenolics	5–30	–	Antibacterial	Álvarez-Chimal et al. (2021)
9	*Lantana camara* flowers	Flower	ZnO NP	Flavonoids	21–27	–	Antimicrobial, Antiviral	Sportelli et al. (2022)
10	*Passiflora foetida* peel	Peel	ZnO NP	Polyphenols	~58	–	Dye degradation	Khan et al. (2021)
11	*Cocos nucifera* leaf	Leaf	ZnO NP	Polyphenols	109–215	Spherical	Antimicrobial, antioxidant, photocatalytic	Rahman et al. (2022)
12	*Aerva javanica* leaf	Leaf	CuO NP	Flavonoids	18–23	–	Antimicrobial	Amin et al. (2021)
13	*Calotropis procera*	Leaf	CuO NP	Saponins, phenolics	20–80; 22	–	Antifungal, antibacterial	Shah et al. (2022)

(continued)

Table 6.1 (continued)

Sr. No.	Plant (Botanical)	Part Used	Nanoparticle Type	Key Bioactive Compounds	Typical Size (nm)	Shape	Applications	Reference
14	*Rubia cordifolia* bark	Bark	CuO NP	Anthraquinones	~50.7	–	Antibacterial, antioxidant	Vinothkanna et al. (2023)
15	*Thespesia populnea*	Leaf	CuO NP	Flavonoids	60–80	–	Antimicrobial	Narayanan et al. (2022)
16	*Calliandra haematocephala*	Leaf	AgNP	Flavonoids	–	–	Antibacterial	Raja et al. (2017)
17	*Ziziphus nummularia* leaf	Leaf	AgNP	Flavonoids	–	–	Antimicrobial, antioxidant	Padalia and Chanda (2021)
18	*Euphorbia maculata* extract	Leaf	AgNP	Various phytochemicals	–	–	Antibacterial	George et al. (2024)
19	*Terminalia catappa* leaf	Leaf	AuNP	Polyphenols	–	–	Wound healing, Antibacterial	Ray et al. (2023)
20	*Psidium guajava* leaf	Leaf	AuNP	Flavonoids	–	–	Antibacterial	Islam et al. (2023)
21	*Amaranthus gangeticus* leaf	Leaf	AuNP	Phenolics	16–40	–	Antidiabetic	Elias et al. (2024)
22	*Kalanchoe pinnata* leaf	Leaf	AgNP	Flavonoids, bufadienolides	–	–	Antibacterial, photocatalytic	Aryan and Mehata (2021)
23	*Coffee arabica* leaf	Leaf	AgNP	Phenolics	–	–	Biosensor (cysteine detection)	"(PDF) Green synthesis of eco-friendly silver nanoparticles using Coffee arabica leaf extract and development of a cost-effective biosensor for cysteine," (n.d.)

24	*Nelumbo nucifera* root	Root	CuO, CoO&FeO	Alkaloids	–	–	Antimicrobial	"(PDF) Green Synthesis and Characterization of Metal Oxide Nanoparticles Using Nelumbo Nucifera Flower and Its Antimicrobial Activity," (n.d.)
25	*Withania somnifera* leaf	Leaf	CuNP	Withanolides, flavonoids	6.28	–	Antioxidant, Antimicrobial	Shanmugapriya et al. (2022)
26	*Hibiscus cannabinus* leaf	Leaf	AgNP	Polyphenols	4.71	Spherical	Antibacterial, photcatalytic	Princy et al. (2023)
27	*Diospyros kaki* leaf	Leaf	AgNP	Reductive small molecules	27.12	Spherical	Antimicrobial, anticancer	Keskin et al. (2023)
28	*Lysiloma acapulcensis* extract	Extract	AgNP	Phenolics	13.85–34.80	Spherical	Antioxidant, antibacterial	Khanal et al. (2022)

Fig. 6.1 Different biological activities of Phytoconstituents

nanoparticles by green synthesis owing to their efficiency of metal salts reduction into nanoparticles while also stabilizing them through physical interactions (Moodley et al. 2018). The functional groups on terpenoids, including hydroxyl and carbonyl groups, are effective in reducing metal ions to their elemental form. Terpenoid- antioxidant, antimicrobial and anti-cancer actions, making them highly versatile in biomedical applications (Karnwal et al. 2024; Nagori et al. 2023b).

6.3.3 Alkaloids

Alkaloids are mainly compounds containing nitrogen present in many plant species, known for their pharmacological effects. Alkaloids such as nicotine, caffeine, and quinine have been implicated in the synthesis of nanoparticles owing to their electron-donating capacity, which helps reduce metal ions into nanoparticles (Shahin et al. 2025). For instance, the alkaloid berberine, found in *Berberis vulgaris*, has been shown to mediate the reduction of silver ions to silver nanoparticles, while also contributing to the antimicrobial action of the material synthesized (Balážová et al. 2024). Alkaloid-based nanoparticles have potential implementation in delivery of drug and cancer chemotherapy because of their bioactivity and compatibility with biological systems (Dhyani et al. 2022).

6.3.4 *Polyphenols*

Polyphenols belongs to group of naturally occurring substances present in plants, with antioxidant, anticancer, and antimicrobial properties. These compounds, including tannins, flavonoids, and phenolic acids, displays a important role in the green chemical synthesis of nanoparticles by reducing metal salts and stabilizing the formed nanoparticles. For example, polyphenols in the extract of *Camellia sinensis* (green tea) have been utilised for silver and gold nanoparticles synthesis, with demonstrated antioxidant and anticancer properties (Onitsuka et al. 2019). Polyphenols are especially effective in stabilizing nanoparticles, as their aromatic rings and hydroxyl groups interact with the nanoparticle surface, preventing aggregation and enhancing their bioactivity (Abuzeid et al. 2023).

6.4 Extraction and Purification of Plant Bioactives

The extraction and purification of bioactive compounds from plants are crucial steps in harnessing their potential for nanoparticle synthesis. These bioactives, which include flavonoids, alkaloids, terpenoids, and polyphenols, serve as a stabilizing agents and reducing agents in nanoparticle formation. The efficiency and effectiveness of the extraction process directly impact the quality and yield of the bioactives, which in turn influence the characteristics of the synthesized nanoparticles. The Fig. 6.2 demonstrates the production of nanoparticles with the help of different phytoconstituents. Several methods are employed for the extraction of plant bioactives, and the selection of method relys on the characteristics of plant material, the nature of the bioactive compounds, and the intended application. The main extraction techniques include solvent extraction, steam distillation, cold press extraction, ultrasound-assisted extraction and supercritical fluid extraction. Moreover, once the bioactives are extracted, purification is required to isolate the desired compounds

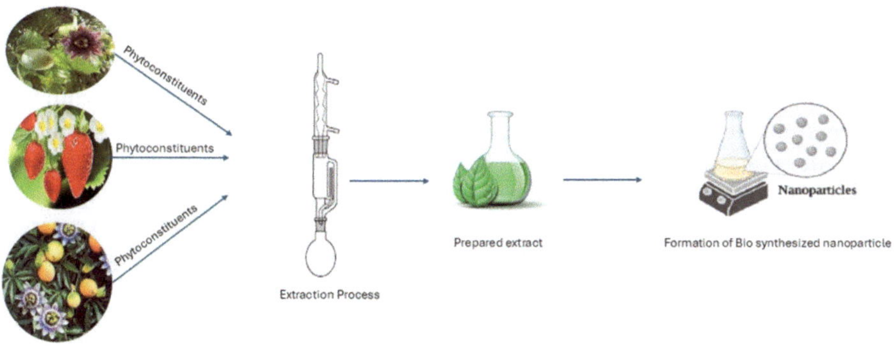

Fig. 6.2 Different phytocostituents involved in the formation of nanoparticles

and remove impurities. Several purification methods are employed, including: column chromatography, thin-layer chromatography, recrystallization, liquid-liquid extraction.

6.5 Applications of Plant Bioactive-Based Nanomaterials

Plant bioactive-based nanomaterials have received widespread recognition and attention in recent current years due to their versatility, biocompatibility, and potential for use in various biomedical, environmental, and industrial applications. These nanomaterials are synthesized using plant-derived bioactive phytoconstituents, which act as stabilizing agents and reducing agent in nanoparticle formation as well as provides unique properties like antioxidant, anticancer and antimicrobial activities. The following outlines and a brief Table no. 6.2 explain the major applications of plant bioactive-based nanomaterials.

6.5.1 Drug Delivery Systems

The use of plant bioactive-based nanomaterials in drug delivery systems has received widespread attention because of their capacity to enhance the efficacy and safety of therapeutic agents. These nanomaterials, derived from plant bioactives, offer numerous benefits advantages, such as biodegradability, biocompatibility, and the potential for targeted drug delivery. By improving solubility, stability, and bioavailability, plant-derived nanoparticles has exhibited ensuring in increasing the performance of various drugs, particularly for diseases such as cancer and microbial infections.

6.5.1.1 Targeted Drug Delivery

The most notable benefit of plant bioactive-based nanomaterials is their ability to target specific tissues or cells, which is crucial in reducing systemic toxicity and enhancing therapeutic outcomes. For example, nanoparticles encapsulated with bioactive compounds like curcumin, quercetin, and resveratrol have been shown to direct drugs specifically to cancer cells, significantly reducing the damage to healthy cells (Ali et al. 2024; Manzari-Tavakoli et al. 2024). Furthermore, these nanoparticles can be functionalized with specific antibodies or ligands that binds to receptor which are overexpressed on the target cells surface, improving the selectivity and precision of drug delivery (Prajapati et al. 2024).

Table 6.2 List of plants, bioactive compounds and their various applications of biosynthesised nanoparticles

Plant Name	Bioactive Compounds	Nanomaterial Type	Application	References
Azadirachta indica (Neem)	Azadirachtin, Flavonoids	Nanoparticles	Nano-pesticides	Hazafa et al. (2022)
Moringa oleifera (Moringa)	Flavonoids, Tannins	Nano-fertilizers	Nutrient delivery, Soil health	Mashamaite et al. (2022)
Cajanuscajan (Pigeon Pea)	Isoflavonoids, Tannins	Nano-fertilizers	Fertilization, Soil enrichment	Garg et al. (2023)
Camellia sinensis (Tea)	Catechins, Flavonoids	Nanoparticles	Antioxidant, Antibacterial	Al-Ghamdi et al. (2022)
Curcuma longa (Turmeric)	Curcumin, Volatile oils	Nanoparticles	Drug delivery, Antioxidant	Rai et al. (2020)
Gossypium hirsutum (Cotton)	Tannins, Flavonoids	Nanoparticles	Drug delivery, wound healing	Kanipandian and Thirumurugan (2014)
Piper nigrum (Black Pepper)	Piperine, Flavonoids	Nanoparticles	Bioenhancer, Drug delivery	Rajput et al. (2022)
Mentha piperita (Peppermint)	Menthol, Flavonoids	Nanoparticles	Anti-inflammatory, Antioxidant	Aftab et al. (2024)
Allium sativum (Garlic)	Allicin, Sulfur compounds	Nanoparticles	Antimicrobial, Anticancer	Subbanna et al. (2020)
Hibiscus rosa-sinensis (Hibiscus)	Anthocyanins, Flavonoids	Nanoparticles	Antioxidant, Anti-inflammatory	Singh et al. (2024)
Ocimum sanctum (Tulsi)	Eugenol, Flavonoids	Nanoparticles	Antibacterial, Anticancer	Jiang et al. (2023), Singhal et al. (2011)
Azadirachta indica (Neem)	Flavonoids, Triterpenoids	Nanoparticles	Pest control, Anticancer	Dutt et al. (2023)
Solanumlycopersicum (Tomato)	Lycopene, Carotenoids	Nanoparticles	Antioxidant, Skin care	Elbrolesy et al. (2023)
Aloe vera	Aloin, Polysaccharides	Nanoparticles	Wound healing, Anti-inflammatory	Nandhini et al. (2023)
Capsicum annuum (Chili)	Capsaicin, Flavonoids	Nanoparticles	Pain relief, Antioxidant	Fathallah (2023)
Pomegranate granatum (Pomegranate)	Polyphenols, Tannins	Nanoparticles	Antioxidant, Anti-aging	Abdellatif et al. (2021)
Ginger officinale (Ginger)	Gingerol, Zingerone	Nanoparticles	Antioxidant, Anticancer	Fadhil Ascar et al. (2019)
Bacopamonnieri (Brahmi)	Bacopasaponins, Alkaloids	Nanoparticles	Cognitive enhancer, Neuroprotective	Nellore et al. (2013)
Linumusitatissimum (Flax)	Lignans, Omega-3 fatty acids	Nanoparticles	Antioxidant, Anticancer	Abbasi et al. (2017)

(continued)

Table 6.2 (continued)

Plant Name	Bioactive Compounds	Nanomaterial Type	Application	References
Lavandula angustifolia (Lavender)	Linalool, Flavonoids	Nanoparticles	Stress relief, Antibacterial	Rather et al. (2022)
Glycyrrhizaglabra (Licorice)	Glycyrrhizin, Flavonoids	Nanoparticles	Anti-inflammatory, Anticancer	Al-Radadi (2024, 2021)
Asparagus officinalis (Asparagus)	Saponins, Flavonoids	Nanoparticles	Antioxidant, Anticancer	Tripathi et al. (2022)
Paeonialactiflora (Peony)	Paeoniflorin, Triterpenoids	Nanoparticles	Anticancer, Anti-inflammatory	Jin et al. (2022)
Tinosporacordifolia (Guduchi)	Tinosporide, Alkaloids	Nanoparticles	Immunomodulatory, Anticancer	Basu et al. (2022)
Nigella sativa (Black Seed)	Thymoquinone, Saponins	Nanoparticles	Antimicrobial	Boke et al. (2024)
Vitis vinifera (Grape)	Resveratrol, Polyphenols	Nanoparticles	Anticancer, antibacterial	Hussein and Mohammed (2021)
Triphala	Tannins, Vitamin C	Nanoparticles	Digestive health, Antioxidant	Ranjani and Hemalatha (2022)
Cinnamomum verum (Cinnamon)	Cinnamaldehyde, Flavonoids	Nanoparticles	Antibacterial, Antioxidant	Ansari et al. (2020), Nemattalab et al. (2022)
Rheum rhabarbarum (Rhubarb)	Anthraquinones, Tannins	Nanoparticles	Anticancer, Antioxidant	Reddy et al. (2016)
Withania somnifera (Ashwagandha)	Withanolides, Alkaloids	Nanoparticles	Anticancer	Meher et al. (2022), Shanmugapriya et al. (2022)
Carthamus tinctorius (Safflower)	Safflor yellow, Flavonoids	Nanoparticles	Antibacterial, antioxidant	Nagaraj et al. (2012), Rodríguez-Félix et al. (2021)

6.5.1.2 Improved Bioavailability

Many plant bioactive compounds suffer from inadequate bioavailability because of their low solubility and quick metabolism in the body. Nanoparticles offer a solution by enhancing the solubility and stability of these compounds. For example, curcumin, which is a phytoconstituent obtained from *Curcuma longa*, has limited bioavailability due to its hydrophobic nature. However, when incorporated into nanoparticles, its bioavailability is significantly improved, allowing for better therapeutic efficacy in conditions such as cancer, inflammation, and oxidative stress (Gera et al. 2017). This enhanced bioavailability is particularly important for the treatment of diseases requiring sustained drug action over time.

6.5.1.3 Sustained and Controlled Release

Nanoparticles made from plant bioactives also enable controlled and sustained drug release, establishing a long time therapeutic efficacy and frequency reduction of drug administration. This is especially favourable for chronic diseases. Nanomaterials like micelles, polymeric nanoparticles and liposomes have been developed using plant-derived compounds to deliver drugs in a control and regulated manner, responding to environmental stimuli such as pH and temperature. Such systems are advantageous for maintaining a consistent drug concentration in the body, thus enhancing the drug's overall effectiveness and patient compliance (Patra et al. 2018).

6.5.1.4 Antimicrobial Drug Delivery

Nanoparticles synthesized from plant bioactives have also shown promise in antimicrobial drug delivery. Plant-derived compounds like flavonoids, polyphenols, and terpenoids have displayed antimicrobial action against a broad variety of pathogens. For example, silver nanoparticles produced from *Azadirachta indica* (neem) extract have exhibited strong antibacterial effects and can be utilized for controlled antibiotic delivery, potentially overcoming the issue of antibiotic resistance (Kumari et al. 2022). Plant bioactive-based nanomaterials can also enhance the effectiveness of antifungal and antiviral drugs, providing a more comprehensive solution for treating microbial infections (Naikoo et al. 2021).

6.5.1.5 Biocompatibility and Safety

One of the pivotal benefits of using plant bioactive-based nanomaterials in drug delivery is their biocompatibility and safety. Unlike synthetic nanoparticles, which may cause toxicity, plant-based nanoparticles are generally well-tolerated by the body. Many plant bioactives possess anti-inflammatory and antioxidant properties, which further contribute to the safety profile of these nanomaterials. For example, *Coriandrum sativum* (coriander)-derived nanoparticles have shown low toxicity and excellent biocompatibility while exhibiting promising anticancer and antimicrobial effects (Khan et al. 2018). Plant bioactive-based nanomaterials represent a promising avenue for advancing drug delivery systems. These nanomaterials not only improve the bioavailability, stability and controlled and targeted release of therapeutic agents but also gives targeted delivery of drugs with enhanced efficacy. The use of plant-derived compounds, with their inherent biocompatibility and therapeutic potential, positions these nanomaterials as key players in the development of next-generation drug delivery systems, offering safer and more effective treatments for a different types of diseases, such as microbial infections and cancer.

6.5.2 Antimicrobial and Anticancer Application

Nanomaterials derived from plant bioactives have shown promising antimicrobial action, that can be exploited in the fight against infections. The growing problem of antibiotic resistance has prompted the exploration of alternative strategies, such as plant-based nanoparticles, to increase the therapeutic efficacy of current antimicrobial agents or provide new therapeutic solutions (Anand et al. 2022). Several plant based compounds, like alkaloids, flavonoids, and terpenoids, possess natural antimicrobial properties. When these compounds are incorporated into nanoparticles, they exhibit enhanced antimicrobial activity, providing better bioavailability and longer-lasting effects compared to their free forms. For example, silver nanoparticles obtained from extracts of *Azadirachta indica* (neem) have shown excellent antibacterial and antifungal activities, potentially reducing the development of resistance and improving treatment outcomes (Khan and Javed 2021). Additionally, plant-based nanoparticles have been used as carriers for conventional antibiotics, improving the drug's stability, solubility, and controlled release. This combination enhances the therapeutic action against a broader spectrum of pathogens, including multi-drug-resistant bacteria. For example, nanoparticles formed from *Coriandrum sativum* (coriander) and *Camellia sinensis* (green tea) extracts have been employed in enhancing the antimicrobial activity of commonly used antibiotics like amoxicillin and tetracycline (Anand et al. 2022). One of the significant advantages of plant bioactive-based nanoparticles in cancer therapy is their ability to target cancer cells specifically, minimizing the side effects typically associated with traditional chemotherapeutic agents. This is achieved through functionalization of the nanoparticles with specific ligands or antibodies that can selectively bind to overexpressed receptors on cancer cells, improving the precision of drug delivery (Ghazal et al. 2024). The antimicrobial and anticancer activities of plant bioactive-based nanomaterials are governed by various mechanisms, including oxidative stress, DNA damage, and modulation of cell signaling pathways. In antimicrobial applications, plant-derived nanoparticles can disrupt microbial cell membranes, cause oxidative damage, and inhibit key enzymes, leading to the death of pathogens (Okeke et al. 2023). Similarly, in cancer therapy, plant bioactive-based nanoparticles can induce apoptosis, inhibit angiogenesis, and prevent metastasis through various molecular mechanisms (Bozzuto et al. 2024).

6.5.3 Antioxidant and Anti-inflammatory Properties

The anti-inflammatory as well as antioxidant potential of plant bioactive-based non-material have gathered notable attention owing to their promising therapeutic applications in a variety of diseases, including chronic conditions like cardiovascular disorder, diabetes, neurodegenerative diseases, and inflammatory conditions (Nagori et al. 2024c). These properties are primarily attributed to the plant bioactive phytoconstituents, such as flavonoids, polyphenols, and terpenoids, found in extracts of plant, which are encapsulated into nanoparticles to enhance their bioavailability and effectiveness. Oxidative stress is a key factor in pathogenesis of numerous diseases and disorders, like cancer, neurodegenerative disorders, and cardiovascular diseases (Valko et al. 2007). Antioxidants display a major contribution in free radicals neutralization and reduction in damage due to oxidative stress. Many plant-derived bioactives, like flavonoids, phenolic acids, and alkaloids, possess strong antioxidant properties, which renders them as potential candidates for the nanoparticles synthesis with enhanced antioxidant activity. Nanoparticles derived from plant bioactives exhibit improved antioxidant effects compared to the free form of the bioactive compounds. For example, nanoparticles synthesized from *Camellia sinensis* (green tea) extract, rich in polyphenols like catechins, have demonstrated promising antioxidant activity, prominently scavenging ROS (reactive oxygen species) and diminishing damage due to oxidative stress (Tran Khac et al. 2023). Similarly, nanoparticles made from *Curcuma longa* (turmeric) extract, which contains curcumin, a well-known antioxidant, have been shown to enhance the body's natural antioxidant defences (Sattar et al. 2024). Chronic inflammation is implicated in various diseases such as arthritis, asthma, inflammatory bowel disease (IBD), and neurodegenerative disorders. The anti-inflammatory activity of plant bioactive compounds, including flavonoids, terpenoids, and polyphenols, has been widely studied, and their incorporation into nanoparticles enhances their ability to modulate inflammatory pathways. For example, nanoparticles synthesized from *Ginger* (*Zingiber officinale*) extract have shown significant anti-inflammatory effects by inhibiting the production of pro-inflammatory cytokines such as TNF-α and IL-1β (Anh Nga et al. 2023). Furthermore, the targeting of specific inflammation-related pathways, coupled with the improved bioavailability and stability of plant phytoconstituents compound in nanoparticle form, permits for further effective and targeted anti-inflammatory treatment. For example, nanoparticles can be tailored to liberate their therapeutic load in response to the inflammatory microenvironment, ensuring that the anti-inflammatory agents arrives at the inflammation site in a regulated manner, thus improving treatment outcomes (Dewangan et al. 2016; Zhu et al. 2021).

6.5.4 Environmental Applications: Pollutant Removal and Water Purification

The increasing levels of environmental pollution have raised significant concerns regarding the degradation of ecosystems and public health. Plant bioactive-based nanomaterials have emerged as effective and sustainable solutions for addressing some of the most pressing environmental issues, particularly in the areas of pollutant removal and water purification. These nanomaterials, derived from plant extracts rich in bioactive compounds, offer a green and eco-friendly alternative to conventional materials used in environmental cleanup processes. Their exceptional and distinctive properties, like larger surface area, biocompatibility, and functional groups capable of interacting with pollutants, renders them promising candidates for water treatment and environmental remediation.

Plant bioactive-based nanomaterials have shown remarkable potential in removing a broad variety of pollutants, such as dyes, heavy metals, organic contaminants and pesticides, from contaminated water and soil. The nanoparticles synthesis from plant bioactives allows for the modification of surface properties, enhancing their adsorption capacity for various pollutants (Dikshit et al. 2021). For example, silver nanoparticles (AgNPs) synthesized using *Azadirachta indica* (neem) extract have been demonstrated to efficiently remove heavy metals, such as lead (Pb) and mercury (Hg), from contaminated water (Kumari et al. 2022). The functional groups present on the surface of the plant-derived nanoparticles interact with the metal ions, forming stable complexes and facilitating their removal through adsorption or precipitation.

In addition to pollutant removal, plant bioactive-based nanomaterials have shown significant potential in the development of advanced water purification systems. The incorporation of plant-derived bioactive compounds into nanomaterials can enhance the removal of microorganisms, including bacteria, viruses, and fungi, making them useful in drinking water purification and wastewater treatment applications. For example, zinc oxide (ZnO) nanoparticles synthesized from *Mentha piperita* (peppermint) extract have been shown to possess strong antimicrobial properties, effectively killing bacteria like *Staphylococcus aureus* and *Escherichia coli* (Ahmad et al. 2023). These nanoparticles can be used in water filtration systems, where their antimicrobial activity helps to eliminate harmful pathogens from drinking water, improving water quality and safety.

6.5.5 Agricultural Applications: Nano-pesticides and Fertilizers

The agricultural industry is one of the largest sectors to benefit from nanotechnology, especially through the development and growth of nanofertilizers and nanopesticides. Plant bioactive-based nanomaterials are increasingly being explored as a sustainable approach to enhancing agricultural productivity and promoting environmental safety. These nanomaterials, synthesized from plant extracts, offer a new frontier in crop protection and soil health improvement, making them vital tools for modern agriculture. Nano-pesticides are pesticide formulations that incorporate nanoparticles or are engineered at the nanoscale to improve their effectiveness, stability, and controlled release. The controlled release feature of nano-pesticides helps in minimizing the frequent application of chemicals, thus reducing the overall pesticide load in the environment. For example, *Azadirachta indica* (neem)-based nanoparticles have been widely studied for their pesticidal properties, providing an eco-friendly replacement to conventional chemical pesticides. Neem-based nano-pesticides have been shown to exhibit significant insecticidal, fungicidal, and anti-bacterial activities, providing effective protection against a range of agricultural pests, including aphids, caterpillars, and fungal pathogens (Chaudhary et al. 2017). Nano-fertilizers are an innovative approach to delivering essential nutrients to plants more efficiently and sustainably. By encapsulating plant nutrients in nanoparticles or by creating nanoparticles that release nutrients in a controlled manner, nano-fertilizers ensure that plants optimally receive the necessary nutrients. The plant bioactive compounds present in these nanoparticles can enhance nutrient uptake, improve soil fertility, and reduce nutrient leaching, which is a major issue with traditional fertilizer use. One example is the use of *Moringa oleifera* (moringa) seed extract in the synthesis of nano-fertilizers. Moringa-based nanoparticles have been developed to enhance nitrogen and phosphorus delivery to plants. These nano-fertilizers have been proved to elevate yields of crop while minimizing the use of chemical fertilizers and reducing nutrient runoff into water bodies (Mohammed and Hawar 2022; Virk et al. 2023). An overview of plant sources, associated bioactive compounds, and the various uses of biosynthesized nanoparticles are given in Table 6.2.

6.6 Techniques for Synthesis of Plant Bioactive Based Nanomaterials

The properties and therapeutic applications of nanoparticles depend on their synthesis method. They are often synthesized using the top-down approach and the bottom-up approach to make nanoparticles, both approaches include physical, chemical, and biological processes. The bulk material under consideration is pulverized into very fine entities utilizing suitable approaches like thermal evaporation, mechanical grinding, ball-milling and sputtering in the top-down approach, while in the bottom-up approach, nano-materials are developed by congregating molecules or atoms into new material having the size in the nano range using biological and chemical methods.

6.6.1 Sol-Gel Method

Two types of components are present in sol-gel as evident from the name itself, 'sol' and 'gel. 'Sols' belongs to subclass colloids in which the solid particles are dispersed in liquid.. Gels constitute continuous chain of particles with liquid filled pores (or polymers containing liquid). A sol-gel phenomenon exhibits establishment of 'sols' in a liquid and then bridging the sol particles (or few subunit with ability of making a porous network) to form a chain like network. It is feasible to

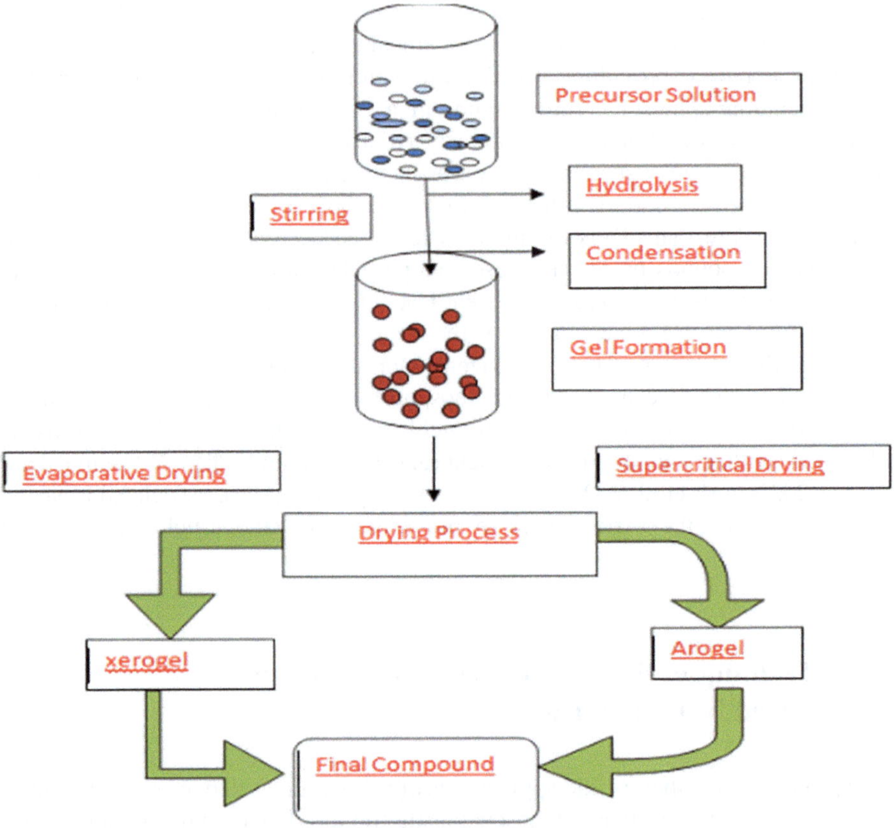

Fig. 6.3 Sol-Gel Method. (Adopted from Patil et al. 2021)

form thin films, powders as well as monolithic solid by drying the liquid (Ummah 2019). This gel is composed of distinct particles or polymers. Synthesis of Sol-gel comprises of following steps such as mixing, gelation, casting, drying, aging as well as densification as depicted in Fig. 6.3. Fe_3O_4 nanoparticles were effectively obtained employing sol–gel methods and simultaneously annealed at various temperatures under vacuum. The morphology studies revealed that the nanoparticles were almost spherical in shape (Kumari et al. 2023). This technique was found to be one of the simplest as well as effective methods to synthesize quality wise genuine nanoparticles. This technique also presents certain advantages over other methods of synthesis like control over the size, surface properties of the materials and texture, less cost, superior quality, ease of implementation, and formation of materials having more surface areas. This simplicity and flexibility features justify its popularity in the formation of nanoscale powders (Navas et al. 2021).

Sol-Gel technique has extensive implementations including waste water management, novel drug delivery, construction appliances, and a wide range of sensors. This technique can be implemented at an industrial platform due to its simplicity and utility (Singh 2022). Accordingly sol-gel constituents can be of various types, such as catalysts, inorganic pigments, drugs, magnetic nanoparticles, metallic nanoparticles and nanocarriers. Additionally, it permits the biological molecules encapsulation like enzymes and proteins which have applications and utilization as biosensors as well as the drugs release in the management of neurodegenerative disorders like Alzheimer's and Parkinson's, for instance. Overall chemistry of sol-gel tends to be specifically sensitive towards following parameters like solvent, pH, time, temperature, agitation and catalysts.

6.6.2 Hydrothermal and Solvothermal Methods

The term "hydrothermal and solvothermal methods" can be termed as ongoing chemical reactions in sealed vessels containing solvents in which the solvents temperature can be guided to nearly around their critical points through heating simultaneously altogether with endogenous pressures. The method is termed as "hydrothermal" where water is employed as the solvent. In few instances, this term also describe processes carried out at ambient state. To make it clear and easy to understand, the approach utilising the solvent as water in the hydrothermal method is particularly mentioned in this chapter; whereas the term "solvothermal method" denotes organic reagents are used as solvents (Aliofkhazraei 2015).

Hydrothermal and solvothermal synthesis is an comprehensive term for methods that grasp advantage of solubility of materials by immersing in extremely hot water under high pressure, which gives crystalline structures. Hydrothermal and solvothermal chemistry have been broadly implemented in the formation of inorganic-organic hybrid compounds, like clusters and coordination polymers. Conversely to other synthesis process, the hydrothermal and solvothermal techniques can elevate the reactants solubility, increase the reactants reactivity, form ideal crystal materials, and generate important and chief metastable phases which are either difficult or sternuous to prepare or even impossible to be prepared by conventional methods (Feng and Li 2017).

6.6.3 Biological Reduction and Phytochemical-Assisted Synthesis

The nanoparticles synthesis has emerged as a subject of great novel interest owing to their optical, catalytic, mechanical, magnetic and electrical properties. Various techniques like the co-precipitation, sol–gel process, hydrothermal techniques, non-aqueous synthesis, Sono chemical method, thermal reduction, microemulsion method and ultrasound irradiation etc. have been developed and optimized to synthesize nanoparticles. The green route paves a way for natural capping agents for the metal nanoparticles stabilization. The green route at the same time prevents the contamination of natural capping agent from hazardous and toxic chemicals. Various types of nanoparticles can be synthesized employing green route method like silver, magnetite, nickel oxide, zinc oxide, copper oxide, gold, manganese dioxide etc (Khatun et al. 2022). The evolution of green synthesis of nanoparticles employing plant based material has happened in the last decade. The plant extracts ability of reducing metal ions has been unravelled during early 1900s but it was utilised dominantly in past 30 years only for the metal salt reduction. Plants contains several bioactive materials like flavonoids, phenols, citric acid, ascorbic acid, polyphenolic, alkaloids, reductase and terpenes which exhibits potential as reducing agents. Synthesis of nanoparticles from plant based bioactives is a very promising and novel nanotechnology approach due to the fact that the plant bioactives itself possess reducing property and acts too as a capping agent. Nanoparticles can be synthesized extracellularly as well as intracellularly from plant based bioactive system (Santhoshkumar et al. 2017).

6.6.3.1 Gold Nanoparticle Synthesis

Extracts obtained from plants has the inherent potential of serving as reducing agents as well as stabilizing agents in the nanoparticles synthesis. Gold nanoparticles are conferred with bactericidal property against food pathogens and bacterial pathogens and it also renders several other pharmacological applications such as tumour destruction through heat, health care, separation science, pharmacokinetic studies, drug delivery, environmental studies, optics, gene delivery, space industry, food industries etc. The primary focus is on gold nanoparticle synthesis mediated by various plant extracts specifically where plant bioactive extracts undergo reaction with metal ions and metal salts leading to reduction. Gold nanoparticles biosynthesis utilizing plant extracts is gaining high popularity owing to the broad spectrum antibacterial action and reducing property. Synthesis of gold nanoparticles in size range of 20–30 nm is arising as a potential agent for cancer therapy. Therapeutic efficacy of nano ranged gold nanoparticles is under evaluation and consideration in treatment of various human carcinoma cells (Ahmed 2015).

6.6.3.2 Silver Nanoparticle Synthesis

Biological methods have been established as cost-effective, dependable, environmental-friendly, sustainable and acceptable for silver nanoparticles synthesis (Ag NPs), subject to toxic chemical elimination. Phytoconstituents (such as soluble polyphenols, aminoacids, carbohydrates, terpenoids and other chemicals) of plant extracts have potential to reduce silver ions. Silver nanoparticles (Ag NPs) obtained from plant extracts can efficiently retard the growth of numerous types of microorganisms because silver nanoparticle (Ag NPs) is conferred with the ability to penetrate bacterial cell wall, silver nanoparticle (Ag NPs) and the silver ions (Ag+ ion) can consequently couples with mercapto groups of numerous vital and essential enzymes. These results in inactivation of different functional biopolymer (enzymes and proteins) in bacterial cell (Raza et al. 2024). Owing to their increased and elevated capacity of adsorption, silver nanoparticles (Ag NPs) gets easily and firmly bound with different parts of microorganisms: (i) the surface of cell, (ii) carbonyl group and disulfide bond of DNA and (iii) proteins. These results in termination of bacterial DNA replication, growth inhibition followed by bacterial cell death (Koduru et al. 2018). Green synthesis utilising extracts of plant bioactives delivers environmental friendly, stable and effective silver nanoparticles (AgNPs) with tailored characteristics (Nagori et al. 2025). Different types of various plants which mediate the synthesis of different metallic nanoparticles are tabulated in Tables 6.3 and 6.4.

6.6.3.3 Zinc Oxide Nanoparticle Synthesis

Nanotechnology is advancing swiftly in the present day scenario owing to its countless applications in agriculture, biomedical studies and research. Zinc oxide nanoparticles (ZnONPs) have implementation in sunscreens, agriculture, food packaging and cosmetic industries. Owing to its enormous capacity of penetrating the cell membrane of microbes, zinc oxide nanoparticles (ZnONPs) are conferred with antibacterial properties. Nanoparticles produced by green synthesis method are a proven eco-friendly and cost effective approach to diminish the environmental pollution (Nagori et al. 2024d; Sharma et al. 2020).

6.6.3.4 Copper Nanoparticles

Copper based materials have broad application and implementation in protecting seasonal crops against disparate moulds and microorganisms, along with those causing fungal and bacterial diseases. Copper possess antibacterial efficacy by generating hydroperoxide free radicals which blocks vital and essential enzymes and modulates integrity of cell membrane. Additionally, copper has the ability to obstruct functional protein groups by replacing the necessary ions. Copper nanoparticles (CuNPs) also facilitates the initiation of antioxidant system, as it is conferred with properties just like antioxidant molecule. This permits them to efficiently neutralize, absorb and douse or remove triplet and singlet oxygen. Various reductive groups, specifically phytoconstituents are present on the outer surface of copper nanoparticles (CuNPs) which renders potential antioxidant activity to copper nanoparticles (Khairul Hanif Mohd Nazri and Sapawe 2020).

Table 6.3 Leaf mediated synthesis of different metallic nanoparticles

S. No.	Scientific name	Family	Common name	Prepared nanoparticle	Characterization	Size (nm)	Shape	Reference
1.	*Aloe barbadensis*	Asphodelaceae (Liliaceae)	Aloe vera	AgNPs	UV–Vis spectrophotometer, FTIR	400	–	Anju et al. (2020)
2.	*Moringa oleifera*	Moringaceae	Drumstick Tree	ZnO NPs	UV-Vis, SEM, XRD, FTIR	20–100	Irregular	Shashikala et al. (2024)
3.	*Azadirachta indica*	Meliaceae	Neem	AgNPs	UV–Vis, (FESEM) field emission scanning electron microscopy, FTIR, DLS, zeta potential	27	Spherical	Alharbi and Alsubhi (2022), Nagori et al. (2023a)
4.	*Lawsoniainermis*	Lythraceae	Henna	AgNPs	UV- Vis, XRD, TEM, FESEM, FTIR	20–70	Spherical	Alhomaidi et al. (2022)
5.	*Eucalyptus globulus*	Myrtaceae	Eucalyptus	Fe_3O_4 NPs	UV-Vis, TEM, FTIR, DLS	12–45	Spherical	Andrade-Zavaleta et al. (2022)
6.	*Camellia Sinensis*	Theaceae	Green Tea	AgNPs	UV- Vis, XRD, High resolution transmission electron microscope (HRTEM), FESEM, FTIR	7	Spherical	Emima Jeronsia et al. (2020)
7.	*Mentha piperita*	Lamiaceae	Peppermint	AgNPs	UV- Vis, TEM, SEM, FTIR, XRD	20	Spherical	Robles-Martínez et al. (2020), Nagori et al. (2024a)
8.	*Carica papaya*	Caricaceae	Papaya	AuNPs	UV-Vis, XRD, TEM, FTIR	25–60	Spherical	Sathya et al. (2024)
9.	*Basil leaves*	Lamiaceae	mint	AgNPs	UV- Vis, HR-TEM, FE-SEM, FTIR, XRD, DLS	48.25	Spherical	Manikandan et al. (2021)
10.	*Carica papaya*	Caricaceae.	Papaya	AgNPs	UV- Vis, TEM, STM, SEM, EDS, XRD, FTIR	10–50	Spherical	Singh et al. (2021)
11.	*Moringa oleifera*	Moringaceae	Drumstick	AgNPs	UV–Vis, FTIR, XRD, SEM-EDS, TEM, DSC	25	Spherical	Younas et al. (2023)
12.	*Eclipta alba*	Asteraceae	Bhringraj	AgNPs	UV–Vis, FTIR, XRD	40–60	Spherical	Mani et al. (2024)
13.	*Ziziphus spina*	Rhamnaceae	Christ's thorn jujube	AgNPs	UV–Vis, FTIR, HE-TEM	26–28	Spherical	Abdelkader (2019)

14.	*Psidium guajava*	Myrtaceae	Guava	CuNPs	UV-Vis, SEM, FTIR, XRD	22–70	Spherical	Sriram (2023)
15.	*Syzygium cumini*	Myrtaceae	Jamun	ZnO NPs	UV-Vis, XRD, FTIR,TEM	18–80	Spherical	Yadav et al. (2025)
16.	*Terminalia catappa*	Combretaceae	Indian Almond	CeO_2 NPs	UV-Vis, SEM, FTIR, XRD	25–90	Oval	Ray et al. (2023)

Table 6.4 Plant mediated synthesis of different metallic nanoparticles

S. No.	Scientific name	Family	Common name	Prepared Nanoparticles	Characterization	Size (nm)	Shape	Reference
1.	*Grewia flavescens*	Malvaceae	Lavender star flower	AgNPs	UV–vis spectroscopy, FT-IR, DLS, XRD, HR-TEM, and TGA	20–30	Spherical	Sana et al. (2023)
2.	*Allium sativum*	Amaryllidaceae	Garlic	SeNPs	UV-Vis, TEM, EDX, FTIR	30–100	Spherical	Mikhailova (2023)
3.	*Lupinus albus and Lupinuspilosus*	Fabaceae	White lupin	AgNPs	UV–Vis, XRD, FTIR, and SEM	19–26	Crystalline	Mirza et al. (2023), Nagori et al. (2024b)
4.	*Cassia sieberiana*	Fabaceae	Drumstick tree	AgNPs	UV–Vis, XRD, FTIR, TEM and SEM	12.9 ± 3.1	Spherical	Kyene et al. (2023)
5.	*Tridax procumbens*	Asteraceae	Coat buttons	ZnO NPs	UV-Vis, XRD, SEM, FTIR	15–90	Rod-shaped	Ahmed et al. (2022)
6.	*Moringa oleifera*	Moringaceae	Drumstick	AgNPs	UV–Vis, EDX, FTIR	119 ± 36	Agglomerated spherical	Rafique et al. (2023), Nagori et al. (2023a)
7.	*Clerodendrum inerme*	Lamiaceae	Glorybower	AuNPs, AgNps	UV-Vis, VSM, SEM, FTIR	12–40	Cubic	Khan et al. (2020)
8.	*Brassica oleracea*	Brassicaceae	Cabbage	AgNPs	UV–Vis, EDX, FTIR	52 nm	Flower like	Manojkumar et al. (2023)
9.	*Ananas comosus*	Bromeliaceae	Pineapple	Cu NPs	UV-Vis, XRD, TEM, FTIR	25–85	Spherical	Sajeev et al. (2025)
10.	*Coriandrum sativum*	Apiaceae	Coriander	CeO_2 NPs	UV-Vis, SEM, FTIR, XRD	30–90	Spherical	Sathiyapriya et al. (2020)
11.	*Cassia alata*	Fabaceae	Ringworm Plant	ZnO NPs	UV-Vis, SEM, FTIR, XRD	18–65	Spherical	Govindarasu et al. (2024)
12.	*Solanum nigrum*	Solanaceae	Black nightshade	CuONPs	UV-Vis, SEM, XRD, FTIR	20–80	Spherical	Venkata et al. (2019)

6.7 Challenges in Developing Plant Bioactive-Based Nanomaterials

There are numerous challenges in the development and optimization of nanostructures, the primary challenge being the proper selection of optimized method for obtaining nanomaterial and secondary, identifying and recognizing the ideal nanomaterial type for a specific bioactive substance under consideration and interest. Additionally, toxicological effects characterization is pursued by distinct regulation and control for safety in environment and human consumption, like the utilization of green synthesis methods. Nanomaterials have been broadly utilized for targeted delivery approach. Nanomaterials are also employed to incorporate proteins and genes into plant tissue and cells (Ansari 2023).

6.7.1 Scalability and Industrial Production

Numerous conventional chemical reaction methods are strenuous to scale up because of the requirement for accurate control over conditions of chemical reactions and the dependency on toxic chemicals. Apart from these, biogenic process are commonly scalable due to the fact that they utilize non-toxic natural materials which permits for effortless scaling of production without demanding strict and precise confinement of hazardous and unsafe by products (Fahim et al. 2024). Researchers are constantly concentrating on the green synthesis of nanoparticles, using an eco friendly and favourable method. Owing to their less cost, nontoxicity, easy availability, and environment favourable nature, substantial research and studies has been performed on plant extract based nanoparticles (NPs) formation and their prospective production in various industries (Hano and Abbasi 2022).

Plant bioactive phytoconstituents encapsulated using nanotechnology techniques has several obvious benefits from processing of food materials to packaging, like enhancement in bioavailability and stability. The future prospective for the nanoencapsulated plant based bioactive compounds emphasize the demand for their global recognition and regulation to ameliorate their safe utilization and marketing (Bazana et al. 2019).

6.7.2 Stability and Shelf-Life of Nanomaterials

Establishing the stability of pharmaceutical nanoparticles is critical in regard to efficacy and safety. These tests tends to suggest perfect storage conditions and to determine the effect of outcomes that falls out of the specified criteria or variations in the process of manufacturing. This will consequently influence the quality, efficacy, or safety of pharmaceutical ingredients in the short, medium and long term

(Santamaría-Aguirre et al. 2018). The shelf life of pharmaceutical products can be determined using various types of stability test. During the initial stages of testing, utmost conditions of humidity and temperature are examined, in order to sort out the degradation products that can be detected in long term storage conditions as well as to designate the utmost promising drug candidates (Phan and Haes 2019). Other studies are also carried out a posteriori under less severe conditions to predict the shelf life of products. In addition to these tests, real-time stability tests are performed over long periods (months to years) (Korshed et al. 2018). These stability testing procedures are employed to forecast shelf life of drug products or for relating with alternative drug formulations. The most commonly employed method is Arrhenius method, which ascertains and determines the deviations of the kinetic constant on behalf of function of temperature; its implementation needs awareness of reactions kinetic order.

6.7.3 Toxicity and Biocompatibility Concerns

Regarding biomedical applications, these nanoparticles (NPs) penetrates in the body and comes in direct contact with cells and tissues; therefore it is essential for investigating their biocompatibility.

Hemocompatibility *Nanoparticles (*NPs) are employed as vectors for the implementation in gene delivery, as biosensors and drug delivery where it comes in direct contact and viscinity with blood. Therefore compatibility behaviour of few nanoparticles with blood are investgated. In order to investigate the in-vitro compatibility of blood with the prepared PHEMA nanoparticles (NPs), hemolysis assay is performed. These assay studies were conducted on the prepared nanoparticles surface. The outcomes suggested that % hemolysis is smallest for nanoparticles (NPs) having 12.37 mM HEMA and 1.06 mM EGDMA. This distinctly indicates a moderate range of biocompatibility (Li et al. 2012).

Toxicity of Nanoparticles Mechanisms of interaction between living systems and nanoparticles (NPs) have not been yet fully recognized. The foremost complexity arises concerning with the nanoparticles (NPs) capability to interact and bind with biological matter and alter the characteristics of their surface, relying on the environment they are present. Scientific information regarding nanoparticle (NPs) and cell interaction, mode of action has been cumulating in recent few years, establishing the fact that cells willingly up take nanoparticles (NPs) through either passive or active mechanisms. However, Intracellular pathway and mechanisms are more strenuous to understand (Witika et al. 2020).

6.8 Conclusion

The development of plant bioactive-based nanomaterials provides a green and sustainable substitute for traditional synthesis techniques. Phytochemicals namely flavonoids, polyphenols, alkaloids, and terpenoids play crucial roles in reducing, capping, and stabilizing nanoparticles. These green-synthesized nanomaterials exhibit excellent biocompatibility, targeted delivery, and multifunctional therapeutic potential. Their applications span drug delivery, antimicrobial and anticancer therapies, environmental remediation, and agriculture. Techniques like sol-gel, hydrothermal, and biological reduction have advanced the precision and scalability of nanoparticle synthesis. However, challenges namely stability, toxicity, and regulatory compliance must need attention. Standardization and thorough safety evaluations are essential for clinical and industrial translation. Overall, plant-based nanotechnology holds immense promise for next-generation biomedical and environmental solutions.

References

Abbasi BH, Anjum S, Hano C (2017) Differential effects of in vitro cultures of Linum usitatissimum L. (Flax) on biosynthesis, stability, antibacterial and antileishmanial activities of zinc oxide nanoparticles: a mechanistic approach. RSC Adv 7:15931–15943. https://doi.org/10.1039/C7RA02070H

Abdelkader H (2019) Green synthesis, characterization and antimicrobial activity of biosynthesized silver nanoparticles using Ziziphusspina-Christi leaf extracts. Adv Microbiol Res 3:1–7. https://doi.org/10.24966/amr-694x/100010

Abdellatif AAH, Alawadh SH, Bouazzaoui A, Alhowail AH, Mohammed HA (2021) Anthocyanins rich pomegranate cream as a topical formulation with anti-aging activity. J Dermatolog Treat 32:983–990. https://doi.org/10.1080/09546634.2020.1721418

Abuzeid HM, Julien CM, Zhu L, Hashem AM (2023) Green synthesis of nanoparticles and their energy storage, environmental, and biomedical applications. Crystals 13:1576. https://doi.org/10.3390/CRYST13111576

Aftab R, Akbar F, Afroz A, Asif A, Khan MR, Rehman N, Zeeshan N (2024) Mentha piperita silver nanoparticle-loaded hydrocolloid film for enhanced diabetic wound healing in rats. J Wound Care 33:XLVIII–LX. https://doi.org/10.12968/jowc.2024.33.Sup3a.xlviii

Ahmad N, Ali S, Abbas M, Fazal H, Saqib S, Ali A, Ullah Z, Zaman S, Sawati L, Zada A, Sohail (2023) Antimicrobial efficacy of Mentha piperata-derived biogenic zinc oxide nanoparticles against UTI-resistant pathogens. Sci Rep 13. https://doi.org/10.1038/S41598-023-41502-W

Ahmed S (2015) iMedPub journals synthesis of gold nanoparticles using plant extract: an overview synthesis of nanoparticles using plant extracts synthesis of gold nanoparticles using plant extracts. Nano Res Appl:1–6

Ahmed SS, Alqahtani AM, Alqahtani T, Alamri AH, Menaa F, Mani RK, Bharathi DR, Kavitha K (2022) Green synthesis, characterizations of zinc oxide nanoparticles from aqueous leaf extract of Tridax procumbens Linn. and assessment of their anti-hyperglycemic activity in Streptozoticin-induced diabetic rats. Materials 15:8202. https://doi.org/10.3390/MA15228202

Akhter SMH, Siddiqui VU, Ahmad S, Husain D, Naeem S, Alam MT (2024) Sustainable synthesis of zinc oxide nanoparticles using Terminalia chebula extract: effect of concentration and temperature on properties and antibacterial efficacy. Nano-Struct Nano-Objects 38:101158. https://doi.org/10.1016/J.NANOSO.2024.101158

Aksu Demirezen D, Yıldız YŞ, Yılmaz Ş, Demirezen Yılmaz D (2019) Green synthesis and characterization of iron oxide nanoparticles using Ficus carica (common fig) dried fruit extract. J Biosci Bioeng 127:241–245. https://doi.org/10.1016/J.JBIOSC.2018.07.024

Al-Ghamdi SA, Alkathiri TA, Alfarraj AE, Alatawi OM, Alkathiri AS, Panneerselvam C, Vanaraj S, Darwish AAA, Hamdalla TA, Pasha A, Khasim S (2022) Green synthesis and characterization of zinc oxide nanoparticles using Camellia sinensis tea leaf extract and their antioxidant, anti-bactericidal and anticancer efficacy. Res Chem Intermed 48:4769–4783. https://doi.org/10.1007/S11164-022-04845-Z

Alharbi NS, Alsubhi NS (2022) Green synthesis and anticancer activity of silver nanoparticles prepared using fruit extract of Azadirachta indica. J Radiat Res Appl Sci 15:335–345. https://doi.org/10.1016/j.jrras.2022.08.009

Alhomaidi E, Jasim SA, Amin HIM, Lima Nobre MA, Khatami M, Jalil AT, Hussain Dilfy S (2022) Biosynthesis of silver nanoparticles using Lawsonia inermis and their biomedical application. IET Nanobiotechnol 16:284–294. https://doi.org/10.1049/nbt2.12096

Ali M, Benfante V, Di Raimondo D, Salvaggio G, Tuttolomondo A, Comelli A (2024) Recent developments in nanoparticle formulations for resveratrol encapsulation as an anticancer agent. Pharmaceuticals 17:126. https://doi.org/10.3390/PH17010126

Aliofkhazraei M (2015) Handbook of nanoparticles. Springer. https://doi.org/10.1007/978-3-319-15338-4

Al-Radadi NS (2021) Facile one-step green synthesis of gold nanoparticles (AuNp) using licorice root extract: antimicrobial and anticancer study against HepG2 cell line. Arab J Chem 14:102956. https://doi.org/10.1016/J.ARABJC.2020.102956

Al-Radadi NS (2024) Physiochemical characterization of Glycyrrhiza mediated bioengineered palladium nanoparticles (PdNPs), their anti-proliferative and other biomedical applications. J King Saud Univ Sci 36:103256. https://doi.org/10.1016/J.JKSUS.2024.103256

Álvarez-Chimal R, García-Pérez VI, Álvarez-Pérez MA, Arenas-Alatorre JÁ (2021) Green synthesis of ZnO nanoparticles using a Dysphania ambrosioides extract. Structural characterization and antibacterial properties. Mater Sci Eng C Mater Biol Appl 118:111540. https://doi.org/10.1016/J.MSEC.2020.111540

Amin F, Fozia, Khattak B, Alotaibi A, Qasim M, Ahmad I, Ullah R, Bourhia M, Gul A, Zahoor S, Ahmad R (2021) Green synthesis of copper oxide nanoparticles using Aerva javanica leaf extract and their characterization and investigation of in vitro antimicrobial potential and cytotoxic activities. Evid Based Complement Alternat Med 2021:5589703. https://doi.org/10.1155/2021/5589703

Anand U, Carpena M, Kowalska-Góralska M, Garcia-Perez P, Sunita K, Bontempi E, Dey A, Prieto MA, Proćków J, Simal-Gandara J (2022) Safer plant-based nanoparticles for combating antibiotic resistance in bacteria: a comprehensive review on its potential applications, recent advances, and future perspective. Sci Total Environ 821:153472. https://doi.org/10.1016/J.SCITOTENV.2022.153472

Anchan S, Pai S, Sridevi H, Varadavenkatesan T, Vinayagam R, Selvaraj R (2019) Biogenic synthesis of ferric oxide nanoparticles using the leaf extract of Peltophorum pterocarpum and their catalytic dye degradation potential. Biocatal Agric Biotechnol 20:101251. https://doi.org/10.1016/J.BCAB.2019.101251

Andrade-Zavaleta K, Chacon-Laiza Y, Asmat-Campos D, Raquel-Checca N (2022) Green synthesis of superparamagnetic iron oxide nanoparticles with Eucalyptus globulus extract and their application in the removal of heavy metals from agricultural soil. Molecules 27:1367. https://doi.org/10.3390/MOLECULES27041367

Anh Nga NT, Sathiyavimal S, A. Al-Humaid L, Al- Dahmash ND, Lee J, Barathi S, Jhanani GK (2023) Deciphering the anticancer, anti-inflammatory and antioxidant potential of Ti nanoparticles fabricated using Zingiber officinale. Environ Res 236:116748. https://doi.org/10.1016/J.ENVRES.2023.116748

Anju TR, Parvathy S, Valiya Veettil M, Rosemary J, Ansalna TH, Shahzabanu MM, Devika S (2020) Green synthesis of silver nanoparticles from Aloe vera leaf extract and its antimicrobial activity. Mater Today Proc 43:3956–3960. https://doi.org/10.1016/j.matpr.2021.02.665

Ansari MA (2023) Nanotechnology in food and plant science: challenges and future prospects. Plants 12. https://doi.org/10.3390/plants12132565

Ansari MA, Murali M, Prasad D, Alzohairy MA, Almatroudi A, Alomary MN, Udayashankar AC, Singh SB, Asiri SMM, Ashwini BS, Gowtham HG, Kalegowda N, Amruthesh KN, Lakshmeesha TR, Niranjana SR (2020) Cinnamomum verum bark extract mediated green synthesis of ZnO nanoparticles and their antibacterial potentiality. Biomolecules 10:336. https://doi.org/10.3390/BIOM10020336

Aryan R, Mehata MS (2021) Green synthesis of silver nanoparticles using Kalanchoe pinnata leaves (life plant) and their antibacterial and photocatalytic activities. Chem Phys Lett 778:138760. https://doi.org/10.1016/J.CPLETT.2021.138760

Asimuddin M, Shaik MR, Adil SF, Siddiqui MRH, Alwarthan A, Jamil K, Khan M (2020) Azadirachta indica based biosynthesis of silver nanoparticles and evaluation of their antibacterial and cytotoxic effects. J King Saud Univ Sci 32:648–656. https://doi.org/10.1016/J.JKSUS.2018.09.014

Balážová Ľ, Bedlovičová Z, Daneu N, Baláž M, Lukáčová Bujňáková Z, Tkáčiková Ľ, Takáčová M, Jacková A (2024) Silver nanoparticles produced in vitro by Berberis vulgaris fruit and their antioxidant, antimicrobial and Ex Ovo irritation potential study. BioNanoScience 14:867–879. https://doi.org/10.1007/S12668-024-01400-5/FIGURES/4

Basu S, Tiwari A, Sharma M, Birla H, Dutta D, Singh M, Jindal D, Pratap Singh S (2022) Recent advancement on phytochemical and medicinal properties of Tinospora cordifolia: an Indian medicinal plant recent advancement on phytochemical and medicinal properties of Tinospora cordifolia: an Indian medicinal plant. 20:3753–3778. https://doi.org/10.14704/NQ.2022.20.12.NQ773702

Bazana MT, Codevilla CF, de Menezes CR (2019) Nanoencapsulation of bioactive compounds: challenges and perspectives. Curr Opin Food Sci 26:47–56. https://doi.org/10.1016/j.cofs.2019.03.005

Boke E, Saygi KO, Erenler R, Kacmaz B, Ergene A (2024) Nigella sativa seed-mediated green biosynthesis of silver nanoparticles and antimicrobial activity. Rev Mex Fis 70:1–7. https://doi.org/10.31349/REVMEXFIS.70.051001

Bozzuto G, Calcabrini A, Colone M, Condello M, Dupuis ML, Pellegrini E, Stringaro A (2024) Phytocompounds and Nanoformulations for anticancer therapy: a review. Molecules 29:3784. https://doi.org/10.3390/MOLECULES29163784

Chau TP, Veeraragavan GR, Narayanan M, Chinnathambi A, Alharbi SA, Subramani B, Brindhadevi K, Pimpimon T, Pikulkaew S (2022) Green synthesis of zirconium nanoparticles using Punica granatum (pomegranate) peel extract and their antimicrobial and antioxidant potency. Environ Res 209:112771. https://doi.org/10.1016/J.ENVRES.2022.112771

Chaudhary S, Kanwar RK, Sehgal A, Cahill DM, Barrow CJ, Sehgal R, Kanwar JR (2017) Progress on Azadirachta indica based biopesticides in replacing synthetic toxic pesticides. Front Plant Sci 8:610. https://doi.org/10.3389/FPLS.2017.00610

Dauthal P, Mukhopadhyay M (2016) Noble metal nanoparticles: plant-mediated synthesis, mechanistic aspects of synthesis, and applications. Ind Eng Chem Res 55:1–10

Deka K, Nongbet RD, Das K, Saikia P, Kaur S, Talukder A, Thakuria B (2025) Understanding the mechanism underlying the green synthesis of metallic nanoparticles using plant extract (s) with special reference to silver, gold, copper and zinc oxide nanoparticles. Hybrid Advances 9:100399. https://doi.org/10.1016/j.hybadv.2025.100399

Dewangan K, Nagori K, Sharma M, Singh A, Chandrakar S, Sahu VD, Sharma G, Gupta S, Solanki H, Majumdar M, Tripathi DK, Alexander A, Ajazuddin (2016) Formulation, evaluation and release studies of indomethacin suppositories. Res J Pharm Technol 9:942–944. https://doi.org/10.5958/0974-360X.2016.00181.5

Dhaka A, Chand Mali S, Sharma S, Trivedi R (2023) A review on biological synthesis of silver nanoparticles and their potential applications. Results Chem 6:101108. https://doi.org/10.1016/J.RECHEM.2023.101108

Dhyani P, Quispe C, Sharma E, Bahukhandi A, Sati P, Attri DC, Szopa A, Sharifi-Rad J, Docea AO, Mardare I, Calina D, Cho WC (2022) Anticancer potential of alkaloids: a key emphasis to colchicine, vinblastine, vincristine, vindesine, vinorelbine and vincamine. Cancer Cell Int 22(1):1–20. https://doi.org/10.1186/S12935-022-02624-9

Dikshit PK, Kumar J, Das AK, Sadhu S, Sharma S, Singh S, Gupta PK, Kim BS (2021) Green synthesis of metallic nanoparticles: applications and limitations. Catalysts 11:902. https://doi.org/10.3390/CATAL11080902

Dutt Y, Pandey RP, Dutt M, Gupta A, Vibhuti A, Raj VS, Chang CM, Priyadarshini A (2023) Silver nanoparticles Phytofabricated through Azadirachta indica: anticancer, apoptotic, and wound-healing properties. Antibiotics 12:121. https://doi.org/10.3390/ANTIBIOTICS12010121

Edo GI, Mafe AN, Ali ABM, Akpoghelie PO, Yousif E, Isoje EF, Igbuku UA, Zainulabdeen K, Owheruo JO, Essaghah AEA, Umar H, Ahmed DS, Alamiery AA (2025) Eco-friendly nanoparticle phytosynthesis via plant extracts: mechanistic insights, recent advances, and multifaceted uses. Nano TransMed 4:100080. https://doi.org/10.1016/J.NTM.2025.100080

Elbrolesy A, Abdou Y, Elhussiny FA, Morsy R (2023) Novel green synthesis of UV-sunscreen ZnO nanoparticles using Solanum Lycopersicum fruit extract and evaluation of their antibacterial and anticancer activity. J Inorg Organomet Polym Mater 33:3750–3759. https://doi.org/10.1007/S10904-023-02744-3

Elias A, Habbu PV, Iliger S (2024) Biosynthesis of gold nanoparticles from Amaranthus gangeticus s. and its anti-diabetic activity on Streptozotocin induced rats. Proceedings 103:86. https://doi.org/10.3390/PROCEEDINGS2024103086

Emima Jeronsia J, Ragu R, Sowmya R, Mary AJ, Jerome Das S (2020) Comparative investigation on Camellia Sinensis mediated green synthesis of Ag and Ag/GO nanocomposites for its anticancer and antibacterial efficacy. Surf Interfaces 21:100787. https://doi.org/10.1016/j.surfin.2020.100787

Fadhil Ascar I, Mazika Ahemd F, Ascar IF, Al-A SB, Alshanon AF (2019) Cytotoxicity and antioxidant effect of ginger gold nanoparticles on thyroid carcinoma cells. J Pharm Sci Res 11(3):1044

Fahim M, Shahzaib A, Nishat N, Jahan A, Bhat TA, Inam A (2024) Green synthesis of silver nanoparticles: a comprehensive review of methods, influencing factors, and applications. JCIS Open 16:100125. https://doi.org/10.1016/j.jciso.2024.100125

Fathallah TZ (2023) Antioxidant activity of silver nanoparticles prepared from capsicum Annuum (Bell Pepper) Extract. E3S Web Conf 391. https://doi.org/10.1051/E3SCONF/202339101126

Fawcett D, Verduin JJ, Shah M, Sharma SB, Poinern GEJ (2017) A review of current research into the biogenic synthesis of metal and metal oxide nanoparticles via marine algae and seagrasses. J Nanosci 2017:8013850. https://doi.org/10.1155/2017/8013850

Feng SH, Li GH (2017) Hydrothermal and solvothermal syntheses. In: Modern inorganic synthetic chemistry: Second Edition. Elsevier. https://doi.org/10.1016/B978-0-444-63591-4.00004-5

Fuloria S, Mehta J, Chandel A, Sekar M, Rani NNIM, Begum MY, Subramaniyan V, Chidambaram K, Thangavelu L, Nordin R, Wu YS, Sathasivam KV, Lum PT, Meenakshi DU, Kumarasamy V, Azad AK, Fuloria NK (2022) A comprehensive review on the therapeutic potential of *Curcuma longa* Linn in relation to its major active constituent curcumin. Front Pharmacol 13:820806. https://doi.org/10.3389/FPHAR.2022.820806

Garg D, Sridhar K, Inbaraj S, Chawla B, Tripathi P, Sharma M, Chew KW, Singhania RR, Garg D, Sridhar K, Inbaraj BS, Chawla P, Tripathi M, Sharma M (2023) Nano-biofertilizer formulations for agriculture: a systematic review on recent advances and prospective applications. Bioengineering 10:1010. https://doi.org/10.3390/BIOENGINEERING10091010

George N, Joy J, Mathew B, Koshy EP (2024) Green mediated synthesis of silver nanoparticle using Euphorbia Maculate leaf extract and their catalytic reduction and antibacterial properties. Mater Today Proc. https://doi.org/10.1016/J.MATPR.2024.01.013

Gera M, Sharma N, Ghosh M, Huynh DL, Lee SJ, Min T, Kwon T, Jeong DK (2017) Nanoformulations of curcumin: an emerging paradigm for improved remedial application. Oncotarget 8:66680. https://doi.org/10.18632/ONCOTARGET.19164

Ghazal H, Waqar A, Yaseen F, Shahid M, Sultana M, Tariq M, Bashir MK, Tahseen H, Raza T, Ahmad F (2024) Role of nanoparticles in enhancing chemotherapy efficacy for cancer treatment. Next Materials 2:100128. https://doi.org/10.1016/J.NXMATE.2024.100128

Govindarasu M, Vaiyapuri M, Kim JC (2024) Protective effect of zinc oxide nanoparticles synthesized using Cassia alata for DSS-induced ulcerative colitis in mice model. Bioprocess Biosyst Eng 47:1393–1407. https://doi.org/10.1007/S00449-024-03047-8

Gupta A, Jain P, Nagori K, Adnan M, Ajazuddin (2024) Treatment strategies for psoriasis using flavonoids from traditional Chinese medicine. Pharmacol Res - Modern Chinese Med 12:100463. https://doi.org/10.1016/j.prmcm.2024.100463

Hano C, Abbasi BH (2022) Plant-based green synthesis of nanoparticles: production, characterization and applications. Biomolecules 12:1–9. https://doi.org/10.3390/biom12010031

Hazafa A, Jahan N, Zia MA, Rahman KU, Sagheer M, Naeem M (2022) Evaluation and optimization of nanosuspensions of Chrysanthemum coronarium and Azadirachta indica using response surface methodology for pest management. Chemosphere 292:133411. https://doi.org/10.1016/J.CHEMOSPHERE.2021.133411

Hussein BY, Mohammed AM (2021) Green synthesis of ZnO nanoparticles in grape extract: their application as anti-cancer and anti-bacterial. Mater Today Proc 42:A18–A26. https://doi.org/10.1016/J.MATPR.2021.03.729

Islam MJ, Khatun N, Bhuiyan RH, Sultana S, Ali Shaikh MA, Amin Bitu MN, Chowdhury F, Islam S (2023) Psidium guajava leaf extract mediated green synthesis of silver nanoparticles and its application in antibacterial coatings. RSC Adv 13:19164–19172. https://doi.org/10.1039/D3RA03381C

Jayaprakash N, Vijaya JJ, Kaviyarasu K, Kombaiah K, Kennedy LJ, Ramalingam RJ, Munusamy MA, Al-Lohedan HA (2017) Green synthesis of Ag nanoparticles using tamarind fruit extract for the antibacterial studies. J Photochem Photobiol B 169:178–185. https://doi.org/10.1016/j.jphotobiol.2017.03.013

Jiang H, Sathiyavimal S, Cai L, Devanesan S, Sayed SRM, Jhanani GK, Lin J (2023) Tulsi (Ocimum sanctum) mediated Co nanoparticles with their anti-inflammatory, anti-cancer, and methyl orange dye adsorption properties. Environ Res 236:116749. https://doi.org/10.1016/J.ENVRES.2023.116749

Jin Y, Li B, Saravanakumar K, Hu X, Mariadoss AVA, Wang MH (2022) Cytotoxic and antibacterial activities of starch encapsulated photo-catalyzed phytogenic silver nanoparticles from Paeonia lactiflora flowers. J Nanostruct Chem 12:375–387. https://doi.org/10.1007/S40097-021-00421-W

Kanipandian N, Thirumurugan R (2014) A feasible approach to phyto-mediated synthesis of silver nanoparticles using industrial crop Gossypium hirsutum (cotton) extract as stabilizing agent and assessment of its in vitro biomedical potential. Ind Crop Prod 55:1–10. https://doi.org/10.1016/J.INDCROP.2014.01.042

Karnwal A, Jassim AY, Mohammed AA, Sharma V, Al-Tawaha ARMS, Sivanesan I (2024) Nanotechnology for healthcare: plant-derived nanoparticles in disease treatment and regenerative medicine. Pharmaceuticals 17:1711. https://doi.org/10.3390/PH17121711

Karuvantevida N, Razia M, Bhuvaneshwar R, Sathishkumar G, Prabukumar S, Sivaramakrishnan S (2022) Bioactive flavonoid used as a stabilizing agent of mono and bimetallic nanomaterials for multifunctional activities. J Pure Appl Microbiol 16:1652–1662. https://doi.org/10.22207/JPAM.16.3.03

Keskin C, Ölçekçi A, Baran A, Baran MF, Eftekhari A, Omarova S, Khalilov R, Aliyev E, Sufianov A, Beilerli A, Gareev I (2023) Green synthesis of silver nanoparticles mediated Diospyros kaki L. (Persimmon): determination of chemical composition and evaluation of their antimicrobials and anticancer activities. Front Chem 11. https://doi.org/10.3389/FCHEM.2023.1187808

Khairul Hanif Mohd Nazri M, Sapawe N (2020) A short review on plants extract mediated synthesis of copper oxide nanoparticles. Mater Today Proc 31:A38–A41. https://doi.org/10.1016/j.matpr.2020.10.966

Khan K, Javed S (2021) Silver nanoparticles synthesized using leaf extract of Azadirachta indica exhibit enhanced antimicrobial efficacy than the chemically synthesized nanoparticles: a comparative study. Sci Prog 104. https://doi.org/10.1177/00368504211012159

Khan MZH, Tareq FK, Hossen MA, Roki MNAM (2018) Green synthesis and characterization of silver nanoparticles using Coriandrum sativum leaf extract. J Eng Sci Technol 13(1):158–166

Khan SA, Shahid S, Lee CS (2020) Green synthesis of gold and silver nanoparticles using leaf extract of Clerodendrum inerme; characterization, antimicrobial, and antioxidant activities. Biomolecules 10:835. https://doi.org/10.3390/BIOM10060835

Khan M, Ware P, Shimpi N (2021) Synthesis of ZnO nanoparticles using peels of Passiflora foetida and study of its activity as an efficient catalyst for the degradation of hazardous organic dye. SN Appl Sci 3. https://doi.org/10.1007/S42452-021-04436-4

Khanal LN, Sharma KR, Paudyal H, Parajuli K, Dahal B, Ganga GC, Pokharel YR, Kalauni SK (2022) Green synthesis of silver nanoparticles from root extracts of Rubus ellipticus Sm and comparison of antioxidant and antibacterial activity. J Nanomater 2022:1832587. https://doi.org/10.1155/2022/1832587

Khatun R, Al Mamun MS, Islam S, Khatun N, Hakim M, Hossain MS, Dhar PK, Barai HR (2022) Phytochemical-assisted synthesis of Fe3O4 nanoparticles and evaluation of their catalytic activity. Micromachines (Basel) 13:1–10. https://doi.org/10.3390/mi13122077

Koduru JR, Kailasa SK, Bhamore JR, Kim KH, Dutta T, Vellingiri K (2018) Phytochemical-assisted synthetic approaches for silver nanoparticles antimicrobial applications: a review. Adv Colloid Interf Sci 256:326–339. https://doi.org/10.1016/j.cis.2018.03.001

Korshed P, Li L, Ngo DT, Wang T (2018) Effect of storage conditions on the long-term stability of bactericidal effects for laser generated silver nanoparticles. Nanomaterials 8. https://doi.org/10.3390/nano8040218

Kumar V, Yadav SK (2012) Characterisation of gold nanoparticles synthesised by leaf and seed extract of Syzygium cumini L. J Exp Nanosci 7:440–451. https://doi.org/10.1080/17458080.2010.543989

Kumari SA, Patlolla AK, Madhusudhanachary P (2022) Biosynthesis of silver nanoparticles using Azadirachta indica and their antioxidant and anticancer effects in cell lines. Micromachines (Basel) 13:1416. https://doi.org/10.3390/MI13091416

Kumari S, Raturi S, Kulshrestha S, Chauhan K, Dhingra S, András K, Thu K, Khargotra R, Singh T (2023) A comprehensive review on various techniques used for synthesizing nanoparticles. J Mater Res Technol 27:1739–1763. https://doi.org/10.1016/j.jmrt.2023.09.291

Kyene MO, Droepenu EK, Ayertey F, Yeboah GN, Archer MA, Kumadoh D, Mintah SO, Gordon PK, Appiah AA (2023) Synthesis and characterization of ZnO nanomaterial from Cassia sieberiana and determination of its anti-inflammatory, antioxidant and antimicrobial activities. Sci Afr 19:e01452. https://doi.org/10.1016/j.sciaf.2022.e01452

Li X, Wang L, Fan Y, Feng Q, Cui FZ (2012) Biocompatibility and toxicity of nanoparticles and nanotubes. J Nanomater 2012. https://doi.org/10.1155/2012/548389

Mani ST, Jayakumar P, Pavithra ME, Saranya K, Rathinavel T, Ammashi S (2024) Green synthesis and characterization of silver nanoparticles from Eclipta alba and its activity against triple-negative breast cancer cell line (MDA-MB-231). Mol Biotechnol 66:3597–3607. https://doi.org/10.1007/s12033-023-00959-w

Manikandan DB, Sridhar A, Krishnasamy Sekar R, Perumalsamy B, Veeran S, Arumugam M, Karuppaiah P, Ramasamy T (2021) Green fabrication, characterization of silver nanoparticles using aqueous leaf extract of Ocimum americanum (Hoary Basil) and investigation of its in vitro antibacterial, antioxidant, anticancer and photocatalytic reduction. J Environ Chem Eng 9:104845. https://doi.org/10.1016/j.jece.2020.104845

Manojkumar U, Kaliannan D, Srinivasan V, Balasubramanian B, Kamyab H, Mussa ZH, Palaniyappan J, Mesbah M, Chelliapan S, Palaninaicker S (2023) Green synthesis of zinc oxide nanoparticles using Brassica oleracea var. botrytis leaf extract: photocatalytic, antimicrobial and larvicidal activity. Chemosphere 323:138263

Manzari-Tavakoli A, Babajani A, Tavakoli MM, Safaeinejad F, Jafari A (2024) Integrating natural compounds and nanoparticle-based drug delivery systems: a novel strategy for enhanced efficacy and selectivity in cancer therapy. Cancer Med 13:e7010. https://doi.org/10.1002/CAM4.7010

Mashamaite CV, Ngcobo BL, Manyevere A, Bertling I, Fawole OA (2022) Assessing the usefulness of Moringa oleifera leaf extract as a biostimulant to supplement synthetic fertilizers: a review. Plants 11:2214. https://doi.org/10.3390/PLANTS11172214

Meher K, Paithankar H, Hosur RV, Lopus M (2022) Ashwagandha-polyphenols-functionalized gold nanoparticles facilitate apoptosis by perturbing microtubule assembly dynamics in breast cancer cells. J Drug Deliv Sci Technol 70:103225. https://doi.org/10.1016/J.JDDST.2022.103225

Mikhailova EO (2023) Selenium nanoparticles: green synthesis and biomedical application. Molecules 28:8125. https://doi.org/10.3390/MOLECULES28248125

Mirza S, Hussaini AA, Öztürk G, Turgut M, Öztürk T, Tugay O, Ulukuş D, Yıldırım M (2023) Photocatalytic and antibacterial activities of ZnO nanoparticles synthesized from Lupinus albus and Lupinus pilosus plant extracts via green synthesis approach. Inorg Chem Commun 155:111124. https://doi.org/10.1016/J.INOCHE.2023.111124

Mohammed GM, Hawar SN (2022) Green biosynthesis of silver nanoparticles from Moringa oleifera leaves and its antimicrobial and cytotoxicity activities. Int J Biomater 2022:4136641. https://doi.org/10.1155/2022/4136641

Moodley JS, Krishna SBN, Pillay K, Sershen, Govender P (2018) Green synthesis of silver nanoparticles from Moringa oleifera leaf extracts and its antimicrobial potential. Adv Nat Sci Nanosci Nanotechnol 9:015011. https://doi.org/10.1088/2043-6254/AAABB2

Nadaf SJ, Jadhav NR, Naikwadi HS, Savekar PL, Sapkal ID, Kambli MM, Desai IA (2022) Green synthesis of gold and silver nanoparticles: updates on research, patents, and future prospects. OpenNano 8:100076. https://doi.org/10.1016/J.ONANO.2022.100076

Nagaraj B, Malakar B, Divya TK, Krishnamurthy NB, Liny P, Dinesh R, Iconaru SL, Ciobanu CS (2012) Synthesis of plant mediated gold nanoparticles using flower extracts of Carthamus tinctorius L. (safflower) and evaluation of their biological activities. Dig J Nanomater Biostruct 7:1289–1296

Nagori K, Sharma M, Agrawal A, Agarwal AK, Sharma A, Verma H, Tripathi DK (2011) On allopathic, ayurvedic and homeopathic medicine in Chhattisgarh, India. Int J Pharm Pharm Sci 3:159–162

Nagori K, Nakhate KT, Yadav K, Ajazuddin, Pradhan M (2023a) Unlocking the therapeutic potential of medicinal plants for Alzheimer's disease: preclinical to clinical trial insights. Future Pharmacol 3:877–907

Nagori K, Pradhan M, Nakhate KT, Badwaik HR, Deshmukh R, Roy A, Sharma R, Srivastava SP, Chawla S, Jain V, Sharma M (2023b) In silico molecular docking analysis of some Terpenoids against 3CLpro of SARS-CoV-2. Res J Pharm Technol 16:4791–4798. https://doi.org/10.52711/0974-360X.2023.00777

Nagori K, Pradhan M, Nakhate KT (2024a) Ethyl gallate ameliorates diabetes-induced Alzheimer's disease-like phenotype in rats via activation of α7 nicotinic receptors and mitigation of oxidative stress. Biochem Biophys Res Commun 737. https://doi.org/10.1016/j.bbrc.2024.150925

Nagori K, Pradhan M, Nakhate KT, Thakur A, Patle K, Thakur A (2024b) Dietary supplements and phytopharmaceuticals for the management of cognitive disorders. In: Medicinal applications of phytopharmaceuticals. Springer Nature Switzerland, pp 135–158. https://doi.org/10.1007/978-3-031-63677-6_9

Nagori K, Pradhan M, Sharma M, Ajazuddin, Badwaik HR, Nakhate KT (2024c) Current progress on central cholinergic receptors as therapeutic targets for Alzheimer's disease. Curr Alzheimer Res. https://doi.org/10.2174/0115672050306008240321034006

Nagori K, Sharma R, Shukla A, Nakhate KT, Dewangan D (2024d) Bioink as integral material for 3d printing. In: 3D printing and microfluidics in dermatology: innovations in drug delivery, pp 1–29. https://doi.org/10.1201/9781032690926-2

Nagori K, Nakhate KT, Yadav K, Thakur A, Ajazuddin, Pradhan M (2025) Innovative strategies for overcoming blood-brain barrier challenges in Alzheimer's disease: a focus on green-synthesized metallic nanoparticles. Inorg Chem Commun 171:113604. https://doi.org/10.1016/j.inoche.2024.113604

Naikoo GA, Mustaqeem M, Hassan IU, Awan T, Arshad F, Salim H, Qurashi A (2021) Bioinspired and green synthesis of nanoparticles from plant extracts with antiviral and antimicrobial properties: a critical review. J Saudi Chem Soc 25:101304. https://doi.org/10.1016/J.JSCS.2021.101304

Nair RS, Morris A, Billa N, Leong C-O (2019) An evaluation of curcumin-encapsulated chitosan nanoparticles for transdermal delivery. AAPS PharmSciTech 20:69. https://doi.org/10.1208/S12249-018-1279-6

Nandhini SN, Sisubalan N, Vijayan A, Karthikeyan C, Gnanaraj M, Gideon DAM, Jebastin T, Varaprasad K, Sadiku R (2023) Recent advances in green synthesized nanoparticles for bactericidal and wound healing applications. Heliyon 9. https://doi.org/10.1016/J.HELIYON.2023.E13128

Narayanan M, Hussain F, Ahamed J, Srinivasan B, Sambantham MT, Al-Keridis LA, AL-mekhlafi FA (2022) Green synthesizes and characterization of copper-oxide nanoparticles by Thespesia populnea against skin-infection causing microbes. J King Saud Univ Sci 34:101885. https://doi.org/10.1016/J.JKSUS.2022.101885

Navas D, Fuentes S, Castro-Alvarez A, Chavez-Angel E (2021) Review on sol-gel synthesis of perovskite and oxide nanomaterials. Gels 7. https://doi.org/10.3390/gels7040275

Nayar AM, Kumar PR, Jayachithra SK (2023) Larvicidal and antibacterial efficacy of green synthesized silver nanoparticles using aqueous leaf extract of Cinnamomum Verum. J Chem Health Risks 13:1006–1012. https://doi.org/10.52783/JCHR.V13.I6.1815

Nellore J, Pauline C, Amarnath K (2013) Bacopa monnieri phytochemicals mediated synthesis of platinum nanoparticles and its neurorescue effect on 1-Methyl 4-Phenyl 1,2,3,6 Tetrahydropyridine. J Neurodegener Dis 2013:972391. https://doi.org/10.1155/2013/972391

Nemattalab M, Rohani M, Evazalipour M, Hesari Z (2022) Formulation of cinnamon (Cinnamomum verum) oil loaded solid lipid nanoparticles and evaluation of its antibacterial activity against multi-drug resistant Escherichia coli. BMC Complement Med Ther 22:1–10. https://doi.org/10.1186/S12906-022-03775-Y/FIGURES/5

Okeke ES, Nweze EJ, Anaduaka EG, Okoye CO, Anosike CA, Joshua PE, Ezeorba TPC (2023) Plant-derived nanomaterials (PDNM): a review on pharmacological potentials against pathogenic microbes, antimicrobial resistance (AMR) and some metabolic diseases. 3 Biotech 13(9):1–25. https://doi.org/10.1007/S13205-023-03713-W

Onitsuka S, Hamada T, Okamura H (2019) Preparation of antimicrobial gold and silver nanoparticles from tea leaf extracts. Colloids Surf B Biointerfaces 173:242–248. https://doi.org/10.1016/J.COLSURFB.2018.09.055

Padalia H, Chanda S (2021) Synthesis of silver nanoparticles using Ziziphus nummularia leaf extract and evaluation of their antimicrobial, antioxidant, cytotoxic and genotoxic potential (4-in-1 system). Artif Cells Nanomed Biotechnol 49:354–366. https://doi.org/10.1080/21691401.2021.1903478

Patil N, Bhaskar R, Vyavhare V, Dhadge R, Khaire V, Patil Y (2021) Overview on methods of synthesis of nanoparticles. Int J Curr Pharm Res 13:11–16. https://doi.org/10.22159/ijcpr.2021v13i2.41556

Patra JK, Das G, Fraceto LF, Campos EVR, Rodriguez-Torres MDP, Acosta-Torres LS, Diaz-Torres LA, Grillo R, Swamy MK, Sharma S, Habtemariam S, Shin HS (2018) Nano based drug delivery systems: recent developments and future prospects. J Nanobiotechnol 16:71. https://doi.org/10.1186/S12951-018-0392-8

(PDF) Green synthesis and characterization of metal oxide nanoparticles using Nelumbo Nucifera flower and its antimicrobial activity [WWW Document] (n.d.). URL https://www.researchgate.net/publication/360623933_Green_Synthesis_and_Characterization_of_Metal_Oxide_Nanoparticles_Using_Nelumbo_Nucifera_Flower_and_Its_Antimicrobial_Activity. Accessed 6.30.25

(PDF) Green synthesis of eco-friendly silver nanoparticles using Coffee arabica leaf extract and development of a cost-effective biosensor for cysteine [WWW Document] (n.d.). URL https://www.researchgate.net/publication/363402819_Green_synthesis_of_eco-friendly_silver_nanoparticles_using_Coffee_arabica_leaf_extract_and_development_of_a_cost-effective_biosensor_for_cysteine. Accessed 6.30.25

Phan HT, Haes AJ (2019) What does nanoparticle stability mean? HHS public access. J Phys Chem C Nanomater Interfaces 123:16495–16507. https://doi.org/10.1021/acs.jpcc.9b00913

Prajapati A, Rangra S, Patil R, Desai N, Jyothi VGSS, Salave S, Amate P, Benival D, Kommineni N (2024) Receptor-targeted nanomedicine for cancer therapy. Receptors 3:323–361. https://doi.org/10.3390/RECEPTORS3030016

Princy SSJ, Hentry C, Alodaini HA, Hatamleh AA, Arokiyaraj S, Bindhu MR (2023) Hibiscus cannabinus seeds assisted spherical silver nanoparticles and its antibacterial and photocatalytic applications. Chem Phys Impact 6:100192. https://doi.org/10.1016/J.CHPHI.2023.100192

Rafique S, Bashir S, Akram R, Jawaid S, Bashir M, Aftab A, Attique A, Awan SU (2023) In vitro anticancer activity and comparative green synthesis of ZnO/Ag nanoparticles by moringa oleifera, mentha piperita, and citrus lemon. Ceram Int 49:5613–5620

Rahman F, Patwary MAM, Siddique MAB, Bashar MS, Aminul HM, Akter B, Rashid R, Anamul HM, Uddin AKMR (2022) Green synthesis of ZnO nanoparticles using *Cocos nucifera* leaf extract: characterization, antimicrobial, antioxidant, and photocatalytic activity. R Soc Open Sci 9(11):220858. https://doi.org/10.1101/2022.10.27.514023

Rai M, Ingle AP, Pandit R, Paralikar P, Anasane N, Santos CAD (2020) Curcumin and curcumin-loaded nanoparticles: antipathogenic and antiparasitic activities. Expert Rev Anti-infect Ther 18:367–379. https://doi.org/10.1080/14787210.2020.1730815

Raja S, Ramesh V, Thivaharan V (2017) Green biosynthesis of silver nanoparticles using Calliandra haematocephala leaf extract, their antibacterial activity and hydrogen peroxide sensing capability. Arab J Chem 10:253–261. https://doi.org/10.1016/J.ARABJC.2015.06.023

Rajak KK, Pahilani P, Patel H, Kikani B, Desai R, Kumar H (2023) Green synthesis of silver nanoparticles using Curcuma longa flower extract and antibacterial activity. https://doi.org/10.48550/arXiv.2304.04777

Rajput A, Mandlik S, Dawre S, Mandlik D, Pingale P, Butani S (2022) Herbal bioenhancers in nanoparticulate drug delivery system. In: Drug delivery technology: herbal bioenhancers in pharmaceuticals, pp 45–86. https://doi.org/10.1515/9783110746808-003

Ranjani S, Hemalatha S (2022) Triphala decorated multipotent green nanoparticles and its applications. Mater Lett 308:131184. https://doi.org/10.1016/J.MATLET.2021.131184

Rather GA, Nanda A, Raj E, Mathivanan N, Nayak BK (2022) Green synthesis of ZnO nanoparticles using the leaf extract of Lavandula angustifolia and evaluation of their antibacterial activity against human pathogens. Int J Health Sci (Qassim):13478–13485. https://doi.org/10.53730/ijhs.v6ns2.8539

Ray L, Das M, Tripathy J (2023) Terminalia catappa leaf extract mediated eco-friendly synthesis of cerium oxide nanoparticles. Mater Today Proc. https://doi.org/10.1016/J.MATPR.2023.12.035

Raza MA, Sharma MK, Nagori K, Jain P, Ghosh V, Gupta U, Ajazuddin (2024) Recent trends on polycaprolactone as sustainable polymer-based drug delivery system in the treatment of cancer: biomedical applications and nanomedicine. Int J Pharm 666:124734. https://doi.org/10.1016/j.ijpharm.2024.124734

Reddy PR, Ganesh SD, Saha N, Zandraa O, Sáha P (2016) Ecofriendly synthesis of silver nanoparticles from garden rhubarb (Rheum rhabarbarum). J Nanotechnol 2016:4964752. https://doi.org/10.1155/2016/4964752

Robles-Martínez M, Patiño-Herrera R, Pérez-Vázquez FJ, Montejano-Carrizales JM, González JFC, Pérez E (2020) Mentha piperita as a natural support for silver nanoparticles: a new anti-Candida albicans treatment. Coll Interface Sci Commun 35:100253. https://doi.org/10.1016/j.colcom.2020.100253

Rodríguez-Félix F, López-Cota AG, Moreno-Vásquez MJ, Graciano-Verdugo AZ, Quintero-Reyes IE, Del-Toro-Sánchez CL, Tapia-Hernández JA (2021) Sustainable-green synthesis of silver nanoparticles using safflower (Carthamus tinctorius L.) waste extract and its antibacterial activity. Heliyon 7:e06923. https://doi.org/10.1016/J.HELIYON.2021.E06923

Sahoo S, Panigrahy SK, Singhamahapatra A, Jena M (2020) Green synthesis of metal nanoparticles using Ocimum sanctum (Tulsi). Int J Pharm Biol Sci-IJPBS TM 10. https://doi.org/10.21276/ijpbs.2020.10.2.42

Sahu N, Jain P, Nagori K (2024) Recent advancement and novel treatment strategies for breast Fibroade- noma: clinical approach and prospects. Curr Cancer Ther Rev 20:1–11. https://doi.org/10.2174/0115733947318171240802100419

Sajeev S, Johnson MS, Nambiar A, Nair RS (2025) Green synthesis of copper nanoparticles using Ananas comosus. In: Signals and communication technology part F76, pp 129–133. https://doi.org/10.1007/978-3-031-68952-9_18

Salem MZM, EL-Hefny M, Ali HM, Abdel-Megeed A, El-Settawy AAA, Böhm M, Mansour MMA, Salem AZM (2021) Plants-derived bioactives: novel utilization as antimicrobial, antioxidant and phytoreducing agents for the biosynthesis of metallic nanoparticles. Microb Pathog 158:105107. https://doi.org/10.1016/J.MICPATH.2021.105107

Sana SS, Vadde R, Kumar R, Arla SK, Somala AR, Krishna Rao KSV, Zhijun Z, Boya VKN, Mondal K, Mamidi N (2023) Eco-friendly and facile production of antibacterial zinc oxide nanoparticles from Grewia flavescens (G. flavescens) leaf extract for biomedical applications. J Drug Deliv Sci Technol 80:1–31. https://doi.org/10.1016/j.jddst.2023.104186

Santamaría-Aguirre J, Alcocer-Vallejo R, López-Fanárraga M (2018) Drug nanoparticle stability assessment using isothermal and nonisothermal approaches. J Nanomater 2018. https://doi.org/10.1155/2018/3047178

Santhoshkumar J, Rajeshkumar S, Venkat Kumar S (2017) Phyto-assisted synthesis, characterization and applications of gold nanoparticles—A review. Biochem Biophys Rep 11:46–57. https://doi.org/10.1016/j.bbrep.2017.06.004

Sathiyapriya R, Balaji M, Rajesh S (2020) Bio-synthesis of cerium oxide nanoparticles from Coriandrum sativum L. leaf extract and their antibacterial activity. Int J Adv Sci Eng 06:1439–1444. https://doi.org/10.29294/ijase.6.3.2020.1439-1444

Sathya TA, Viswanathan S, Kolar AB, Jahirhussain G, Alagumanian S, Sobana S, Arumugam N (2024) Environmental profiling of gold nanoparticles by flavonoids fractionalization from carrica papaya leaf extract for photocatalytic debasement of organic contaminants and it's cytotoxic analysis. Environ Res 259:119445. https://doi.org/10.1016/J.ENVRES.2024.119445

Sattar ZA, Mishaal Mohammed A, Khalaf YH (2024) Green synthesis of Curcuma longa-mediated palladium nanoparticles as antioxidant, antimicrobial, and anticancer agents. Results Chem 9:101675. https://doi.org/10.1016/J.RECHEM.2024.101675

Shah IH, Ashraf M, Sabir IA, Manzoor MA, Malik MS, Gulzar S, Ashraf F, Iqbal J, Niu Q, Zhang Y (2022) Green synthesis and characterization of copper oxide nanoparticles using Calotropis procera leaf extract and their different biological potentials. J Mol Struct 1259:132696. https://doi.org/10.1016/J.MOLSTRUC.2022.132696

Shahin N, Jain P, Nagori K (2025) Advancements in natural alkaloid-loaded drug delivery systems for enhanced peptic ulcer treatment: a review. Curr Drug Ther 20:988–1000. https://doi.org/10.2174/0115748855312609240628110440

Shanmugapriya J, Reshma CA, Srinidhi V, Harithpriya K, Ramkumar KM, Umpathy D, Gunasekaran K, Subashini R (2022) Green synthesis of copper nanoparticles using Withania somnifera and its antioxidant and antibacterial activity. J Nanomater 2022:7967294. https://doi.org/10.1155/2022/7967294

Sharma J, Sweta Thakur C, Vats M, Sharma SK (2020) Green synthesis of zinc oxide nanoparticles using neem extract. In: AIP Conf Proc, vol 2220. https://doi.org/10.1063/5.0002093

Shashikala BS, Ashwini KR, Lavanya DR, Jessica Y, Ushasri P, Sunitha DV, Shobha G, Praveen Prakash Dsouza P, Chandra Babu Naidu K (2024) Moringa Oleifera leaf-based zinc oxide nanoparticles: green synthesis, characterization, and their antibacterial investigations. Inorg Chem Commun 170:113355. https://doi.org/10.1016/J.INOCHE.2024.113355

Shen X, Sheng H, Zhang Y, Dong X, Kou L, Yao Q, Zhao X (2024) Nanomedicine-based disulfiram and metal ion co-delivery strategies for cancer treatment. Int J Pharm X 7:100248. https://doi.org/10.1016/J.IJPX.2024.100248

Singh NB (2022) Green synthesis of nanomaterials. In: Handbook of microbial nanotechnology. Elsevier, pp 225–254. https://doi.org/10.1016/B978-0-12-823426-6.00007-3

Singh SP, Mishra A, Shyanti RK, Singh RP, Acharya A (2021) Silver nanoparticles synthesized using Carica papaya leaf extract (AgNPs-PLE) causes cell cycle arrest and apoptosis in human prostate (DU145) cancer cells. Biol Trace Elem Res 199:1316–1331. https://doi.org/10.1007/s12011-020-02255-z

Singh A, Bharti R, Thakur A, Verma M, Sharma R (2024) Leaf extract mediated green synthesis of iron-oxide nanoparticles (FeO -NPs) by using Hibiscus rosa-sinensis (China rose): a potential approach and its biological application. Orbital: Electronic J Chem 16:271–278. https://doi.org/10.17807/ORBITAL.V16I4.21174

Singhal G, Bhavesh R, Kasariya K, Sharma AR, Singh RP (2011) Biosynthesis of silver nanoparticles using Ocimum sanctum (Tulsi) leaf extract and screening its antimicrobial activity. J Nanopart Res 13:2981–2988. https://doi.org/10.1007/S11051-010-0193-Y/METRICS

Sportelli MC, Gaudiuso C, Volpe A, Izzi M, Picca RA, Ancona A, Cioffi N (2022) Biogenic synthesis of ZnO nanoparticles and their application as bioactive agents: a critical overview. Reactions (Basel) 3:423–441. https://doi.org/10.3390/REACTIONS3030030

Sriram T (2023) Synthesis of copper nanoparticles from the leaf extract of Psidium guajava L. and its antioxidant, anticancer, antifungal activities. Ann Phytomed Int J 12. https://doi.org/10.54085/ap.2023.12.2.55

Subbanna S, Ts G, Basalingappa KM (2020) Biogenic nanoparticles from Allium Sativum and its bioactives applications. Eur J Mol Clin Med 7(8):212–232

Tran Khac K, Hoang Phu H, Tran Thi H, Dinh Thuy V, Do Thi H (2023) Biosynthesis of silver nanoparticles using tea leaf extract (camellia sinensis) for photocatalyst and antibacterial effect. Heliyon 9:e20707. https://doi.org/10.1016/J.HELIYON.2023.E20707

Tripathi D, Modi A, Smita SS, Narayan G, Pandey-Rai S (2022) Biomedical potential of green synthesized silver nanoparticles from root extract of Asparagus officinalis. J Plant Biochem Biotechnol 31:213–218. https://doi.org/10.1007/S13562-021-00684-Y/METRICS

Ummah MS (2019) No 主観的健康感を中心とした在宅高齢者における 健康関連指標に関する共分散構造分析Title. Sustainability (Switzerland) 11:1–14

Valko M, Leibfritz D, Moncol J, Cronin MTD, Mazur M, Telser J (2007) Free radicals and antioxidants in normal physiological functions and human disease. Int J Biochem Cell Biol 39:44–84. https://doi.org/10.1016/J.BIOCEL.2006.07.001

Venkata ALK, Anthony SP, Muthuraman MS (2019) Synthesis of Solanum nigrum mediated copper oxide nanoparticles and their photocatalytic dye degradation studies. Mater Res Express 6:125402. https://doi.org/10.1088/2053-1591/AB52A6

Verma ML, Dhanya BS, Thakur M, Jeslin J, Jana AK (2021) Plant derived nanoparticles and their biotechnological applications. Compr Anal Chem 94:331–362. https://doi.org/10.1016/BS.COAC.2021.01.011

Vinothkanna A, Mathivanan K, Ananth S, Ma Y, Sekar S (2023) Biosynthesis of copper oxide nanoparticles using Rubia cordifolia bark extract: characterization, antibacterial, antioxidant, larvicidal and photocatalytic activities. Environ Sci Pollut Res 30:42563–42574. https://doi.org/10.1007/S11356-022-18996-4/METRICS

Virk P, Awad MA, Saleh Abdu-llah Alsaif S, Hendi AA, Elobeid M, Ortashi K, Qindeel R, El-Khadragy MF, Yehia HM, Ferkry Serag EL-DIN M, Ali Salama H (2023) Green synthesis of Moringa oleifera leaf nanoparticles and an assessment of their therapeutic potential. J King Saud Univ Sci 35:102576. https://doi.org/10.1016/J.JKSUS.2023.102576

Witika BA, Makoni PA, Matafwali SK, Chabalenge B, Mwila C, Kalungia AC, Nkanga CI, Bapolisi AM, Walker RB (2020) Biocompatibility of biomaterials for nanoencapsulation: current approaches. Nanomaterials 10:1–40. https://doi.org/10.3390/nano10091649

Yadav A, Kumar H, Kumar P, Rani G, Maken S (2025) Syzygium cumini leaf extract mediated green synthesis of ZnO nanoparticles: a sustained release for anticancer, antimicrobial, antioxidant, and anti-corrosive applications. J Mol Struct 1325:141017. https://doi.org/10.1016/J.MOLSTRUC.2024.141017

Younas M, Rasool MH, Khurshid M, Khan A, Nawaz MZ, Ahmad I, Lakhan MN (2023) Moringa oleifera leaf extract mediated green synthesis of silver nanoparticles and their antibacterial effect against selected gram-negative strains. Biochem Syst Ecol 107:104605. https://doi.org/10.1016/J.BSE.2023.104605

Zhu FD, Hu YJ, Yu L, Zhou XG, Wu JM, Tang Y, Qin DL, Fan QZ, Wu AG (2021) Nanoparticles: a hope for the treatment of inflammation in CNS. Front Pharmacol 12:683935. https://doi.org/10.3389/FPHAR.2021.683935/XML/NLM

Chapter 7
Techniques for Characterization of Plant Bioactive-Based Nanomaterial

Jagadesh Astagirinathan, Abimannan Arulkumar, Arthi Carounanidhi, and F. Andrea Mary

7.1 Introduction

The rapid advancement of nanotechnology in recent years has led to the development and application of a diverse array of nanomaterials. Among these, plant bioactive-based nanomaterials have garnered significant attention due to their unique properties and vast potential across various sectors, including the food industry, clothing, medicine, cosmetics, agriculture, electronics, bioengineering, and environmental remediation (Okeke et al. 2023).

The synthesis of these nanomaterials using plant-derived natural compounds offers a sustainable and environmentally friendly alternative to traditional chemical synthesis methods, which often involve the use of toxic reagents and processes that can be harmful to the environment (Soni et al. 2021).

The inherent advantages of plant bioactive-based nanomaterials stem from the diverse bioactive properties exhibited by the phytochemicals extracted from various plant species (Sohail et al. 2022). When these bioactive compounds are incorporated into nanostructures, they can serve to enhance the functionality and efficacy of the resultant materials, as the unique properties of the nanoscale can amplify the beneficial effects of the plant-derived compounds. Moreover, the biocompatibility

J. Astagirinathan · F. Andrea Mary (✉)
Department of Human Genetics, Sri Ramachandra Institute of Higher Education and Research, Chennai, Tamil Nadu, India
e-mail: andreamary@sriramachandra.edu.in

A. Arulkumar
Department of Biotechnology, Achariya Arts and Science College, Puducherry, India

A. Carounanidhi
Department of Biotechnology, Sri Ramachandra Institute of Higher Education and Research, Chennai, Taminadu, India

V. Ghosh et al. (eds.), *Applications of Plant Bioactive Compounds-Based Nanoformulations*, Nanotechnology in the Life Sciences,
https://doi.org/10.1007/978-3-032-13086-0_7

of these plant derived compounds makes them suitable for use in drug delivery systems and therapeutic applications that involve direct contact with biological systems (Di Gioia et al. 2020).

The interaction of these nanomaterials with biological systems, including bio distribution and cellular uptake, is highly governed by factors such as particle size, morphology, chemical composition, surface charge, and thermal stability. The nanomaterials are thus, not only important for reproducibility and stability, but are equally important for ensuring that nanomaterials perform as intended (Leong et al. 2019).

In this review, both traditional and advanced techniques are sought to elucidate the characterization of plant bioactive-based nanomaterials. UV visible spectroscopy, Fourier transform IR spectroscopy (FTIR), Dynamic Light Scattering (DLS) and zeta potential measurements are presented to describe the formation size distribution and stability of particles. Next, integrate imaging techniques, including Transmission Electron Microscopy (TEM) and Scanning Electron Microscopy (SEM), for high resolution visualization of nanoparticle morphology and structure. The interaction between bioactive compounds and nanoparticles is further deepened into structural analysis methods including X-ray Diffraction (XRD), X-ray Photoelectron Spectroscopy (XPS), and Atomic Force Microscopy (AFM).

The characterization methods developing for nanomaterials are emerging trends that are revolutionizing the way it is understood and control the properties of nanoparticles. Currently high throughput techniques, in situ monitoring and game changing speed and precision on these techniques is becoming the popular. Real time insights into nanoparticle synthesis are gained using techniques like Nanoparticle Tracking Analysis (NTA) and Surface enhanced Raman spectroscopy (SERS).

Alongside, instruments like the Scanning Mobility Particle Sizer (SMPS) and Brunauer-Emmett-Teller (BET) monitoring tool help fine tune precision by examining property such as particle size distribution and surface area. With these development, plant bioactive based nanomaterials are becoming more and more capable of satisfying the stiff requirements of modern applications where speedy advancement and release are a must. In essence, it's not just observing nanoparticles; think of it as nanoparticle alchemy, actively sculpting them to fit specific purposes.

This points the importance of plant bioactive based nanomaterials in various fields, their characterizations and the methods adopted for such analyses. In future, a comprehensive understanding of these materials and their properties will be crucial for exploiting their full potential and resolving the difficulties that the current state of technological development imposes.

7.2 Traditional Techniques for Characterization of Nanomaterials

Traditional characterization techniques remain fundamental in analyzing nanomaterials, offering critical insights into their size, shape, structure, and composition. These methods ensure the reliability and reproducibility of nanomaterial synthesis, guiding their application in diverse fields. As shown in Fig. 7.1, techniques like UV-Vis, FTIR, NMR, DLS, SEM, TEM, XRD, TGA, XPS, and AFM collectively provide a comprehensive understanding of nanomaterial properties.

7.2.1 UV-Visible Spectroscopy

UV-visible spectroscopy has been used for the metallic nanoparticles like silver and gold synthesis using plant bioactive based nanomaterials as an indispensable tool real-time tracking of nanoparticle formation by detecting Surface Plasmon

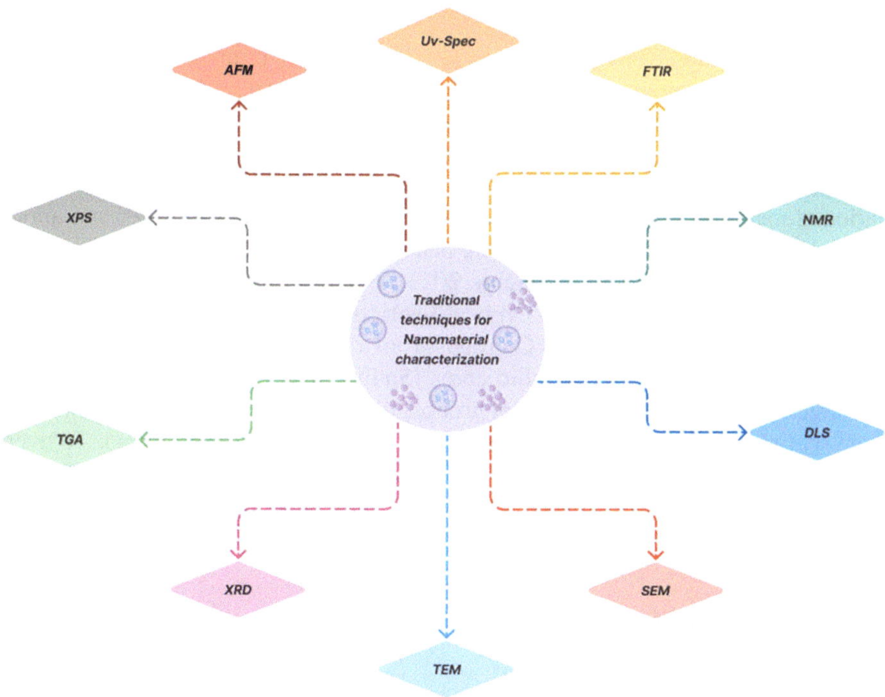

Fig. 7.1 This illustration highlights key traditional techniques used for the characterization of plant bioactive-based nanomaterials. Each method ranging from spectroscopic and microscopic analysis to thermal and surface characterization contributes uniquely to understanding the physicochemical properties of synthesized nanomaterials

Resonance (SPR). This latter wavelength specificity happens by virtue of nanoparticles overlapping light to cause conductive electron oscillations that results in a peak SPR. The position and strength of this peak can provide useful information on the size, shape and degree of particle aggregation (Pechyen et al. 2024). The majority of silver nanoparticles have a SPR peak at about 420 nm, whereas for gold nanoparticles it is roughly around 530 nm. Blue direction of shift is commonly correlated to a decrease in particle size and therefore increased availability whereas any additional redshift upholds aggregation or increasing the particle. Real-time monitoring of these shifts allows researchers to optimize synthesis conditions and ensure nanoparticle stability (Khlebtsov et al. 2022). Recent advancements in automated UV-visible spectroscopy systems enable high-throughput monitoring, for example surface enhanced UV spectroscopy have greatly accelerating the characterization process by allowing us to analyse molecular interactions at nanoscale level with improved stability (Gupta et al. 2025).

7.2.2 Fourier Transform Infrared (FTIR) Spectroscopy

The functional group at surfaces of nanomaterials are mainly examined by Fourier Transform Infrared (FTIR) Spectroscopy. FTIR can reveal the different functional groups that are being attached to nanoparticles by measuring vibrational modes of chemical bonds (Eid 2022). Important for wood bioactive nanomaterials; plant biomass will also reduce metal ions and passivate nanoparticle surfaces, so the proven strategies would be highly relevant (Barhoum et al. 2020). FTIR analysis revealed various functional groups like hydroxyl (-OH) group, carbonyl (C=O), and amide which are very important for stabilizing nanoparticle property as well as to alter its biocompatibility and reactivity (Gebre 2023). For instance, The FTIR analysis of nanoparticles fabricated using polyphenol-enriched extracts can offer insight into how the presence of -OH and C=O groups from a biomolecule leads to interactions between this molecule and metallic cores (Sadhasivam et al. 2020). While FTIR effectively detects surface functional groups, it provides limited information about their spatial distribution on the nanoparticle surface (Wang et al. 2021). To overcome this limitation, FTIR mapping has been developed, allowing for the visualization of functional group distribution over larger sample areas (Siddique 2024).

7.2.3 Nuclear Magnetic Resonance (NMR)

NMR spectroscopy is an indispensable method for analysing molecular structures with the use of resonance signals generated by magnetic field which reveals intricate details about atomic arrangements. It exploits the magnetic characteristics of certain nuclei, mainly hydrogen (H) and carbon (C), which reacts to the radiofrequency pulses which in turn provides information about their chemical surroundings (Tampieri et al. 2021). Due to this characteristics NMR is widely used in analysis of bioactive compounds, helpful in identifying functional groups, and even to quantify the concentration of compounds (De Vita et al. 2020). Furthermore, NMR provide distinctive advantage for nanoparticle characterization, especially those incorporating bioactive plant compounds, secondary metabolites such as phenols, flavonoids, terpenoids, and alkaloids, exhibit unique NMR signals that helps in identifying and verifying the functional groups within nanoparticles (Patil 2020). In addition to these advanced techniques, including 1D and 2D NMR modalities like Correlation Spectroscopy (COSY) and Heteronuclear Single Quantum Correlation (HSQC), provide comprehensive information on molecular interactions and particle stability, enabling to analyse nanoparticles used for drug delivery and therapeutic purposes (De Vita et al. 2020). These nanomaterials can now be better characterized to improvements in sensitivity and resolution brought about by high-field NMR technology. However, there are drawbacks to NMR, particularly with regard to signal overlap in more complex samples, and the expense of the equipment may prevent broad usage in regular analysis (Hatzakis 2019). Despite these drawbacks, NMR is still a vital tool for the structural verification and in-depth analysis of plant bioactive-based nanomaterial, providing non-destructive, incredibly insightful information that is crucial for applications in nano medicine and biotechnology.

7.2.4 Dynamic Light Scattering (DLS)

Dynamic Light Scattering (DLS) is a standard technique that measures the size distribution and zeta potential of nanoparticles in suspension. It measures the hydrodynamic radius (nanoparticle size- with solvent and any surface coatings) rather than dry core size (Rahdar et al. 2019). This is done through intelligently studying the Brownian motion of particles in solution which causes light scattering (Knysh et al. 2023). DLS is essential to assess the size uniformity and colloidal stability of plant-based nanoparticles, as it provides information on surface charge using zeta potential measurements (Bagheri et al. 2023). Higher absolute zeta potential values represent improved colloidal stability while lower may indicate formation of significant nanoparticle agglomeration (Pochapski et al. 2021). For these thermally labile nanomaterial, temperature-controlled DLS can explain how size and stability of nanoparticles are influenced by temperature (Darabi n.d.), while recent developments with multi-angle DLS systems have taken the measurement accuracy to an even higher level in wide-size-distribution samples.

7.2.5 Scanning Electron Microscopy (SEM)

Scanning Electron Microscopy (SEM) serves as an indispensable analytical tool for elucidating the complex surface architecture and compositional intricacies of plant-derived nanomaterial. The atomic resolution capabilities of SEM make it ideal for observing surface topography, particle size distribution and any contour features that can be crucial determinants in biological efficacy, stability and functionality within these materials (Puri et al. 2024). SEM helps us to visualize the crystallinity of the material, particle aggregation, and surface roughness which all affect cell uptake, and bioavailability of the bioactive compounds derived from plants (Falsafi et al. 2020). Advancements in SEM imaging such as Focused Ion Beam (FIB) and Electron Beam Lithography (EBL), which offers deep insights about structure of the bioactive by enabling cross-sectional analysis. Combined with Energy Dispersive X-ray Spectroscopy (EDS) (Yang et al. 2019; Khonina et al. 2024). SEM can further analyse the elemental composition and help to identify the key phytochemicals across the nanoparticle matrix. This helps us to understand the arrangement of cells and their interaction with target cells or tissues, which has great implications in therapeutics (Talha et al. 2022). In addition, SEM also provides crucial data on nanoparticle stability such as surface modifications and degradation patterns under physiological pH, temperature and ionic strength (Ngouangna et al. 2022). In the case of plant based bioactive nanomaterial which are sensitive to environmental shifts, SEM has the ability to monitor these changes (in situ), thus helping us to predict performance, efficacy and controlled release in clinical application, with these mentioned multi-dimensional attributes, SEM is an essence for advancing nanoparticle-based therapeutics and delivery system of plant-based nanomaterial (Lv et al. 2019; Karnwal et al. 2024; Kaushik et al. 2023).

7.2.6 Transmission Electron Microscopy (TEM)

Transmission Electron Microscopy (TEM) provides the highest-resolution imaging, allowing for atomic-scale observation of nanoparticles (Lin et al. 2021). TEM can be used to create detailed information about the size, shape and internal structure of nanoparticles by transmitting a beam of electron through an extremely thin sample. It is also used to particle-shaped assessments; TEM helps to identify whether nanoparticles are spherical, rod or triangular shapes which greatly affect their biological activities in the application of plant bioactive-based nanomaterial (Bals et al. 2022). Additionally, TEM can detect the formation of core-shell structures, where a metallic core is encased with bioactive compounds from plant extracts, enhancing the stability and functionality of the nanoparticles (Villalobos-Noriega et al. 2021). Recent advancements in cryo-TEM have enabled the visualization of nanoparticles in their natural, hydrated state, preserving the biological interactions between the nanoparticles and plant bioactive (Lin et al. 2024).

7.2.7 X-Ray Diffraction (XRD)

The X-ray Diffraction (XRD) technique can be used to assess the crystallinity of nanoparticles. These diffraction patterns suggest the phase composition and crystal structure of nanomaterials from analysis by XRD method, since this is considered as a benchmark technique to depict all over information about atomic planes in crystals (Ali et al. 2022). Silver nanoparticles synthesized from plant derived bioactives often exhibit Face Centred Cubic (fcc) type diffraction peaks (Sharma et al. 2022; Jain et al. 2021). Moreover, broader diffraction peaks might suggest the existence of amorphous structures or smaller crystal sizes (Lan et al. 2021).

7.2.8 Thermo Gravimetric Analysis (TGA)

Thermo Gravimetric Analysis (TGA) is required in both case of plant-based nanomaterials (Cherkashina et al. 2024). TGA can provide useful information on the decomposition of organic compounds bonded to nanoparticle surfaces by measuring mass changes during heating (Kunc et al. 2022; Clausen et al. 2019). At a steady rate, TGA can identify thermal stability by tracking shifts in weight as a function of temperature and offers both quantitative and qualitative insights about physical changes, such as sample gain or loss in response to regulated atmospheres (rich in oxygen or inert). The sample's thermal stability, and/or its ability to retain properties nearly unaltered when subjected to heat, is the key insight obtained from the TGA curve. The curve, which shows weight loss under heat, also provides information on the composition of the sample which is very much crucial in understanding the steps involved in decomposition (Saadatkhah et al. 2020).

Melting and Decomposition are combined into one factsheet as TGA is universally used alongside Differential Scanning Calorimetry (DSC) allowing the heat flow mass loss to simultaneously provide a whole picture of material thermal properties (Kijo-Kleczkowska et al. 2022). Using recent trends, there are TGA systems available that have been automated to run numerous samples at one time so the sample volume has also dramatically increased (Farivar et al. 2021). In addition, TGA-FTIR systems enable determining the type of gases evolved upon thermal decomposition and help to interpret better on chemical resistance mechanism for plant-based nanomaterial (Barhoum et al. 2020).

7.2.9 X-Ray Photoelectron Spectroscopy (XPS)

X-ray Photoelectron Spectroscopy (XPS), is able to quantify the elemental composition and chemical state of atoms located at nanomaterial surfaces (Stevie and Donley 2020). It is essential for exploring the bioactive compound-nanoparticles interaction that could help to understand whether surface functionalization depends on the different oxidation states (Harish et al. 2023). For the very same reasons, XPS is often employed to monitor changes in binding energies of core electrons that reflect oxidation states of metals such as silver (Ag) or gold (Au) (Oliveira et al. 2021). In general, XPS can also be used to scan surface contaminants which may affect nanoparticle performance as well whether they are being considered for use in drug delivery or within an environmental sensing application (Thakur and Kumar 2022). Recent improvements in XPS technology have increased the sensitivity and resolution to a level that enables detailed analysis of new ultrathin films or nanoscale surface coatings. The addition of depth profiling capabilities enables comprehensive examination of multi-layered systems, such as core-shell nanostructures created with plant bioactives (Greczynski et al. 2023).

7.2.10 Atomic Force Microscopy (AFM)

Atomic Force Microscopy (AFM) is a vital methodology for the nanoscale investigation of surface topography and mechanical characteristics of plant bioactive-based fabrics (Stan and King 2020). This method employs a sharp tip which is rastered on the surface of the sample for generating high-resolution 3-dimensional (3D) images of surfaces morphology and structure. This is critical for eco-friendly plant-derived nanomaterial with numerous rich and complex surface attributes that result from interactions with the diverse compounds within plants interacting onto metal cores (Akhtar et al. 2019). However, AFM can be used for high resolution surface mapping and hence it might still be useful to provide important structural information on nanoparticle aggregation and therefore colloidal stability/reactivity of these materials under application conditions such as drug delivery devices; biosensor systems or environmental remediation (Nguyen-Tri et al. 2020).

The force spectroscopy performed by AFM makes it feasible to explore the mechanical characteristics at nano level of various materials including adhesion, elasticity and surface hardness through controlled forces with a sharp tip under constant contact area. This is especially critical for assessing how plant-based nanoparticles behave in biological systems, or survive varied environmental settings (Stan and King 2020). For example, AFM has been employed in the investigations of adhesive interactions between polysaccharide-based nanomaterial (such as cellulose Nano crystals) and plant bio actives yielding crucial information related to stability/functionality of such materials within aqueous media (Jana and Jana 2022). Recent advancements including high-speed AFM further enable imaging dynamic processes such as nanoparticle degradation or aggregation in real-time, making it more applicable for research on plant bioactive-based nanomaterial (Xu and Wang 2022).

Advanced modes, as well as Conductive AFM (CAFM) and Magnetic Field dependent Measurements (MFM) can be performed hence making this technique available for investigation of multifunctional nanomaterial in a whole. Taken together, these properties render AFM an irreplaceable instrument for studying structural, mechanical and functional characteristics of diverse types of plant-derived nanomaterial to promote applications in a myriad fields from biotechnology to medicine and environmental science (Joo et al. 2024).

7.3 Emerging Techniques for Characterizing Plant Bioactive-Based Nanomaterials

To fully comprehend the potential use of plant bioactive based nanomaterial and to understand its characteristics, and behaviour in a variety of industries, such as agriculture and medicines, advancement in characterisation is essential (Begum et al. 2022). Techniques such as Nanoparticle Tracking Analysis (NTA), Surface-Enhanced Raman Spectroscopy (SERS), Scanning Mobility Particle Sizer (SMPS) and Brunauer-Emmett-Teller (BET) are those of the techniques that have gained prominence in their advancement. These methods offer enhanced precision and sensitivity, facilitating a deeper understanding of nanomaterials derived from plant bioactives (Mourdikoudis et al. 2018).

7.3.1 Nanoparticle Tracking Analysis (NTA)

Biological activity and interactions of nanoparticles with cellular systems can be greatly influenced by its size distribution (Yusuf et al. 2023). Nanoparticle Tracking Analysis (NTA) that allows for the quantitative measurement of size distribution and concentration of nanoparticles in colloidal systems (Foreman-Ortiz et al. 2022). By combining the principles of Brownian motion and light scattering, it can analyse individual particles within a sample (Wells et al. 2024). In terms of its advantages, one of the main benefits is that it has the ability to be real-time viewed, which is allows for researchers to monitor changes in the size and distribution of particles over a period of time, which is useful to evaluate the stability of colloidal dispersions where continuous monitoring can provide data on particle behaviour under diverse conditions (Griffiths et al. 2020; Briffa et al. 2020). NTA can precisely measure size by tracking how nanoparticles move under a laser beam and diffusion rates are needed for valid fit measurements and to design materials that have the desired therapeutic activity (Nissen et al. 2022).

Additionally, NTA quantifies nanoparticle concentration, which is extremely useful for determining the release kinetics of bioactive molecules from nanocarriers. Recent finding can optimize formulations for improved therapeutic efficacy when they understand how plant-derived bioactive are released from their carriers (Kowkabany and Bao 2024). Beyond these applications, NTA is useful for ascertaining the encapsulation efficiency of bioactive compounds within nanoparticles and evaluating the stability of these formulations under a variety of environmental conditions. This information is essential for developing plant-derived therapeutic delivery systems that enhance bioavailability and effectiveness (Ortiz et al. 2021).

7.3.2 Surface-Enhanced Raman Spectroscopy (SERS)

On the other hand, Surface-Enhanced Raman Spectroscopy (SERS) is an advanced tactic for amplification of Raman scattering signals from molecules attached to nanostructure surfaces (Eskandari et al. 2022). Its principal strength lies in its superlative detection mechanism, making it possible to analyse bioactive molecules at exceedingly low concentrations; usually sub-picogram amounts. It is important to have this sensitivity for characterizing plant bioactives that might only be present in minute amounts as part of complex biological mixtures (Tahir et al. 2021). Raman spectroscopy demonstrate that the power as a transformative tool in such instances and use SERS to do ‘chemical fingerprints’: not only does the molecule-specific nature of the Raman spectra show individual molecular constitutions at high details but also specific plant-derived-molecule identification and quantification. This will advance the study of their therapeutic benefits and action mechanisms (Cialla-May et al. 2021; Yavuz et al. 2024).

Moreover, when it comes to following the bioactives release kinetics from nanocarriers SERS is also of a great importance (Malekjani and Jafari 2022). The nature of the Raman spectra can change, over time upon monitoring and leads to understand stability as well release profile (release kinetics) encapsulated bioactive compounds needed for improving drug delivery systems (Sabaghi et al. 2022). SERS is always combined with NTA and other characterization techniques in SERS to give an all-rounded understanding of the plant bioactive-based nanomaterial (Li et al. 2022). As NTA quantifies the size distribution and concentration, SERS provides information regarding chemical composition of bioactive molecules in a complex biological system to establish two-to-three-dimensional analysis that can further guide study on plant-derived nanomaterials design with improved efficacy. The use of such an integrated approach would permit a thorough investigation there involves both physical and chemical characteristics plant-originated nanomaterials to promote its most efficient applications towards drug delivery, bio sensing etc., (Ding et al. 2024; Wang et al. 2022).

7.3.3 Scanning Mobility Particle Sizer (SMPS)

Scanning Mobility Particle Sizer (SMPS) is designed for the accurate sizing of nanoparticles around 2 nm to 1000 nm (Duelge et al. 2022). Employing an ionizer, SMPS classifies aerosolized particles according to their electrical mobility in the electric field and thereby enabling higher resolution for smaller particles as compared with conventional methods such as Dynamic Light Scattering (DLS) (Rahman et al. 2024). It operates by ionizing particles in an aerosol flow, followed by passage through an electrostatic classifier where the particle is separated on the basis of their electrical mobility (Duelge n.d.). The precipitated particles are air routed counts in classified mode, to a Condensation Particle Counter (CPC) that provides number concentration of certain size selected less than a micro meter aerosol and generates Particulate Matter (PM) size distribution graphics (Xu et al. n.d.).

SMPS turned out to be a very powerful tool for the characterization of plant bioactive based nanomaterials especially in complex, heterogeneous systems due to large differences in particle size nature driven by composition complexity of plants extracts (Khan et al. 2022). For instance, SMPS monitoring the size of silver nanoparticles synthesized by plant extracts in different reaction stages show how bioactive compounds influence nanoparticle nucleation and growth dynamics. By introducing this technique, it benefits the researchers to maximize their reaction conditions and guarantee that they can synthesize nanoparticles within an application defined size window essential for applications in drug delivery as well environmental sensing & remediation technologies (Restrepo and Villa 2021; Mabrouk et al. 2021).

It also has the capability of detecting submicron particles while scanning aerosol, which is especially useful in nanoparticle research. Additionally, its ability to provide real-time monitoring of nanoparticle size during synthesis allows for greater control over nanoparticle production, ensuring consistency and optimizing the final product for targeted applications (Harish et al. 2023).

7.3.4 Brunauer-Emmett-Teller (BET) Method

The Brunauer-Emmett-Teller is a gas adsorption technique used to measure the surface area of solid or porous material. Using nitrogen as gas, it analyses the quantity of gas adsorbed on a material's surface under different pressures and subsidizes further information such as pore volume or average pore size (Irwansyah et al. 2024). This model is an extension of the Langmuir Monolayer adsorption that corresponds to Multi-layered adsorption, thus better in assessing nanoporous materials (like plant-derived nanomaterial) (Zhang et al. 2021).

The plant bioactive-based supported nanomaterials can be evaluated for catalytic, adsorptive or drug delivery applications by BET analysis thus making them suitable for environmental applications where high effectiveness in terms of pollutant adsorption or catalytic degradation is essential (Bhosale et al. 2023; Bibi et al. 2023). Higher surface areas are always desired as they result in greater reactivity and adsorption capacity so that BET is a critical tool for designing these materials (Sinha et al. 2019). It is also a very high potential in the pore structure analysis of materials, typically used to characterize biopolymers or cellulose-based nanomaterials that are often microporous or mesoporous. Pore size distribution is critical for developing improved controlled release systems and adsorption technologies (Singh et al. 2022).

7.4 Synergism Between the Characterization Techniques

Plant-based nanomaterials are often complex and sensitive, with unique thermal and molecular properties that affect their functionality in real-world applications. This synergistic method allows us to ensure that these materials are stable under various conditions, while confirming the presence and integrity and different aspects of bioactive compounds. Each pairing leverages the strengths of individual techniques thus providing more comprehensive data that are accurate and reliable in characterization. The above-mentioned Fig. 7.2 reflects how these and other complementary techniques together offer a robust analysis for developing reliable, application-specific nanomaterials.

7.4.1 SMPS and BET

Combining the SMPS and BET methods could enable complete characterization of particle size distribution, as well as specific surface area which are important for evaluating plant bioactive-based nanomaterials. The use of additional routine and advanced characterization methods is an option to enhance the understanding from different perspectives and accordingly, assists in potential tailoring strategies on utilization regarding various conditions (Arvaniti et al. 2015).

In one of the works which reported synthesis of plant bioactive based silica nanoparticle, SMPS was used to observe particle size distribution and BET analysis showed specific surface area and porosity properties of the particles. In that way, this linear approach allowed to the formation of nanomaterials with better reaction properties for proper catalysis and cure carriage antics (Basu et al. 2024).

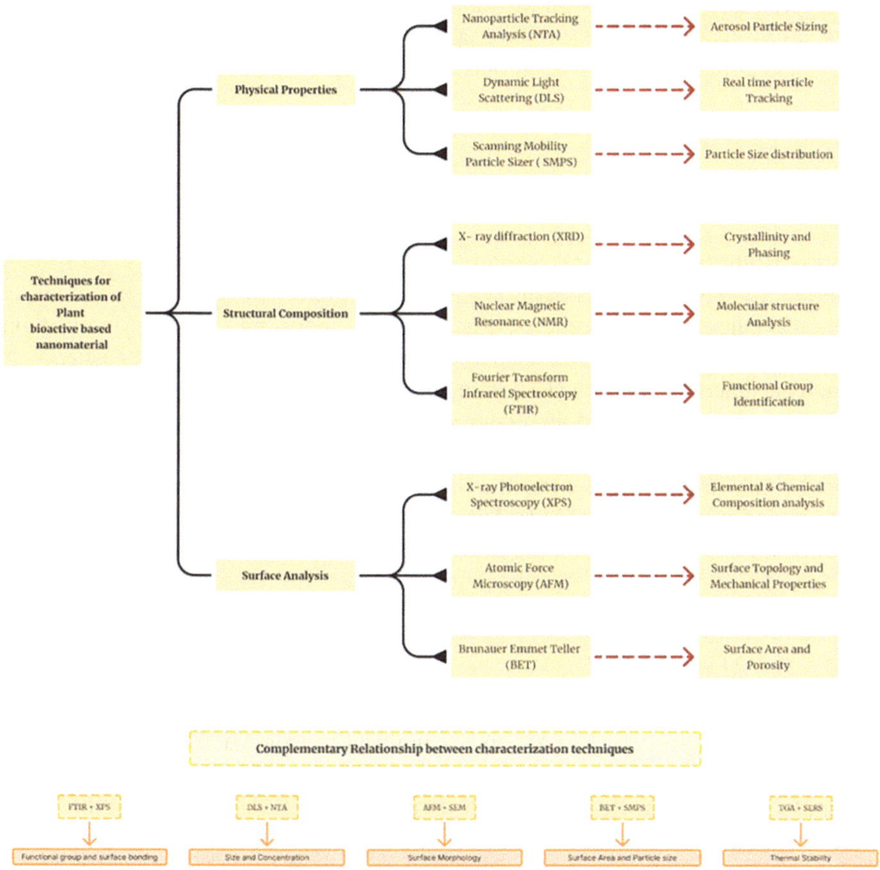

Fig. 7.2 Flowchart illustrating the various characterization techniques for plant bioactive-based nanomaterial and their complementary relationship between those techniques. The chart categorizes techniques into physical, chemical, and biological characterization methods, highlighting their complementary relationships. This visual representation aids in understanding the diverse approaches for analysing the properties and functionalities of plant-based nanomaterial, facilitating informed selection of techniques based on specific research objectives

7.4.2 FTIR and XPS

Fourier Transform Infrared Spectroscopy (FTIR) and X-ray Photoelectron Spectroscopy (XPS) work synergistically to analyse the chemical composition of nanomaterials. FTIR is primarily used to identify functional groups on the surface of nanoparticles by detecting vibrations in chemical bonds, while XPS provides elemental composition and oxidation state information through core-level binding energy analysis. For example, in a study on the surface chemistry of plant-extract-mediated silver nanoparticles, FTIR was used to detect functional groups like hydroxyl and carbonyl groups, and XPS confirmed the oxidation state of silver, showing reduced silver ions on the nanoparticle surface (Ajitha et al. 2015). This combination ensures precise surface characterization, crucial for understanding bioactivity and compatibility with biological systems.

7.4.3 DLS and NTA

Dynamic Light Scattering (DLS) and Nanoparticle Tracking Analysis (NTA) are commonly used together to measure nanoparticle size distribution and colloidal stability. DLS calculates the average particle size by analysing the scattering of light, whereas NTA provides individual particle tracking, giving precise measurements for each particle in a solution (Rahdar et al. 2019; Foreman-Ortiz et al. 2022). This dual approach was applied in characterizing polyphenol-loaded nanoparticles, where DLS assessed overall particle stability, and NTA verified individual particle size and concentration, demonstrating improved precision compared to using a single method (Garavand et al. 2022). These techniques combined offer high accuracy in sizing and distribution, essential for stability in applications like drug delivery.

7.4.4 AFM and SEM

Atomic Force Microscopy (AFM) and Scanning Electron Microscopy (SEM) provide complementary insights into surface morphology. While SEM offers high-resolution imaging to observe surface structure and particle aggregation, AFM captures topographical and mechanical data such as surface roughness and elasticity (Puri et al. 2024; Akhtar et al. 2019). A study characterizing cellulose-based bionanocomposites demonstrated that SEM visualized the uniformity of particles, and AFM further analysed nanoscale roughness and mechanical properties (Mishra et al. 2024). Together, they offer a complete view of particle morphology and stability, which is critical for applications where surface interactions influence performance, such as biosensing (Li et al. 2016).

7.4.5 TGA and SERS

For the combination of Thermogravimetric Analysis (TGA) and Surface-Enhanced Raman Spectroscopy (SERS), studies demonstrate that these techniques together provide a comprehensive analysis of the thermal and structural stability of plant bioactive-based nanomaterials. TGA helps monitor weight loss as the material is heated, which provides insights into thermal stability and decomposition characteristics. Meanwhile, SERS enhances the detection of molecular structures on the

Table 7.1 Comparison of characterization techniques for plant bioactive-based nanomaterials

S.No	Technique	Size Range	Sensitivity	Specific Application	Key Strengths	Limitations	Data Output	Sample Requirements	Ref
1	*UV-visible spectroscopy*	N/A (optical properties)	Moderate	Monitoring nanoparticle formation; concentration and surface Plasmon resonance (SPR)	Simple, rapid, cost-effective	Limited to optically active nanoparticles	Quantitative (absorbance)	Liquid samples, clear solutions	Akash et al. (2020)
2	*FTIR spectroscopy*	N/A (molecular vibrations)	High	Identifying functional groups on nanoparticle surfaces	Non-destructive, provides molecular fingerprint	Cannot measure particle size	Qualitative (spectra)	Solid, liquid, colloidal samples	White (2020)
3	*NMR spectroscopy*	N/A (atomic interactions)	High	Structural elucidation of organic coatings and surface interactions	Detailed molecular-level information	Requires large amounts of sample	Quantitative & Qualitative	Requires pure sample (solid or liquid)	Reif et al. (2021)
4	*Dynamic light scattering (DLS)*	1 nm—10 µm	Moderate	Particle size distribution in colloids	Quick size distribution analysis	Sensitive to agglomeration, assumes spherical shape	Quantitative (size distribution)	Colloidal, liquid suspensions	Babick (2020)
5	*Zeta potential analysis*	N/A (surface charge)	High	Surface charge, colloidal stability analysis	Useful for assessing colloidal stability	Affected by conductivity and pH	Quantitative (charge values)	Liquid, colloidal samples	Pochapski et al. (2021)
6	*X-ray diffraction (XRD)*	< 1 nm—100 nm	High	Determining crystallographic structure	High precision in identifying crystal phases	Requires crystalline samples	Quantitative (crystalline structure)	Solid, crystalline samples	Ali et al. (2022)

(continued)

Table 7.1 (continued)

S.No	Technique	Size Range	Sensitivity	Specific Application	Key Strengths	Limitations	Data Output	Sample Requirements	Ref
7	*Thermogravimetric analysis (TGA)*	N/A (thermal stability)	High	Thermal stability and composition analysis (organic vs. inorganic content)	Precise thermal stability data	Destructive, thermal-only	Quantitative (weight change vs. temp)	Solid samples	Gross-Rother et al. (2020)
8	*Scanning electron microscopy (SEM)*	10 nm—1 μm	High	Surface morphology and topography analysis	Provides high-resolution images of surface morphology	Limited depth of field, sample preparation required	Imaging	Solid samples, thin films	Karakuş and Coşkun (2023)
9	*Transmission electron microscopy (TEM)*	1–100 nm	Very high	Direct imaging of nanoparticle size, morphology, and structure	High resolution, direct visualization	Complex sample preparation, time consuming	Imaging	Thin solid films, dry samples	Malatesta (2021)
10	*X-ray photoelectron spectroscopy (XPS)*	1 nm (surface depth analysis)	High	Surface chemistry, elemental composition, oxidation states	Quantitative surface analysis	Limited to surface analysis, vacuum needed	Quantitative (surface chemistry)	Solid, thin films	Greczynski et al. (2023)
11	*Nanoparticle tracking analysis (NTA)*	10 nm—1 μm	High	Particle size distribution and concentration in a liquid medium	Direct visualization, tracks individual particles	Difficult with polydisperse samples	Quantitative (particle size & count)	Liquid, colloidal suspensions	Gross-Rother et al. (2020)

12	*Surface-enhanced raman spectroscopy (SERS)*	N/A (molecular detection)	Very high	Ultra-sensitive detection of surface-bound molecules, including bioactives	Allows trace detection and molecular identification	Requires metal nanostructures for enhancement	Qualitative & Quantitative (spectra)	Solid, liquid, with metal surfaces	Fu et al. (2024)
13	*Scanning mobility particle sizer (SMPS)*	2 nm—1 μm	Very high	Measuring particle size and distribution in aerosols	High accuracy in particle sizing	Limited to aerosol particles	Quantitative (size distribution)	Aerosols	Shankar et al. (2023)
14	*Brunauer-emmett-teller (BET) analysis*	N/A (surface area)	High	Surface area and porosity analysis of nanomaterials	Accurate surface area measurement	Only solid samples, no liquids	Quantitative (surface area)	Solid, powder samples	Almaghrabi et al. (2023)

Nanoparticle due to their unique physical and chemical properties they need a diverse and fine techniques for analysis. The above given table depicts a wide overview of commonly used methods for characterization, enumerating their detailed attribute such as size range, sensitivity, applications, strength and limitations providing a detailed attributes, potential advantages and disadvantages which are critical for application in biomedical, environmental, and industrial fields surface by amplifying Raman signals, allowing for precise identification of bioactive compound (Clausen et al. 2019; Eskandari et al. 2022).

For instance, research into non-noble metal nanomaterials applied to biosensing shows that SERS is effective in analysing biomolecular structures with high sensitivity, while TGA confirms material stability at elevated temperatures. In one study, TGA-SERS was applied to analyse turmeric-derived nano capsules, where TGA tracked thermal decomposition, and SERS provided details on encapsulated curcumin molecules, illustrating their stability under physiological conditions. This combination is particularly valuable for applications in biomedicine where both the structural integrity and active bioavailability of plant-derived compounds are critical (Chen et al. 2024; Kundu and Nithiyanantham 2013).

7.5 Challenges and Limitations in Characterization

7.5.1 *Complex Mixtures of Bioactive*

The complexity arises from the myriad phytochemicals that can be employed in their synthesis presenting a challenge to accurately characterizing plant-based nanomaterials. Plant extracts often contain many bioactive molecules, which are chemically diverse and can make the analysis difficult. Methods like NMR, FTIR can help to identify these compounds. Nonetheless, isolating and even identifying the bioactive present within it involves a major challenge (Shaikh et al. 2021).

7.5.2 *Inconsistency in Nanoparticle Synthesis*

A further drawback is the heterogeneity of nanoparticle synthesis when plant bioactive are used. Diverse plant extracts as well as different concentration and reaction conditions can result in variability of particle size, shape or stability. The highly variable diameter of the AuNPs to be analysed makes it nearly impossible to standardize and compare characterizations using techniques like DLS or TEM, which are extremely sensitive whether measure is characterized (Ahmad et al. 2021).

7.5.3 *Integration of Multiple Techniques*

The complexity of plant bioactive based nanomaterials lies in its multifaceted nature, for which single technique is incomplete to characterize it on complete scale. In many cases, employing a combination of techniques (for example, DLS for size distribution; TEM for morphology and XPS for surface chemistry) is essential to better understand the properties of nanomaterials. This method can be laborious and expensive, emphasizing the necessity of multi-modal characterization (Mohammadidargah et al. 2024).

7.6 Implications for Future Research

Particularly in large-scale production settings, future studies should concentrate on integrating automated technologies to expedite the characterisation process. Furthermore, the creation of uniform characterization procedures will improve repeatability in labs and sectors. More research into in-situ methods will yield real-time information on the development of nanoparticles, enhancing synthesis control and guaranteeing more stable material characteristics.

7.7 Conclusion

From UV-Visible Spectroscopy for real-time nanoparticle monitoring to sophisticated techniques like XPS for surface chemistry characterization, the main methods used to analyse plant bioactive-based nanomaterial in this chapter are discussed. Each of these techniques provides different insights into important properties such as size, shape, surface charge, chemical composition and thermal stability, and must be used in combination with one another to maximize the performance of nanomaterials in a wide range of applications. In-situ monitoring and high-throughput characterisation are providing new tools that have the power to radically change how scientists study nanomaterials, making measurements more accurate and more efficient. By addressing issues present today such as the need for standardized procedures and differences in the composition of plant extracts, these sophisticated characterisation techniques will allow green nanotechnology to be widely leveraged.

References

Ahmad T, Iqbal J, Bustam MA, Irfan M, Asghar HM (2021) A critical review on phytosynthesis of gold nanoparticles: issues, challenges and future perspectives. J Clean Prod 309:127460

Ajitha B, Reddy YA, Reddy PS (2015) Green synthesis and characterization of silver nanoparticles using Lantana camara leaf extract. Mater Sci Eng C Mater Biol Appl 49:373–381

Akash MS, Rehman K, Akash MS, Rehman K (2020) Ultraviolet-visible (UV-VIS) spectroscopy. In: Essentials of pharmaceutical analysis, pp 29–56

Akhtar I, Rehman MA, Choi W, Kumar S, Lee N, Cho SJ, Park HH, Park KH, Seo Y (2019) Three-dimensional atomic force microscopy for ultra-high-aspect-ratio imaging. Appl Surf Sci 469:582–592

Ali A, Chiang YW, Santos RM (2022) X-ray diffraction techniques for mineral characterization: a review for engineers of the fundamentals, applications, and research directions. Minerals 12(2):205

Almaghrabi M, Alqurshi A, Jadhav SA, Mazzacuva F, Cilibrizzi A, Raimi-Abraham B, Royall PG (2023) Evaluating thermogravimetric analysis for the measurement of drug loading in mesoporous silica nanoparticles (MSNs). Thermochim Acta 730:179616

Arvaniti EC, Juenger MC, Bernal SA, Duchesne J, Courard L, Leroy S, Provis JL, Klemm A, De Belie N (2015) Determination of particle size, surface area, and shape of supplementary cementitious materials by different techniques. Mater Struct 48:3687–3701

Babick F (2020) Dynamic light scattering (DLS). In: Characterization of nanoparticles. Elsevier, pp 137–172

Bagheri AR, Aramesh N, Hasnain MS, Nayak AK, Varma RS (2023) Greener fabrication of metal nanoparticles using plant materials: a review. Chemical Physics Impact 7:100255

Bals J, Loza K, Epple P, Kircher T, Epple M (2022) Automated and manual classification of metallic nanoparticles with respect to size and shape by analysis of scanning electron micrographs. Mater Werkst 53(3):270–283

Barhoum A, Jeevanandam J, Rastogi A, Samyn P, Boluk Y, Dufresne A, Danquah MK, Bechelany M (2020) Plant celluloses, hemicelluloses, lignins, and volatile oils for the synthesis of nanoparticles and nanostructured materials. Nanoscale 12(45):22845–22890

Basu A, Sae-Be A, Namporn T, Suriyaphan O, Sithisarn P, Leanpolchareanchai J, Plommaithong P, Chatsukit A, Sa-Ngiamsuntorn K, Naruphontjirakul P, Ruenraroengsak P (2024) Delivery of avocado seed extract using novel charge-switchable mesoporous silica nanoparticles with galactose surface modified to target Sorafenib-resistant hepatocellular carcinoma. Int J Nanomedicine 31:10341–10365

Begum SJ, Pratibha S, Rawat JM, Venugopal D, Sahu P, Gowda A, Qureshi KA, Jaremko M (2022) Recent advances in green synthesis, characterization, and applications of bioactive metallic nanoparticles. Pharmaceuticals 15(4):455

Bhosale SR, Jagadhane KS, Bhosale RR, Vanalakar SA, Qureshi MH, Patil DN, Dhavale RP, Shimpale VB, Kolekar GB, Anbhule PV (2023) Mesoporous carbon encapsulated zinc oxide nanorods derived from plant species 'Argyreia sharadchandrajii' for live cell imaging of drug delivery and multimodal bioactivities. New J Chem 47(27):12739–12751

Bibi M, Batool SA, Iqbal S, Zaidi SB, Hussain R, Akhtar M, Khan A, Alqahtani MS, Abbas M, Rehman MA (2023) Synthesis and characterization of mesoporous bioactive glass nanoparticles loaded with peganum harmala for bone tissue engineering. Heliyon 9(11)

Briffa SM, Sullivan J, Siupa A, Carnell-Morris P, Carboni M, Jurkschat K, Peters RJ, Schultz C, Seol KH, Kwon SJ, Park S (2020) Nanoparticle tracking analysis of gold nanoparticles in aqueous media through an inter-laboratory comparison. JoVE 20(164)

Chen L, Liu H, Gao J, Wang J, Jin Z, Lv M, Yan S (2024) Development and biomedical application of non-noble metal nanomaterials in SERS. Nanomaterials 14(20):1654

Cherkashina NI, Pavlenko ZV, Domarev SN, Ruchiy AY, Solgalov VV (2024) Creation of a composite material based on plant-based components. Nanotekhnologii v Stroitel'stve 16(1):67–76

Cialla-May D, Krafft C, Rösch P, Deckert-Gaudig T, Frosch T, Jahn IJ, Pahlow S, Stiebing C, Meyer-Zedler T, Bocklitz T, Schie I (2021) Raman spectroscopy and imaging in bioanalytics. Anal Chem 94(1):86–119

Clausen PA, Kofoed-Sørensen V, Nørgaard AW, Sahlgren NM, Jensen KA (2019) Thermogravimetry and mass spectrometry of extractable organics from manufactured nanomaterials for identification of potential coating components. Materials 12(22):3657

Darabi M. Experimental and theoretical investigation of mechanical responses of bacteria under hypoosmotic pressure

De Vita S, Terracciano S, Bruno I, Chini MG (2020) From natural compounds to bioactive molecules through NMR and in silico methodologies. Eur J Org Chem 40:6297–6317

Di Gioia S, Hossain MN, Conese M (2020) Biological properties and therapeutic effects of plant-derived nanovesicles. Open Medicine 15(1):1096–1122

Ding F, Sánchez-Villasclaras S, Pan L, Lan W, García-Martín JF (2024) Advances in vibrational spectroscopic techniques for the detection of bio-active compounds in virgin olive oils: a comprehensive review. Foods 13(23):3894
Duelge K New Nanoparticle Characterization Techniques by Differential Mobility Analysis. Doctoral dissertation, University of Maryland, College Park
Duelge K, Mulholland G, Zachariah M, Hackley VA (2022) Accurate nanoparticle size determination using electrical mobility measurements in the step and scan modes. Aerosol Sci Technol 56(12):1096–1113
Eid MM (2022) Characterization of nanoparticles by FTIR and FTIR-microscopy. In: Handbook of consumer nanoproducts. Springer Singapore, Singapore, pp 1–30
Eskandari V, Sahbafar H, Zeinalizad L, Marashipour R, Hadi A (2022) A review of paper-based substrates as surface-enhanced Raman spectroscopy (SERS) biosensors and microfluidic paper-based SERS platforms. J Comput Appl Mech 53(1):142–156
Falsafi SR, Rostamabadi H, Assadpour E, Jafari SM (2020) Morphology and microstructural analysis of bioactive-loaded micro/nanocarriers via microscopy techniques; CLSM/SEM/TEM/AFM. Adv Colloid Interf Sci 280:102166
Farivar F, Yap PL, Hassan K, Tung TT, Tran DN, Pollard AJ, Losic D (2021) Unlocking thermogravimetric analysis (TGA) in the fight against “fake graphene” materials. Carbon 179:505–513
Foreman-Ortiz IU, Ma TF, Hoover BM, Wu M, Murphy CJ, Murphy RM, Pedersen JA (2022) Nanoparticle tracking analysis and statistical mixture distribution analysis to quantify nanoparticle–vesicle binding. J Colloid Interface Sci 615:50–58
Fu Y, Liu T, Wang H, Wang Z, Hou L, Jiang J, Xu T (2024) Applications of nanomaterial technology in biosensing. J Sci Adv Material Devices 2:100694
Garavand F, Cacciotti I, Vahedikia N, Rehman A, Tarhan Ö, Akbari-Alavijeh S, Shaddel R, Rashidinejad A, Nejatian M, Jafarzadeh S, Azizi-Lalabadi M (2022) A comprehensive review on the nanocomposites loaded with chitosan nanoparticles for food packaging. Crit Rev Food Sci Nutr 62(5):1383–1416
Gebre SH (2023) Bio-inspired synthesis of metal and metal oxide nanoparticles: the key role of phytochemicals. J Clust Sci 34(2):665–704
Greczynski G, Haasch RT, Hellgren N, Lewin E, Hultman L (2023) X-ray photoelectron spectroscopy of thin films. Nat Rev Methods Primers 3(1):40
Griffiths D, Carnell-Morris P, Wright M (2020) Nanoparticle tracking analysis for multiparameter characterization and counting of nanoparticle suspensions. In: Nanoparticles in biology and medicine: methods and protocols. Springer US, New York, pp 289–303
Gross-Rother J, Blech M, Preis E, Bakowsky U, Garidel P (2020) Particle detection and characterization for biopharmaceutical applications: current principles of established and alternative techniques. Pharmaceutics. 12(11):1112
Gupta A, Juyal VK, Anjum B, Gupta S (2025) Vis spectroscopy in drug discovery: illuminating insights, vol 10. Modern Spectroscopic Techniques for Drug Discovery and Environmental Sustainability, p 97
Harish V, Ansari MM, Tewari D, Yadav AB, Sharma N, Bawarig S, García-Betancourt ML, Karatutlu A, Bechelany M, Barhoum A (2023) Cutting-edge advances in tailoring size, shape, and functionality of nanoparticles and nanostructures: a review. J Taiwan Inst Chem Eng 149:105010
Hatzakis E (2019) Nuclear magnetic resonance (NMR) spectroscopy in food science: a comprehensive review. Compr Rev Food Sci Food Saf 18(1):189–220
Irwansyah FS, Amal AI, Diyanthi EW, Hadisantoso EP, Noviyanti AR, Eddy DR, Risdiana R (2024) How to read and determine the specific surface area of inorganic materials using the Brunauer-Emmett-teller (BET) method. ASEAN J Sci Eng 4(1):61–70
Jain AS, Pawar PS, Sarkar A, Junnuthula V, Dyawanapelly S (2021) Bionanofactories for green synthesis of silver nanoparticles: toward antimicrobial applications. Int J Mol Sci 22(21):11993
Jana S, Jana S (eds) (2022) Micro-and Nanoengineered gum-based biomaterials for drug delivery and biomedical applications. Elsevier

Joo S, Jeong U, Gong C, Hong S (2024) Bridging mechanical and electrical analyses in AFM: advances, techniques, and applications. Acc Mater Res

Karakuş E, Coşkun B (2023) A brief overview of basic surface morphology characterization techniques in thin films. J Mater Electron Devices 3(1):32–37

Karnwal A, Jassim AY, Mohammed AA, Sharma V, Al-Tawaha AR, Sivanesan I (2024) Nanotechnology for healthcare: plant-derived nanoparticles in disease treatment and regenerative medicine. Pharmaceuticals. 17(12):1711

Kaushik A, Singh RK, Tyagi PK (2023) Green synthesized nanoparticle based drug delivery: recent trends and future prospects. Precis Nanomed 6(4):1109–1131

Khan Y, Sadia H, Ali Shah SZ, Khan MN, Shah AA, Ullah N, Ullah MF, Bibi H, Bafakeeh OT, Khedher NB, Eldin SM (2022) Classification, synthetic, and characterization approaches to nanoparticles, and their applications in various fields of nanotechnology: a review. Catalysts 12(11):1386

Khlebtsov NG, Dykman LA, Khlebtsov BN (2022) Synthesis and plasmonic tuning of gold and gold–silver nanoparticles. Russ Chem Rev 91(10):RCR5058

Khonina SN, Kazanskiy NL, Butt MA (2024) Grayscale lithography and a brief introduction to other widely used lithographic methods: a state-of-the-art review. Micromachines 15(11):1321

Kijo-Kleczkowska A, Szumera M, Gnatowski A, Sadkowski D (2022) Comparative thermal analysis of coal fuels, biomass, fly ash and polyamide. Energy 258:124840

Knysh A, Sokolov P, Nabiev I (2023) Dynamic light scattering analysis in biomedical research and applications of nanoparticles and polymers. J Biomed Photonics Eng 9(2):020203

Kowkabany G, Bao Y (2024) Nanoparticle tracking analysis: an effective tool to characterize extracellular vesicles. Molecules 29(19):4672

Kunc F, Gallerneault M, Kodra O, Brinkmann A, Lopinski GP, Johnston LJ (2022 Jun) Surface chemistry of metal oxide nanoparticles: NMR and TGA quantification. Anal Bioanal Chem 414(15):4409–4425

Kundu S, Nithiyanantham U (2013) In situ formation of curcumin stabilized shape-selective Ag nanostructures in aqueous solution and their pronounced SERS activity. RSC Adv 3(47):25278–25290

Lan S, Zhu L, Wu Z, Gu L, Zhang Q, Kong H, Liu J, Song R, Liu S, Sha G, Wang Y (2021) A medium-range structure motif linking amorphous and crystalline states. Nat Mater 20(10):1347–1352

Leong HS, Butler KS, Brinker CJ, Azzawi M, Conlan S, Dufès C, Owen A, Rannard S, Scott C, Chen C, Dobrovolskaia MA (2019) On the issue of transparency and reproducibility in nanomedicine. Nat Nanotechnol 14(7):629–635

Li Q, Zhang T, Pan Y, Ciacchi LC, Xu B, Wei G (2016) AFM-based force spectroscopy for bioimaging and biosensing. RSC Adv 6(16):12893–12912

Li P, Zhou B, Ge M, Jing X, Yang L (2022) Metal coordination induced SERS nanoprobe for sensitive and selective detection of histamine in serum. Talanta 237:122913

Lin Y, Zhou M, Tai X, Li H, Han X, Yu J (2021) Analytical transmission electron microscopy for emerging advanced materials. Matter 4(7):2309–2339

Lin X, Zhu C, Chen M, Gonzalez PR, Chen X, Zhao Z, Danino D, Corke H (2024) Development of gastric digestion-tolerant plant protein-based nanoparticles: fabrication, characterization, antioxidant activity, and stability. Food Hydrocoll 151:109815

Lv J, Christie P, Zhang S (2019) Uptake, translocation, and transformation of metal-based nanoparticles in plants: recent advances and methodological challenges. Environ Sci Nano 6(1):41–59

Mabrouk M, Das DB, Salem ZA, Beherei HH (2021) Nanomaterials for biomedical applications: production, characterisations, recent trends and difficulties. Molecules 26(4):1077

Malatesta M (2021) Transmission electron microscopy as a powerful tool to investigate the interaction of nanoparticles with subcellular structures. Int J Mol Sci 22(23):12789

Malekjani N, Jafari SM (2022) Intelligent and probabilistic models for evaluating the release of food bioactive ingredients from carriers/nanocarriers. Food Bioprocess Technol 15(7):1495–1516

Mishra B, Panda J, Mishra AK, Nath PC, Nayak PK, Mahapatra U, Sharma M, Chopra H, Mohanta YK, Sridhar K (2024) Recent advances in sustainable biopolymer-based nanocomposites for smart food packaging: a review. Int J Biol Macromol 11:135583

Mohammadidargah M, Pedram P, Cabrera-Barjas G, Delattre C, Nesic A, Santagata G, Cerruti P, Moeini A (2024) Biomimetic synthesis of nanoparticles: a comprehensive review on green synthesis of nanoparticles with a focus on Prosopis farcta plant extracts and biomedical applications. Adv Colloid Interf Sci 15:103277

Mourdikoudis S, Pallares RM, Thanh NT (2018) Characterization techniques for nanoparticles: comparison and complementarity upon studying nanoparticle properties. Nanoscale 10(27):12871–12934

Ngouangna EN, Jaafar MZ, Norddin MM, Agi A, Oseh JO, Mamah S (2022) Surface modification of nanoparticles to improve oil recovery mechanisms: a critical review of the methods, influencing parameters, advances and prospects. J Mol Liq 360:119502

Nguyen-Tri P, Ghassemi P, Carriere P, Nanda S, Assadi AA, Nguyen DD (2020) Recent applications of advanced atomic force microscopy in polymer science: a review. Polymers 12(5):1142

Nissen M, Förster R, Wieduwilt T, Lorenz A, Jiang S, Hauswald W, Schmidt MA (2022) Nanoparticle tracking in single-antiresonant-element fiber for high-precision size distribution analysis of mono-and polydisperse samples. Small 18(38):2202024

Okeke ES, Nweze EJ, Anaduaka EG, Okoye CO, Anosike CA, Joshua PE, Ezeorba TP (2023) Plant-derived nanomaterials (PDNM): a review on pharmacological potentials against pathogenic microbes, antimicrobial resistance (AMR) and some metabolic diseases. 3 Biotech. 13(9):291

Oliveira C, Chaves CR, Bargiela P, da Rocha MD, da Silva AF, Chubaci JF, Boström M, Persson C, Malta M (2021) Surface studies of the chemical environment in gold nanorods supported by X-ray photoelectron spectroscopy (XPS) and ab initio calculations. J Mater Res Technol 15:768–776

Ortiz AC, Yañez O, Salas-Huenuleo E, Morales JO (2021) Development of a nanostructured lipid carrier (NLC) by a low-energy method, comparison of release kinetics and molecular dynamics simulation. Pharmaceutics. 13(4):531

Patil AS (2020) Plant secondary metabolites: isolation, characterization and biological properties. Studera Press

Pechyen C, Tangnorawich B, Toommee S, Marks R, Parcharoen Y (2024) Green synthesis of metal nanoparticles, characterization, and biosensing applications. Sensors Int 31:100287

Pochapski DJ, Carvalho dos Santos C, Leite GW, Pulcinelli SH, Santilli CV (2021) Zeta potential and colloidal stability predictions for inorganic nanoparticle dispersions: effects of experimental conditions and electrokinetic models on the interpretation of results. Langmuir 37(45):13379–13389

Puri A, Mohite P, Ansari Y, Mukerjee N, Alharbi HM, Upaganlawar A, Thorat N (2024) Plant-derived selenium nanoparticles: investigating unique morphologies, enhancing therapeutic uses, and leading the way in tailored medical treatments. Material Adv 5(9):3602–3628

Rahdar A, Amini N, Askari F, Susan MA (2019) Dynamic light scattering: a useful technique to characterize nanoparticles. J Nanoanalysis 6(2):80–89

Rahman M, Ou Q, Pui D (2024) Electrospray-scanning mobility particle sizer (ES-SMPS) technique: superior sizing and multimodal characterization of colloidal nanoparticles compared to NTA and DLS. Anal Chem 96(48):18946–18956

Reif B, Ashbrook SE, Emsley L, Hong M (2021) Solid-state NMR spectroscopy. Nat Rev Methods Primers 1(1):2

Restrepo CV, Villa CC (2021) Synthesis of silver nanoparticles, influence of capping agents, and dependence on size and shape: a review. Environ Nanotechnol Monitor Manag 15:100428

Saadatkhah N, Carillo Garcia A, Ackermann S, Leclerc P, Latifi M, Samih S, Patience GS, Chaouki J (2020) Experimental methods in chemical engineering: Thermogravimetric analysis—TGA. Can J Chem Eng 98(1):34–43

Sabaghi M, Tavasoli S, Hoseyni SZ, Mozafari MR, Degraeve P, Katouzian I (2022) A critical review on approaches to regulate the release rate of bioactive compounds from biopolymeric matrices. Food Chem 382:132411

Sadhasivam S, Vinayagam V, Balasubramaniyan M (2020) Recent advancement in biogenic synthesis of iron nanoparticles. J Mol Struct 1217:128372

Shaikh WA, Chakraborty S, Owens G, Islam RU (2021) A review of the phytochemical mediated synthesis of AgNP (silver nanoparticle): the wonder particle of the past decade. Appl Nanosci 11(11):2625–2660

Shankar SN, Mital K, Le E, Lewis GS, Eiguren-Fernandez A, Sabo-Attwood T, Wu CY (2023) Assessment of Scanning Mobility Particle Sizer (SMPS) for online monitoring of delivered dose in an in vitro aerosol exposure system. Toxicol In Vitro 92:105650

Sharma NK, Vishwakarma J, Rai S, Alomar TS, AlMasoud N, Bhattarai A (2022) Green route synthesis and characterization techniques of silver nanoparticles and their biological adeptness. ACS Omega 7(31):27004–27020

Siddique IM (2024) Exploring functional groups and molecular structures: a comprehensive analysis using FTIR spectroscopy. Development 28:1–2

Singh G, Ramadass K, Sooriyakumar P, Hettithanthri O, Vithange M, Bolan N, Tavakkoli E, Van Zwieten L, Vinu A (2022) Nanoporous materials for pesticide formulation and delivery in the agricultural sector. J Control Release 343:187–206

Sinha P, Datar A, Jeong C, Deng X, Chung YG, Lin LC (2019) Surface area determination of porous materials using the Brunauer–Emmett–Teller (BET) method: limitations and improvements. J Phys Chem C 123(33):20195–20209

Sohail M, Rabbi F, Younas A, Hussain A, Yu B, Li Y, Iqbal S, Ullah KH, Qadeer A, Aquib M, Iqbal H (2022) Herbal bioactive–based nano drug delivery systems. In: Herbal bioactive-based drug delivery systems. Academic Press, pp 169–193

Soni V, Raizada P, Singh P, Cuong HN, Saini A, Saini RV, Van Le Q, Nadda AK, Le TT, Nguyen VH (2021) Sustainable and green trends in using plant extracts for the synthesis of biogenic metal nanoparticles toward environmental and pharmaceutical advances: a review. Environ Res 202:111622

Stan G, King SW (2020) Atomic force microscopy for nanoscale mechanical property characterization. J Vac Sci Technol B 38(6)

Stevie FA, Donley CL (2020) Introduction to x-ray photoelectron spectroscopy. J Vac Sci Technol A 38(6)

Tahir MA, Dina NE, Cheng H, Valev VK, Zhang L (2021) Surface-enhanced Raman spectroscopy for bioanalysis and diagnosis. Nanoscale 13(27):11593–11634

Talha M, Islam NU, Zahoor M, Sadiq A, Nawaz A, Khan FA, Gulfam N, Alshamrani SA, Nahari MH, Alshahrani MA, Mahnashi MH (2022) Biological evaluation, phytochemical screening, and fabrication of Indigofera linifolia leaves extract-loaded nanoparticles. Molecules 27(15):4707

Tampieri A, Szabó M, Medina F, Gulyás H (2021) A brief introduction to the basics of NMR spectroscopy and selected examples of its applications to materials characterization. Phys Sci Rev 6(1):20190086

Thakur A, Kumar A (2022) Recent advances on rapid detection and remediation of environmental pollutants utilizing nanomaterials-based (bio) sensors. Sci Total Environ 834:155219

Villalobos-Noriega JM, Rodríguez-León E, Rodríguez-Beas C, Larios-Rodríguez E, Plascencia-Jatomea M, Martínez-Higuera A, Acuña-Campa H, García-Galaz A, Mora-Monroy R, Alvarez-Cirerol FJ, Rodríguez-Vázquez BE (2021) Au@ Ag Core@ Shell nanoparticles synthesized with rumex hymenosepalus as antimicrobial agent. Nanoscale Res Lett 16:1–9

Wang HL, You EM, Panneerselvam R, Ding SY, Tian ZQ (2021) Advances of surface-enhanced Raman and IR spectroscopies: from nano/microstructures to macro-optical design. Light Sci Appl 10(1):161

Wang D, Du X, He L (2022) Investigation of the impact of a pesticide adjuvant on dimethoate persistence, penetration, and stability on apples using surface-enhanced Raman spectroscopy. ACS Agric Sci Technol 2(4):788–795

Wells MJ, Chen JY, Bodycomb J, Wolgemuth D, Stretz HA, Zacheis GA, Bautista M, Bell KY (2024) Multi-laser nanoparticle tracking analysis (NTA): a unique method to visualize dynamic (shear) and dynamic (Brownian motion) light scattering and quantify nonliving natural organic matter (NNOM) in environmental water. Sci Total Environ 949:174985

White R (2020) Chromatography/Fourier transform infrared spectroscopy and its applications. CRC Press

Xu K, Wang B (2022) Recent trends in AFM imaging speed and improvement methods. Curr Nanosci 18(3):277–290

Xu Z, Pan K, Liu L, Tang S, Zhang Z Condensation particle counter. In: Condensation particle counting technology and its applications. CRC Press, pp 96–130

Yang L, Wei J, Ma Z, Song P, Ma J, Zhao Y, Huang Z, Zhang M, Yang F, Wang X (2019) The fabrication of micro/nano structures by laser machining. Nanomaterials 9(12):1789

Yavuz E, Sakir M, Onses MS, Salem S, Yilmaz E (2024) Advancements in reusable SERS substrates for trace analysis applications. Talanta 31:126640

Yusuf A, Almotairy AR, Henidi H, Alshehri OY, Aldughaim MS (2023) Nanoparticles as drug delivery systems: a review of the implication of nanoparticles' physicochemical properties on responses in biological systems. Polymers 15(7):1596

Zhang L, Ma P, Dai L, Li S, Yu W, Guan J (2021) In situ crystallization and growth of TiO 2 nanospheres between MXene layers for improved adsorption and visible light photocatalysis. Catal Sci Technol 11(11):3834–3844

Chapter 8
Application of Plant Based Nanomaterial for Treating Dementia

Abraham Olufemi Asuku, Priscilla Omobonke Adebayo, Ayodeji Johnson Ajibare, and Saheed Abiodun Ayoola

8.1 Introduction

Dementia is a term for several numbers of illnesses that impair memory, thinking, and the capacity to carry out daily tasks. Over time, the illness worsens. The likelihood of a person having dementia increases with aging but there are some exceptions to this (Asuku et al. 2024a). Dementia is known as a syndrome that can be caused by a various number of diseases that gradually damage the brain and kill the nerve cells which eventually causes a decline in cognition (Ayinla et al. 2020). For example, the dementia syndrome that is associated with a gradual loss of language abilities occurs from different diseases including frontotemporal lobar degeneration, tumour in the language cortex and Alzheimer disease (AD). Based on worldwide estimates, dementia affects up to 7% of individuals above the age 65 with a higher prevalence of 8–10% in developed countries due to their higher life expectancies (Baumgart et al. 2015). Based on the report of Diagnostic and Statistical Manual of Mental Disorders-5 (DSM-5), the term dementia is now known as major neurocognitive disorder (MND) (Bennett et al. 2020).

A. O. Asuku (✉) · S. A. Ayoola
Bioresources Development Centre, National Biotechnology Research and Development Agency, Ogbomoso, Nigeria

P. O. Adebayo
Department of Physiology, Faculty of Basic Medical Sciences, College of Health Sciences, University of Ilorin, Ilorin, Nigeria

A. J. Ajibare
Department of Physiology, Neuro-Reproductive and Metabolism Unit, Faculty of Basic Medical and Health Sciences, College of Medicine, Lead City University, Ibadan, Oyo State, Nigeria

V. Ghosh et al. (eds.), *Applications of Plant Bioactive Compounds-Based Nanoformulations*, Nanotechnology in the Life Sciences,
https://doi.org/10.1007/978-3-032-13086-0_8

Dementia is a remarkable health condition that is facing the public. Approximately about 47 million individuals are suffering from dementia globally, this is proposed to rise to 131 million by the year 2050 (World Alzheimer Report 2015). The common dementia symptoms includes continuous deterioration in reasoning, memory, learning, language, and judgement, along with difficulties in performing familiar tasks and focusing, confusion about time and location, and mood swings (Basheikh et al. 2021; Asuku et al. 2025). Due to the loss of neuronal cells in dementia, no cure exists at this time. As a result, symptomatic management remains the primary approach: addressing behavioural issues, modifying the environment to support function, and providing safety counselling.

The main objective of dementia management is to reduce the patient's cognitive decline and slow cognitive deterioration. Depending on the type of dementia, different treatments are currently available. Memantine and inhibitors of cholinesterase are the two main medications recommended for dementia treatment. Acetylcholinesterase inhibitors (AChEIs), like rivastigmine, donepezil and galantamine can be used to treat dementia associated with AD. The dosage of the drugs varies based on the severity of the disease. Memantine could also be administered. Those drugs are very costly and presented with a lot of adverse effects.

Nanomedicine is the area of biomedical sciences that applies nanotechnology to prevent, cure, and treat different disorders with the aid of nanoscale materials such as nanorobots and biocompatible nanoparticles for diverse purposes such as diagnosis, delivery, and treatment in a living organism (Asuku and Ogungbangbe 2024). Nanotechnology is referred to as the production, modification and analysis of materials that are smaller than 100 nm in size called nanomaterials. Nanomaterials have been greatly utilized in the healthcare and medical fields in the previous 20 years due to their exceptional properties and minuscule size (Hassan et al. 2015). The interest in these materials occurs as a result of their distinct physiochemical features which arise from their nanoscale dimensions including electrical conductivity, magnetism, altered chemical reactivity and optical qualities (Adisa et al. 2020). These nanoscale materials have been created into a variety of BBB-crossing nanoparticles (NPs). These nanoparticles have the potential to act on both the cellular substances and molecular structures. These structures, which might include proteins, nucleic acids, peptides and cellular membrane components can alter the physiology of the biological activities in tissues and cells in an unexpected ways. The NPs-based therapy strategy is steadily attracting a lot of interest.

8.2 Overview of Plant Based Nanomaterials

There are about 35,000 different species of natural herbs that are used as medicine all around the world at the moment. The components of these natural medications are the extracts or active ingredients of these plants which include polyphenols, terpenes and flavonoids among others. Based on the work of Zhang et al. (2019), these compounds possess diverse functions and uses, including being used as antipsychotics, antifatigue, antioxidants, antidepressants, anxiolytics, anti-inflammatory, and anticancer agents. The herbs with the antioxidant, antidepressant, and anti-inflammatory properties may produce a promising result in AD management. However, the following drawbacks limit the clinical application of these medications, toxicity, insolubility, and extremely high molecular weight. Such limitation may be eliminated by the appropriate combination of natural extracts and the application of nanotechnology. Here, we outline a few of the most widely used herbal extracts that target dementia, including curcumin, triptolide, resveratrol, quercetin, berberine, and epigallocatechin gallate. Their administration is through the nasal route, respiratory system or the gastrointestinal tract to cross the blood brain barrier (BBB) and get to the central nervous system (CNS). They can be loaded on lipids nanoparticles, nanosuspension and nanoemulsion like liposomes.

8.3 Types of Plant-Based Nanomaterials

Nanomaterial may be further categorized into nanoparticles, nanofibers, and nanosheets.

Plant-based nanoparticles represent a significant role in medical fields and have been proven to have several applications in biomedicine. These nanoparticles have an ideal shape and size that makes them very appropriate for safe delivery of drugs, dementia treatment, and cancer treatment, among others. Because of their nano size, NPs possess the ability to pass through tiny capillaries and allow the efficient absorption of loaded drugs into cells and enable their accumulation at specific sites. Additionally, the drug can be retained at the active location for an extended duration. It is feasible to regulate the dimensions of nanoparticles utilized in delivery of drugs, offering effective protection for the medication. The primary benefits of utilizing nanoparticle (NP) approaches include targeted drug delivery and improved safety profiles (Patra et al. 2018). Phytochemical-rich herbal extracts, comprising flavonoids, alkaloids, and other metabolites, therefore, present an attractive alternative source for nanomaterials preparation by greener methods.

Various secondary metabolites in these extracts are acting as agents that stabilize or/and reduce for the structural development of NPs, for various applications across many fields, especially in nanomedicine (Parveen et al. 2015). A major factor governing the physiochemical characteristics of nanoparticles is the increasing size of particles while the ratio of surface area-to-volume enters the domain where the impact of quantum becomes significant. Nanomaterials have different properties compared with bulk materials, since they possess a much larger surface area to volume ratio, which increases greatly as the particle size reduces. The nanoparticles are classified based on the synthesis method employed for their preparation. The production of NPs from plants has a number of merits over chemical and physical methods of synthesis. It is highly effective, faster, and less expensive than chemical or physical methods and involves lower energy with moderate conditions. In general, roots, stems, latex, leaves, and even the bark of almost every part of the plant

Table 8.1 Comparison of conventional drugs and nanoparticles-based therapies

Nanoparticle-based therapies	Conventional drugs
High bioavailability	Low bioavailability
Reduced side effects	Increased side effects
Overcome barrier posed by the blood–brain barrier due to its physiological properties like reactivity, strength, and surface area	Difficulty in penetrating the blood–brain barrier
Enhance drug target	Limited drug target
Enhance therapeutic efficacy	Reduced therapeutic efficacy
High absorption	Poor absorption

can be used for NP synthesis. Even though the exact mechanisms involved in the synthesis of NPs by plants are not known, it has been confirmed that phenols, flavonoids, and several other bioactive compounds found in plants act as reducing agents to the metal ions and, in turn, result in the formation of NPs. It includes a variety of valuable properties, together with several different bioactive reducing agents, so biological syntheses of nanomaterials-actually silver nanoparticles-have received considerable interest.

The biomedical application of the various nanoparticles gotten from plants occurs by reformulating medications into smaller particle sizes; this is nanotechnology, and it can greatly enhance oral bioavailability (Table 8.1) and thus replace many traditional oral formulations, which are inefficient due to low bioavailability as a therapeutic choice. Also, the different systems are non-invasive, hence much safer and more comfortable to the patients, enhance the poor pharmacokinetics of the peptides and proteins that can easily be degraded in the body and limit the drugs toxic effects (Rashmi and Mohanty 2022).

8.4 Mechanisms of Action of Plant-Based Nanomaterials

With their capability for the therapeutic agents delivery at active sites, the nanocarriers are used immensely in medical settings, providing for targeted and controlled administration. The potential mechanism of action includes:

8.5 Roles of Nanoparticles-Based Therapies in the Brain

8.5.1 Enhancing the Delivery of Drugs to the Brain

The treatment of neurodegenerative diseases (NDs) like AD and Parkinson's disease (PD) is always not effective because of low brain penetration of therapeutic compounds via the BBB (Asuku et al. 2024b). The CNS has diverse physical barriers that prevent it from neurotoxic substances, circulating blood cells, and infectious

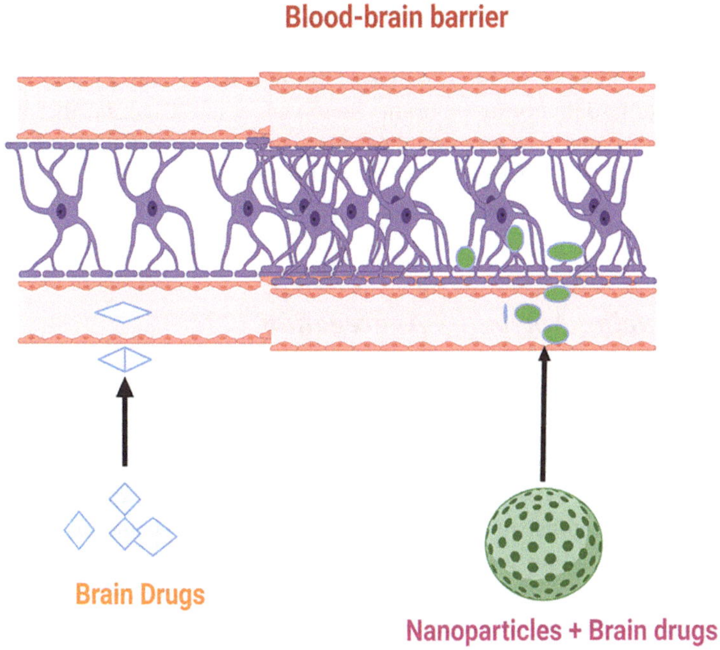

Fig. 8.1 Drug delivery to the brain with and without nanoparticles

micro-organisms, which also prevent most therapeutic methods (Fig. 8.1). These include the BBB, blood-cerebrospinal fluid barrier, blood-retinal barrier, and blood-spinal cord barrier (Asuku et al. 2024b; Nady et al. 2023). In a condition in which the body is functioning normally, molecules may pass through the BBB by diverse transport mechanisms: transport proteins, lipophilic diffusion, transcellular-paracellular tight junctions, adsorptive transcytosis, and receptor-mediated endocytosis (González-Sanguino et al. 2020). A plethora of researches have investigated these transport mechanisms in improving CNS drugs delivery. Considering the inadequacy of the traditional means of delivery, there is an obvious need for specific and active strategies in enhancing therapies transport across the BBB to attenuate these diseases. Nanocarrier-based systems have recently made significant advancement in overcoming limitations of traditional therapies simply because of their nanometre-scale size, allowing for the unique functional characteristics (Hano and Abbasi 2021). Its physiological properties like stability, surface area, reactivity, strength, sensitivity, and solubility greatly allow the traversing of nanomedicine across the BBB so easily, standing in opposition to major hurdles for therapies pertinent to the CNS and also the therapeutic agent delivery (Saeedi et al. 2021). Besides, there are three major pathways for the nanomaterial transport through the BBB, which include alternative, non-invasive, and invasive drug delivery pathways. Of all these, the non-invasive techniques are widely employed and rely on endogenous cellular processes that facilitate movement of drugs across the BBB by transcellular pathways (Xie et al. 2019). Combining intranasal with CPP-modified PLGA NPs for brain insulin delivery in AD, this has been developed; it significantly enhances the efficiency of brain delivery through nasal administration and thus potentially insulin delivery to the brain (Steinbach et al. 2016). Furthermore, drug-carrying NPs are capable of crossing the BBB via receptor-mediated and adsorption-mediated endocytosis, other than carrier-mediated pathways utilized by various endogenous substances for brain delivery (Zhou et al. 2018).

8.5.2 *Reducing Amyloid-β Aggregation*

The altered breakdown of amyloid precursor protein into ordered beta-sheet oligomers, Aβ, and insoluble fibrils disrupts synaptic signalling, leading to neurotoxicity, neurodegeneration, and eventually dementia (Bagyinszky et al. 2017). Such processes further lead to hyperphosphorylation of tau protein. Memory loss and cognitive impairment are consequences of reduced choline uptake and decreased release of acetylcholine, leading to significant presynaptic cholinergic cell death. Thus, nanocarriers with tailored functions can be designed to improve drug delivery by anchoring AD medications onto the surfaces of nanoparticles that can surmount the CNS-related obstacles highlighted earlier. Molecules that can bind Aβ have been attached to nanoparticles for therapeutic purposes in AD. Among the currently utilized nanoparticles, PEG-PLA and lipososmes nanoparticles are favoured due to their recognized safety, low immunogenic response, and complete biodegradability.

8.5.3 *Neuroinflammation Modulation*

Nanomaterials interact with the brain and help modulate neuroinflammation. For instance, quercetin, a flavonoid bioactive compound which is found in vegetables and fruits, possess anti-inflammatory properties; it helps in blocking the inflammatory response, and it can be encapsulated into nanoparticles to improve its brain delivery.

8.6 Applications of Plant Based Nano-Material in Dementia Treatment

8.6.1 *Drug Delivery Systems*

The absence of a foolproof treatment and diagnostic method for this challenging disease prompts researchers to investigate various therapeutic strategies aimed at improving both diagnosis and treatment for dementia (Zvěřová 2018). Over the past

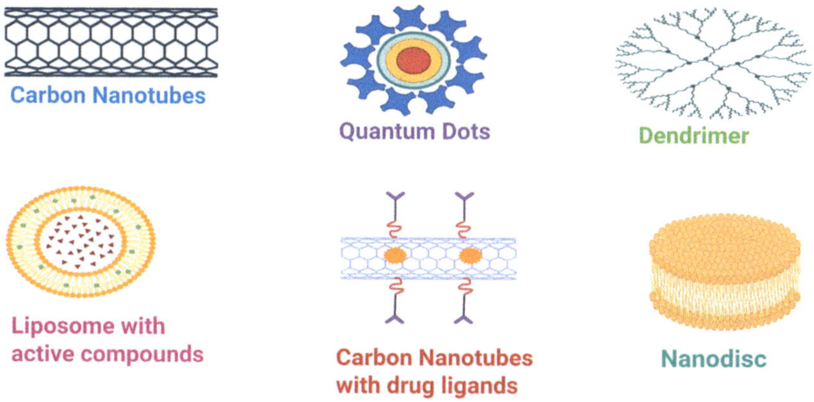

Fig. 8.2 Nanomaterial-based carriers

ten years, a diverse range of inorganic and organic nanomaterial-based carriers (Fig. 8.2), including dendrimers, carbon nanotubes, fullerenes, quantum dots (QDs), liposomes (LIPs), nanodiscs and magnetic NPs, have been explored for their possible applications in biosensors, regeneration of tissues, diagnostics, targeted drug delivery, and cell culture, among other areas in biomedical sciences (Zdrojewicz et al. 2016). These nanocarriers have the ability to cross the BBB so as to get to the targeted area of AD patients brain. Consequently, these nanosized vehicles can exhibit superior clinical efficacy.

(A) Nanotubes

Nanotubes take the shape of a cylinder created from rolled graphite with buckyballs positioned at one or both ends (Ménard-Moyon 2018). Nanotubes can cross cell membranes by endocytosis, direct insertion, or diffusion. In order to make these tube-like nanostructures more accessible to cells, ammonium or carboxylic functional groups could be attached to them. Thus, nanotubes can act as carriers for the cellular delivery of nucleic acids, peptides, and other therapeutic substances. Yang et al. (2013) used carbon nanotubes (CNTs) to target brain cells. They were orally given to mice for 10 days. Electron microscopy examination showed that there are significant improvements in learning, memory, and overall cognitive function due to enhanced acetylcholine transport by CNTs in induced AD mouse model (Yang et al. 2013).

(B) Quantum dots: Quantum dots are from 2 to 10 nm in size, comprising a core-shell structure of inorganic material, while an aqueous organic substance acts as the coating to which biomolecules can attach, enabling targeting of various biomarkers. Quan et al. (2016) developed a quantum dot nano-vehicle capable of targeting surface cells, playing a pivotal function in the detection of AD. Therefore, QDs enhance the sensitivity for detecting amyloid-β in AD progression (Quan et al. 2016).

(C) Liposomes: Regarded as the classic drug delivery vehicles, liposomes (LIP) are spherical structures that have a lipid bilayer membrane, which can be unilamellar or multilamellar in nature, and contain an aqueous interior. These could be a promising approach for AD treatment. LIPs can encapsulate water-dissolving drugs in the aqueous core while incorporating lipophilic drugs within the membrane. This allows for rapid delivery and improved efficacy of the drugs (Gregoriadis 2016).

This approach offers advantages like minimizing adverse impacts on normal cells, enhancing specific drug delivery to target sites, and enhancing the overall safety and efficacy profile of the medication (Andey et al. 2016). Given the shortcomings of conventional medications or biomolecules in reaching clinical trials, targeted nano-LIPs are emerging as promising drug delivery systems for AD (Ross et al. 2018). Though they have not yet advanced to clinical trials, they have shown to be flexible and biocompatible, with great capabilities for transporting many therapeutic agents across the BBB to reach brain tissues. Liposomes can be used for either one target or many targets. A vast number of modification using crossing peptides via the BBB, as well as complexes of integrated liposome-ligand featuring retroinverted peptides, phosphatidic acid, and curcumin have been produced, with the intention of targeting and inhibiting aggregation of β-amyloid (Ross et al. 2018). Liposomes are in the group of promising delivery systems which could cross the BBB, and shield the nervous system from aggregation of amyloid plaque.

(D) Carbon dots (CDs): They are 0D fluorescent, carbon-based nanomaterials, having sizes less than 10 nm and are commonly classified as carbonized polymer dots (CPDs), carbon quantum dots (CQDs), carbon nitride dots (CNDs), and carbonized graphene quantum dots (GQDs) (Tejwan et al. 2020). They possess special properties like modified surface functional groups, less toxicity, great photoluminescence, great controlled size distribution and biocompatibility; they have the ability to cross the BBB (Zhang et al. 2021). The CDs were surface modified with carboxyl and amine groups to complex with various drugs for the CNS, and these act as carriers to carry the medications into the CNS to treat dementia (Du et al. 2019).

This chapter highlights a few commonly used herbal extracts aimed at treating dementia. These herbal extracts include quercetin, epigallocatechin gallate, berberine, triptolide, curcumin, and resveratrol. Those extracts were in several nanoparticle forms such as liposomes, nanoemulsion, immunoliosomes, nanosuspension, carbon nanotubes, and solid lipid nanoparticles, which were administered via gastrointestinal route, respiratory system, or nasal route to enable the herbal extract to pass through the BBB more easily to the CNS.

8.6.2 Curcumin

Curcumin, a polyphenolic compound, is extensively utilized in Asia for treating inflammatory conditions, due to its ability to combat inflammation (Abdulmalek et al. 2021). Curcumin is a naturally occurring phenolic compound known to have positive effects on several chronic health issues such as obesity, diabetes, and depression (Rinwa and Kumar 2013). Notably, these chronic conditions can be risk factors for AD and associated with its causes or consequences. Also, curcumin is a common spice in cooking. Curcumin is promising in the management and treatment of diverse neurodegenerative disorders as a result of its ability to pass through the BBB. Curcumin can help slow down AD symptoms through moderating its various risk factors (Reitz et al. 2011). Curcumin is found to strengthen the memory capacity by increasing synaptic plasticity and neurogenesis in aged animals (Belviranlı et al. 2013). It may also raise the production of docosahexaenoic acid, which would improve plasma membrane integrity and, therefore, normal functions of synapses and mitochondrial (Wu et al. 2020). On the other hand, curcumin lowered alanine aminotransferase activity, a common marker of hepatic damage, but increased plasma myeloperoxidase, which is linked to inflammatory process (DiSilvestro et al. 2012). Despite having so much variety of biological activity, poor solubility, rapid elimination, and instability have been found in the clinical application of curcumin (Barbara et al. 2017). Consequently, a few nano formulation-based approaches have been developed in the last few decades to address the challenges in preclinical and clinical applications.

Various types of nano formulations are nanoparticles, liposomes/lipids, micelles, and nanogels/micro/water (Abbas 2021). By encapsulating curcumin in nanoparticles, its chemical stability is enhanced while protecting it from pH and enzymatic-induced disintegration. Moreover, incorporating curcumin into nanoparticles boosts its body circulation (Abbas 2021). Cox et al. (2015) showed that chronic supplementation with lipid curcumin preparation enhanced cognition, lowered fatigability and decreased the debilitating impact of psychological stress on mood, contributing to improved quality of life for the aged. Thus, dietary intake of curcumin could decrease the incidence of AD and improve cognitive functions and/or hinder/halt the process of aging/neurodegenerative disorders. The BBB shields the brain from large molecules and also obstructs the entry of therapeutic agents. After being formulated into nanoparticles, curcumin improves its bioavailability and can effectively pass through the BBB to the brain, where it shows its therapeutic impacts (Zhang et al. 2024). Enhancing the penetration of BBB is among the ways through which neurological conditions can be treated effectively. One promising approach is nanostructured lipid carriers, wherein the drug may be embedded or layered within a solid colloidal drug delivery system of a lipid core. Due to great biocompatibility, low toxicity, and biodegradability, this method has been widely used for drug nanotechnology delivery vehicles.

Sadegh Malvajerd et al. (2019) successfully developed the curcumin encapsulation into nanostructured lipid carriers (NLCs) to improve accumulation of drugs in the brain. Among all the carriers, NLCs are more effective for the permeation of BBB. Diverse nanoparticle delivery techniques have been designed for loading curcumin. This aims to reduce the dosage of the exact therapeutic molecule and enhance outcomes with reduced harm and damage to the system. Curcumin decreases the production of Aβ, prevents its accumulation, and enhances its clearance. It also blocks inflammatory signaling pathways and reduces inflammatory cytokines production. Besides, it neutralizes the free radicals and reduces oxidative stress (Liu et al. 2023).

8.6.3 Quercetin

Quercetin is a flavonoid that is commonly seen in many vegetables and fruits like asparagus, onions, apples, and red leaf lettuce, and it has been documented to have protective attributes against oxidative damage by decreasing peroxidation of lipids (Pinheiro et al. 2020a). Furthermore, it assists in removing Aβ, decreasing lysis of cell and preventing inflammation. But, in spite of its diverse pharmacological effects, it has low clinical applications because of its limited solubility, inadequate absorption, and swift metabolism (Pinheiro et al. 2020b). Moreover, quercetin aglycone demonstrates decreased effectiveness after passing through the BBB. Consequently, clinical research on quercetin and its analogues as compounds which have protective impacts on the neurons need to focus on improving their bioavailability, which could enhance clinical outcomes. Many methods and strategies like production of molecules with improved BBB permeability and the use of nanoparticles have been used to address these issues. Some researchers have encapsulated quercetin-modified sulfur nanoparticles (Qc@SNPs) in microbubbles (MB) to produce the Qc@SNPs-MB system (Liu et al. 2020). The system and ultrasound has the ability to open the BBB rapidly. The fast build-up of Qc@SNPs in the brain efficiently decreases the apoptosis of the neuron, inflammatory reactions, oxidative stress of endoplasmic reticulum, and calcium imbalance, thereby protecting the nervous system and fulfilling the objective of AD treatment.

8.6.4 Resveratrol

Polyphenol-rich natural substances are always used to prevent diseases associated with aging because of their neuroprotective attributes and ability to combat inflammation. Resveratrol (RES), a polyphenol, is mostly found in nuts, berries, and grapes. It has many promising properties like antiproliferative, anti-inflammatory, neuroprotective, and antioxidant (Guan et al. 2016). Due to the aforementioned properties and its ability to bind to Aβ1–42, RES was also suggested to have potential for AD management and treatment (Labban et al. 2021). The easiest way of administering RES is the oral route, but its oral administration has many disadvantages, like low bioavailability, poor solubility, poor liver metabolism, fast elimination, and lack of chemical stability (Komorowska et al. 2020). The problems of poor solubility and lack of stability could be addressed by encapsulating RES with nanocarriers, this will improve the bioavailability. The techniques entail nanoemulsions, lipid core nanocapsules, selenium nanoparticles, solid lipid nanoparticles, and transfer bodies.

Salem et al. (2019) designed two types of transfer bodies which are resveratrol-loaded and nanoemulsions, and they enhanced passage through the BBB. The transfer bodies demonstrated better penetration in to the BBB compared to the nanoemulsion and caused a remarkable greater improvement in the amnestic rats' memory than nanoemulsion. As mentioned earlier, nanoparticles from selenium possess unique properties for absorbing Aβ. Yang et al. (2018) encapsulated RES into selenium NPs to develop a Res@SeNP structure. This gave rise to a significant and better penetration of the BBB compared to the use of pure drugs. The transport of resveratrol through the brain is enabled by solid lipid and antibody-functionalized nanoparticles, extending its in vivo metabolism (Loureiro et al. 2017). The encapsulated resveratrol in nanobody demonstrated effective permeation through the BBB and was consequently discovered to be highly accumulated in the brain, which might enhance its bioavailability as well as clinical outcomes.

8.6.5 Epigallocatechin Gallate (EGCG)

EGCG is a product of green tea. It is a polyphenol compounds. It possesses great antioxidant quality. It is a non-toxic compound that has demonstrated healing potential against many diseases, including neurological diseases, tumours, and Down's syndrome (Granja et al. 2017). EGCG can hinder accumulation of tau, decreases aggregation of Aβ, and suppresses the expression of inflammatory cytokines in AD persons. Nevertheless, there is an issue of instability with EGCG (Krupkova et al. 2016). A general approach to decrease the clearance of drugs is to deliver them using nanostructured approaches. It has been discovered and reported that EGCG/AA NPs, dual drug-loaded PEGylated PLGA nanomaterials showed a significant decrease in the drug clearance and increase in brain delivery (Cano et al. 2018). The administration (oral) of EGCG/AA NPs in APP/PS1 transgenic mice caused a substantial synapses increase, reduced brain inflammation, and burden of Aβ (Cano et al. 2019). Preclinical study has reported that long-term aluminum exposure produces neurological alterations which hamper the abilities of rats to learn (Prema et al. 2017). Besides this, administration of aluminium leads to neurodegeneration, decline in cognition, and loss of neurons (Granja et al. 2017). Drugs targeting aluminium have shown promising results for AD treatment. Nanoparticles loaded with epigallocatechin gallate have demonstrated remarkable significant neuroprotective impacts in the animal model of AD induced by aluminium chloride (Singh et al. 2018). Furthermore, it was documented that EGCG-stabilized selenium nanomaterials impede aggregation of Aβ, and break down the already produced Aβ aggregates into safe ones (Yu et al. 2014).

8.6.6 Berberine

Berberine could be obtained from the Chinese medicinal materials of Cortex Phellodendri, Sangean, and Coptis Rhizoma. Study has reported that berberine possesses multiple pharmacological properties, such as controlling lipid and blood glucose levels, and antiviral as well as anti-inflammatory properties (Živančević et al.

2022). Moreover, berberine improves AD model rats' cognitive performance, and has demonstrated potency to inhibit hyperphosphorylation of tau and aggregation of Aβ, making it a promising natural substance for AD prevention and management (Fang et al. 2020; Raju et al. 2021). It has been reported that MSNs-BBR-L/berberine nanoparticles demonstrated an eight-fold increase in passing through the BBB compared to the administration (oral) of only berberine (Singh et al. 2021). The study demonstrated that MSNs-BBR-L showed promise as drug delivery system for the brain. Berberine selective ability to pass through the BBB is a limitation to its clinical use.

8.6.7 Ginkgo Biloba Leaf Extracts

Ginkgo biloba leaf extracts possess diverse bioactive molecules such as terpenoids and flavonoids and they are recognised for their neuroprotective activities. Ginkgo biloba leaf extracts has been reported to be effective for the treatment of AD and dementia (Asuku et al. 2024c). Nanoparticle formulations of ginkgo biloba have been investigated for their attributes to enhance cognitive function and reduce oxidative stress in AD.

8.6.8 Other Compounds

Certain herbal compounds are faced with challenges when attempting to cross the BBB due to their notable harmful and toxic contents. Triptolide, an anti-inflammatory and immunosuppressive compound obtained from *Tripterygium wilfordii*, has demonstrated neuroprotective potential for treating Alzheimer's disease. However, its clinical usage in Alzheimer's patients is constrained as a result of its toxicity. A multicoated nanoparticle of triptolide was designed by Jia et al. (2021). This formulation remarkably decreased cytotoxicity and prevented cell death caused by Aβ1-42 in PC12 cells. *Andrographis paniculata,* is one of the phytoconstituents being researched as an alternative therapy for persons suffering from AD due to its abilities to combat inflammatory process, fights oxidative stress, and protects the neurons (Bilia et al. 2017). It possess poor permeability through the BBB, hence compromising its clinical efficacy against neurological disorders. Delivering andrographolide through nanoparticles significantly improved its permeability across the BBB. Other herbal extracts that have been incorporated into nanoparticles in attempts to improve their accumulation in the brain include piperine (piperine) (Elnaggar et al. 2015), rhynchophylline (Uncaria) (Wu et al. 2020), and Calamus (Saini et al. 2021). The aforementioned results suggest that herbal monomeric compounds encapsulated into nanoparticles can improve their bioavailability through the enhancement of BBB permeability, toxicity reduction, and/or biodistribution modulation. Other nanoparticles are those incorporating rosemary or sage extracts and nanoparticles loaded with omega-3 fatty acids.

8.6.9 *Nanomedicine for the Clearance of Tau Aggregates*

As discussed above, the misfolding and aggregation of Tau protein are major contributors to neuronal death in AD, and hence most of the nanoparticle-based therapies target Tau aggregation. Recent reports have suggested that methylene blue prevents or inhibits Tau aggregation in tauopathies. Okuda et al. synthesized derivatives of curcumin that could function as dual inhibitors for the aggregation of tau and Aβ (Okuda et al. 2016). Furthermore, Gao et al. (2020) reported neuron tau-targeting biomimetic nanoparticles for curcumin delivery, which might impede the progression of AD by the inhibition of tau accumulation (Gao et al. 2020).

8.7 Challenges and Future Direction

Nanoparticles are a promising approach for treatment of dementia since they can minimize enzymatic breakdown, reduce clearance by endothelial cells, and decrease peripheral side effects, all while enhancing targeting, bioavailability, and overcoming barriers posed by the BBB. It is thus crucial that more work should be done on the development of nanoparticles for bioactive compounds recognized to have potent anti-apoptotic, anti-inflammatory and antioxidant properties to ensure high permeability through the BBB. In the future, multi-component and multifunctional nanoparticles will also be prepared by incorporating different sizes, shapes, and functionalities in order to optimize their beneficial properties. Nonetheless, the transition of nanomedicines into clinical use for dementia treatment still presents numerous obstacles. Safety remains a significant concern for the clinical application of nanomaterial for treating dementia.

8.8 Conclusion

Dementia is one of the most perilous conditions that pose a threat to both the mental and physical conditions of the aged. Due to aging of the population, the numbers of aged individuals are growing steadily. However, existing modes of treatments are too meager to meet the therapeutic needs. Traditional herbal preparations, especially from the Traditional Chinese medicine, have gained increasing attention. There is limitation to clinical applications because of their numerous deficiencies. Moreover, in Alzheimer's disease, many conventional drugs do not successfully enter the cells to be treated. The application of nanomaterials is promising in the management of dementia. Nanoparticle-loaded compounds can easily cross the BBB and reach the affected areas. The use of nanotechnology for better delivery of effective natural herbs and bioactive molecules can greatly enhance the efficacy and make us more confident with the cure for dementia.

References

Abbas M (2021) Potential role of nanoparticles in treating the accumulation of amyloid-Beta peptide in Alzheimer's patients. Polymers 13:1051. https://doi.org/10.3390/polym13071051

Abdulmalek S, Nasef M, Awad D, Balbaa M (2021) Protective effect of natural antioxidant, curcumin nanoparticles, and zinc oxide nanoparticles against type 2 diabetes-promoted hippocampal neurotoxicity in rats. Pharmaceutics 13:1937. https://doi.org/10.3390/pharmaceutics13111937

Adisa IO, Rawat S, Pullagurala VLR, Dimkpa CO, Elmer WH, White JC, Gardea-Torresdey JL (2020) Nutritional status of tomato (Solanum lycopersicum) fruit grown in Fusarium-infested soil: impact of cerium oxide nanoparticles. J Agric Food Chem 68(7):1986–1997

Andey H, Rani R, Agarwal V (2016) Liposome and their applications in cancer therapy. Braz Arch Biol Technol 59:e16150477

Asuku AO, Ogungbangbe GO (2024) Applications of nanomaterials in biomedicine. In: Kaur H, Phanden RK, Singh RK, Sikarwar BS (eds) Discovery, disruption, and future implications of nanomaterials. IGI Publisher. https://doi.org/10.4018/979-8-3693-4397-5

Asuku AO, Ayinla MT, Gbonjubola OO, Babatunde SS, Olajide TS, Oyerinde TO (2024a) Neurocognitive aspects of dementia. In: Moradikor N, Chatterjee I, Mohamed W (eds) Nutrition in brain aging and dementia. Nutritional neurosciences. Springer, Singapore. https://doi.org/10.1007/978-981-97-4117-5_

Asuku AO, Ayinla MT, Aransiola SA, Ajibare AJ, Adeyemo MB, Adeyemo RO (2024b) Chitosan-based nanoparticles to bypass the blood-brain barrier for the treatment of neurological diseases: a review. Pac J Med Sci 25(1):1072–1625

Asuku AO, Ajibare AJ, Ajagun EJ, Babatunde SS, Olajide TS, Oyerinde TO, Ogungbangbe GO, Odugbesan OA (2024c) Herbal extracts in antiaging products. In: Yingngam B (ed) Eco-friendly skin solutions for natural cosmeceuticals. IGI Publisher. https://doi.org/10.4018/979-8-3693-5473-5.ch012

Asuku AO, Ogungbangbe GO, Olajide TS (2025) Effect of happiness and positive emotions on dementia. In: Moradikor N, Haghipanah M, Chatterjee I, Hamdar H (eds) Neuroscience of feelings. Progress in Brain Research. Elsevier. https://doi.org/10.1016/bs.pbr.2025.03.001

Ayinla MT, Asuku AO, Bayo-Olugbami A, Ayeni AO, Abiola AA, Owoyele BV (2020) Melatonin and vitamin C modulate cholinergic neurotransmission and oxidative stress in scopolamine–induced rat model of memory impairment. J Afr Assoc Physiol Sci 8(1):20

Bagyinszky E, Giau VV, Shim K, Suk K, An SSA, Kim S (2017) Role of inflammatory molecules in the Alzheimer's disease progression and diagnosis. J Neurol Sci 376:242–254

Barbara R, Belletti D, Pederzoli F, Masoni M, Keller J, Ballestrazzi A, Grabrucker AM (2017) Novel curcumin loaded nanoparticles engineered for blood-brain barrier crossing and able to disrupt Abeta aggregates. Int J Pharm 526:413–424

Basheikh M, AlQayidi A, Addas M, Al-Zanbaqi M, Abumelha A, Basheikh M (2021) Prevalence of systolic heart failure in patients with dementia in Saudi Arabia: single-center retrospective data review. Middle East J Fam Med 7:6–10

Baumgart M, Snyder HM, Carrillo MC, Fazio S, Kim H, Johns H (2015) Summary of the evidence on modifiable risk factors for cognitive decline and dementia: a population-based perspective. Alzheimers Dement 11(6):718–726

Belviranlı M, Okudan N, Atalık KEN, Öz M (2013) Curcumin improves spatial memory and decreases oxidative damage in aged female rats. Biogerontology 14:187–196

Bennett J, Das JM, Emmady PD (2020) Spinal cord injuries. StatPearls Publishing, Treasure Island

Bilia AR, Piazzini V, Guccione C, Risaliti L, Asprea M, Capecchi G, Bergonzi MC (2017) Improving on nature: the role of nanomedicine in the development of clinical natural drugs. Planta Med 83(05):366–381

Cano A, Ettcheto M, Espina M, Auladell C, Calpena AC, Folch J, Barenys M et al (2018) Epigallocatechin-3-gallate loaded PEGylated-PLGA nanoparticles: a new anti-seizure strategy for temporal lobe epilepsy. Nnomedicine 14(4):1073–1085

Cano A, Ettcheto M, Chang JH, Barroso E, Espina M, Kühne BA (2019) Dual-drug loaded nanoparticles of epigallocatechin-3-gallate (EGCG)/ascorbic acid enhance therapeutic efficacy of EGCG in a APPswe/PS1dE9 Alzheimer's disease mice model. J Control Release 301:62–75. https://doi.org/10.1016/j.jconrel.2019.03.010

Cox DBT, Platt RJ, Zhang F (2015) Therapeutic genome editing: prospects and challenges. Nat Med 21(2):121–131

DiSilvestro RA, Joseph E, Zhao S, Bomser J (2012) Diverse effects of a low dose supplement of lipidated curcumin in healthy middle aged people. Nutr J 11:1–8

Du J, Xu N, Fan J, Sun W, Peng X (2019) Carbon dots for in vivo bioimaging and theranostics. Small 15:1805087

Elnaggar Y, Etman SM, Abdelmonsif DA, Abdallah OY (2015) Intranasal Piperine-loaded chitosan nanoparticles as brain-targeted therapy in Alzheimer's disease: optimization, biological efficacy, and potential toxicity. J Pharm Sci 104:3544–3556. https://doi.org/10.1002/jps.24557

Fang Z, Tang Y, Ying J, Tang C, Wang Q (2020) Traditional Chinese medicine for anti-Alzheimer's disease: berberine and evodiamine from Evodia rutaecarpa. Chin Med 15:82. https://doi.org/10.1186/s13020-020-00359-1

Gao C, Chu X, Gong W, Zheng J, Xie X, Wang Y, Yang M, Li Z, Gao C, Yang Y (2020) Neuron tau-targeting biomimetic nanoparticles for curcumin delivery to delay progression of Alzheimer's disease. J Nanobiotechnol 18(1):71

González-Sanguino C, Ausín B, Castellanos MÁ, Saiz J, López-Gómez A, Ugidos C, Muñoz M (2020) Mental health consequences during the initial stage of the 2020 coronavirus pandemic (COVID-19) in Spain. Brain Behav Immun 87:172–176

Granja A, Pinheiro M, Reis S Epigallocatechin gallate nanodelivery systems for cancer therapy. Nutrients 2017:8:307. https://doi.org/10.3390/nu8050307

Gregoriadis G (2016) Liposomes in drug delivery: how it all happened. Pharmaceutics 8:19

Guan P, Lu Y, Qi J, Wu W (2016) Readily restoring freeze-dried probilosomes as potential nanocarriers for enhancing oral delivery of cyclosporine a. Colloids Surf B Biointerfaces 144:143–151. https://doi.org/10.1016/j.colsurfb.2016.04.006

Hano C, Abbasi BH (2021) Plant-based green synthesis of nanoparticles: production, characterization and applications. Biomolecules 12(1):31

Hassan S, Nadzim SZA, Shiratuddin N (2015) Strategic use of social media for small business based on the AIDA model. Procedia Soc Behav Sci 172:262–269

Jia L, Nie XQ, Ji HM, Yuan ZX, Li RS (2021) Multiple-coated PLGA nanoparticles loading Triptolide attenuate injury of a cellular model of Alzheimer's disease. Biomed Res Int:1–9. https://doi.org/10.1155/2021/8825640

Komorowska J, Wątroba M, Szukiewicz D (2020) Review of beneficial effects of resveratrol in neurodegenerative diseases such as Alzheimer's disease. Adv Med Sci 65:415–423. https://doi.org/10.1016/j.advms.2020.08.002

Krupkova O, Ferguson SJ, Wuertz-Kozak K (2016) Stability of (−)-epigallocatechin gallate and its activity in liquid formulations and delivery systems. J Nutr Biochem 37:1–12. https://doi.org/10.1016/j.jnutbio.2016.01.002

Labban A, Ma Y, Dube L (2021) A neurobehavioral account of differential consumer responses to price and in-store display between un/healthy food. Eur J Mark 55(11):2988–3009

Liu Y, Gong Y, Xie W, Huang A, Yuan X, Zhou H (2020) Microbubbles in combination with focused ultrasound for the delivery of quercetin-modified sulfur nanoparticles through the blood brain barrier into the brain parenchyma and relief of endoplasmic reticulum stress to treat Alzheimer's disease. Nanoscale 12:6498–6511. https://doi.org/10.1039/C9NR09713A

Liu N, Ruan J, Li H, Fu J (2023) Nanoparticles loaded with natural medicines for the treatment of Alzheimer's disease. Front Neurosci 17:1112435

Loureiro JA, Andrade S, Duarte A, Neves A, Queiroz J, Nunes C (2017) Resveratrol and grape extract-loaded solid lipid nanoparticles for the treatment of Alzheimer's disease. Molecules 22:277. https://doi.org/10.3390/molecules22020277

Ménard-Moyon C (2018) Applications of carbon nanotubes in the biomedical field. In: Smart Nanoparticles for Biomedicine, vol 354, pp 83–101

Nady DS, Bakowsky U, Fahmy SA (2023) Recent advances in brain delivery of synthetic and natural nano therapeutics: reviving hope for Alzheimer's disease patients. J Drug Deliv Sci Technol:105047

Okuda M, Hijikuro I, Fujita Y, Teruya T, Kawakami H, Takahashi T, Sugimoto H (2016) Design and synthesis of curcumin derivatives as tau and amyloid β dual aggregation inhibitors. Bioorg Med Chem Lett 26(20):5024–5028

Parveen A, Parveen B, Parveen R, Ahmad S (2015) Challenges and guidelines for clinical trial of herbal drugs. J Pharm Bioallied Sci 7(4):329–333

Patra JK, Das G, Fraceto LF, Campos EVR, Rodriguez-Torres MDP, Acosta-Torres LS (2018) Nano based drug delivery systems: recent developments and future prospects. J Nanobiotechnol 16:71

Pinheiro R, Granja A, Loureiro JA, Pereira MC, Pinheiro M, Neves AR (2020a) RVG29-functionalized lipid nanoparticles for quercetin brain delivery and Alzheimer's disease. Pharm Res 37:139. https://doi.org/10.1007/s11095-020-02865-1

Pinheiro R, Granja A, Loureiro JA, Pereira MC, Pinheiro M, Neves AR (2020b) Quercetin lipid nanoparticles functionalized with transferrin for Alzheimer's disease. Eur J Pharm Sci 148:105314. https://doi.org/10.1016/j.ejps.2020.10531

Prema A, Justin Thenmozhi A, Manivasagam T, Mohamed Essa M, Guillemin GJ (2017) Fenugreek seed powder attenuated aluminum chloride-induced tau pathology, oxidative stress, and inflammation in a rat model of Alzheimer's disease. J Alzheimers Dis 60:S209–S220. https://doi.org/10.3233/JAD-161103

Quan L, Wu J, Lane LA, Wang J, Lu Q, Gu Z (2016) Enhanced detection specificity and sensitivity of Alzheimer's disease using amyloid-β-targeted quantum dots. Bioconjug Chem 27:809–814

Raju M, Kunde SS, Auti ST, Kulkarni YA, Wairkar S (2021) Berberine loaded nanostructured lipid carrier for Alzheimer's disease: design, statistical optimization and enhanced in vivo performance. Life Sci 285:119990. https://doi.org/10.1016/j.lfs.2021.119990

Rashmi R, Mohanty SK. Examining chronic disease onset across varying age groups of Indian adults using competing risk analysis. Sci Rep, 2022, 13(1), 5848

Reitz C, Brayne C, Mayeux R (2011) Epidemiology of Alzheimer disease. Nat Rev Neurol 7(3):137–152

Rinwa P, Kumar A (2013) Quercetin suppress microglial neuroinflammatory response and induce antidepressent-like effect in olfactory bulbectomized rats. Neuroscience 255:86–98

Ross C, Taylor M, Fullwood N, Allsop D (2018) Liposome delivery systems for the treatment of Alzheimer's disease. Int J Nanomedicine 13:8507–8522

Sadegh Malvajerd S, Azadi A, Izadi Z, Kurd M, Dara T, Dibaei M (2019) Brain delivery of curcumin using solid lipid nanoparticles and nanostructured lipid carriers: preparation, optimization, and pharmacokinetic evaluation. ACS Chem Neurosci 10:728–739. https://doi.org/10.1021/acschemneuro.8b00510

Saeedi A, Saeedi M, Maghsoudi A, Shalbaf A (2021) Major depressive disorder diagnosis based on effective connectivity in EEG signals: a convolutional neural network and long short-term memory approach. Cogn Neurodyn 15(2):239–252

Saini S, Kaur N, Pati PK (2021) Phytohormones: key players in the modulation of heavy metal stress tolerance in plants. Ecotoxicol Environ Saf 223:112578

Salem HF, Kharshoum RM, Abou-Taleb HA, Naguib DM (2019) Brain targeting of resveratrol through intranasal lipid vesicles labelled with gold nanoparticles: in vivo evaluation and bioaccumulation investigation using computed tomography and histopathological examination. J Drug Target 27:1127–1134. https://doi.org/10.1080/1061186X.2019.1608553

Singh J, Dutta T, Kim KH, Rawat M, Samddar P, Kumar P (2018) 'Green' synthesis of metals and their oxide nanoparticles: applications for environmental remediation. J Nanobiotechnol 16:1–24

Singh AK, Singh SS, Rathore AS, Singh SP, Mishra G, Awasthi R (2021) Lipid-coated MCM-41 mesoporous silica nanoparticles loaded with Berberine improved inhibition of acetylcholine esterase and amyloid formation. ACS Biomater Sci Eng 7:3737–3753. https://doi.org/10.1021/acsbiomaterials.1c00514

Steinbach K, Vincenti I, Kreutzfeldt M, Page N, Muschaweckh A, Wagner I, Merkler D (2016) Brain-resident memory T cells represent an autonomous cytotoxic barrier to viral infection. J Exp Med 213(8):1571–1587

Tejwan N, Saha SK, Das J (2020) Multifaceted applications of green carbon dots synthesized from renewable sources. Adv Colloid Interf Sci 275:102046

World Alzheimer Report. The Global Impact of Dementia: An analysis of prevalence, incidence, cost and trends. Alzheimer's Disease International, 2015. www.alz.co.uk/worldreport2015corrections

Wu Y, Xu X, Chen Z, Duan J, Hashimoto K, Yang L, Yang C (2020) Nervous system involvement after infection with COVID-19 and other coronaviruses. Brain Behav Immun 87:18–22

Xie J, Shen Z, Anraku Y, Kataoka K, Chen X (2019) Nanomaterial-based blood-brain-barrier (BBB) crossing strategies. Biomaterials 224:119491

Yang ZZ, Zhang YQ, Wang ZZ, Wu K, Lou JN, Qi XR (2013) Enhanced brain distribution and pharmacodynamics of rivastigmine by liposomes following intranasal administration. Int J Pharm 452(1–2):344–354

Yang L, Wang W, Chen J, Wang N, Zheng G (2018) A comparative study of resveratrol and resveratrol-functional selenium nanoparticles: inhibiting amyloid β aggregation and reactive oxygen species formation properties. J Biomed Mater Res A 106:3034–3041. https://doi.org/10.1002/jbm.a.36493

Yu H, Schlüter AD, Zhang B (2014) Synthesis of high generation dendronized polymers and quantification of their structure perfection. Macromolecules 47:4127–4135

Zdrojewicz Z, Waracki M, Bugaj B, Pypno D, Cabała K (2016) Medical applications of nanotechnology. Postepy Hig Med Dosw 69:1196–1204

Zhang C, Wu X, Xian Y, Zhu L, Lin G, Lin ZX (2019) Evidence on integrating pharmacokinetics to find truly therapeutic agent for Alzheimer's disease: comparative pharmacokinetics and disposition kinetics profiles of stereoisomers Isorhynchophylline and Rhynchophylline in rats. Evid Based Complement Alternat Med:4016323. https://doi.org/10.1155/2019/4016323

Zhang W, Sigdel G, Mintz KJ, Seven ES, Zhou Y, Wang C (2021) Carbon dots: A future Blood–Brain Barrier penetrating nanomedicine and drug nanocarrier. Int J Nanomedicine 16:5003

Zhang P, Liu H, Yu Y, Peng S, Zhu S (2024) Role of Curcuma longae Rhizoma in medical applications: research challenges and opportunities. Front Pharmacol:1430284. https://doi.org/10.3389/fphar.2024.1430284

Zhou Y, Peng Z, Seven ES, Leblanc RM (2018) Crossing the blood-brain barrier with nanoparticles. J Control Release 270:290–303

Živančević K, Baralić K, Bozic D, Miljaković EA, Djordjevic AB, Ćurčić M (2022) Involvement of environmentally relevant toxic metal mixture in Alzheimer's disease pathway alteration and protective role of berberine: bioinformatics analysis and toxicogenomic screening. Food Chem Toxicol 161:112839. https://doi.org/10.1016/j.fct.2022.112839

Zvěřová M (2018) Alzheimer disease and blood-based biomarkers and ash; potential contexts of use. Neuropsychiatr Dis Treat 14:1877–1882

Chapter 9
Plant-Based Smart Nanomaterials

Lakshmi Sai Varshini Yedavalli, Saloni Semwal, Vedika Kartha, Masilamani Karthikeyan, Mohandass Ramya, and Pasupathi Rathinasabapathi

9.1 Introduction to Nanomaterials

Nanotechnology and nanoparticles smaller than 100 nm are a rapidly developing transdisciplinary area. Nanomaterials, which are built from atoms, can be organic, inorganic, or else hybrids with diverse shapes/structures. Nanoparticles, which are particularly relevant in industrial applications because of their increased stiffness and ductility, are produced by techniques such as chemical vapor decomposition and sol-gel synthesis. Nanomaterials address various challenges, especially in cancer therapy. Their potential for cancer targeting, tumor microenvironment, and immune system makes them good candidates for improving drug delivery, toxicity reduction, and bioavailability. Nanoparticles are also employed in biomedical fields, for imaging applications such as magnetic resonance imaging (MRI), computed tomography (CT), positron emission tomography (PET), and ultrasound examinations. The particular characteristics of nanomaterials, including mechanical, chemical, and electrical properties, make them useful for environmental cleanup and medicine, and technology. Present research is also directed at the responsible and sustainable exploitation of nanomaterials, as nanomaterials influence the development of technology and society (Khan et al. 2021).

L. S. V. Yedavalli · S. Semwal · V. Kartha · M. Karthikeyan · M. Ramya
P. Rathinasabapathi (✉)
Department of Genetic Engineering, SRM Institute of Science and Technology, Kattankulathur, Chengalpattu, Tamil Nadu, India
e-mail: rathinap1@srmist.edu.in

V. Ghosh et al. (eds.), *Applications of Plant Bioactive Compounds-Based Nanoformulations*, Nanotechnology in the Life Sciences,
https://doi.org/10.1007/978-3-032-13086-0_9

9.1.1 Plant-Based Nanomaterials

Plant extracts derived from flowers, seeds, stems, and roots have been utilized to generate nanoparticles. Secondary metabolites of medicinal plants show a wide range of biological activities, which can be used as valuable biomarkers in nanoparticle synthesis. Cultivation of plant biomolecules for the production of plant-based proteins, including collagen, gelatin, elastin, monoclonal anti-cancer antibodies, and recombinant anti-cancer vaccines, has also been explored, although metal nanoparticles are much in the news. Enzymes and polysaccharides play a key role in various biological processes. Numerous catabolizing enzymes, and metabolic biomolecules such as amino acids, vitamins, and alkaloids are also used in forming and growing nanoparticles. It is to be surmised that less toxic, cheaper, more biocompatibility, and better size controlling aspects offered a higher prominence to produce nanomaterials through greener ways over physical and chemical methods (Hano and Abbasi 2021).

Plant-derived Nanomaterials have the potential to transform certain fields without adverse environmental consequences. Their combination of renewable origins, functional versatility, and alignment with green technology positions them as a cornerstone for a more sustainable and environmentally friendly ecosystem. However, the synthesis of plant-based nanomaterials can be more efficient, and their synthetic methods are highlighted through smart acknowledged processes and newly developed techniques like biosensing, bioimaging, therapeutics, etc. (Jadoun et al. 2020)

9.2 Types of Plant-Based Nanomaterials

9.2.1 Cellulose Nanomaterials

Cellulose Nanomaterials are composed of both crystalline and amorphous domains of 100 nm in length, roughly a micron. They can be derived from cellulose resources, including cottonseed, rice, wheat, wood, bast fibers, etc. Cellulose-based Nanomaterials can be isolated using several different methods and techniques (such as cellulose nanofibrils), which are generally produced using the approaches of phase drying, solvent extraction, and template synthesis and which can subsequently have their structural properties fine-tuned to guide their performance. For example, ionizing the material or directing charges to the material surface can help improve its dispersibility in solvents, or by introducing nonpolar or hydrophobic compounds, compatibility with other composites can be achieved. Due to the progress made in the synthesis and functionalization of cellulose Nanomaterials, these NMs possess the capability to be accurately engineered, thus improving their versatility and efficiency in various fields (Shayan et al. 2022).

Bacterial cellulose (BC) and cellulose nanocrystals (CNCs) are extensively researched for biomedical applications, drug delivery, and targeted therapy.

BC-based nanocomposites, featuring fibrous morphology and durability that provide mechanical stability, are employed for tissue engineering, cartilage regeneration, artificial skin, and wound dressing. However, their small pore size (14.3 nm) and slow degradation hinder bone tissue regeneration, prompting modifications to increase pore size and degradation rates for hydrogels and nanofiber development. Rod-shaped crystalline material, acid hydrolysis-released CNCs, microcrystalline cellulose/cotton-based agents as a source of supply, high specific surface area, and biocompatibility nature. Extraction techniques, like ultrasonic-assisted hydrolysis and deep eutectic solvents, improve CNC yield and properties. The high aspect ratio and thermal stability of CNCs make them suitable for hydrogels, enzyme immobilization, and drug delivery systems, enabling applications in cosmetics and biomedicine via biodegradability and low toxicity.

9.2.2 Lignin-Based Nanomaterials

Lignin can be depolymerized from lignocellulosic biomass by a process wherein lignin is fragmented into annular analogs and should be chemically converted into products of smaller molecular weight. Some variables influence the extraction reaction (e.g., temperature, the ability of solute/solvent, and microenvironment pH). The phenolic and aliphatic moieties of lignin can be chemically modified and can be grafted into polymeric matrices to decrease production costs. In many cases, lignin does not have to be altered and its native form can be applied directly to the same. Lignin-derived nanomaterials can serve as chemical reagents, fuels, and polymers. Lignin may also enable the release of fertilizers and herbicides during agricultural production by serving as a sizing or dispersing agent. Lignin is one of the most naturally occurring polymers and can be used to create sustainable recycling mechanisms. In general, Nanomaterials possess outstanding features such as UV absorption and biodegradability and can be combined into high-performance composite materials and eco-friendly industrial processes (Gigli et al. 2022; Chen et al. 2018).

9.2.3 Plant-Derived Proteins and Polysaccharides

Protein polymers derived from plants are abundant, biodegradable, biocompatible, and metabolically decomposable, making them ideal for various applications. Their conformational structure allows them to be used as agents in food and drug delivery systems. Protein-based nanostructures enhance the stability of nutraceuticals by immobilizing them in protective environments. For nanocomposite synthesis, factors such as protein physical and chemical properties, biocompatibility, and immunogenicity must be considered to ensure in vivo biodegradability. Polypeptide cross-linking during encapsulation increases delivery effectiveness, and bio-polymer-based systems (proteins and polysaccharides) have been most commonly used for encapsulation of biomaterials (Jafarzadeh et al. 2022).

Polysaccharides, as biomolecules including proteins, DNA/RNA, starch, and cellulose, play important roles in cell signaling, communication, and structural integrity. These renewable and biodegradable compounds exhibit therapeutic potential, adhesion to cell surfaces, and metabolic compatibility. Their hydrophilicity and enzymatic degradation allow them to be used for nanoparticle synthesis and drug transport. Protein-polysaccharide nanoconjugates enhance the stability, bioavailability, and delivery of bioactive nutraceuticals, with advances in therapeutic uses. Plant-based nutraceuticals, also known as polyphenols, alkaloids, terpenoids, glycosides, tannins, and sterols, have important medical uses. Polyphenols like resveratrol (grapes), curcumin (turmeric), and catechins (green tea) have anti-inflammatory and neuroprotective effects. These compounds mitigate arthritis, bowel disorders, Alzheimer's, and Parkinson's, and protect against cognitive decline through antioxidant activity. Additionally, curcumin and resveratrol have demonstrated potential in cancer therapy, while polyphenols and flavonoids aid in diabetes management by enhancing insulin sensitivity and glucose metabolism. Plant polyphenols and sterols contribute to cardiovascular health by improving lipid profiles, preventing atherosclerosis, and promoting better well-being (Jafarzadeh et al. 2022).

9.2.4 Green Synthesis of Nanoparticles Using Plant Extracts

Enhancing the green synthesis of nanoparticles focuses on green solvents, nontoxic reducing agents, and nontoxic stabilizing agents. Unlike chemical methods, green synthesis creates nanoparticles suitable for applications in biomedicine, healthcare, and marketing. Plant extracts from leaf, root, stem, and seed are widely used, allowing nanoscale material fabrication with minimal health risks to humans. These plant-based nanomaterials are readily fabricated and have uses in areas ranging from food science to agriculture, bioengineering, and nanomedicine. Silver, gold, copper, and nickel metals are commonly used in the synthesis of nanoparticles. Silver nanoparticles (AgNPs) are synthesized with the use of silver metal ions, biological reducing agents, and stabilizing biomolecules, including vitamins, proteins, and alkaloids. Almost all plants can facilitate AgNP synthesis. For example, gold nanoparticles (AuNPs) possess attractive traits such as surface modification, low toxicity, and high biocompatibility. Selective reducing agents such as phenols, flavonoids, and proteins reduce gold ions, leading to stable nanoparticle structures. Zinc oxide nanoparticles (ZnO NPs) are one of the most popular biomedicine, electronics, and cosmetics agents. They are inexpensive and can be produced from plant material, i.e., seeds and leaves. ZnO NPs have a large excitation binding energy (60 meV) and a band gap of 3.37 eV, which gives semiconductor properties. Plant extract-mediated copper nanoparticles (CuNPs) reduce aqueous cuprous ions, which are a cost-effective alternative to the use of gold and silver nanoparticles.

Nevertheless, they still pose biosafety concerns. Furthermore, metals nickel, manganese, titanium, platinum, and cerium have also been investigated in nanoparticle synthesis for plant-derived applications in biomedical and industrial sectors due to their biocompatibility, cheapness, and environmental friendliness (Jadoun et al. 2020).

9.3 Smart Nanomaterials

Smart Nanomaterials are composed of materials that can be influenced by some stimuli such as mechanical stress, pH, and temperature (Fig. 9.1). Such particles share the capability of integrating with other molecules and introducing changes in order to perform better and more efficiently in the system.

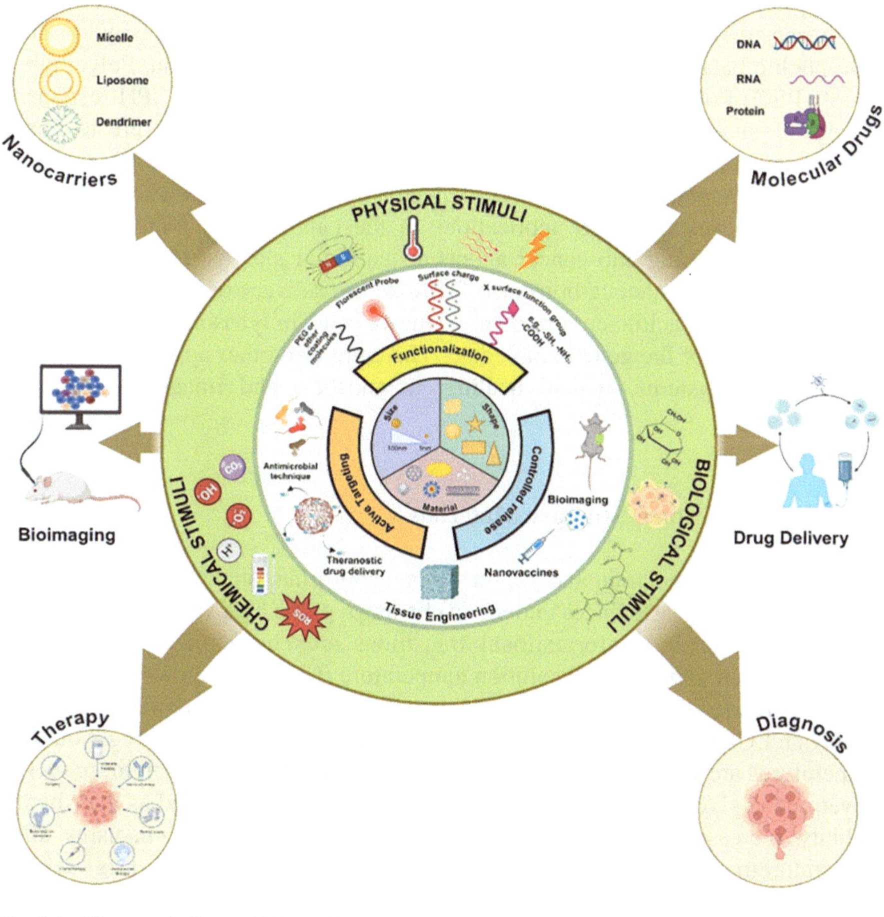

Fig. 9.1 Characteristics and biomedical applications of plant-based smart nanomaterials

9.3.1 Smart Nanomaterials Can Be Drafted as Different Types

9.3.1.1 pH-Responsive Materials

Nanoparticles can release drugs as a response to certain pH levels during the processes of treating tumor microenvironments where H+ dissociations occur, and the pH varies. In these acidic conditions, polymeric micelles and other nanogels can trap drugs and stimulate the release when placed in an acidic medium. pH-responsive nanomaterials usually have distinct functional groups that vary and can change depending on different pH conditions, resulting in conformational changes, changes in hydrophobicity, and solubility changes. Some of these functional groups are amino groups, giving protons in acidic environments, and carboxyl groups, deprotonating ions in basic environments; some groups, imidazole and amine-containing groups are all very sensitive to pH variations and may result in a large variation of structural conformation and solubility of nanoparticles. Polymeric micelles are a type of pH-responsive nanomaterials that contain a hydrophobic core that helps in drug encapsulation, and changes in the pH can lead to changes in the hydrophilic-hydrophobic balance of the media, for example, in an acidic media, Polyethylene glycol (PEG)-Polyacrylic acid (PAA) stimulate release of drugs. PH-responsive hydrogels, such as Poly (N-vinylpyrrolidone-co-acrylic acid), create three-dimensional networks that can take up water and swell in an acidic medium, which, consequently, enables drug release (i.e., from a hydrogel matrix). These properties have a significant effect on applications in the cancer field, for example, pH-responsively releasing anti-cancer agents at the tumor site, as tumor sites can be acidic. PH-sensitive nanocarbons also facilitate the delivery of DNA/RNA content to cells, where the acidic environment of the vesicles in lysosomes is exploited to force the release of the genetic molecule. There are also other applications such as drug delivery systems, wound healing, diagnostics, and imaging (Figueiredo et al. 2017).

9.3.1.2 Temperature-Sensitive Materials

Thermo-reversible gelation and polymer chain conformational changes. In thermo-reversible gelation, substances undergo a change in phase by a variation in temperature. Polymer chain conformational transitions are characterized by polymers having either an upper critical solution temperature (UCST) or a lower critical solution temperature (LCST). Polymers like poly (N-isopropyl acrylamide) (PNIPAM) display an LCST at around 32 °C, and they undergo denaturation/aggregation if the temperatures are increased beyond this limit. On the other hand, polymers such as polyethylene glycol (PEG) exhibit UCST behavior and are swelled as their hydrophilicity increases above the threshold temperature. These mechanisms allow temperature-responsive nanomaterials to be used efficiently in drug delivery systems.

Cytocompatibility encapsulated drugs can be released by phase transitions, such as for PNIPAM-functionalized nanoparticles and nanogels. An example is nanogels, which for their phase transitions change their shape and structural characteristic conformation to release drugs in a programmable manner. This method holds great potential in thermal cancer therapy by allowing nanoparticles to be engineered such that they are targeted to tumor cells for selective drug loading and release.

Besides drug delivery, temperature-responsive nanomaterials also have applications in biomedical and cosmetic areas, which reflect the utility and prospects of temperature-responsive nanomaterials to improve healthcare technologies (Karimi et al. 2016).

9.3.1.3 Chemical Responsive Materials

Polydopamine-based systems can alter their molecular structure based on the activity of specific chemicals. Such chemical response may be used not only for therapeutic but also for environmental sensing applications. Their response mechanism is generally based on the surface properties of the nanomaterial or on a change of conformation, which promotes drug desorption from the encapsulated drugs.

These reactions are driven by

1. pH sensitivity: Due to changes in the environmental pH, the nanoparticles undergo conformational changes in shape and cause the release of the materials.
2. Redox sensitivity: When oxidative and reductive reactions occur, it promotes the release of enclosed drugs.
3. Ionic sensitivity: Nanoparticles might express sensitivity to the ions calcium, chloride, and sodium and, subsequently, induce the release of drugs upon ionic alteration of the media.
4. Enzyme sensitivity: Enzymes (e.g., matrix metalloproteinases, MMPs) or lysozyme may cause structural conformation change, leading to the release of encapsulated drugs.

Chemical-sensitive nanoparticles are of ultimate importance in biomedical applications, releasing drugs in response to target chemical stimuli, including pH changes or enzymatic activity. For instance, pH-responsive nanoparticles release agents in acidic tumor microenvironments, or enzyme-responsive nanoparticles provide targeted enzymatic pathway activation or tissue targeting, which provides highly accurate targeted delivery, reduces side effects, and improves drug efficacy. Polymeric micelles and enzyme-responsive hydrogels are commonly used to deliver drugs over time, hence, by increasing treatment efficacy and minimizing damage to healthy cells. These nanoparticles are also critical in the field of imaging and diagnostics, serving as contrast agents or biosensors for the identification of target biomarkers or physiological status. For example, enzyme-activated fluorescent nanoparticles can provide improved imaging resolution as well as real-time diagnosis. Furthermore, they enable gene therapy by supporting the translocation and controlled delivery of genetic material, thereby increasing the safety and effectiveness of the delivery system. In addition to the above, chemical-sensing nanoparticles are multi-functionalized for bio-sensing and therapeutic devices, including hydrogels for wound healing. Their versatility underscores their potential to revolutionize medicine by integrating precision, efficiency, and advanced diagnostic and therapeutic capabilities (Gulfam et al. 2018).

9.3.1.4 Magneto-Responsive Materials

Magnetic or ferrous materials, such as Iron Oxide (Fe3O4), may respond to magnetic field variations and therefore may be capable of detecting such changes. They can also be applied in hyperthermia therapy for cancer treatment, drug release, and magnetic resonance imaging (MRI). Magneto-responsive materials act according to a surrounding environment by using techniques such as magnetization—in which magnetism is influenced by the configuration of magnetic domains in which a force detected by the magnet is employed for motion and deformation of the material —magnetic hyperthermia—in which magnetic field variations are based on the thermal effect created by the material. Magneto-responsive materials are composites involving ferromagnetic nanoparticles, such as iron oxide nanoparticles, cobalt/nickel nanoparticles, or utilizing magneto-responsive hydrogels and composites. Examples of their use are targeted drug delivery, where an alternate magnetic field drives the movement of drugs by magnetic nanoparticles to their target site, further decreasing side effects by increasing specificity. Superparamagnetic iron oxide nanoparticles (SPIONs) can be applied in magnetic hyperthermia for cancer therapy, in which the magnetic nanoparticles are heated to generate heat and destroy cancer cells, and therefore, are more susceptible to chemotherapies. Magneto-responsive materials are also suitable for actuators in drug delivery and invasive surgery. They can also be applied in tissue culture and engineering, MRI, biosensing, and diagnostics (Baglioni and Bonini 2014).

9.3.1.5 Photo-Responsive Nanomaterials

Nanoparticles extracted from titanium oxide (TiO2) trap light energy and generate heat from it. These kinds of nanoparticles also have the potential for use in solar conversions and photocatalysis. When exposed to light, photo-responsive materials can change their physical or chemical properties, and these changes can include structural changes, photochemical reactions through light energy to induce chemical reactions, and photo-isomerization, where a light force can stimulate the conformational change in the material and allow it to go through reversible isomerization. Gold Nanoparticles are used in photothermal therapy as they can absorb light energy and use it to extract nanomaterials through heat energy, and by using light energy in ranges like visible and infrared light, they can be used to kill tumor cells. Not only carbon-based nanomaterials, graphene oxides and carbon nanotubes can also be used in combination with light energy to lyse tumor cells and targeted cell treatment. Photochromic molecules based on spiropyrans and azobenzene can undergo isomerization in the presence of UV or visible light, respectively, and have been employed in sensors, digital electronics, and actuators. Employing these properties, photo-active materials can be applied to drug delivery purposes as, at specific wavelengths, the contents encapsulated may be released to achieve greater therapeutic efficacy with reduced therapeutic effort. Their optical characteristics enable them to be employed in monitoring devices for monitoring pollutants, temperature, pH, and other parameters in sensors (Paul et al. 2021).

9.3.1.6 Electro-Responsive Nanomaterials

Electro-responsive nanoparticles such as Carbon nanotubes (CNTs) can be utilized as materials and implemented into storage devices, sensors, and smart textiles. For these materials, when an electric field or currents are applied, the materials can react and change their conformation. There are some cases where nanomaterials such as electroactive polymers (EAPs) are deformed under stress and bending/stretching of the material and some materials undergo electrochemical reactions through ion of catalyst and the generation of electric potential in the media and ionic gels and hydrogels allow ion dissociation for these nanoparticles and can control the materials shape or induce swelling by the force of an electrical current/field. Examples and applications of these materials include Carbon Nanotubes (CNTs), which are used in energy storage devices, actuators, and sensing purposes; graphene oxide, which can be used for drug delivery applications and in energy storage devices because of its high conductivity and versatility. Volume, dimension, and shape change of the Polypyrrole (PPy) is controllable through its applied voltage and can be applied as an actuator or sensor. Analogous to this idea, Poly (N-isopropyl acrylamide) (PNIPAAm) hydrogels expand and contract in the presence of electric fields and can be used in actuators and tissue engineering (Kolosnjaj-Tabi et al. 2019).

9.3.1.7 Hydrogels with Nanoscale Structures

The swelling characteristic of hydrogels in response to temperature and heat can be observed in gold/silver nanoparticles and can be used in drug delivery systems, therapeutic mechanisms, etc.. Hydrogel-responsive materials can respond to stimuli and behave in a desired manner, e.g., they can be responsive to environment pH or temperature variations via temperature-sensitive hydrogels, and the structure of the hydrogel or volume of the hydrogel can change or vary in case the temperature environment varies. Responsive hydrogels to light energies and magneto-responsiveness have also been developed and, relying on light energies or magnetic fields, respectively, provide structural deformations that can be harnessed for drug encapsulation and release. These exogenous stimuli can evoke a regulated and directional drug release, and these nanomaterials can be applied for tissue engineering, biosensing, as well as applications in the biomedical fields of bioimaging diagnostics and wound healing (Lem et al. 2024; Maiz-Fernández et al. 2020).

9.3.1.8 Stimuli-Responsive Liposomes

Nanoparticles can be implemented into liposomes through genetically engineered methods and allow the release of encapsulated drugs under slight changes in temperature or light, and these liposomes can be used for gene therapy and drug delivery for chemotherapy. Hence, it can be concluded that such smart nanomaterials, which are genetically engineered and can respond to certain changes in temperature, pressure, and pH, can be used in various fields in the medical, electronics, and other industries (Lem et al. 2024).

9.3.2 Smart Plant Sensors

In agricultural industries, selecting the desired plant trait with the optimal feature was a major challenge back then, so through the production of recent technological advances, plant-based sensors have been developed that control, optimize, and improve plant productivity and plant yield, These plant sensors have digitized sensors implemented onto the plant itself and can monitor changes in the plant model and translate the chemical signals into digitized information that can be accessed easily through a computer, Scientists have hereby used this technique to assess the performance parameters of these plant-based sensors based on their resolution, accuracy, and durability. Plant-based sensing is also judged by the plant signaling molecule related to health states, which used to be communicated via electrical, wireless, and optical signals to agricultural and phenotyping devices. Hence plant plant-based smart sensors can be enabled to link with actuated electronic devices for research purposes and to assess the optimal conditions and necessary parameters that need to be controlled for the overall productivity and growth of the plant (Singh et al. 2024; Giraldo et al. 2019; Singh and Verma 2019).

9.3.3 Smart Nano Agrochemicals

Functionalized plant-derived nanomaterials have been designed for use in agrochemical applications as nano pesticides, fertilizers, and plant growth replacements. For instance, pyrethrum insecticides can be formulated as nano pesticides, which, when sprayed, may inadvertently impact non-target species such as honeybees by negatively affecting their population. Nano agrochemicals are subjected to chemical degradation pathways with several steps, including filtration, catalytic degradation, adsorption, and biological treatments. These materials exhibit environmentally responsive behaviors; for example, smart nano sorbents respond to stimuli like pH and temperature changes by binding positively charged agrochemicals. Electrochemical interactions between chemicals and nanoparticles are further strengthened at high temperatures when the kinetic energy of particles is increased because it contributes to high reactivity and degradation efficiency.

However, certain nanoparticles and nanoagrochemicals have enormous toxic effects on non-target species. For instance, low-dose exposure to nanocapsules loaded with atrazine is toxic to aquatic species (e.g., Prochilodus lineatus), and low-dose exposure to simazine packed in lipid-based and solid lipid nanoparticles (SNLs) can be lethal and suppress the body length and increase the mortality in Caenorhabditis elegans animals within 48 h of exposure. Furthermore, the non-specific active ingredient release may cause both cellular and molecular damage, such as lysosomal disruption and changes in proteins, pigments, and DNA, which are generally signaled through pH-dependent conformational change. Recently, the binding affinity of nanocomposites has also affected their distribution, bioavailability, and toxicity within plant cells. The formation of nano pesticide complexes has been associated with protein degradation and cytogenetic abnormalities due to DNA conformational changes. The extent of nano pesticide toxicity is influenced by factors such as the size of active ingredients, biodegradability rates, and ingredient concentrations. For instance, atrazine-induced nano capsules display lower toxicity over 48 h and do not show significant effects on carbonic anhydrase and antioxidant activities in the *Prochilodus lineatus*. On the other hand, treatment with nano pesticides in *Enchytraeus crypticus* has been linked to increased toxicity, decreased hatching, and delayed plant cell growth. This evidence highlights the need for a critical assessment of the ecological and biological implications of plant-derived nano-pesticides to permit their sustainable and protective use (Sharma et al. 2022).

9.4 Properties of Plant-Based Smart Nanomaterials

Plant-based smart nanomaterials are novel types of nanomaterials that exhibit a response to external stimuli, including temperature, pH, light, magnetic field, and other environmental factors. These nanomaterials integrate intrinsic characteristics of plant-derived components, such as cellulose, starch, proteins, and lipids, and apply tailor-made functions to allow for clever interactions with the environment. The functionality of plant-derived smart nanomaterials is largely determined by drug delivery, environmental sensing, and responsive coating properties (Table 9.1).

Table 9.1 Characteristics of plant-based smart nanomaterials

Characteristic	Description	Examples/ materials	Applications	Key benefits	Reference
Environmental stimuli	React to pH, temperature, moisture, or light, enabling controlled drug release, pollutant removal, or real-time sensing.	Cellulose, chitosan, and anthocyanin films	Drug delivery, environmental sensors, smart coatings	Precise control, multifunctional capabilities	Ebrahimi Tirtashi et al. (2019), Gasti et al. (2021) and Chen et al. (2019)
Biodegradability	Decomposes naturally into non-toxic products like CO_2 and water, reducing environmental impact.	Cellulose nanocrystals, chitosan	Biodegradable food packaging, sustainable agriculture	Eco-friendly, supports a circular economy	Zheng et al. (2024) and Jagtiani (2022)
Biocompatibility	Safe for medical and food applications; minimal toxicity and immune reactions.	Chitosan nanoparticles, lignin composites	Wound healing, drug delivery, biosensors	Safe for human use, no harmful side effects	Tian and Liu (2023) and Cui et al. (2017)
Cost-effectiveness	Derived from renewable plant sources; minimizes the use of synthetic chemicals.	Lignin nanoparticles, plant-derived films	Affordable alternatives to synthetic nanomaterials	Low production costs, supports green technology	Piryaei et al. (2025)
Functional versatility	Can be tailored for specific uses by functionalizing with enzymes, ligands, or other bioactive molecules.	Functionalized CNCs, chitosan nanoparticles	Biosensors, diagnostics, and smart agricultural systems	Wide range of applications, increased efficiency	Capello et al. (2021) and Wang et al. (2019)

9.4.1 Responsiveness to Environmental Stimuli

Plant-based smart nanomaterials show high sensitivity and response to environmental stimuli, including pH, temperature, moisture, and light. These represent some of the highest values of versatility for drug delivery systems, sensing, and coatings. Materials that also fall within this category are biopolymers such as cellulose, chitosan, and alginate, whose structure in response to pH changes (e.g., swelling or degradation) can be exploited and applied to controlled-release systems. These nanomaterials grow or shrink in length on exposure to temperature transitions. With these materials, temperature-responsive drug delivery devices are fabricated. These materials are known to be stable under humid conditions, and the scale and mechanical properties of the materials can be scaled by moisture presence and humidity.

Light-sensitive plant-based nanomaterials exhibit photothermal effects as their solar-absorbing nature leads to the generation of heat, which initiates drug release or photochemical alterations resulting in molecular transformations. Certain of these materials can catalyze a reaction upon light irradiation, and that type of reaction has potential applications to environmental remediation tasks. In general, the sensitivity of plant-based nanomaterials to different environmental conditions increases their ability to perform and thus enables their use in a broader spectrum of state-of-the-art technologies (Pooja Saraswat et al. 2023).

9.4.2 *Biodegradability and Sustainability*

Plant-based smart nanomaterials exhibit various potential advantages from a biodegradability and sustainability perspective, offering a means to address the issue of environmental pollution. In contrast to synthetic nanomaterials that remain in the environment, nanomaterials derived from plants tend to decompose into innocuous products (e.g., CO_2 and water), with negligible environmental impact, mediated by bacteria, fungi, and environmental factors.

Cellulose and chitosan, two of the naturally occurring plant polymers, which provide themselves as eco-friendly substitutes to polymers from a biodegradability point of view, are extensively utilized. But also, the production of these materials comes from CO_2-consuming agricultural waste, thus reducing the carbon emission, instead petroleum products produce greenhouse gases during production. This is also in line with building a more sustainable circular economy and reducing its environmental footprint while promoting its environmental applications in packaging, medicine, in agriculture as well as environmental applications. For example, cellulose nanocrystals are considered biodegradable and strong alternatives to plastics as “packaging”, and chitosan-based nanoparticles are very good at absorbing heavy metals in wastewater; thus, they offer an eco-friendly solution for wastewater pollution (Turasan et al. 2023; Vieira et al. 2022) (Fig. 9.2).

9.4.3 *Biocompatibility and Safety*

Plant-based smart nanomaterials are gaining recognition for their biocompatibility and safety, making them suitable for drug delivery, wound healing, and food packaging. Cellulose nanoparticles from plant sources are very popular owing to their

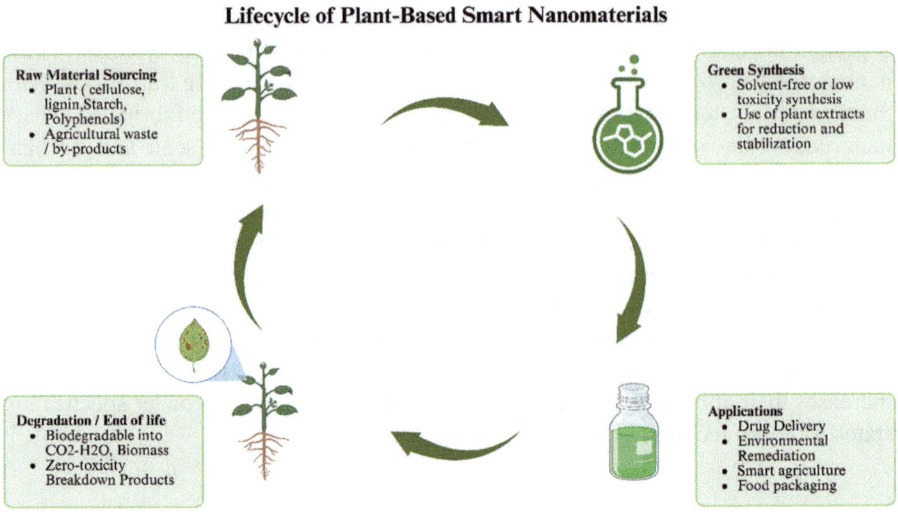

Fig. 9.2 Lifecycle of Plant-Based Smart Nanomaterials highlighting sustainable sourcing, green synthesis, functional application, and eco-friendly degradation

low toxicity and drug delivery without adverse effects. Chitosan-based nanoparticles (crab shells or fungi origin) accelerate tissue regeneration, it is non-toxic, and hence very useful for wound healing applications. Although these novel technologies also have excellent safety profiles, regulatory (FDA, EMA, and EFSA) agencies are actively watching their utilization, and they need to be extensively characterized qualitatively and quantitatively by rigorous toxicological investigations to evaluate the possible impact on human health and the environment. As adoption ensued, regulatory agencies took steps to assure and protect safety and efficacy in many aspects (Vieira et al. 2022).

9.5 Applications of Plant-Based Smart Nanomaterials

The plant-intelligent nanomaterial has wide practical applications in the real world (Table 9.2).

9.5.1 Drug Delivery and Therapeutics

Plant-based smart nanomaterials have progressed to the next generation of powerful platforms for highly selective and regulated drug delivery, overcoming critical limitations of conventional approaches. These nanomaterials, obtained from natural polymers such as cellulose, lignin, and plant proteins, are biocompatible and

Table 9.2 Applications of plant-based smart nanomaterials

Application Area	Mechanism	Example/material	Applications	Key benefits	Reference
Drug delivery	pH-responsive materials release drugs in acidic environments (e.g., tumor microenvironments).	Cellulose Nanofibrils (CNFs), lignin-based nanoparticles	Targeted cancer therapy, precise drug delivery, reduced side effects.	Enhanced efficacy, reduced toxicity	Figueiredo et al. (2017), Zhao et al. (2020) and Imlimthan et al. (2019)
Environmental monitoring	Functionalized nanoparticles detect pollutants via adsorption, photocatalysis, and signal transduction.	Lignin nanoparticles, cellulose nanocrystals (CNCs)	Oil spill cleanup, heavy metal detection in water, and gas sensors for air quality monitoring.	Eco-friendly, high sensitivity to pollutants	Imlimthan et al. (2019) and Carrion et al. (2021)
Agriculture	Controlled release of nutrients and pesticides through stimuli-responsive mechanisms.	Lignin-based nanofertilizers, neem-based nanopesticides	Increased crop yields, reduced chemical runoff, improved soil health, and efficient pest control.	Sustainable farming, reduced environmental impact.	Carrion et al. (2021)
Food packaging	Smart films detect spoilage with color changes or maintain freshness through antimicrobial properties.	Anthocyanin-based films, nanostarch, lipid nanoparticles	Real-time freshness indicators, extended shelf life of products, and reduced spoilage in perishable goods.	Reduced food waste, improved consumer safety	Ebrahimi Tirtashi et al. (2019) and Capello et al. (2021)

environmentally friendly. Their sensitivity to exogenous cues, i.e., pH, temperature, or enzymes, makes them perfect for customized therapy, thereby allowing the delivery of therapeutic agents in a site-specific manner. In this way, not only, the therapeutic effect can be improved, but also less systemic toxicity can be produced. In contrast to artificial analogs, plant-derived nanomaterials offer a natural, cost-effective, environmentally friendly, and fully compatible with green chemistry principles solution (Gulia et al. 2022; Barzin et al. 2023; Zhang et al. 2021; Hajialyani et al. 2018).

9.5.1.1 pH-Responsive Plant-Based Nanomaterials

Plant-based nanomaterials hydrogels, such as cellulose nanofibrils (CNFs), exhibit pH-sensitive properties that allow them to respond to the acidic conditions found in tumor microenvironments. Materials that are biocompatible and undergo conformational/structural modification when exposed to low pH in cancerous tissue that effectively release entrapped therapeutic agents (e.g., chemotherapies, nucleic acids) are described. These hydrogel-based materials are mainly employed for cancer treatment, in which their ability to release drugs in a specific manner at the tumor site lessens off-target toxicity and increases the efficacy of the treatment (Shayan et al. 2022; Figueiredo et al. 2017; Hajialyani et al. 2018).

9.5.1.2 Thermo-Responsive Plant-Based Nanomaterials

Here, thermo-responsive plant-based nanomaterials (e.g., lignin or starch nanoparticles) undergo self-adaptation into alternative phases at a specific temperature. These materials are capable of releasing their encapsulated therapeutic agents in the presence of high temperatures, typically achieved by treating them using a hyperthermia local treatment. The heat brings about a structural rearrangement of the nanomaterials that allows them to control the delivery of the pharmaceutical to the site of interest in a regulated manner. Such nanomaterials are combined with treatment by hyperthermia in the hopes of localized drug delivery to provide anticancer therapy, increase local drug concentration at the tumor site, and reduce systemic toxicity (Liu et al. 2022; Guo et al. 2022).

9.5.1.3 Antimicrobial and Photodynamic Properties of Lignin-Based Nano Sprays

Nano sprays containing lignin are characterized by antimicrobial and photodynamic activities, and it is highly feasible to sterilize an object for disinfection. The photosensitizing action of the nanospray on singlet oxygen production as a result of illumination gives it this biocidal action. The Photoluminescence of the LNSR is exploited for bioimaging, thus combining therapeutic capabilities with diagnostic capabilities. Lignin-derived coatings are promising from an antimicrobial perspective, in particular, in applications involving light-induced disinfection, (virtually any) very useful on medical devices and environmental surfaces. Plant-origin nanomaterials are inherently biodegradable and bio-compatible (i.e., they decompose in vivo over time, which reduces the risk of long-term toxicity) in nature. This function is particularly of concern for applications, e.g., for controlled drug delivery, in which the safety of the degradation of the material needs to be ensured once the intended function of the material has been executed. These materials have potential use in areas such as drug delivery, wound healing, and tissue engineering, where biocompatibility and biocompatible degradation, without any toxicity, are crucial. The effectiveness of plant-based smart nanomaterials for drug delivery and drug therapy is shed light on through the progress towards the frontline because of the capability to intelligibly respond to stimuli in the environment or the body to accurately and gradually release drugs, enhance bioavailability, and decrease side effects. Their sustainable synthesis and versatility make them a great alternative to synthetic carriers. Based on sustained scientific investigations and technological developments, plant-derived intelligent nanomaterials are heading to transform present medicine by offering safer, more efficient, and more accurate therapeutic modalities and interventions (Paganini et al. 2021a; Vanlalveni et al. 2021; Okeke et al. 2023).

9.5.2 Environmental Applications of Plant-Based Smart Nanomaterials

Given the increased environmental pollution, there is a significant demand for eco-friendly, disposable technologies for water, soil, and air sewage control. Bio-smart nanomaterials implanted in the soil can be considered eco-friendly alternatives to synthetic materials for efficiently remediating and removing heavy metals, organic pollutants, and microbial pollutants through specific remediation and real-time monitoring. Their intelligent mechanisms, including adsorption and photocatalysis, enable pollutant removal while reducing environmental footprints. CNCs functionalized with tannic acids or dyes selectively bind heavy metal ions like Fe^{3+}, Hg^{2+}, Pb^{2+}, or Cu^{2+}, producing distinct colorimetric or fluorescence changes. For example, tannic acid-encapsulated CNCs switch from pale gray to black upon complexation with Fe^{3+}, and fluorescently labeled CNCs selectively quench their emission toward Hg^{2+} or Pb^{2+}. These materials can also be used for water quality monitoring and toxic metal sensing due to their high sensitivity and selectivity even at low concentrations (Raja et al. 2023; Bradford et al. 2021).

Gas-responsive nanomaterials of CNCs for gas sensing with high surface area and re-programmable properties are employed as probes and sensors for tropane ammonia and carbon monoxide. Optical properties, i.e., alterations of reflection wavelength and/or alteration of electrical conductivity that are produced by adsorption of a gas. These sensors apply to urban air quality control, worker safety, and real-time hazardous gas emissions monitoring and have high sensitivity and greenness. For soil remediation, lignin nanoparticles and cellulose composites are preferential pollutants, selective adsorbents which capture pollutants such as pesticides, herbicides, and heavy metals through hydrogen bonding, van der Waals bonding, and ionic exchange. The porous structure of their materials increases both the pollutant retention and decreases in leaching into groundwater, and the restoration of soil health through aeration and fertility. These materials provide contamination protection for groundwater sources, reduce toxic loadings, and, for subsistence purposes, sequester contaminants in the soil, thereby reducing contamination loading. By improving soil quality, they have contributed to sustainable agriculture and its long-term viability. Plant-based smart nanomaterials have tremendous potential to solve environmental problems, achieve sustainability, and create a healthy ecosystem (Raja et al. 2023; Aswathi et al. 2022).

9.5.3 Agriculture

Plant-derived, bio-adapted smart nanomaterials, isolated from natural polymers of origin, including cellulose, lignin, and plant extract biomolecules, tackle a variety of crucial aspects of agricultural applications through controlled agrochemical delivery strategies as well as in situ soil and plant monitoring that enhance the efficiency of resource utilization and reduce the risk of environmental pollution. Specifically, neem-based nano pesticides, e.g., release bioactive molecules, such as azadirachtin, in a highly efficient mode that depends on environmental stimuli, e.g., moisture or pH, and release the pesticide at a targeted and spatialized time. Their UV stability allows for sustained effectiveness with lower application rates and corresponding reduction of chemical leaching into the environment, protecting aquatic ecosystems and supporting regular pest control to achieve healthier crops (Gigli et al. 2022; Pooja Saraswat et al. 2023).

Lignin-based nano fertilizers provide controlled, slow-release delivery of nutrients and result in higher crop yield and soil fertility, along with nutrient loss and greenhouse gas release. These nano fertilizers offer precision agriculture (reduction of fertilization rate and prevention of surface water contamination by leaching). Cellulose nanocrystal (CNC)- based sensors enable real-time soil analysis by monitoring defined soil properties (i.e., soil moisture, pH value, and nutrient content) to optimize irrigation and fertilization. These sensors can optimize the use of resources, minimize drought stress, and control deficiencies in nutrients, resulting in healthier plants with higher yields. Green options of plant-derived nanomaterials are presented that can improve the efficiency and usability of resources while promoting ecological stability. Advances in the fields of smart sensors, controlled-release delivery systems, and soil remediation technologies will boost their role in sustainable agriculture (Gigli et al. 2022; Baliyarsingh and Pradhan 2023).

9.5.4 Food Packaging and Preservation

Plant-based smart nanomaterials are transforming the landscape of food packaging and processing by providing active, intelligent functions to counteract microbial contamination, evapotranspiration, and/or off-odor/off-flavor.

Artificial packaging bio-based biodegradable reformulations have been developed to respond to stimuli of the natural changes in the pH, in the gas emissions, and in the development of microorganisms, and hence to offer the possibility of real-time food quality control and extension of the shelf life. They are the basis of natural polymers (cellulose, lignin, and plant bio-extracts) for decreasing present food losses and for creating novel sustainable packaging solutions. For example, pH-responsive films containing anthocyanins from red cabbage or blackberries exhibit color transformation in response to spoilage-related pH changes, i.e., ammonia vapors associated with animal meat or fish spoilage. These movies provide consumers and shopkeepers with a simple, real-time sign of freshness to reduce food waste and enhance safety. Anthocyanin-containing films change, perceptibly, from purple to yellow when exposed to ammonia, i.e., indicating spoiled food (Baneshi et al. 2024; Shankar et al. 2021; Xu et al. 2024).

Nanostarch, which is extracted from natural sources such as corn and cassava via hydrolysis, yields nanoparticles that not only improve mechanical properties but also impart moisture barrier properties to packaging films. It can adsorb antimicrobial agents to prevent spoilage and extend the shelf life of fresh produce and dairy products. Being biodegradable, nanostarch-based packaging is green and functional. Cellulose nanocrystals (CNCs) extracted using acid hydrolysis of plant fibers improve the stiffness, stretchability, and barrier properties of the packaging materials. CNCs function as oxygen scavenger agents to prevent oxidative deterioration and maintain food-grade quality. They are efficient in the storage of fresh food and can be combined with hygroscopic materials to maintain optimal storage conditions. These bio-based nanomaterials are now paving the way towards sustainable, smart food packaging, resulting in an improved ecotoxicity of food consumption without compromising food safety and quality (Paganini et al. 2021b; Fang et al. 2022; Travičić et al. 2023; Bangar et al. 2023).

9.5.5 Biosensors and Diagnostics

Plant-derived smart nanomaterials have been extensively employed for designing biosensors and diagnostic platform devices for disease diagnosis, health monitoring, and environmental surveillance purposes. Because of biocompatibility, large surface area, and sensitivity to stimuli, they suit many applications requiring accuracy and sensitivity. These materials are natural polymers (i.e., cellulose, lignin) and plant extracts and are eco-friendly alternatives for synthetic nanomaterials, particularly for point-of-care diagnostics, and hence, can enhance the performance of biosensors in terms of faster speed and higher sensitivity, and at a lower cost.

Tunable functionalized cellulose nanocrystals (CNCs), functionalized with metal nanoparticles or enzymes, are used as a label for the sensors of glucose and enable non-invasive glucose tracking in clinical and laboratory settings. The modified CNCs containing glucose oxidase or containing Fe_2O_3 NPs display electrochemical glucose sensing or color change and thus are of use in continuous glucose control in diabetes. In addition, anthocyanin-encapsulated visual colorimetric biological sensors show a shift in the visible color upon contact with an analyte, so early detection of bacterial adulteration in food is assured. These inexpensive, easily accessible sensors provide convenient possibilities for food safety diagnostics (Singh and Verma 2019).

DNA (linear and branched) hybrid nanostructures encapsulated within lignin-based nanomaterials are also synthesized for wearable nanomaterials for the detection of physiological parameters, including temperature, pH, and sweat composition. Because of their biocompatibility and mechanical properties, they are very suited for flexible devices, which could be useful to athletes and patients suffering from chronic diseases. Plant-derived nanomaterials are transforming biosensing by providing high sensitivity, low cost, and rapid assay performance. Present research efforts are directed toward creating efficient, portable, and environmentally friendly diagnostic systems that will positively impact the medical field, food safety, and environmental monitoring (Turasan et al. 2023).

9.6 Challenges and Future Directions

9.6.1 Limitations and Challenges Associated with Plant-Based Smart Nanomaterials

Plant-derived smart nanomaterials have been developed as a green and sustainable successor to conventional synthetic materials with a multifunctionality potential in many areas. Nevertheless, their generalized use is subject to several issues such as scalability, stability, regulatory policies, and so on. It is critical to overcome these barriers to realize their full potential in areas including healthcare, agriculture, food packaging, and environmental sensing (Kashyap et al. 2023; Bhushan et al. 2024).

9.6.1.1 Production Scalability

Despite the great application prospects of plant-derived smart nanomaterials, production scalability is one of the key obstacles limiting wide-scale practicality. Although laboratory-scale synthesis approaches have shown the possibility and viability of such materials, it is a complicated issue to bridge the gap from industrial-scale production. Scalability issues primarily stem from the variability in plant-derived raw materials, the complexity of extraction processes, and the high cost of scaling up environmentally friendly synthesis methods. These difficulties constrain the general commercial use of plant-based smart nanomaterials, especially in areas such as health care, food packaging, and environmental use (Letchumanan et al. 2021; Aflori 2021).

Plant-derived nanomaterials are produced from raw materials, including cellulose, lignin, and bioactive plant extracts. Yet, the speciation and amount of plant bioactive compounds are very different in the presence of environmental factors, like soil type, temperature, or seasonality. This inherent variability puts in place a discrepancy in the quality, quantity, and efficacy of the nanomaterials being fabricated. I.e., plants growing in nutrient-deficient soils can produce lower concentrations of lignin and cellulose, respectively, thus lowering nanoparticle synthesis efficiency. Controlling raw material supply uniformity in an industrial setting is crucial for maintaining consistent product quality. Standardized agricultural and extraction protocols are key to overcoming this gap, but they invariably come with significant investments in equipment and quality control. One of the challenges for this is variability in the amount of lignin present in plant species and in plantation settings, which has resulted in inconsistency across modalities of NP size and function, making it difficult to scale up production (Kashyap et al. 2023; Bhushan et al. 2024).

The complexity and cost of green synthesis approaches employed to synthesize plant-biobased smart nanomaterials are one of the main bottlenecks. Although this approach removes the use of toxic reagents, it typically requires lengthy extraction, purification, and functionalization of plant extracts. For instance, the green synthesis of silver nanoparticles using plant extracts necessitates careful regulation of pH, temperature, and reaction time to allow reproducibility and high yield. Scaling such steps towards the industrial scale raises operational costs as well as demands sophisticated monitoring systems to control process consistency (Bhushan et al. 2024).

Furthermore, the high energy and material consumption for plant-based material processing are a barrier to further scale-up. E.g., the synthesis of cellulose nanofibrils (CNFs), and mechanical, chemical, or enzymatic processing are all energy-consuming steps. Physical processes (e.g., high-pressure homogenization or ball milling) are energy-intensive, and this is often the case even in scaling up to high-volume production. Chemical and enzymatic treatments, while effective at reducing energy consumption, often require expensive reagents or enzymes, further increasing production costs. Energy saving vs. cost-effectiveness is an ongoing challenge for upscaling these processes. For instance, the mechanical production of cellulose nanofibrils through high-pressure homogenization proved energy-intensive, highlighting the need for more sustainable large-scale processes (Letchumanan et al. 2021).

The creation of standardized and economical production processes for plant-based smart nanomaterials that are scalable is essential to solve the scalability issue. Strategies including automation integration, process optimization improvement, and the realization of hybrid systems that integrate green synthesis and traditional approaches are under investigation. Hybrid approaches can still provide greater control over material properties while achieving environmentally friendly production. For instance, the integration of chemical and biological synthesis methods can not only increase the nanoparticle yield but also shorten the production time and cost.

Additionally, scalable supply chains for inputs are required to be uninterrupted. Partnerships with agricultural industries can give reliable routes to plant-based feedstocks, which in turn reduce variability and production costs. Industrials can tap into low-cost, environmentally benign resources—e.g., agricultural residues (corn husks, wheat straw, or wood waste) for nanomaterial production on a larger scale. One instance of this methodology is the use of agricultural byproducts such as wheat straw as source material for cellulose recovery and bio-separation, which has lowered materials costs and enhanced supply chain security for cellulose-based nanomaterials (Bangar et al. 2023).

9.6.1.2 Stability and Performance

A key issue to address that limits the practical implementation of plant-derived smart nanomaterials is their stability and functionality under real-world conditions. Although such materials show fairly favorable properties in controlled laboratory conditions, their performance and structural soundness are typically compromised when exposed to the outside environment, including temperature change, humidity, light irradiation, and mechanical loading. These restrictions limit their accuracy and applicability, especially in applications related to drug delivery, food packaging, and environmental sensing, in which stable performance over prolonged periods is an essential factor. For example, cellulose nanofibre (CNF)-based materials, whose synthesis for potential use in hydrogels and packaging films has been widely investigated, are very hygroscopic. Due to the hydrophilic character of cellulose, it is susceptible to swelling, and swelling may impact the mechanical properties of cellulose and the functions of cellulose as a whole. In drug delivery systems, this instability negatively affects the controlled-release performance of therapeutic agents, especially in humid or dynamic environments. Another issue is that thermal stability is still a subject to be addressed for most plant-based nanomaterials. For instance, structural modifications and degradation of lignin nanoparticles at high temperatures occur, thereby restricting their application in high-temperature environments, including food processing or industrial uses. Thermal aging also impacts their function as drug vehicles or environmental monitors, where temperature variations may impact the responsiveness and performance of the drug carriers or sensors. One of the challenges is the diminished stability and functional performance of lignin nanoparticles at elevated temperatures, which limits the broad application of lignin nanoparticles in high-temperature applications (Gupta et al. 2023; Mandal et al. 2021).

Mechanical stability is another of the main factors that govern the ability of plant-based smart nanomaterials to perform. In applications such as flexible sensors and coatings, materials derived from cellulose or lignin often exhibit limited mechanical durability, especially under repeated stress or deformation. This deficiency of robustness degrades their long-term utility in devices that need to be both flexible and continuously operated, such as wearable biosensors for healthcare monitoring or soil sensors for precision farming. The mechanical toughness of these materials has been an active research topic, while maintaining their biodegradability is still one of the focuses. To overcome these challenges, material modifications and hybrid systems are under investigation to enhance the stability and functionality of plant-based nanomaterials. For example, chemical crosslinking or compatibilization with synthetic polymers can improve the moisture barrier and mechanical integrity of cellulose-based films to make them more suitable for wet applications. For instance, surface functionalization of silver nanoparticles with stabilizing agents can be used to prevent agglomeration and enhance their photo-stability towards exposure to light. Hybrid systems that combine plant-based nanomaterials with metal oxides or other nanocomposites have also shown promising results in enhancing thermal and mechanical stability (Mandal et al. 2021; Saraiva et al. 2024).

9.6.2 Future Trends

The trend of plant-based smart nanomaterials is directed toward the resolution of existing challenges by an amelioration of current materials, hybrid systems, and sustainable techniques for material production, and a combination with new technologies such as the Internet of Things (IoT) and artificial intelligence (AI). Plant-based components (e.g., cellulose or lignin) and synthetic components (e.g., metal oxides or graphene) are hybrid nanomaterials and are currently being developed to increase the robustness, strength, and function of the hybrid nanomaterials. These developments facilitate water purification, food packaging, and drug delivery applications. Further, green synthesis, e.g., microwave-assisted and ultrasound-assisted approaches, is being fine-tuned to optimize their scalability, chemistry, economic feasibility, and environmental footprint. The integration of IoT and AI further enhances the potential of plant-based nanomaterials, enabling real-time environmental monitoring, predictive analytics in healthcare, and more efficient production processes (Bhushan et al. 2024; Anjum et al. 2019).

Another important trend is the commercialization of plant-based smart nanomaterials with an emphasis on cost reduction and production at the scale of agriculture and forestry residues. Through this method, production costs are decreased and circular economy models can be supported. Researchers are also developing multifunctional materials, for example, antimicrobial and stimuli-responsive nanomaterials, for use in health care and food packaging. Innovations such as self-repairing nanomaterials are enhancing the resilience and performance of these systems and fulfilling sustainable-oriented objectives. With continuous development, however, these materials are foreseen to change industries by providing environmentally benign, efficient, and multi-functional solutions in areas ranging from medicine and agriculture to environmental surveillance and packaging. Plant-based smart nanomaterials of the future have the potential to fuel innovation in a sustainable and environmentally responsible way (Kaushik et al. 2023; Trivedi et al. 2022).

9.7 Conclusion

Plant-derived smart nanomaterials are a new frontier of materials science that provides an eco-friendly and green substitute for synthetic materials. These smart materials based on natural plant material are extremely sensitive to stimuli in the environment, including pH, temperature, light, and magnetic field, and thus, are promising for a wide range of applications in medicine, agriculture, environmental sustainability, and advanced technology. The combination of their unique properties, such as biocompatibility, biodegradability, and functional diversity, makes plant-based smart nanomaterials the key to global challenges in sustainably targeted drug delivery and biosensors for precision agriculture and environmental remediation. These materials enable transformative capabilities with negligible ecological footprints. Their capability to integrate green chemistry concepts and high functionality makes them a pioneer of sustainable development. With research advances, plant-derived smart nanomaterials will remain at the forefront of an era of high-performance materials integrated with environmental and societal requirements. Their incorporation into scalable systems and state-of-the-art technologies will create an equilibrium between technological advancement and ecological conservation, which sets the stage for a more sustainable and intelligent ecosystem (Hajialyani et al. 2018; Aflori 2021; Gupta et al. 2023).

References

Aflori M (2021) Smart nanomaterials for biomedical applications—a review. Nanomaterials 11(2):396

Anjum S, Anjum I, Hano C, Kousar S (2019) Advances in nanomaterials as novel elicitors of pharmacologically active plant specialized metabolites: current status and future outlooks. RSC Adv 9(69):40404–40423

Aswathi VP, Meera S, Maria CGA, Nidhin M (2022) Green synthesis of nanoparticles from biodegradable waste extracts and their applications: a critical review. Nanotechnol Environ Eng 24:8

Baglioni P, Bonini M (2014) Magneto-responsive nanocomposites: preparation and integration of magnetic nanoparticles into films, capsules, and gels. Adv Colloid Interface Sci 207:3–13

Baliyarsingh B, Pradhan CK (2023) Prospects of plant-derived metallic nanopesticides against storage pests—a review. J Agric Food Res 14:100687–100687

Baneshi M, Aryee ANA, English M, Mkandawire M (2024) Designing plant-based smart food packaging solutions for prolonging consumable life of perishable foods. Food Chem Adv 5:100769

Bangar SP, Kajla P, Ghosh T (2023) Valorization of wheat straw in food packaging: a source of cellulose. Int J Biol Macromol 227:762–776

Barzin M, Bagheri AM, Ohadi M, Abhaji AM, Salarpour S, Dehghannoudeh G (2023) Application of plant-derived exosome-like nanoparticles in drug delivery. Pharm Dev Technol 28(5):383–402

Bhushan D, Shoran S, Kumar R, Gupta R (2024) Plant biomass-based nanoparticles for remediation of contaminants from water ecosystems: recent trends, challenges, and future perspectives. Chemosphere:143340

Bradford SA, Shen C, Kim H, Letcher RJ, Rinklebe J, Ok YS et al (2021) Environmental applications and risks of nanomaterials: an introduction to CREST publications during 2018–2021. Crit Rev Environ Sci Technol 52(21):3753–3762

Capello C, Leandro GC, Gagliardi TR, Valencia GA (2021) Intelligent films from chitosan and biohybrids based on Anthocyanins and Laponite®: physicochemical properties and food packaging applications. J Polym Environ 29(12):3988–3999

Carrion CC, Nasrollahzadeh M, Sajjadi M, Jaleh B, Soufi GJ, Iravani S (2021) Lignin, lipid, protein, hyaluronic acid, starch, cellulose, gum, pectin, alginate and chitosan-based nanomaterials for cancer nanotherapy: challenges and opportunities. Int J Biol Macromol 178:193–228

Chen L, Zhou X, Shi Y, Gao B, Wu J, Kirk TB et al (2018) Green synthesis of lignin nanoparticle in aqueous hydrotropic solution toward broadening the window for its processing and application. Chem Eng J 346:217–225

Chen C, Yao W, Sun W, Guo T, Hangya L, Wang X et al (2019) A self-targeting and controllable drug delivery system constituting mesoporous silica nanoparticles fabricated with a multi-stimuli responsive chitosan-based thin film layer. Int J Biol Macromol 122:1090–1099

Cui X, Guan X, Zhong S, Chen J, Zhu H, Li Z et al (2017) Multi-stimuli responsive smart chitosan-based microcapsules for targeted drug delivery and triggered drug release. Ultrason Sonochem 38:145–153

Ebrahimi Tirtashi F, Moradi M, Tajik H, Forough M, Ezati P, Kuswandi B (2019) Cellulose/chitosan pH-responsive indicator incorporated with carrot anthocyanins for intelligent food packaging. Int J Biol Macromol 136:920–926

Fang S, Guan Z, Su C, Zhang W, Zhu J, Zheng Y et al (2022) Accurate fish-freshness prediction label based on red cabbage anthocyanins. Food Control 138:109018

Figueiredo P, Ferro C, Kemell M, Liu Z, Kiriazis A, Lintinen K et al (2017) Functionalization of carboxylated lignin nanoparticles for targeted and pH-responsive delivery of anticancer drugs. Nanomedicine 12(21):2581–2596

Gasti T, Dixit S, D'souza OJ, Hiremani VD, Vootla SK, Masti SP et al (2021) Smart biodegradable films based on chitosan/methylcellulose containing Phyllanthus reticulatus anthocyanin for monitoring the freshness of fish fillet. Int J Biol Macromol 187:451–461

Gigli M, Fellet G, Pilotto L, Sgarzi M, Marchiol L, Crestini C (2022) Lignin-based nano-enabled agriculture: a mini-review. Front Plant Sci 26:13

Giraldo JP, Wu H, Newkirk GM, Kruss S (2019) Nanobiotechnology approaches for engineering smart plant sensors. Nat Nanotechnol 14(6):541–553

Gulfam M, Sahle FF, Lowe TL (2018) Design strategies for chemical-stimuli-responsive programmable nanotherapeutics. Drug Discov Today 24(1):129–147

Gulia K, James A, Pandey S, Dev K, Kumar D, Sourirajan A (2022) Bio-inspired smart nanoparticles in enhanced cancer theranostics and targeted drug delivery. J Funct Biomater 13(4):207

Guo Y, Sun L, Wang Y, Wang Q, Jing D, Liu S (2022) Nanomaterials based on thermosensitive polymer in biomedical field. Front Chem 21:10

Gupta G, Khan J, Singh NK (2023) Plants based nanomaterials for water treatment—a review on sustainable approach for water conservation. NanoWorld J 9:27

Hajialyani M, Tewari D, Sobarzo-Sánchez E, Nabavi SM, Farzaei MH, Abdollahi M (2018) Natural product-based nanomedicines for wound healing purposes: therapeutic targets and drug delivery systems. Int J Nanomedicine 13:5023–5043

Hano C, Abbasi BH (2021) Plant-based green synthesis of nanoparticles: production, characterization and applications. Biomolecules. 12(1):31

Imlimthan S, Correia A, Figueiredo P, Lintinen K, Balasubramanian V, Airaksinen AJ et al (2019) Systematic in vitro biocompatibility studies of multimodal cellulose nanocrystal and lignin nanoparticles. J Biomed Mater Res A 108(3):770–783
Jadoun S, Arif R, Jangid NK, Meena RK (2020) Green synthesis of nanoparticles using plant extracts: a review. Environ Chem Lett 13:19
Jafarzadeh S, Forough M, Amjadi S, Javan Kouzegaran V, Almasi H, Garavand F et al (2022) Plant protein-based nanocomposite films: a review on the used nanomaterials, characteristics, and food packaging applications. Crit Rev Food Sci Nutr 6:1–27
Jagtiani E (2022) Cellulose based smart sensors. Adv Biochips 3(1):50–70
Karimi M, Sahandi Zangabad P, Ghasemi A, Amiri M, Bahrami M, Malekzad H et al (2016) Temperature-responsive smart nanocarriers for delivery of therapeutic agents: applications and recent advances. ACS Appl Mater Interfaces 8(33):21107–21133
Kashyap BK, Singh V, Solanki MK, Kumar A, Ruokolainen J, Kesari KK (2023) Smart nanomaterials in cancer Theranostics: challenges and opportunities. ACS Omega 8(16):14290–14320
Kaushik A, Singh RK, Tyagi PK (2023) Green synthesized nanoparticle based drug delivery: recent trends and future prospects. Precision Nanomed 6(4)
Khan S, Mansoor S, Rafi Z, Kumari B, Shoaib A, Saeed M et al (2021) A review on nanotechnology: properties, applications, and mechanistic insights of cellular uptake mechanisms. J Mol Liq 348:118008
Kolosnjaj-Tabi J, Gibot L, Fourquaux I, Golzio M, Rols MP (2019) Electric field-responsive nanoparticles and electric fields: physical, chemical, biological mechanisms and therapeutic prospects. Adv Drug Deliv Rev 138:56–67
Lem O, Gangurde P, Koivuniemi A, Keskinen A, Efimov A, Durandin N et al (2024) Far-red light-triggered cargo release from liposomes bound to a photosensitizer-cellulose nanofiber hydrogel. Carbohydr Polym 336:122134–122134
Letchumanan D, Sok SPM, Ibrahim S, Nagoor NH, Arshad NM (2021) Plant-based biosynthesis of copper/copper oxide nanoparticles: an update on their applications in biomedicine, mechanisms, and toxicity. Biomolecules 11(4):564
Liu X, Li M, Li X, Ge M, Liu S, Li S et al (2022) A photosensitive sustainable lignin nanoplatform for multimodal image-guided mitochondria-targeted photodynamic and photothermal therapy. Mater Today Chem 4(26):101000
Maiz-Fernández S, Pérez-Álvarez L, Ruiz-Rubio L, Vilas-Vilela JL, Lanceros-Mendez S (2020) Polysaccharide-based in situ self-healing hydrogels for tissue engineering applications. Polymers 12(10)
Mandal S, Chatterjee B, Buddhadev Layek (2021) Cellulose-based nanomaterials in drug delivery applications. In: Elsevier eBooks, pp 57–86
Okeke ES, Nweze EJ, Anaduaka EG, Okoye CO, Anosike CA, Joshua PE et al (2023) Plant-derived nanomaterials (PDNM): a review on pharmacological potentials against pathogenic microbes, antimicrobial resistance (AMR) and some metabolic diseases. 3 Biotech 13(9)
Paganini CC, Laroque DA, Augusto B, Falcão M (2021a) Shelf-life extension of meat products by cellulose acetate antimicrobial film incorporated with oregano's essential oil. Res Soc Develop 10(16):e271101623335
Paganini CC, Laroque DA, Augusto B, Falcão M (2021b) Shelf-life extension of meat products by cellulose acetate antimicrobial film incorporated with oregano's essential oil. Research Society and Development. 10(16):e271101623335–e271101623335
Paul S, Thakur NS, Chandna S, Reddy YN, Bhaumik J (2021) Development of a light activatable lignin nanosphere based spray coating for bioimaging and antimicrobial photodynamic therapy. J Mater Chem B 9(6):1592–1603
Piryaei M, Abolghasemi MM, Raked MS (2025) Eco-friendly and smart biopolymer with green synthesis nanoparticles, anthocyanins, sodium alginate, pectin, and checking the degree of greenness of the method. Results Surf Interfaces:100433

Pooja Saraswat, Singh S, Prasad M, Misra R, Rajput VD, Ranjan R (2023) Applications of bio-based nanomaterials in environment and agriculture: a review on recent progresses. Hybrid Adv 4:100097–100097

Raja MA, Ahmad MA, Md Daniyal, Husen A (2023) Management of wastewater and other environmental issues using smart nanomaterials. In: Elsevier eBooks, pp 489–503

Saraiva DV, Remiëns SN, Ethan, Vermaire IR, Tran L (2024) Flexible, photonic films of surfactant-functionalized cellulose nanocrystals for pressure and humidity sensing. Small Struct 5(7)

Shankar S, Khodaei D, Lacroix M (2021) Effect of chitosan/essential oils/silver nanoparticles composite films packaging and gamma irradiation on shelf life of strawberries. Food Hydrocoll 117:106750

Sharma S, Kumar A, Choudhary A, Harish BM, Karmakar P, Sharma P et al (2022) Recent developments in smart nano-agrochemicals: a promise for revolutionizing present-day agriculture. Mater Today Proc 69:530–534

Shayan M, Gwon J, Meen Sung Koo, Lee D, Adhikari A, Wu Q (2022) pH-responsive cellulose nanomaterial films containing anthocyanins for intelligent and active food packaging applications. Research Square. 2022 Jul 13

Singh AK, Verma N (2019) 4. Plants and plant-derived materials used for biosensor development. In: De Gruyter eBooks, vol 24, pp 73–88

Singh S, Bhardwaj S, Tiwari P, Dev K, Ghosh K, Maji PK (2024) Recent advances in cellulose nanocrystals-based sensors: a review. Material Adv 5(7):2622–2654

Tian B, Liu J (2023) Smart stimuli-responsive chitosan hydrogel for drug delivery: a review. Int J Biol Macromol 235:123902

Travičić V, Cvanić T, Ćetković G (2023) Plant-based nano-emulsions as edible coatings in the extension of fruits and vegetables shelf life: a patent review. Foods 12(13):2535–2535

Trivedi R, Upadhyay TK, Mujahid MH, Khan F, Pandey P, Sharangi AB et al (2022) Recent advancements in plant-derived nanomaterials research for biomedical applications. Processes 10(2):338–338

Turasan H, Rouf TB, Yilmaz T, Kokini JL (2023) Nanomaterials with plant or animal origin for greener biodelivery and biosensor applications: a review. Recent Progress in Materials 05(02):1–36

Vanlalveni C, Lallianrawna S, Biswas A, Selvaraj M, Changmai B, Rokhum SL (2021) Green synthesis of silver nanoparticles using plant extracts and their antimicrobial activities: a review of recent literature. RSC Adv 11(5):2804–2837

Vieira IRS, de Carvalho APA, de Conte-Junior CA (2022) Recent advances in biobased and biodegradable polymer nanocomposites, nanoparticles, and natural antioxidants for antibacterial and antioxidant food packaging applications. Compr Rev Food Sci Food Saf 21(4):3673–3716

Wang J, Guo Z, Xiong J, Wu D, Li S, Tao Y et al (2019) Facile synthesis of chitosan-grafted beta-cyclodextrin for stimuli-responsive drug delivery. Int J Biol Macromol 125:941–947

Xu Y, Wu Z, Li A, Chen N, Rao J, Zeng Q (2024) Nanocellulose composite films in food packaging materials: a review. Polymers 16(3):423

Zhang N, Jia C, Ma X, Li J, Wang S, Yue B et al (2021) Hierarchical core–shell Fe_3O_4@ $mSiO_2$@Chitosan nanoparticles for pH-responsive drug delivery. J Nanosci Nanotechnol 21(5):3020–3027

Zhao J, Zheng D, Tao Y, Liu Y, Wang L, Liu J et al (2020) Self-assembled pH-responsive polymeric nanoparticles based on lignin-histidine conjugate with small particle size for efficient delivery of anti-tumor drugs. Biochem Eng J 156:107526–107526

Zheng D, Cao S, Li D, Wu Y, Duan P, Liu S et al (2024) Fabrication and characterization of chitosan/anthocyanin intelligent packaging film fortified by cellulose nanocrystal for shrimp preservation and visual freshness monitoring. Int J Biol Macromol 264:130692–130692

Chapter 10
Plant-Based Nanocomposite Formulation and Applications

Manoja Das

10.1 Introduction

Composites are made by mixing two or more dissimilar materials. A new type of unique material known as nanocomposites is created by combining one or more different materials at the nanoscale level, which is of great value (Malhotra and Ali 2017). The term nanocomposite refers to the combination of nanosized biodegradable polymers with polymer composites, which have at least 100 nm in size. Nanomaterials have a larger surface area as compared to mass for interaction, leading to advantageous qualities. The morphology and interface features of nanocomposites affect their qualities in addition to the individual components employed for preparation. Nanocomposites are appealing for numerous applications due to their superior attributes over individual materials, including enhanced abrasion resistance, fouling resistance, extreme hydrophobicity, fantastic hydrophilicity, thermal energy transport, electronic and ionic transport, and fluid transport (Rhim et al. 2013). Nanocomposites are predicted to reach a US$9 billion business by 2025, with quantities approaching five million tons (Rai et al. 2024). The declining costs of nanomaterials will drive growth in this industry, rising production levels, and the resolution of technical problems about their mass-scale manufacturing as nanofiber-based composites replace other conductive materials in a significant part of the market, particularly in electrical and electronics sectors, which become more popular in the upcoming years.

M. Das (✉)
Department of Biotechnology, Gandhi Institute of Engineering and Technology University (GIET University), Gunupur, Odisha, India
e-mail: manojadas@giet.edu

V. Ghosh et al. (eds.), *Applications of Plant Bioactive Compounds-Based Nanoformulations*, Nanotechnology in the Life Sciences,
https://doi.org/10.1007/978-3-032-13086-0_10

The domain of nanocomposites has attracted more attention from many people in the recent past, with considerable attempts to regulate the nanostructures using appropriate synthetic techniques. The high surface-to-volume ratios of nanomaterials and nanolayers make them perfect for use with polymeric materials. Polymeric materials are easily processed and have a low dielectric constant. However, the most popular high dielectric constant ceramic materials are made at high temperatures and are found to be fragile. Thus, the fibre-based products are integrated into the matrix of polymers as reinforcing agents to increase the mechanical strength of the matrix. Apart from that, adding fibre-based materials to the composite improves mechanical qualities like toughness, compressive strength, tensile strength, tensile modulus, and twisting strength. The agricultural waste fibers are suitable candidates to combine with various polymers, monomers and as additives, including polyethylene, polypropylene, low-density polyethylene, unsaturated polyester resin, epoxy resin, melamine, methyl or ethyl acrylate, starch, phosphoric acid, urea, urethane, natural dye, artificial dye, gelatine, etc., to prepare commercial composites (Pai et al. 2023). Moreover, a few additives are added to composites that produce free radicals that alter the characteristics of the composites. The cellulosic fibre prepared from agricultural wastes also contains hemicellulose, lignin, pectin, and other substances.

10.2 Classification of Nanocomposites

The nanocomposites can be classified into various types based on the kind of reinforcements or fillers used during the preparation of the materials. A few of them are described as below:

10.2.1 Ceramic Matrix Nanocomposites

These are a specific class of nanocomposites that have drawn much attention due to their exceptional thermal and mechanical properties. This nanocomposite is comprised of ceramic fibres, particles, and ceramic materials embedded in a ceramic matrix. These composites are potentially used in several fields, such as clinical sciences, aerospace, and energy production. Further, these nanocomposites are utilized in artificial joint implants to address the fracture failure issues to increase patient mobility, which avoids the need for expensive surgery. The nanocomposite alumina-toughened zirconia has the most popular application in the field.

10.2.2 Metal and Its Oxide-Based Nanocomposites

Bimetallic nanocomposites are being employed as alloys or in shells and possess a wide range of optical and catalytic properties. In different sensors and biosensors, metal oxide nanoparticles and nanowires with semiconducting properties are used as filler materials for gas sensing. The advantages of oxides of semiconductor metals to be used in sensors are that they are less costly, easily available, and stable in high gas density. There are several applications of metals and metal alloys, such as the preparation of orthopaedic materials, dental structures, and other clinical applications (Allali et al. 2022).

10.2.3 Polymer-Based Nanocomposites

There are numerous types of polymer-based nanocomposites prepared using different kinds of compounds based on their uses, including inorganic & organic, organic &organic, polymer & ceramic, inorganic and organic amalgams, polymer & polymer, and bio-composite, etc. For instance, nanocomposites made up of clay and coffee nanoparticles in a particular ratio are commonly applied as fillers in many instruments.

10.2.4 Carbon-Based Nanocomposites

Carbon nanotubes are commonly used in the preparation of composites, chemical sensors, biosensors, and nanoelectronics devices due to their specific mechanical and electrical characteristics. They are used in conductive nanocomposites and the components that need to discharge the electrostatic potentials during their uses. The electric appliances are added with carbon nanotube-based nanocomposites for proper electrostatic potentials to be discharged during their operation. The frequently used carbon nanotubes in the electronics field are single-walled nanotubes (SWNTs) and multi-walled nanotubes (MWNTs). The MWNTs comprise a coaxial configuration of concentric single nanotubes, whereas SWNTs do. In contrast, SWNTs are made up of cylindrical single graphite nanostructure sheets that are rolled into tubes with a diameter of 1–3 nm only.

10.2.5 Noble Metal–Based Nanocomposites

Different types of metals are used for the preparation of nanocomposites. During the preparation of nanocomposites, more than two metals are melted and mixed in a proper ratio to develop a polymeric matrix. The nanocomposites prepared using different porous metal oxides have tenable porosity, more chemical stability, minimal bulge, high mechanical strength, and decomposable stability. These nanocomposites have better sensitivity for detecting reducing and oxidizing gases at lower working temperatures.

10.3 Characterization of Nanocomposites

It is vital to characterize the structural geomorphology of nanocomposites to comprehend the structural properties of the materials effectively. The examination of shape, size, structure, configuration, thermal stability, optical, magnetic, electrical, and thermal characteristics of nanocomposites is possible using various techniques (Fazal-ur-Rehman et al. 2024). Among the methods employed to describe the nanocomposites are.

10.3.1 UV-Vis Analysis

A scientific method called ultraviolet-visible (UV-Vis) analysis determines a sample's chemical properties by measuring its reaction to visible and ultraviolet light. Measurement of light absorption in the visible and ultraviolet spectra is done using UV-Vis spectroscopy. Electronic transitions are described by UV-Vis, which is frequently employed in quantitative analysis.

10.3.2 FT-IR Analysis

Fourier Transform Infrared Spectroscopy determines the uniqueness of organic, polymeric, and inorganic materials. In this instrument, infrared light is used to analyze the chemical properties of compounds. The Fourier Transformation Infrared spectrometer confirmed the functional group of materials employing the KBr pellet in the transmittance mode between 4000 and 400 cm^{-1}. In addition to XRD, FTIR can be used to assess the sample's structural characteristics (Zhao et al. 2022).

10.3.3 Microscopic Methods

The methods primarily comprise three forms of microscopy: transmission electron microscopy (TEM), atomic force microscopy (AFM), and scanning electron microscopy (SEM). In transmission electron microscopy (TEM), an electron beam provides high-resolution pictures of tiny objects. TEM analysis is used in numerous disciplines, including materials science, microbiology, nanotechnology, cancer research, virology, and palaeontology. This microscopic examination helps to forecast the inner structure of nanocomposites, morphology, particle size, gelling agent impact, nanoparticle dispersion, structural stacking, surface roughness, etc. Using a concentrated electron beam, SEM analysis generates a high-resolution image of the surface of the sample. SEM resolution is approximately 1–2 nm. The surface needs to be electrically conductive to analyze the surface morphology of the nanocomposites with SEM. A thin layer of carbon or gold is applied to the non-conductive sample to make it electrically conductive.

10.3.4 AFM Analysis (or Atomic Force Microscopy)

Scanning probe microscopy is another name for AFM. Atomic force microscopy can measure the surface roughness of samples at the Angstrom (A^0) level. AFM detects the tensions between the atoms in a sample to image its surface at the atomic resolution level for conducting and non-conducting materials.

10.3.5 X-Ray Diffraction (XRD)

This instrument is used to determine the orientation, crystallinity, lattice parameters, purity, and size of the nanocomposite. Moreover, it is utilized to determine the phase and unit cell information of the powder nanocomposites. It works on Scherrer's formula to calculate the particle size (D). $D = K\lambda/\beta \cos\theta$, where D is the particle's size, K is referred to as Scherer's constant, λ is the wavelength of X-rays, β represents the diffraction peak's full width at half maximum, and Bragg's angle (angle of diffraction) is denoted by θ. This formula is applied to nanocomposites having a distinct crystalline structure.

10.3.6 Powder X-Ray Diffraction (PXRD)

One of the common techniques for describing materials at the nanoscale is this one. Powder X-ray diffraction (PXRD) is a non-destructive analytical technique that is applied to determine the phase composition, crystal structure, and orientation of different forms of samples. This technique can identify a crystalline material's phase, which also provides information on the unit cell dimensions. In the account of spectroscopic and microscopic methods, powder XRD analysis yields essential information about the samples, such as phase documentation, sample transparency, crystalline structure, etc. (Umavathi et al. 2020).

10.3.7 EDX Analysis

An EDX spectrometer is used to perform the energy-dispersive X-ray spectroscopy to ascertain the atomic composition of nanocomposites. The elemental composition of materials is recognizable and measured using energy-dispersive X-ray spectroscopy, sometimes called EDX analysis. This non-destructive technique is frequently combined with scanning electron microscopy (SEM). Using EDX analysis, analysts and researchers can learn important details about a sample's chemical makeup at the macro, micro, or nanoscale. A concentrated electron beam is used to expose a sample during the EDS analysis procedure, which causes the sample's atoms to emit distinctive X-rays. At particular energy levels peculiar to that element, each element in the sample releases X-rays. Using energy-dispersive X-ray spectroscopy, the emitted X-rays can be identified and evaluated to determine the existence and comparative quantity of elements present in the sample.

10.3.8 Thermal Stability

Thermal analysis encompasses many techniques, such as thermal-mechanical analysis (TMA), dynamic mechanical thermal analysis (DMTA), differential scanning calorimetry (DSC), and thermogravimetric analysis (TGA). One method used in thermal analysis, a field of materials science that examines how a material's properties vary with temperature, is thermomechanical analysis (TMA). Dynamic mechanical, thermal analysis (DMTA) can be performed to analyze a polymer nano-composite mechanical characteristic. A rotational rheometer can be used to perform DMTA in torsion. The temperature is continuously fluctuating while the material is subjected to oscillatory shear. Besides, DMTA is used to assess the mechanical performance of solid materials having significant application-related characteristics like stiffness, brittleness, damping, or impact resistance. Thermogravimetric analysis (TGA) is carried out using an instrument called a thermogravimetric analyzer. The instrument constantly measures the mass of the samples at different temperatures and varies over time.

10.4 Plant-Based Nanocomposites

Numerous phytochemicals, including flavonoids, alkaloids, tannins, saponins, and other metabolites, are found in plants that significantly impact the generation of nanomaterials. Green synthesis has emerged as a popular method for creating nanoparticles and composites in recent years. Natural biomolecules can actively contribute to the synthesis of nanoparticles of various sizes and shapes, which can serve as a catalyst for the development of harmless, eco-friendly, and easy procedures for the synthesis of nanocomposites (NCs). Plant-based nanocomposites combine biopolymers from plants with nanostructures to produce useful materials with improved qualities (Siddiqui et al. 2023). These materials are becoming increasingly popular since they are sustainable, biodegradable, and can substitute petroleum-based and synthetic products. Plant waste is a valuable source of biopolymers, including proteins, polysaccharides, and lipids, which can be converted into nanomaterials. These nanocomposites generate hybrid materials by fusing the inherent properties of biopolymers, including cellulose, lignin, and starch, with the functional properties of nanoparticles. One of the best active components of biological nanomaterials is plant-based nanocomposites, which may have antidiabetic, anti-inflammatory, antibacterial, anti-cancer, and antioxidant effects (Anwer et al. 2023). Types of plant-based nanocomposites are shown in Table 10.1 along with their composition, properties, and specific uses.

10.4.1 Materials for Plant-Based Composites

These days, various plant-based biopolymeric chemicals are employed to create nanocomposites. The majority of compounds consist of:

10.4.1.1 Cellulose Nanocomposites

Cellulose is the most common biopolymer on the planet. It is a polysaccharide comprising more than a thousand glucose molecules. Cellulose is prevalent in all naturally occurring organic substances, making up over 33% of all vegetable matter and serving as the essential structural constituent of plant cell walls. The rigidity and strength of plant cell walls are attributed to the presence of cellulose. The common sources of cellulose are agricultural waste, fruits, vegetables, whole grains, legumes, nuts, and seeds (Chopra and Manikanika 2021). Agricultural waste, including pineapple peel, rice and wheat stalks, oat husks, straw, and leaves, can extract cellulose.

Table 10.1 Types of plant-based nanocomposites: composition, characteristics and application

Sl no.	Types of nanocomposites	Composition	Characteristics	Application	References
1.	Cellulose-based nanocomposites	1. Cellulose (base materials)-derived from plant 2. Nanofillers (reinforcing agents) 3. Polymer matrix (optional) 4. Additives (if needed)	1. Biodegradability and biocompatibility 2. Excellent barrier properties 3. Thermal stability 4. High mechanical strength 5. Large surface area and reactivity	1. Packaging material 2. Strengthening in bioplastics 3. Water purification 4. Biomedical 5. Electronics and sensors	Joseph et al. (2020)
2.	Starch-based nanocomposites	1. Starch (base materials)-derived from a plant 2. Nanofillers (reinforcing agents) 3. Plasticizers 4. Crosslinking agents(optional)	1. Film coating ability 2. Biodegradability 3. High biocompatibility 4. Improve mechanical properties 5. Antimicrobial potential	1. Biodegradable packaging 2. Edible films 3. Agricultural mulch 4. Biomedical 5. Water purification	Tibolla et al. (2020)
3.	Chitosan-based nanocomposites	1. Chitosan (base materials)-derived from fungi or animal 2. Nanofillers (reinforcing agents) 3. Plasticizer 4. Crosslinking agents(optional) 5. Solvents or acid medium	1. Biodegradability 2. Antimicrobial activity 3. Film-forming ability 4. Biocompatibility 5. High mechanical strength	1. Biomedical 2. Food packaging 3. Environmental application 4. Agriculture application 5. Cosmetic and personal care	Guo et al. (2024)
4	Protein-based nanocomposites	Protein (base materials)-derived from plant 2. Nanofillers (reinforcing agents) 3. Plasticizer 4. Crosslinking agents(optional) 5. Solvents or medium (for film formation)	1. Good film-forming ability, 2. High mechanical strength 3. Biodegradability 4. Biocompatibility 5. Barrier properties	1. Edible coating 2. Biosensor 3. Wound healing and tissue engineering 4. Drug delivery 5. Food packing	Krishna et al. (2024)

(continued)

Table 10.1 (continued)

Sl no.	Types of nanocomposites	Composition	Characteristics	Application	References
5	Lignin-based nanocomposites	1. Lignin derived from plant biomass 2. Nanofillers 3. Polymer matrix (optional) 4. Plasticizer (if required)	1. UV resistance 2. Antioxidant activity 3. Antimicrobial properties 4. Improved mechanical strength 5. Biodegradability and sustainability	1. UV-shielding films 2. Biomedical field 3. Food packaging 4. Water treatment 5. Agricultural applications	Parvathy et al. (2021)

Cellulose is esterified to produce bioplastics, which include cellulose acetate (CA), cellulose acetate propionate (CAP), and cellulose acetate butyrate (CAB). Because CA is more durable and has superior optical clarity, it is exciting. Cellulose and cellulose-derived nanocomposites are now appropriate for use in food wrapping, preparation of cellulose nanocomposite film with a good tensile strength, and agricultural uses.

10.4.1.2 Cellulose Nanocrystals (CNCs)

Cellulose nanocrystals (CNCs) are special nanostructures made from cellulose, the most prevalent and unlimited natural polymer. The rod-shaped nanoparticles known as cellulose nanocrystals (CNCs) are made from cellulose. Typically, CNCs measure 100 nm in length and 2–5 nm in diameter. Much attention has been paid to these nanostructures' mechanical, optical, chemical, and rheological properties. Because of their tiny size, they cannot reflect or refract light, which can produce transparent and colourless materials. CNCs have outstanding mechanical strength, stiffness, and thermal stability and are renewable, biodegradable, and non-toxic (Rana et al. 2021). Potential applications for CNCs include tissue engineering, medication delivery, medical devices, cosmetics, and protein scaffolds.

10.4.1.3 Cellulose Nanofibers (CNFs)

Cellulose nanofibers (CNFs) are usually 5–20 nm in diameter and many micrometres long. Although cellulose nanofibers have good mechanical qualities, they are typically too short for ecologically friendly composites like CNF-reinforced composites. When nanofibers are so precisely and delicately aligned, their mechanical qualities improve. Apart from spinning, mechanical stretching is another effective method for obtaining cellulose nanofibers with enhanced alignment and high mechanical qualities (Surendran and Sherje 2022).

10.4.1.4 Microcrystalline Cellulose

It is processed wood pulp. It is an inert chemical that does not break down during digestion or absorb much. When consumed in high amounts, it gives food bulk and may have a laxative effect. It is utilized in solid dosage forms, such as tablets, and has outstanding compressibility qualities. It is also present in many processed foods and can be employed as a stabilizer, texture adjuster, suspending agent, or anti-caking agent, among other things.

10.4.1.5 Bacterial Cellulose

Bacteria ferment to produce bacterial cellulose (BC), a naturally occurring organic substance. The biomaterial known as bacterial cellulose has exceptional mechanical qualities, a high water-holding capacity, and a distinctive structure. The substance is also crystalline, biocompatible, biodegradable, and extremely pure and obtained from bacteria like *Sarcina ventriculi*, *Acetobacter*, *Komagataeibacter*, and *Agrobacterium (*Pandey et al. 2023*)*.

10.4.1.6 Starch Nanocomposites

They comprise two different kinds of molecules: branched amylopectin and linear and helical amylose. Pure starch is a white, tasteless, and odourless powder, and alcohol and cold water do not dissolve it. They are renewable polymers of carbohydrates made from rice, corn, or potatoes. A polymeric carbohydrate, starch, or amylum comprises several glucose units connected by glycosidic linkages. In the industrial sector, starch is frequently transformed into sugars, for instance, through malting. These sugars can undergo fermentation to yield ethanol, which is used to make biofuel, whisky, and beer (Abdullah and Dong 2018). Furthermore, a lot of processed meals contain sugars made from processed starch. Depending on their botanical origin and manufacturing conditions, starch-based films and coatings can have many properties. Usually, when making films for coating and food packaging applications, starch is modified to enhance its functions. One of the best ways to enhance their physicochemical and functional characteristics is through crosslinking or reinforcing, among other activities. Additionally, CNFs were found to be a significant reinforcing material for starch-based nanocomposite materials. A nanocomposite film is prepared with CNF as a reinforcing filler, sucrose in oxidized form as a cross-linker, and starch as a biopolymer matrix. Using sucrose to crosslink the nanocomposite film during production increases its mechanical strength while decreasing its permeability to oxygen and water vapour (Sun et al. 2024). Esterification and etherification are two further chemical processes that alter starch by substituting other chemical groups for its hydroxyl groups, which reduces the starch's affinity for water molecules. A promising substitute for many applications involving synthetic plastic is thermoplastic starch (TPS) (Kshitij et al. 2024). It is environmentally friendly because it contains naturally occurring and biodegradable ingredients. Thermoplastic starch is a possible substitute for synthetic plastic. TPS is brittle and has poor mechanical qualities because it is inherently prone to retrogradation with its limitations. Additionally, it has a high-water solubility, low water resistance, and poorer performance at higher humidities due to its inherent hydrophilia.

10.4.1.7 Agar Nanocomposites

Marine algae (family: Rhodophyceae) and certain seaweeds are the common sources of agar production, which is made up of agarose and agaropectin (Malagurski et al. 2017). Agaropectin is a non-gelling, sulphated, branching, and charged compound, whereas agarose is a neutral, linear molecule comprising β-1,3-linked-D-galactose and α-1,4-linked 3,6-anhydro-L-galactose units. Agar is one of the most promising polysaccharides for the preparation of films and coatings due to its transparency, low hydrophilicity, and better film-forming properties. However, agar films have limited use in food packaging due to their poor mechanical qualities, lower thermal stability, and weak antibacterial activity. It has been concentrated on enhancing these characteristics by using nano-fillers, such as metallic and bimetallic nanoparticles, nano clay, and nanocellulose, to reinforce agar films (Atef et al. 2014). It has been found that the addition of less clay to the nanocomposites results in a higher grade of intercalation and clay dispersion. This raises the mechanical strength of nanocomposites and lowers the water vapour permeability (WVP). Moreover, various properties of agar nanocomposite film, such as mechanical, optical, and thermal properties, can be enhanced by adding Zn and Cu elements. The shape, physicochemical properties, and film functioning were also affected by adding Zn and Cu minerals to agar.

10.4.1.8 Zein Nanocomposites

Zein is a naturally occurring protein that is soluble in alcohol and is commercially derived from corn kernels. Leucine, proline, alanine, and phenylalanine are among the hydrophobic amino acid residues abundant in zein. It has film-forming qualities and is hydrophobic, biodegradable, biocompatible, and widely accessible (Tadele et al. 2024). Because of these characteristics, it is a popular choice for creating films and coatings made of biopolymer-based nanocomposite materials. However, there are drawbacks to zein-based films, including low elongation, poor thermal characteristics, poor processability, and brittleness. Therefore, plasticizers like ethers, aldehydes, long-chain fatty acids, and glycerol are frequently utilized in zein to improve the quality. Zein is converted into silky, oil-resistant, greasy films and coatings by using plasticizers for food packaging. It can preserve food value by serving as an oxygen barrier and preventing decomposition. The preparation and application of different nano-fillers in zein-based nanocomposites and organic active agents for antibacterial food packaging have been documented (Wang et al. 2024).

10.4.1.9 Wheat Gluten (WG) Nanocomposites

It consists of two essential plant proteins, such as gliadins and glutenins. Gliadins are prolamin proteins, whereas glutenins are linked proteins joined by disulfide bonds. Various interactions, including covalent bonds, non-covalent interactions, and entanglements, stabilize the gluten system, a special network of polymers. Since gluten is a heteropolymer, it can interact with other molecules through chemical grafting or cross-linking. In contrast to glutenins, gliadins are soluble in aqueous alcohol; however, glutenins become soluble in aqueous alcohols when their disulfide links are reduced. WG is being used in non-food applications, such as food packing, with the preparation of biodegradable films and coatings. Making fresh or minimally processed fruits and vegetables last longer after harvest and controlling respiratory exchange appears to be very promising aspects.

10.4.1.10 Lignin Nanocomposites

Lignin comprises three complicated polymeric molecules and is a phenolic chemical (containing an aromatic ring with a —OH group attached). The main component of wood is lignin, a complex organic polymer that contains oxygen. It is the second most widespread organic compound found on Earth after cellulose. The cell walls of wood contain lignin, a secondary metabolite that accounts for 24–35% dry weight of softwoods and 17–25% of hardwoods. There are several industrial applications for lignin, including as a soil conditioner, a glue for linoleum, a binder for laminated particleboard or composite wood products, and a filler or active ingredient in phenolic resins. Lignin is also used to make dimethyl sulfoxide and vanillin, a synthetic vanilla (Parvathy et al. 2020).

10.4.1.11 Nanocomposites Made from Soy Protein

Soy protein (SP), widely accessible worldwide, is derived from soybean seeds. Two essential proteins called albumin and globulin make up soy proteins. They are sold commercially as soy flour (SF), soy protein concentrate (SPC), and soy protein isolate (SPI). Since SPI has a larger protein concentration than the other SP compounds, it has a greater capacity to form films, making it ideal for coatings and films. Because of soy protein's biodegradability, biocompatibility, and film-forming properties, investigators are interested in using it as a basic material to create edible coatings and films.

10.4.1.12 Plant-Protein Nanocomposites

Proteins are biocompatible, biodegradable, and extensively distributed in nature. Protein nanocomposites have recently attracted much interest because of the advantageous properties of both proteins and nanotechnology. Protein nanocomposites, protein-nanoparticle conjugates, and protein nanoparticles have so become a replacement nano-carrier medium. Protein nanoparticles are smaller, non-toxic, and quickly eliminated from the human body compared to traditional nanoparticles derived from metals and other inorganic compounds (Wang et al. 2020). Proteins are basically amphiphilic by nature, so they can interact with both hydrophobic and hydrophilic parts of the compounds. In addition to protein nanoparticles, scientists have effectively used protein-nanoparticle conjugates in several applications.

10.4.1.13 Pectin Nanocomposites

Pectin is a naturally occurring, benign, and versatile heteropolysaccharide crucial to the food, pharmaceutical, and cosmetics industries. It is one of the primary constituents of the cell walls of all higher plants and occupies up to 0.5–4.0% of the total fresh weight of a plant. Particularly in fruits like apples, oranges, and lemons, pectin, a structural fibre, is present in the plant cells' intracellular layer and main cell wall. Promoting cell adhesion and propagation and improving tissue remodelling through creating an extracellular matrix *in vivo*, pectin and pectin-based composites function as interactive matrices or scaffolds.

10.4.1.14 Alginate Nanocomposites

Alginate, a polymer that is naturally obtained from brown seaweed, is well-known for stabilising and producing gels. Alginate's remarkable flexibility, mechanical strength, and sensory and visual clarity are crucial for food applications. Combining alginate biopolymers with other polymers and nanomaterials can improve their functional properties, leading to the development of smart alginate hybrid nanomaterials and nanocomposites (Zia et al. 2020). They are perfect for creating novel, multifunctional smart nanomaterials because of their programmable physicochemical and biological characteristics.

10.4.1.15 Chitosan Nanocomposites

Chitins are present in fungal cell walls and crustaceans. Both include chitin, and an alkali to treat chitin (deacetylation) yields chitosan. The units that make up this polysaccharide are β-(1 → 4) D glucosamine and β-(1 → 4)-N-acetyl-D glucosamine. Chitosan is remarkably flexible, strong mechanically, and thermally stable. It is chemically capable of forming compounds with ions and transition metals.

Chitosan possesses antibacterial activity and is also non-toxic, biodegradable, and biocompatible. Additionally, it forms a great barrier against radiation, gases, and moisture. Adding nano additives to chitosan improves its biological activity, mechanical strength, and thermal stability. With their improved physicochemical and biological characteristics, these chitosan-based nanocomposites have various uses in pulp and paper, textile, food packaging, medical, wastewater treatment, cosmetics, and tissue engineering industries (Guo et al. 2024). Chitosan nanocomposites come in a variety of forms, such as carbon nanotube chitosan, chitosan-metal nanocomposites, which are composed of chitosan and metal nanoparticles, and nanocomposites of chitosan polymers.

10.4.1.16 Nanocomposites Based on Carnauba Wax

Commercial applications are made of wax-based coatings for various fruits and vegetables to reduce dehydration, shrivelling, and shrinking. Beeswax, carnauba, candelilla, paraffin, and rice bran wax are among the waxes that have long been used successfully in coatings. The wax coating has been applied to a variety of fruits and vegetables, including oranges, lemons, grapefruits, apples, tomatoes, rutabaga, cherries, cucumbers, apricots, bananas, dates, grapes, guavas, mangoes, peaches, nectarines, and pears. Carnauba, plant-based wax is obtained from the leaves of the Brazilian palm tree (*Copernica cerifera*). Since the 1930s, carnauba has been widely used to coat fruits and vegetables to keep them shiny and prevent weight loss. Natural antimicrobial agents and nanoparticles have been successfully used with carnauba wax coating as substitutes for these chemical fungicides. The carnauba wax on application to Indian jujube fruits after harvest reduces respiration rate and ethylene production.

10.4.2 Formulation of Plant-Based Nanocomposites

Nanocomposites can be created using the same techniques as nanoparticles, with a few modifications. There are various methods for producing nanocomposites, and each has pros and cons (Fig. 10.1).

The creation of nanocomposites is accomplished using both chemical and mechanical techniques. It was simple to expand the process of mechanical alloying to the industrial level for the synthesized materials, but maintaining structural homogeneity and purity is difficult. Here is a brief overview of some of the techniques for the synthesis of plant-based nanocomposites are covered:

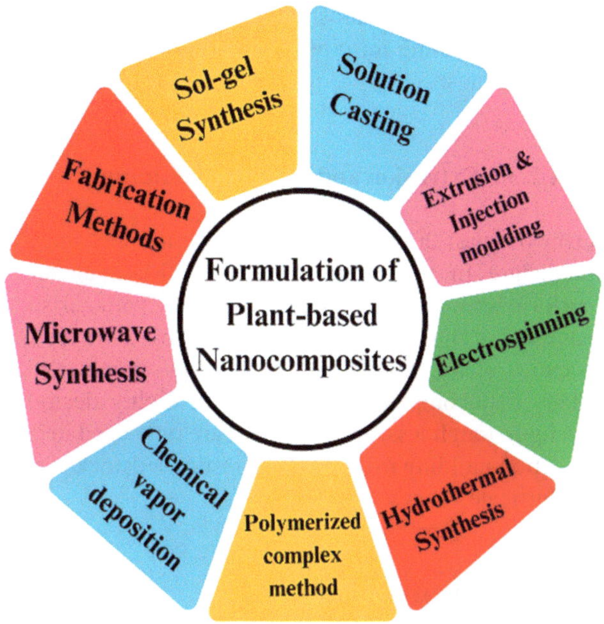

Fig. 10.1 Different methods of preparation of plant-based nanocomposites

10.4.2.1 Sol-Gel Synthesis: Sol-Gel

synthesis is the process of creating a chemical with a metal or metalloid component encircled with suitable ligands (gel) in a colloidal solution (sol). Nanocomposites are frequently prepared using the sol-gel process. For instance, high-porosity aerogels are perfect for use in nanocomposites. A second component is necessary for the assembly of aerogel nanocomposites. Additionally, reactive gas treatments may impact the chemical modification of the aerogel particles. The second component might be added before supercritical drying in the sol-gel process of the metal oxides, or it might be included after supercritical drying in the vapor phase. As a result of this procedure, a range of nanocrystalline metal/semiconductor oxides, alloys, and elements are created.

10.4.2.2 Solution Casting

The solution casting process is commonly used for creating composite films. For the produced film-forming solution to have the appropriate dispersion and homogeneity, the polymer composites and nanomaterials must be vigorously stirred or ultrasonicated in the organic solvent or water. The foundation of the solution-casting approach is Stokes' law. The prepolymer and polymer are uniformly blended and soluble in the proper solution by this technique. Although the nanocomposites were distributed in either the same or a separate solution, the matrix phase and the polymer disintegrated easily in the medium. Both eventually become blended. For instance, the solution must swell to clay during the creation of clay-based nanocomposites, indicating miscibility. Casting the film-forming solution onto glass sheets or petri plates.

10.4.2.3 Extrusion and Injection Molding

Over the past 20 years, ceramic injection molding (CIM) has become a common procedure in the ceramics industry. Material science is well-versed in high-performance ceramics with structural characteristics at the submicron or even nanoscale. The nanocomposites made by CIM are not yet commercially available because of the challenging processing characteristics of such ultrafine particles. Many nanocomposite ceramics with highly desirable qualities, such as extreme strength, toughness, and hardness, are only used in expensive function applications. Injection molding using thermoplastic feedstocks appears to be an appealing option for intricate shapes.

10.4.2.4 Electrospinning

Although the word electrospinning, which comes from "electrostatic spinning," was first coined in 1994, the basic concept goes back more than 60 years. Electrospinning is a productive method for creating polymer nanofibers. In recent years, a variety of polymers, mostly in a solvent solution but some in molten form, have been successfully electro-spurned into ultrafine fibers. Potential uses for these fibers, particularly as reinforcement in creating nanocomposite materials, have been identified. A polymer solution, like cellulose acetate, was added to the electric field. The fluid was sandwiched between two electrodes with oppositely polarized electrical charges to generate the polymer filaments. One electrode was attached to a collector, and the other was submerged in the solution. The charged solution jets were expelled from a metal spinneret with a tiny hole, where they evaporated to form fibers gathered on the collector.

10.4.2.5 Hydrothermal Synthesis

Hydrothermal synthesis is a widely used technique for precipitating metal or semiconductor metal oxides in single or multiple phases straight from the relevant homogeneous or heterogeneous solution. Hydrothermal synthesis is the favored one-step method for the synthesis of various single or multiphase oxides and phosphates. Moreover, single crystals of emeralds, rubies, quartz, and alexandrite are grown via hydrothermal synthesis due to their versatility and convenience of use. Additionally, the process is being used to create nanomaterials for energy and environmental uses like catalysis and dye-sensitized solar cells.

10.4.2.6 Polymerized Complex Method

The Pechini process-based polymeric precursor wet chemical approach has been used to manufacture various ceramic oxides. The method has a number of benefits for processing ceramic powders, including homogeneity, homogeneous mixing of many components at the molecular level, and direct and accurate control of stoichiometry. High-temperature superconductors, catalysts, dielectric, fluorescent, and magnetic materials are all made using this technique. Additionally, the technique is favored for the application of coatings made of oxide films, such as Lithium-ion batteries and dye-synthesized solar cells using the nanostructured electrode.

10.4.2.7 Chemical Vapor Deposition

In this process use of a chemical reaction in the vapour phase to add a solid layer on a heated surface. Nano or microstructure coatings, powders, fibers, and multiphase compounds made of metals, metal oxides, and non-metallic components like silicon and carbon can all be produced *via* this flexible technique. One benefit of CVD is its ability to fabricate single or multiphase nanomaterials and achieve high throughput because of its rapid deposition rate.

10.4.2.8 Microwave Synthesis

The use of microwave dealing out the applicable in many different domains, including food preparation, medicine, and chemistry. Microwave processing is crucial in the areas of design of microwave apparatus, novel material development, sintering, joining, modeling, and microwave material interaction. Microwave processing is being utilized more and more to create carbon fibers and ceramics at low temperatures and timeframes since it offers reliable heating at a reduced temperature.

10.4.2.9 Fabrication Methods

Even though mixing two or more polymers is not new, the introduction of nanofillers has enabled the creation of complex high-performance materials. Polymer blend nanocomposites can exhibit several technologically beneficial properties. The use of polymer mixed nanocomposites in electromagnetic interference (EMI) shielding, sensors, batteries, fuel cells, capacitors, vehicles, and packaging is now widely applied (Hou et al. 2024).

One or more techniques are used to create a nanocomposite. For instance, there are several ways to manufacture chitosan-based nanocomposites, including the sol-gel method, solution casting, electro-spinning, freeze-drying, *in si*tu synthesis, layer-by-layer assembly, and emulsion processes. Chitosan is a polymer of great concern because of its antibacterial, biodegradable, cost-effective, and biocompatible qualities. Similarly, pulping is regarded as a revolutionary way to produce cellulose nanofibers. These are made from wood chips that are 0.02–0.04 nm wide and 1–3 mm long and comprise numerous cellulose microfibrils. The main source of cellulose nanofibers, known to have a width of less than 100 nm, are cellulose microfibrils. To create films, membranes, and coatings, chitosan and nanoparticles are dissolved in a solvent, cast into a mold, and then dried to create nanocomposites. They are also made by *in situ* polymerization, which disperses the nanoparticles in the solution while forming a polymer matrix. Nanofiber is created using the electrostatic spinning of chitosan and nanoparticles. The sol-gel technique is used to produce a colloidal suspension with chitosan and nanoparticles that gel to create a solid matrix.

10.4.3 Applications of Plant-Based Nanocomposites

In the last few years, natural polymer-based nanocomposites have attracted a lot of attention from both scientific and industrial users (Fig. 10.2).

They can help eliminate the drawbacks of employing petroleum-derived polymeric materials and the related environmental problems. As a substitute for

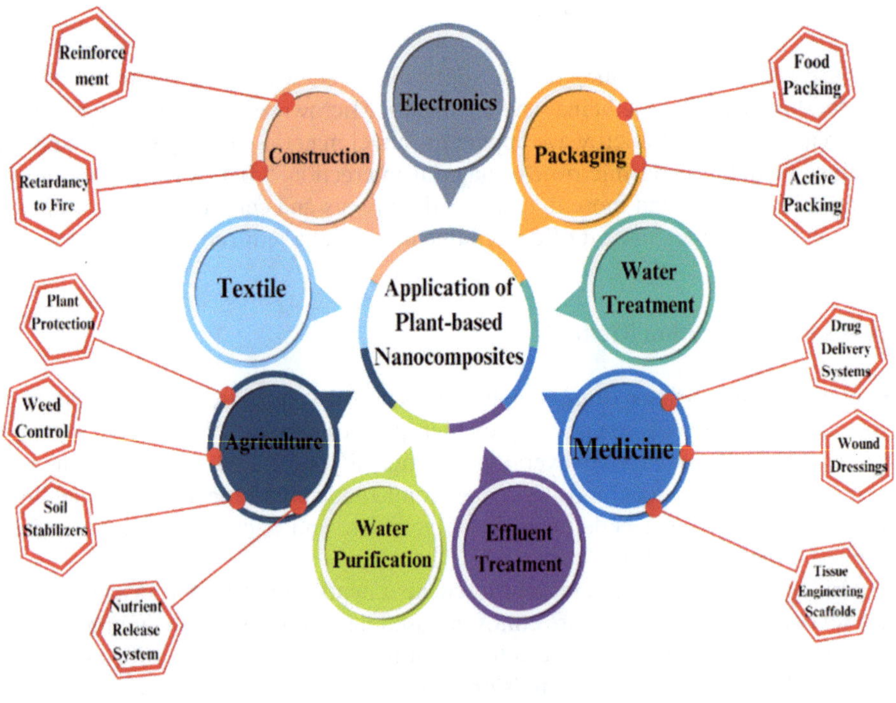

Fig. 10.2 Various applications of plant-based nanocomposites

packaging materials that are not biodegradable, natural polymers like proteins, polysaccharides, and lipids, as well as their mixtures or composites with nanoparticles, can be utilized as edible and/or biodegradable films in food packaging.

10.4.3.1 Food Packaging

The packaging materials made of synthetic compounds have a negative effect on the environment, mostly on the marine trophic chain. In place of plastic, biopolymers could be utilized to create novel materials like edible and antimicrobial films appropriate for food packaging. For example, starch is one of the natural polymers that can make edible films. Food packaging has not made extensive use of starch-based materials because of their hydrophilic nature, low gas barrier, and sensitivity to water. The worldwide packaging industry has been interested in plant-based proteins as a substitute for synthetic polymers for the previous several years due to their sustainability, biodegradability, positive health effects, and environmental benefits (Kumar et al. 2021). The active architecture of proteins makes it simple to alter them chemically and physically to create films, porous systems, composites, hybrids, and fibers. Plant-protein-based packaging and films typically exhibit various advantageous thermal, mechanical, barrier, and sensory qualities due to their distinct composition.

10.4.3.2 Active Packaging

Protecting and extending the duration of food products' shelf life requires active and smart food packaging, which serves as a barrier against environmental factors like as humidity, air, and bacterial infection. When coupled with bio-polymer resins, natural fibers including jute, banana, pineapple leaf, bamboo, hemp, sisal, and kenaf can enhance the mechanical properties of composite materials that decompose naturally, thereby improving the quality and integrity of food packaging (Hadidi et al. 2022).

10.4.3.3 Nutrient Release Systems

Plant-based nanocomposite materials are applied to create a smooth delivery system for nutrients, herbicides, and fertilizers. The Smart Fertilizers system uses encapsulated nanocomposites to manage the delivery of nutrients. The most abundant and widely dispersed polysaccharide found in nature is natural cellulose. It is a naturally occurring biodegradable substance that is insoluble in water because of hydrogen bonds between and within molecules.

10.4.3.4 Soil Stabilizers

Physical and chemical stabilizers and regular soil irrigation have been in use to enhance soil strength and surface resistance and prevent wind erosion. The use of biopolymers for soil stability and wind erosion mitigation has been studied recently. By providing sustainable and affordable alternatives to traditional materials, carbohydrate biopolymers have great promise for transforming the mitigation of soil erosion, enhancing soil fertility, and developing the building and construction industries. The scientific community is paying more and more attention to creating alginate-based nanocomposites in agriculture to fight drought and nutrient loss for sustainable development.

10.4.3.5 Plant Protection

Agricultural waste composites, such as jute, coir, or bagasse, can be utilized as biodegradable mulching sheets. Natural crop covers made of fiber utilizing plant-based composites, such as jute or coir textiles, can be made to shield plants from pests, harsh weather, and ultraviolet light. Biopesticides or beneficial bacteria can be added to coir or hemp mats. Serve as systems for the controlled discharge of pests and diseases (Malik and Rawat 2021). Plant-based composite guards can shield young plants and saplings from mechanical harm or grazing animals.

10.4.3.6 Reinforcement

Plant-based natural fiber reinforced materials have other potential applications in the building sector, aside from traditional cement composites. More renewable resources must be used in the construction industry due to the drive for sustainability in the application of building materials. Plant-based natural fibers are inexpensive, renewable resources that are widely available on the Earth. Plant-based fibers include cotton, hemp, jute, flax, ramie, sisal, and bagasse, and fibers derived from timber are being used as reinforced materials.

10.4.3.7 Retardancy to Fire

Natural fibers can adjust the mechanical, thermal, and flammability behavior of polymeric matrices, such as thermosets and thermoplastics. Bamboo and flax can contribute some intrinsic flame-retardant properties to the composite due to their silica concentration and insulating porous structures. However, lignin, cellulose, and hemicellulose are inherently combustible and can cause ignitability and fire spread in natural fibre-reinforced polymer composites (NFPCs). Additionally, the border plays a part in addressing the flammability of natural fibre-reinforced polymer composites. Thus, modifying the inherent characteristics of natural fibres and their interfacial interactions is now an emphasis area of study (Guo et al. 2024).

10.4.3.8 Drug Delivery Systems

In recent years, it has been shown that the biomolecules obtained from plants such as proteins, carbohydrates, lipids, cell inclusions, and phospholipids, can be used to build specialized medication delivery methods. In this delivery method, the vitamins, pharmaceuticals, and nutraceuticals are encapsulated properly with suitable nanocomposites so that the closed compounds can be preserved and released slowly in the body, thereby increasing their efficacy and bioavailability. Additionally, the negative effects of the medicated compounds are reduced greatly by this delivery system. These medication formulations are now used in a range of administration methods, including oral, nasal, cutaneous, and inhalation, to administer pharmaceutical components in a controlled or targeted manner (Karimah et al. 2021). Drug formulations made with colloidal delivery methods derived from plants can alter their pharmacokinetic and pharmacodynamic properties by altering their size, content, and structure.

10.4.3.9 Wound Dressings

The next phase in wound care treatment is using natural polymers and herbal ingredients in novel dressing materials. Plant compounds are added to wound dressings to provide them with antibacterial and antioxidant properties that aid in vascularization, epithelization, and wound contraction. However, their regulated release may be crucial in promoting angiogenic pathways, suppressing inflammation, and stimulating collagen synthesis, all of which can hasten the healing process of wounds (Khizar et al. 2022). Finally, polyphenols can use hydrogen bonds to cross-link natural polymers, especially collagen and gelatin, to improve their mechanical strength and thermal stability.

10.4.3.10 Tissue Engineering Scaffolds

Scaffolds can be designed using naturally occurring cellulose-based scaffolds that are highly biocompatible, readily available, easy to produce, sustainable and renewable, and environmentally benign. Promising prospects for tissue engineering applications, these cellulose-based scaffolds have demonstrated distinct mechanical properties, better biocompatibility or cytocompatibility, multifunctionality, and improved cell attachment or proliferation (Eskilson et al. 2023).

10.4.3.11 Conductive Nanocomposites

The use of materials like nanocellulose (NC) in the design of electromagnetic interference (EMI) shielding materials is growing in popularity. The development of EMI shielding materials has shown much attention in cellulose because of its superior mechanical and thermal qualities, biocompatibility, and biodegradability (Zielińska et al. 2023). The three important types of nanocellulose are cellulose nanoparticles, sometimes referred to as cellulose nanowhiskers (CNWs), cellulose nanocrystals (CNC) or nanocrystalline cellulose (NCC), and cellulose nanofibrils (CNFs). Despite being naturally occurring, cellulose and its derivatives are non-conductive, current can nevertheless flow between the polymer matrix and these conducting elements by adding conducting fillers.

10.4.3.12 In Textiles

Biocomposites are generally defined as manufactured composite materials that include two or more constituents, with one or more derived elements from natural or bio-based resources. A broad definition of biocomposites is when natural fibers that are not derived from wood, including cotton, jute, sisal, and hard and softwood, are combined with biopolymers, like polylactic acid, and sisal are combined. Developing a new range of fibre composites to take the place of synthetic fibres and plastics that are sustainable during the manufacturing process and beyond their useful lives is one of the main goals of creating biocomposites (Idumah 2021).

10.4.3.13 Water Purification

Polymers provide customizable characteristics, enhanced processability, exceptional stability, high surface area for quick decontamination, selectivity to eliminate various contaminants, and water treatment cost savings (Elella et al. 2024). Therefore, one of the future paths to achieving environmental water standards and meeting the water requirements of the expanding population is the development of polymeric materials.

10.4.4 Limitations of Plant-Based Nanocomposites

Plant-based composites, also known as bio-composites, have the possibility of substituting synthetic polymer-based composites in the future due to their appealing, unique qualities, affordability, ease of manufacturing, environmental friendliness, and post-use recycling capabilities (Khodakarami and Bagheri 2021). Natural fiber nanocomposites have enormous potential as a sustainable substitute for traditional materials, despite their restricted uses and ongoing development. The performance and quality of plant fiber-based composites can be further improved by employing the appropriate technical techniques (Ilyas et al. 2024). Plant-based fibers offer numerous benefits, but they also have certain drawbacks. One main problem with plant fibers is that they are extremely polar, which causes a number of problems with incompatibility with most thermosetting and thermoplastic matrices (Ramesh et al. 2017). Therefore, altering the constituents' characteristics to meet the composite material's requirements is essential to creating high-quality and high-performing bio-composites, which will provide a product with consistent, uniform, predictable, and reproducible properties. Strong interfacial bonds must be formed between the binder and the lignocellulosic substrates to make these modifications (Balakrishnan et al. 2024). Therefore, the development of bio-composites requires a comprehensive understanding of how these alterations impact the properties of the composite. Natural fiber nanocomposites are constructed from resources that are renewable and consist of a matrix, ideally produced from biomass, which may or may not be biodegradable, and reinforcements composed of natural fibers (Akter et al. 2022). Plant fibers are hydrophilic due to the hydroxyl groups in lignocellulose, which inhibits their compatibility with thermoplastics that are hydrophobic and makes them vulnerable to moisture degradation. These limitations make using plant fibers as reinforcement for polymers more difficult. Surface changes are usually necessary to decrease moisture absorption and increase fiber adherence to the polymer matrix. Several techniques have been developed to increase composite quality and fiber-matrix compatibility, including alkaline, silane, and other chemical treatments. Hydrophilic biopolymers and hydrophobic nanomaterials do not interact well when left unmodified (Gurunathan et al. 2015). Achieving uniform nanostructure dispersion in biopolymers, establishing cost-effective large-scale production methods, and preserving the balance between long-term stability and biodegradability are some of the major obstacles in the manufacture and use of plant-based nanocomposites.

10.5 Conclusion

Globally, sustainable development is widely acknowledged, although there are differences in methods and approaches. The concept of sustainable development emphasises taking into account both the present and future generations. Currently, studies are being conducted under the umbrella of eco-materials, defined as materials intended to lessen the adverse consequences of human activities on the environment. Eco-friendly materials can play a leadership role in encouraging sustainable development. Subsequently, traditional composite materials are being replaced by lighter, environmentally friendly materials like biocomposites. Natural fiber-reinforced nanocomposite materials are the subject of exploration because they possess low density and strong mechanical qualities. These compounds have several advantages, including low cost, biodegradability, and availability as natural resources. Many natural fiber-based composites possess desirable mechanical qualities suitable for various applications. Since cellulose-based textiles offer superior insulation, they have long been used to reinforce automobile polymers. Plant nanocomposite materials are nowadays used in various industries, such as the automobile sector, food industry, textile industry, wastewater treatment, and water purification. However, there are limitations to using natural plant fibers as reinforcement in polymer products because of various drawbacks, such as water absorption. The hydrophilic lignocellulose found in plant fibers is incompatible with hydrophobic thermoplastics like polyolefins due to its hydroxyl groups, which cause poor moisture resistance. Furthermore, difficult processing techniques, limited dimensional stability, and non-reproducible physical characteristics can all result in less-than-ideal final attributes for natural fibre nanocomposites. Similarly, another drawback of employing natural fibre composites for reinforcement is that they are not compatible with the various metal matrices. Most commercially available polymers, like unsaturated polyester, have non-polar qualities, so they are incompatible with the polar and hydrophilic nature of natural fibres. Composite materials deteriorate in various ways under external loads, which lowers their mechanical qualities. Furthermore, natural fibres' complex surfaces can make it difficult to adhere to the polymer matrix, creating compatibility issues between the fibres and the matrix. The durability and long-term performance of natural fibers may be limited due to their increased susceptibility to moisture and degradation. Despite numerous challenges, natural fiber nanocomposite can now be optimized for performance and properties through continuous research and development with new formulations and techniques. This opens up a promising new area of research with several exciting new applications in various industries, most notably the automotive sector, food industry, agriculture, and food packaging.

References

Abdullah ZW, Dong Y (2018) Recent advances and perspectives on starch nanocomposites for packaging applications. J Mater Sci 53(22):15319–15339. https://doi.org/10.1007/s10853-018-2613-9

Akter M, Uddin MH, Tania IS (2022) Biocomposites based on natural fibers and polymers: a review on properties and potential applications. J Reinf Plast Compos 41(17–18):705–742. https://doi.org/10.1177/07316844211070609

Allali M, Dahamni MA, Ghamnia M, Boukhachem A, Boukrédimi D, Tonneau D et al (2022) Synthesis and investigation of pure and CU-doped NIO nanofilms for future applications in wastewater treatment rejected by the textile industry. Catalysts 12(9):931. https://doi.org/10.3390/catal12090931

Anwer AH, Ahtesham A, Shoeb M, Mashkoor F, Ansari MZ, Zhu S et al (2023) State-of-the-art advances in nanocomposite and bio-nanocomposite polymeric materials: a comprehensive review. Adv Colloid Interf Sci 318:102955. https://doi.org/10.1016/j.cis.2023.102955

Atef M, Rezaei M, Behrooz R (2014) Preparation and characterization agar-based nanocomposite film reinforced by nanocrystalline cellulose. Int J Biol Macromol 70:537–544. https://doi.org/10.1016/j.ijbiomac.2014.07.013

Balakrishnan TS, Sultan MTH, Shahar FS, Nayak SY, Shah AUM, Sebaey TA et al (2024) Plant fiber reinforcements as alternatives in pultruded FRP composites manufacturing: a review. J Nat Fibers 21(1). https://doi.org/10.1080/15440478.2023.2298396

Chopra L, Manikanika N (2021) Extraction of cellulosic fibers from the natural resources: a short review. Mater Today Proc 48:1265–1270. https://doi.org/10.1016/j.matpr.2021.08.267

Elella MHA, Mohamed F, Gawad OFA, Abdallah HM (2024) Polymer nanocomposite films and coatings in the textile industry. Elsevier eBooks, pp 631–662. https://doi.org/10.1016/b978-0-443-19139-8.00008-5

Eskilson O, Zattarin E, Berglund L, Oksman K, Hanna K, Rakar J et al (2023) Nanocellulose composite wound dressings for real-time pH wound monitoring. Mater Today Bio 19:100574. https://doi.org/10.1016/j.mtbio.2023.100574

Fazal-ur-Rehman ZH, Adeel S, Özomay Z, Mia R (2024) Properties and performance relationship of biopolymers in textile industry. In: Ahmed S, Shabbir M (eds) Biopolymers in the textile industry. Springer, Singapore. https://doi.org/10.1007/978-981-97-0684-6_4

Guo Y, Qiao D, Zhao S, Zhang B, Xie F (2024) Advanced functional chitosan-based nanocomposite materials for performance-demanding applications. Prog Polym Sci 157:101872–101872. https://doi.org/10.1016/j.progpolymsci.2024.101872

Gurunathan T, Mohanty S, Nayak SK (2015) A review of the recent developments in biocomposites based on natural fibers and their application perspectives. Compos Part A Appl Sci Manuf 77:1–25. https://doi.org/10.1016/j.compositesa.2015.06.007

Hadidi M, Jafarzadeh S, Forough M, Garavand F, Alizadeh S, Salehabadi A et al (2022) Plant protein-based food packaging films; recent advances in fabrication, characterization, and applications. Trends Food Sci Technol 120:154–173. https://doi.org/10.1016/j.tifs.2022.01.013

Hou X, Feng XR, Jiang K, Zheng YC, Liu JT, Wang M (2024) Recent progress in smart electromagnetic interference shielding materials. J Mater Sci Technol 186:256–271. https://doi.org/10.1016/j.jmst.2024.01.008

Idumah CI (2021) Novel trends in conductive polymeric nanocomposites, and bionanocomposites. Synth Met 273:116674. https://doi.org/10.1016/j.synthmet.2020.116674

Ilyas RA, Asyraf MRM, Rajeshkumar L, Awais H, Siddique A, Shaker K et al (2024) A review of bio-based nanocellulose epoxy composites. J Environ Chem Eng 12(5):113835. https://doi.org/10.1016/j.jece.2024.113835

Joseph B, Sagarika VK, Sabu C, Kalarikkal N, Thomas S (2020) Cellulose nanocomposites: fabrication and biomedical applications. J Bioresour Bioprod 5(4):223–237

Karimah A, Ridho MR, Munawar SS, Adi DS, Ismadi N, Damayanti R et al (2021) A review on natural fibers for development of eco-friendly bio-composite: characteristics, and utilizations. J Mater Res Technol 13:2442–2458. https://doi.org/10.1016/j.jmrt.2021.06.014

Khizar S, Alrushaid N, Khan FA, Zine N, Jaffrezic-Renault N, Errachid A et al (2022) Nanocarriers-based novel and effective drug delivery system. Int J Pharm 632:122570. https://doi.org/10.1016/j.ijpharm.2022.122570

Khodakarami M, Bagheri M (2021) Recent advances in synthesis and application of polymer nanocomposites for water and wastewater treatment. J Clean Prod 296:126404. https://doi.org/10.1016/j.jclepro.2021.126404

Krishna A, Arya VT, Geethanjali D, Joji J, John N (2024) Protein based bio nanocomposites: synthesis, properties, and applications. In: Advances in Bio nanocomposites. Elsevier, pp 211–224

Kshitij M, Dhal MK, Banerjee A, Prasannavenkadesan V, Katiyar V, Kumar A (2024) Thermoplastic starch-based melt processed films for packaging applications: experimental and computational studies. Mater Today Commun:111475. https://doi.org/10.1016/j.mtcomm.2024.111475

Kumar M, Tomar M, Punia S, Dhakane-Lad J, Dhumal S, Changan S et al (2021) Plant-based proteins and their multifaceted industrial applications. LWT 154:112620. Available from: https://doi.org/10.1016/j.lwt.2021.112620

Malagurski I, Levic S, Nesic A, Mitric M, Pavlovic V, Dimitrijevic-Brankovic S (2017) Mineralized agar-based nanocomposite films: potential food packaging materials with antimicrobial properties. Carbohydr Polym 175:55–62. https://doi.org/10.1016/j.carbpol.2017.07.064

Malhotra BD, Ali MA (2017) Nanocomposite materials. Elsevier eBooks, pp 145–159. https://doi.org/10.1016/b978-0-323-44923-6.00005-4

Malik T, Rawat S (2021) Microbial valorization of Coir pith for development of compost and bioethanol production. In: Environmental and microbial biotechnology, pp 1–20. https://doi.org/10.1007/978-981-15-9696-4_1

Pai SHS, Pandey SK, Samuel EJJ, Jang JU, Nayak AK, Han H (2023) Recent advances in NiO-based nanostructures for energy storage device applications. J Energy Storage 76:109731. https://doi.org/10.1016/j.est.2023.109731

Pandey A, Singh MK, Singh A (2023) Bacterial cellulose: a smart biomaterial for biomedical applications. J Mater Res 39(1):2–18. https://doi.org/10.1557/s43578-023-01116-4

Parvathy G, Sethulekshmi AS, Jayan JS, Raman A, Saritha A (2020) Lignin based nano-composites: synthesis and applications. Process Saf Environ Prot 145:395–410. https://doi.org/10.1016/j.psep.2020.11.017

Parvathy G, Sethulekshmi AS, Jayan JS, Raman A, Saritha A (2021) Lignin based nano-composites: synthesis and applications. Process Saf Environ Prot 145:395–410

Rai R, Shetty AR, Shree M, A G, H SC, H G et al (2024) exploring the antimicrobial and antioxidant potential of CuO doped NiO nanocomposites: synthesis, characterization, and melamine sensing evaluation. Nano 10(1):417–431. https://doi.org/10.1080/20550324.2024.2414137

Ramesh M, Palanikumar K, Reddy KH (2017) Plant fiber-based bio-composites: sustainable and renewable green materials. Renew Sust Energ Rev 79:558–584. https://doi.org/10.1016/j.rser.2017.05.094

Rana AK, Frollini E, Thakur VK (2021) Cellulose nanocrystals: pretreatments, preparation strategies, and surface functionalization. Int J Biol Macromol [Internet] 182:1554–1581. https://doi.org/10.1016/j.ijbiomac.2021.05.119

Rhim JW, Park HM, Ha CS (2013) Bio-nanocomposites for food packaging applications. Prog Polym Sci 38(10–11):1629–1652. https://doi.org/10.1016/j.progpolymsci.2013.05.008

Siddiqui S, Surananai S, Sainath K, Khan MZ, Kuppusamy RRP, Suneetha YK (2023) Emerging trends in the development and application of 3D printed nanocomposite polymers for sustainable environmental solutions. Eur Polym J 196:112298. https://doi.org/10.1016/j.eurpolymj.2023.112298

Sun J, Yang X, Bai Y, Fang Z, Zhang S, Wang X et al (2024) Recent advances in cellulose nanofiber modification and characterization and cellulose nanofiber-based films for eco-friendly active food packaging. Foods 13(24):3999. https://doi.org/10.3390/foods13243999

Surendran G, Sherje AP (2022) Cellulose nanofibers and composites: an insight into basics and biomedical applications. J Drug Delivery Sci Technol 75:103601. https://doi.org/10.1016/j.jddst.2022.103601

Tadele DT, Islam MS, Mekonnen TH (2024) Zein-based nanoparticles and nanofibers: co-encapsulation, characterization, and application in food and biomedicine. Trends Food Sci Technol:104809. https://doi.org/10.1016/j.tifs.2024.104809

Tibolla H, Czaikoski A, Pelissari FM, Menegalli FC, Cunha RL (2020) Starch-based nanocomposites with cellulose nanofibers obtained from chemical and mechanical treatments. Int J Biol Macromol 161:132–146

Umavathi S, Mahboob S, Govindarajan M, Al-Ghanim KA, Ahmed Z, Virik P et al (2020) Green synthesis of ZnO nanoparticles for antimicrobial and vegetative growth applications: a novel approach for advancing efficient high-quality health care to human wellbeing. Saudi J Biol Sci 28(3):1808–1815. https://doi.org/10.1016/j.sjbs.2020.12.025

Wang Z, Kang S, Cao S, Krecker M, Tsukruk VV, Singamaneni S (2020) Protein-based functional nanocomposites. MRS Bull 45(12):1017–1026. https://doi.org/10.1557/mrs.2020.302

Wang X, Xue J, Wang Y, Zhu H, Chen S, Xiao Z et al (2024) Development and characterization of zein/gum Arabic nanocomposites incorporated edible films for improving strawberry preservation. Adv Compos Hybrid Mater 7(6). https://doi.org/10.1007/s42114-024-01051-w

Zhao C, Liu G, Tan Q, Gao M, Chen G, Huang X et al (2022) Polysaccharide-based biopolymer hydrogels for heavy metal detection and adsorption. J Adv Res 44:53–70. https://doi.org/10.1016/j.jare.2022.04.005

Zia F, Salman M, Ali M, Iqbal R, Rasul A, Najam-Ul-Haq M et al (2020) Alginate-based bio nanocomposites. Elsevier eBooks, pp 173–205. https://doi.org/10.1016/b978-0-12-816751-9.00008-8

Zielińska A, Karczewski J, Eder P, Kolanowski T, Szalata M, Wielgus K et al (2023) Scaffolds for drug delivery and tissue engineering: the role of genetics. J Control Release 359:207–223. https://doi.org/10.1016/j.jconrel.2023.05.042

Chapter 11
Controlled Release of Antimicrobial Biomolecules by Plant-Based Nanomaterials

Saheed Abiodun Ayoola, Abiodun Kamoldeen Ajijolakewu, Ayobami Elias Ajadi, Abraham Olufemi Asuku, and Saidu Bio Adamu

11.1 Introduction

Antimicrobial resistance (AMR) arises when microorganisms like fungi, bacteria, parasites, and viruses evolve to resist or evade antimicrobial substances, including antibiotics, designed to treat infections (World Health Organization (WHO) 2023). This phenomenon, which has become one of the most pressing global health challenges of the twenty-first century, is as a result of the increasing prevalence of AMR infections and the paucity in the development of newer antimicrobial agents. AMR not only compromises the efficacy of existing treatments but also contributes to increased morbidity, mortality, and healthcare costs (Karnwal et al. 2023). If there is no prompt and adequate management and intervention, by 2050, deaths toll as a result of AMR could surpass deaths from all other causes of mortality globally (Sahu et al. 2022).

The limitations of conventional antimicrobial therapies, including rapid degradation, low bioavailability and unintended off-target effects, necessitate the urgent need for advanced delivery systems that guarantee the controlled and targeted release of bioactive molecules. Nanotechnology offers a promising solution to these challenges by enabling the encapsulation of therapeutic molecules within nanoparticles and hydrogels, and thereafter achieving controlled release profiles

S. A. Ayoola (✉) · A. O. Asuku · S. B. Adamu
Bioresources Development Centre, National Biotechnology Research and Development Agency, Ogbomosho, Nigeria

A. K. Ajijolakewu
Department of Microbiology, Faculty of Life Sciences, University of Ilorin, Ilorin, Nigeria

A. E. Ajadi
Department of Microbiology, Faculty of Sciences, Federal University of Health Sciences, Ila-Orangun, Nigeria

V. Ghosh et al. (eds.), *Applications of Plant Bioactive Compounds-Based Nanoformulations*, Nanotechnology in the Life Sciences,
https://doi.org/10.1007/978-3-032-13086-0_11

and enhanced antimicrobial efficacy (Morelli et al. 2017). Plant-based nanomaterials possess an excellent platform, due to their abundant bioactive compounds, environmental sustainability, and biocompatibility (Lan 2022; Sharma et al. 2025).

The application of nanotechnology in drug delivery has greatly enhanced the bioavailability of therapeutic agents, enabling more efficient treatment while minimizing adverse effects (Lan 2022). Plant-derived nanomaterials such as cellulose nanocrystals, lignin nanoparticles, and phytochemical-based systems, offer a renewable and cost-effective alternative to synthetic materials. These materials not only align with the growing emphasis on sustainability but also exhibit low toxicity and biodegradability, making them particularly suitable for controlled-release of antimicrobial therapies (Mann et al. 2018). Apart from their environmental benefits, plant-based nanomaterials possess outstanding physico-chemical properties that facilitate their use in biomedical applications. For instance, carbon nanotubes and other biocompatible nanoparticles derived from plants are highly useful in drug delivery and therapeutic interventions because of their favorable interactions with biological systems (Jeevanandam et al. 2021). Therapeutic potentials of plant-based bioactive compounds could be further enhanced by incorporating them into nanoparticles and thus integrating biological activity with targeted delivery.

The development of multifunctional nanomaterials capable of serving diverse purposes, including imaging, targeting, and therapeutic delivery, on a single platform were made possible as a result of advances in nanotechnology (Karthik et al. 2024). These competences and advances are particularly relevant for monitoring complex diseases, such as cancer, where precise targeting of tumor cells is essential for effective treatment (Ghahramani et al. 2022). Plant-based materials were targeted with nanotechnology, and this allows researchers to create innovative therapeutic systems that are both efficient and environmentally sustainable (Yuan et al. 2019).

The uniqueness of nanomaterials enhances their methods in antimicrobial resistance capabilities and also their relatively smaller sizes have made their solubility, release and movement in mixtures to be more robust for effectiveness (Díaz-Puertas et al. 2023). A number of potent antimicrobials have been extracted from plants. Vegetables possess excellent antimicrobial substances having antimicrobial, antioxidant, anticancer and also a good candidate for food preservation (Cao et al. 2023). Silver nanoparticles synthesized by the polysaccharides of *Leucaena leucocephala* seeds have been found to possess anticancer, antifungal and preservative potentials (Taher et al. 2022). However, due to the disadvantages witnessed in antimicrobial and pharmacological formulations, nanomaterial technological applications have solved these problems and hence come with a robust design that not only release drugs' functional biologically active ingredients but also protect the substances and thus give rise to synergism in the drugs' effectiveness (Quichaba et al. 2022).

Furthermore, the eco-friendly nature of plant-based nanomaterials contributes to reducing environmental pollution often associated with synthetic materials. The combination of biocompatibility, biodegradability, and the ability to incorporate bioactive compounds underscores the potential of plant-derived nanomaterials to advance controlled-release systems in antimicrobial therapies. This approach not only offers a sustainable alternative but also paves the way for novel strategies to combat AMR and improve therapeutic outcomes (Sharma et al. 2025).

With a focus on their potential to transform antimicrobial therapy, this chapter examines the function of plant-based nanomaterials in the regulated release of antimicrobial biomolecules. It explores how nanomaterials and biomolecules interact, and how these systems might be used practically to treat microbial diseases. This strategy offers a creative and sustainable response to the world's healthcare problems by utilizing the complementary properties of antimicrobial biomolecules and plant-derived nanotechnology.

11.2 A Brief History on Nanomaterials

The discovery of antibiotics by Alexander Flemings marked the beginning of a new era in the history of chemotherapy by mankind as difficult to treat diseases were easily treated. However, the sudden development of antibiotics resistance necessitated the quests for various other alternatives and methodologies to solve this menace. The discovery and development of scanning tunneling microscopes was a turning point towards the birth of nanotechnology due to the fact that, this microscope made it possible for atoms and molecules to be manipulated by researchers in any way they so desired. This new development of nanotechnology later paved way for nanoparticles to be developed for antimicrobials starting with silver nanoparticles due to the nature of higher understandings that scientists have of its antimicrobial activities. This understanding from silver nanoparticles, coupled with technological innovations later culminated into researches in other areas like zinc oxide, gold and numerous other nanoparticles that existed today. Additionally, several methods of experimentation into antimicrobial applications of nanoparticles have been done both in vitro and in vivo spanning several fields that transcend beyond medical uses (Mondal et al. 2024).

11.3 General Method of Controlled Release of Antimicrobial Biomolecules

Release mechanism of antimicrobial biomolecules is the way in which bioactive molecules in a drug or an antimicrobial are constantly and moderately released in a manner that enhances effectiveness, even distribution at given periods and also reducing negative consequences. This release technique can be monitored and controlled in order to achieve functionally efficient release. This could be done by manipulating the ratio and type of the nanocarrier materials. Encapsulation of nanocarriers through the use of freeze dying technique could be used to achieve controlled release (Cao et al. 2023). Summarily, nano-release techniques allow a robust water solubility, stability improvements and removal of toxicity, all of which are factors that prevented bioactive compounds from actively performing their optimum antimicrobial activities (Sims et al. 2020). This controlled release system finds applications in various control of antimicrobials in areas that include but not limited to pharmaceuticals and antimicrobial delivery systems, microbial inhibition and preservation (Gonçalves et al. 2018; Bahrami et al. 2019; Sharma et al. 2021), dairy products, prevention of microbial attacks in food and food packaging materials (Deka et al. 2016; Joung et al. 2016), as well as biomedicals in wound dressings to prevent microbial attack on wounds (El-Wakil et al. 2019; Ternullo et al. 2019; Liang et al. 2021).

Because of the problems associated with majority stability of bioactive molecules which could subsequently hinder their efficacies, suitable nanocarriers are used in order to foster efficient and controlled release into the receptor sites for optimum bioactivity. In order to achieve this controlled release, nanoencapsulation techniques are used. These techniques not only guarantee controlled release, but also provides solubility and controlled stability of the bioactive molecules. For antimicrobial materials, nanoencapsulation techniques are numerous but could be broadly categorized into 3 viz: Organic nanocarriers (Starches, proteins and polysaccharides), inorganic nanocarriers (carbon nanotubes, gold nanoparticles and quantum dots) and hybrid nanocarriers which could be hydrogel or metal-organic frameworks (Cao et al. 2023; Zhang et al. 2021).

11.3.1 Organic Nanocarriers

Organic nanocarriers are nanoparticles of proteins, polysaccharides and starches like soluble proteins, corn alcohol, modified starch, pectin, wheat, alcohol-soluble protein and various other materials. The nature of proteins as biological molecules having diverse physical and chemical and structural compositions which can combine and intermingle with various components has made proteins excellent candidates in delivery mechanisms both in food and medical applications. Proteins used in nanocarrier include plant proteins, protein peptides and animal proteins (Rashidi 2021). Polysaccharide (Chitosan), the second most abundant natural biopolymer, on the other hand is a derivative of chitin. Chitosan nanoparticles are excellent bioactive nanoparticles with antimicrobial activities, robust release potentials and efficient loading characteristics (Imran et al. 2019). The pharmaceutical and antimicrobial uses of starch is numerous due to its cost effectiveness, biocompatibility and relative availability as a result of its numerous sources. Direct application of starch has a few applications because of its poor enzymatic resistance, lack of hydrophobic groups, and relatively large molecular weight (Tibolla et al. 2020). However, effective encapsulation and antimicrobial activities can be enhanced by modifying the molecular structure of starch to form nanoparticles, nanocrystals, nanofibers and nanostructured starches which are nanocarriers derivatives (Wang et al. 2022).

11.3.2 Inorganic Nanocarriers

Inorganic nanomaterials are nanomaterials of metallic origin like gold nanomaterial, silver nanomaterials, carbon nanomaterials, quantum dots and many others. These nanomaterials can be physically or chemically synthesized and have several structural and size ranges. In fact, inorganic nanomaterials are the easiest to produce and the most compatible of all nanomaterials. These nanocarriers allows for both drugs bounding and modification of surfaces which thus leads to efficient release and effective antimicrobial activities (Kaur et al. 2020).

Gold nanoparticle (AuNP) which is an example of inorganic nanoparticle carriers has an efficient delivery mechanism and because of its excellent properties, especially the optical nature, they find good applications in medical and antimicrobial uses. Gold nanoparticles is highly effective against broad spectrum of microorganisms including both Gram-negative and Gram-positive bacteria. This is as a result of the forces of attraction and stronger bonds that are created by the gold nanoparticles' surfaces (Thakor et al. 2011; Ling et al. 2018; Attia et al. 2016). Carbon nanotubes, another inorganic form of nanocarriers, possess excellent properties that make them potent antimicrobials. Carbon nanotubes are hollow and tubular graphene sheets that are folded at intervals. They have various size ranges and classes which are dependent on the number of graphene sheets they possess. Carbon nanotubes are very potent antimicrobials and this is due to their flexible and stable nature (Chad et al. 2010). Quantum dots on the other hand, are atoms which are colloidal crystals with ability to donate energy. Quantum dots are effective nanocarriers and has a great release effectiveness and an excellent antimicrobial (Ahmad et al. 2021).

11.3.3 Hybrid Nanocarriers

These are combinations of two or multiple inorganic or organic nanocarriers which usually result in an improved hybrid nanocarriers with excellent properties of the two combined nanomaterials and thus give better synergistic effects. This combination comes in different forms and could be combinations of organic nanocarriers or organic combination with inorganic nanocarriers or inorganic with inorganic nanocarriers (Cao et al. 2023). Hydrogels which are three dimensional solid networks and can absorb liquids and other bioactive fluids are examples of hybrid nanocarriers Hydrogels nanoparticles' formation occurs through combination of charged polymeric materials that eventually come together and thus become potent antimicrobial. The product so formed (hydrogel) can engulf bioactive molecules and also efficiently release the engulfed antimicrobials effectively. A number of hydrogel combinations have been designed which not only improve delivery and availability of molecules but also increase antimicrobial activities of the hybrid (Cao et al. 2023). Another hybrid nanocarriers are the MOFs (Metal-organic framework materials) which are crystallized molecules that are formed as a result of combinations of organic and inorganic nanocarriers. MOFs have been developed from combinations of several nanomaterials that gave superb antimicrobial activities, excellent release of bioactive molecules and a high level of biocompatibility (Moussa et al. 2016; Abedi et al. 2021).

11.4 Types of Nanomaterials

Types of nanomaterials, their examples, including the description and their properties were represented in Table 11.1.

11.5 Plant-Based Nanoparticles, Their Synthesis and Mechanism of Action

Numerous plant species and their various parts are excellent precursors of nanoparticle production. This is as a result of the presence of active secondary metabolites like tannins, saponins, alkaloids, flavonoids, terpenes, phenols, steroids and a host of other useful products (Bukhari et al. 2021; García-Pérez et al. 2021).

Silver nanoparticles are the best studied nanoparticles and their excellent interactions with numerous species of bacteria and fungi have been observed and studied. A number of physical and chemical methods have been used in the synthesis of various nanoparticles. However, majority of these methods are either very expensive and sometimes not safe. Alternative methods which are synthetic in nature have been developed to solve the menace associated with physical and chemical synthesis. Among these synthetic methods, biological approaches are the best because they are highly reliable, non-toxic, simple and environmental-friendly (Bandekar et al. 2020; Yılmaz et al. 2023). Several authors have demonstrated synthesizing nanoparticles by using various plant species and parts. This has yielded fantastic results which are not only economically friendly but also environmentally friendly. Biosynthesis of nanoparticles through the use of plants that are rich in bioactive metabolites are cost effective, environmentally friendly, biocompatible and highly sustainable as compared to the physical and chemical nanoparticles (Ahmad et al. 2024; Balaji et al. 2023; Khan et al. 2020). A number of metals and their oxides (Zinc, silver, copper, etc) coupled with bioactive plant extracts have been used in the biosynthesis of nanoparticles which have come out with excellent antimicrobials to treat several diseases caused by microorganisms of medical importance (Tilahun et al. 2023; Rahman et al. 2022). Also, several reports have highlighted the use of antimicrobial nanoparticles in the treatment of microbes including their resistant types likes bacteria, fungi, viruses (Lin and Tujios 2023; Berini et al. 2021; Yadeta and Lealem 2024) and even biofilms (Kaur et al. 2020).

Table 11.1 Types of nanomaterials

Types of nanomaterials	Example based	Description	References
Carbon-based nanoparticles	Fullerens	Spherical molecule made of sp^2 hybridized carbon atoms Globular hollow cages	Liu et al. (2021)
	Graphenes	A two-dimensional planar hexagonal composed of carbon atoms in a honeycomb lattice	Zhao and Zhu (2015)
	Carbon nanotubes (CNT)	Formed from graphene nanofoils with honeycomb structures Hollow cylindrical structure with single, double or multiple walls (SWNTs, DWNTs, MWNTs)	Smagulova et al. (2020)
	Carbon nanofiber	Produced similarly to graphene nano foil and CNT but wound into a cone shape instead of cylindrical tubes	Trabelsi et al. (2021)
	Carbon black	Amorphous carbon material with a spherical shape Particle aggregate due to rapid interaction	Raju et al. (2022)
Metallic nanoparticles	Metallic oxide	Synthesized from metals using destructive or constructive processes to achieve nanometric sizes Typically, spherical with diverse colour and sensitivity to external factor	Patil (2023)
	Quantum dots	QDs have a core-shell structure to enhance stability and efficiency Spherical, rod-shaped, or other shapes depending on synthesis	Maleki et al. (2022)
Dendrimers		They have a uniform, tree-like structure, highly branched, spherical and symmetrical They are made from synthetic polymers (e.g. PANAM, PPI)	Maleki et al. (2022)
	Polymeric nanoparticles	Organic-based nanoparticles made Made of natural or synthetic polymers including polylactic acid (PLA), polyethylene glycol (PEG) and chitosan Typically, nano-spherical or nano-capsular in shape	Zielińska et al. (2020)
Composite nanoparticles		Combination of a matrix (e.g. polymer, metal, ceramic) and nanofillers (e.g. CNTs, graphene nanoparticles) Diverse, depending on filler type: spherical, tubular or planar	Panneerselvam et al. (2017)

Table 11.2 Types of plant based nanomaterials

Types of plant based nanomaterials	Sources	References
Cellulose nanocrystal	Derived from cotton, wood pulp, tunicates, wheat straw, ramie, and other lignocellulosic materials	Putri et al. (2020)
	Same sources as CNCs but processed differently to retain longer fibrillar structure	Benítez et al. (2013)
Starch nanoparticles	Derived from starch-rich plants such as maize, potato, wheat, rice, and cassava	Peressini et al. (2020)
Pectin nanoparticles	Extracted from the primary cell wall of citrus peel or apple pomace	Gutierrez-Alvarado et al. (2022)
Lignin nanoparticles	Found in plant cell walls, especially in wood and bark	Wang et al. (2022)
Plant extract nanoparticles	Derived from plant extracts, including leaves, flowers, roots, seeds, and bark. Examples: Green tea, neem, turmeric, aloe vera, and ginger	Arif et al. (2022)
Polyphenolics nanoparticles	Derived from natural polyphenols such as flavonoids, tannins, catechins, and curcumin. Common sources include tea, grapes, berries	Katongole et al. (2020)

Mechanism of antimicrobial action of nanoparticles have extensively been discussed by several authors. Part of the mechanisms on microorganisms are permeability of the cell membrane, cell disruption and death (Spirescu et al. 2021). Additionally, polysaccharides and proteins of bacteria have also been reported to be encapsulated by nanoparticles and also biofilm are also easily penetrated by antimicrobial nanoparticles (Bharadishettar et al. 2021; Pushpamalini et al. 2021).

Plant-based nanomaterials have several types. In Table 11.2, the types of plant-based nanomaterials with their sources, structural description, methods of production and applications were highlighted. Also, Fig. 11.1 depict the flow chart of general method for the synthesis of antimicrobial nanoparticle.

11.6 Applications of Nanomaterials

Plant-based nanomaterials, like every other types of nanomaterials, have their unique applications in several fields of studies and activities of human endeavors. Antimicrobial diseases and infections caused by microbes could be effectively treated using antimicrobial agents, some of which are plant extracts. The effectiveness of these plant extracts is as a result of bioactive metabolites present in them. The effectiveness of plants bioactive compounds against diseases-causing microbes is further improved through the use of nanotechnology by using nanoparticles. Plant extracts are combined with nanomaterials in order to proffer better results (Freeland et al. 2021; Singh and Mijakovic 2022).

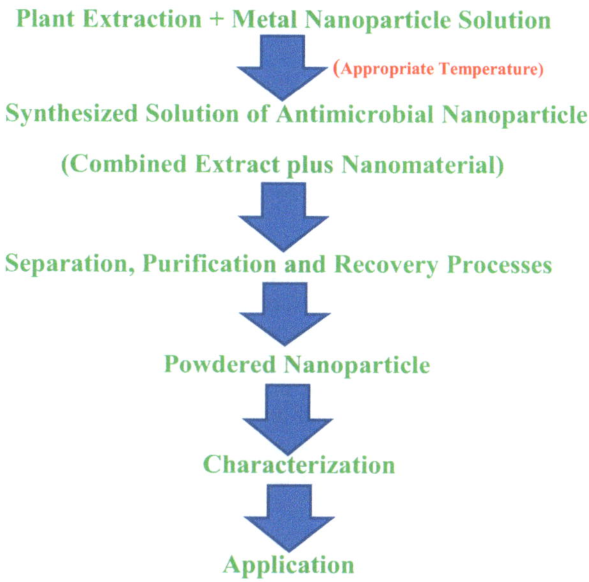

Fig. 11.1 General flowchart for biosynthesis of antimicrobial nanoparticles

Nanoparticles offer efficient utilization of resources due to the fact that they are very active and highly potent against infectious microbes even when applied at smaller doses and low concentrations. Also, their use in various fields of human endeavors cannot be underestimated. A number of antimicrobial nanomaterials have been developed as active against medically important microbes. Additionally, biologically active surface coatings and sterilants have been developed against microbial attack and invasion on surfaces and wounds (Sirotkin et al. 2022; Taha 2022). Good examples of these are coatings used in food packaging materials, plasters, gauzes and other wound dressing materials. These coatings have served as excellent alternatives to antibiotics and thus reduce the consumption of antibiotics by patients. It has also helped in food packaging applications as the effects of these coatings remain on surfaces of packaging materials days after packaging and thus prevent subsequent reinfection by spoilage microbes (Imani et al. 2020; Barabadi et al. 2021). These are used in several applications like food and beverage industries, dental and medical treatments. Nanoparticles has garnered applications in numerous fields of human endeavors like agriculture, medicine, pharmaceutics, food packaging, cosmetics, lubricant and others (Nandhini et al. 2023; Rizki and Klaypradit 2023).

Table 11.3 highlighted the specific applications of plant-based nanomaterials while Table 11.4 presented the general application of nanomaterials as regards their field-specific (industries-related) uses.

Table 11.3 Applications of plant-based nanomaterials

Nanoparticle	Properties	Applications	References
Silver Nanoparticles (Ag-NPs)	Broad-acting, potent antibacterial activity; efficient disinfection of biological pollutants (viruses, bacteria, fungi)	Disinfecting medical devices, home appliances, water treatments, nanomedicine; antimicrobial coatings for medical gear	Nasim et al. (2021)
Gold Nanoparticles (AuNPs)	Non-cytotoxic, high surface area, size-dependent chemical, electronic, and optical properties	Photothermal therapy, drug delivery, gene therapy, biosensing, biolabeling, and molecular imaging; antimicrobial coatings	Mehdipour et al. (2022)
Quantum dots	Effective in targeted drug delivery	Cancer treatment, bio-imaging, biosensing	Chary (2023)
Iron Oxide (Fe_3O_4)	Magnetic properties	Targeted drug delivery	Shen et al. (2018)
Zinc Oxide (ZnO)	Effective in drug delivery; antimicrobial coatings	Biosensing; agriculture for precise delivery of agrochemicals	Wang et al. (2020)
Titanium Nanoparticles (TiO_2)	High aspect ratio, reactivity, controlled pore volume, electrostatic, hydrophilic, and hydrophobic interactions	Water treatment, adsorption catalysis, sensing, and optoelectronics; antimicrobial coatings for medical gear	Ates et al. (2020)
Copper Oxide (CuO), Magnesium Oxide (MgO), Selenium Nanoparticles (Se)	Efficient antimicrobial properties	Antimicrobial coatings for medical gear	Mousaa et al. (2022)
Nanoparticles and Nanocapsules	Ability to control and release agrochemicals (fertilizers, insecticides, herbicides) precisely	Agriculture: site-specific delivery of insecticides and fertilizers with minimal collateral damage	Sharma (2024)
Nanocatalysts	High surface-to-volume ratio; geometric surface effect; electronic impact; quantum size effect	Catalysis: high activity, selectivity, and productivity; materials like aluminum, iron, TiO_2, clays, and silica used	Giannakourou and Tsironi (2021)
Nanomaterials for biosensors	High sensitivity enabled by nanotechnology	Creation of high-performance electrochemical sensing devices for environmental/food safety monitoring and medical diagnostics	Beaver and Dantanarayana (2021)

Table 11.4 General applications of nanomaterials

S/N	Applications	Specific Uses	References
1	Food and agricultural Applications	Nutraceuticals Texture improvement Delivery of nutrient Food delivery Nano-packaging Nano-fertilizers Nano-herbicides	Ali et al. (2020)
2	Medicinal and Pharmaceutical applications	Control of pathogens Healing of wounds Antimicrobial (antiviral, antifungal and antibacterial) Anticancer	Ahmed et al. (2020), Ibrahim et al. (2020) and Anand et al. (2022)
3	Industrial applications	Superplastic ceramics Nano-coatings for papers Nano-cosmetics Nano-composites Nano-pigmentation Antifouling coating Electronics Optoelectronics Photochemicals	Ali et al. (2020), Yan et al. (2022)
4	Environmental applications	Ultraviolet protection Environmental catalysts Scavenging for pollutants Nano-remediation Effluent treatment Nano-degradation of polymers	Ali et al. (2020)

11.7 Shortcomings of Nanomaterials

The integration of plant-based nanomaterials into agricultural processes holds immense potential for enhancing productivity, yet it is accompanied by significant challenges that demand careful consideration. Key limitations can be categorized into synthesis, uptake, and environmental impacts. The environmentally friendly synthesis of monodispersed nanoparticles, often using plant extracts, is crucial for ensuring their functional applications. However, issues such as enzymatic degradation during synthesis hinder reproducibility and lead to inconsistent performance (Das et al. 2021). Despite advances with biocompatible materials like polymers and lipids, challenges in achieving consistent yields and reliable properties persist, complicating the development of dependable nanomaterials for agricultural use (Österberg et al. 2023). Furthermore, the uptake and transport of nanomaterials within plants are influenced by their size, as nanoparticles typically need to be smaller than 20 nm in at least one dimension to penetrate plant cells effectively (Zhi et al. 2022). Because of the little understanding of the mechanisms that surround uptake of nanoparticles and their localization with plant tissues, their optimization efforts have been complicated (Wong et al. 2016). Also, plant's health and yield can negatively be affected as a result of toxicity that arise from plants' physiology and nanomaterials interaction (Ruttkay-Nedecký et al. 2017). Apart from interactions of plant, environmental consequences are also of great concern. Nanomaterials has the potential to accumulate within the soil-plant interphase and this poses serious concerns on their toxicity (Ferdous and Nemmar 2020), long term ecological impacts and aggregation of carbon-based nanoparticles, which aggravates their efficacy and management (Makvandi et al. 2020). For this reasons, ecological implications and safety of nanomaterials in agricultural settings need to be assessed in order to ensure effective integration (Meher 2024).

11.8 Conclusion

The controlled release of antimicrobial biomolecules by plant-based nanomaterials has achieved a landmark innovation and advances in the history of chemotherapy. In order to seek for superb alternatives that are friendly to nature, nanotechnology has helped in this regard through the development and use of plant derived nanomaterials. Plant-based nanomaterials majorly originate from renewable sources and are largely used in agricultural, environmental and medicinal processes to solve series of problems like environmental and microbial resistance issues. Apart from the cost effectiveness of these plant-based nanomaterials, they are also biocompatible and possess a great deal of bioactivities. Also, their uniqueness and ability to continuously target and release antimicrobial biomolecules have greatly improved their therapeutic effectiveness and also lowering associated side effects. However, there are some setbacks associated with these materials like scaling of production, cytotoxicity and environmental concerns during synthesis and disposal. Even though, these associated problems are reduced when plant-based biosynthesis is used in producing nanomaterials as compared to other conventional ones like physical and chemical methods, further research studies are still needed in order to completely proffer solutions to these problems and thus reduce threat posed on animals, man and environment.

References

Abedi M, Abolmaali SS, Heidari R, Mohammadi SS, Tamaddon AM (2021) Hierarchical mesoporous zinc-imidazole dicarboxylic acid MOFs: surfactant-directed synthesis, pH-responsive degradation, and drug delivery. Int J Pharm 602:120685. https://doi.org/10.1016/j.ijpharm.2021.120685

Ahmad B, Hafeez N, Bashir S, Rauf A, Mujeeb UR (2021) Phytofabricated gold nanoparticles and their biomedical applications. Biomed Pharmacother 89:414–425. https://doi.org/10.1016/j.biopha.2017.02.058

Ahmad A, Khan M, Osman SM, Haassan AM, Javed MH, Ahmad A et al (2024) Benign-by-design plant extract-mediated preparation of copper oxide nanoparticles for environmentally related applications. Environ Res 247:118048. https://doi.org/10.1016/j.envres.2023.118048

Ahmed T, Shahid M, Noman M, Niazi MBK, Mahmood F, Manzoor I et al (2020) Silver nanoparticles synthesized by using Bacillus cereus SZT1 ameliorated the damage of bacterial leaf blight pathogen in rice. Pathogens 9:160

Ali MA, Ahmed T, Wu W, Hossain A, Hafeez R, Masum MI et al (2020) Advancements in plant and microbe-based synthesis of metallic nanoparticles and their antimicrobial activity against plant pathogens. Nanomaterials 10:1146. https://doi.org/10.3390/nano10061146

Anand U, Carpena M, Kowalska-Góralska M, Garcia-Perez P, Sunita K, Bontempi E, Dey A et al (2022) Safer plant-based nanoparticles for combating antibiotic resistance in bacteria: a comprehensive review on its potential applications, recent advances, and future perspective. Sci Total Environ 821:153472. https://doi.org/10.1016/j.scitotenv.2022.153472

Arif M, Ullah R, Ahmad M, Ali A, Ullah Z, Ali M et al (2022) Green synthesis of silver nanoparticles using euphorbia wallichii leaf extract: its antibacterial action against citrus canker causal agent and antioxidant potential. Molecules 27(11):3525. https://doi.org/10.3390/molecules27113525

Ates B, Koytepe S, Ulu A, Gurses C, Thakur VK (2020) Chemistry, structures, and advanced applications of nanocomposites from biorenewable resources. Chem Rev 120:9304–9362

Attia YA, Vázquez-Vázquez C, Blanco MC, Buceta D, López-Quintela MA (2016) Gold nanorod synthesis catalysed by Au clusters. Faraday Discuss 191:205–213. https://doi.org/10.1039/C6FD00015K

Bahrami A, Delshadi R, Jafari SM, Williams L (2019) Nanoencapsulated nisin: an engineered natural antimicrobial system for the food industry. Trends Food Sci Technol 94:20–31. https://doi.org/10.1016/j.tifs.2019.10.002

Balaji V, Perumal S, Palanisamy S, Karuppaiah M, Asaithambi S, Velauthapillai D et al (2023) Bio-inspired synthesis of silver nanoparticles and their nanocomposites for antibacterial and anticancer activity: a comparative study. J Alloys Compd 966:171503. https://doi.org/10.1016/j.jallcom.2023.171503

Bandekar SS, Kerur SS, Kore SK, Hegde PG (2020) Synthesis and antimicrobial activity of plant based silver nanoparticles – a review. Biosc Biotechnol Res Commun 13(13):122–130. https://doi.org/10.21786/bbrc/13.13/17

Barabadi H, Mojab F, Vahidi H, Marashi B, Talank N, Hosseini O et al (2021) Green synthesis, characterization, antibacterial and biofilm inhibitory activity of silver nanoparticles compared to commercial silver nanoparticles. Inorg Chem Commun 129:108647

Beaver K, Dantanarayana A (2021) Materials approaches for improving electrochemical sensor performance. J Phys Chem B 125(43):11820–11834. https://doi.org/10.1021/acs.jpcb.1c07063

Benítez A, Torres-Rendon J, Poutanen M, Walther A (2013) Humidity and multiscale structure govern mechanical properties and deformation modes in films of native cellulose nanofibrils. Biomacromolecules 14(12):4497–4506. https://doi.org/10.1021/bm401451m

Berini F, Orlandi VT, Gamberoni F, Martegani E, Armenia I, Gornati R et al (2021) Antimicrobial activity of nanoconjugated glycopeptide antibiotics and their effect on Staphylococcus aureus biofilm. Front Microbiol 12:657431. https://doi.org/10.3389/fmicb.2021.657431

Bharadishettar N, Bhat KU, Bhat PD (2021) Coating technologies for copper based antimicrobial active surfaces: a perspective review. Metals 11:711

Bukhari A, Ijaz I, Gilani E, Nazir A, Zain H, Saeed R et al (2021) Green synthesis of metal and metal oxide nanoparticles using different plants' parts for antimicrobial activity and anticancer activity: a review article. Coatings 11:1374

Cao J, Gao M, Wang J, Liu Y, Zhang X, Ping Y et al (2023) Construction of nano slow-release systems for antibacterial active substances and its applications: a comprehensive review. Front Nutr:1109204. https://doi.org/10.3389/fnut.2023.1109204

Chad D, Vecitis KRZ, Kang S, Elimelech M (2010) Electronic-structure-dependent bacterial cytotoxicity of single-walled carbon nanotubes. ACS Nano 4:5471–5479. https://doi.org/10.1021/nn101558x

Chary K (2023) Carbon quantum dots in healthcare: a promising solution for sustainable healthcare and biomedical practices. E3S Web Conf 453:01017. https://doi.org/10.1051/e3sconf/202345301017

Das N, Kumar A, Rayavarapu R (2021) The role of deep eutectic solvents and carrageenan in synthesizing biocompatible anisotropic metal nanoparticles. Beilstein J Nanotechnol 12:924–938. https://doi.org/10.3762/bjnano.12.69

Deka C, Aidew L, Devi N, Buragohain AK, Kakati DK (2016) Synthesis of curcumin-loaded chitosan phosphate nanoparticle and study of its cytotoxicity and antimicrobial activity. J Biomater Sci Polym 27:1659–1673. https://doi.org/10.1080/09205063.2016.1226051

Díaz-Puertas R, Álvarez-Martínez FJ, Falco A, Barrajón-Catalán E, Mallavia R (2023) Phytochemical-based nanomaterials against antibiotic-resistant bacteria: an updated review. Polymers 15:1392. https://doi.org/10.3390/polym15061392

El-Wakil NA, Hassan EA, Hassan ML, Abd El-Salam SS (2019) Bacterial cellulose/phytochemical's extracts biocomposites for potential active wound dressings. Environ Sci Pollut Res Int 26:26529–26541. https://doi.org/10.1007/s11356-019-05776-w

Ferdous Z, Nemmar A (2020) Health impact of silver nanoparticles: a review of the biodistribution and toxicity following various routes of exposure. Int J Mol Sci 21:2375

Freeland G, Hettiarachchy N, Atungulu GG, Apple J, Mukherjee S (2021) Strategies to combat antimicrobial resistance from farm to table. Food Rev Int 39:27–40

García-Pérez P, Miras-Moreno B, Lucini L, Gallego PP (2021) The metabolomics reveals intraspecies variability of bioactive compounds in elicited suspension cell cultures of three Bryophyllum species. Ind Crop Prod 163:113322. https://doi.org/10.1016/j.indcrop.2021.113322

Ghahramani Y, Mokhberi M, Mousavi SM, Hashemi SA, Nezhad FF, Chiang W et al (2022) Synergistically enhancing the therapeutic effect on cancer, via asymmetric bioinspired materials. Molecules 27(23):8543. https://doi.org/10.3390/molecules27238543

Giannakourou MC, Tsironi TN (2021) Application of processing and packaging hurdles for fresh-cut fruits and vegetables preservation. Foods 10:830. https://doi.org/10.3390/foods10040830

Gonçalves RFS, Martins JT, Duarte CMM, Vicente AA, Pinheiro AC (2018) Advances in nutraceutical delivery systems: from formulation design for bioavailability enhancement to efficacy and safety evaluation. Trends Food Sci Technol 78:270–291. https://doi.org/10.1016/j.tifs.2018.06.011

Gutierrez-Alvarado K, Chacón-Cerdas R, Starbird R (2022) Pectin microspheres: synthesis methods, properties, and their multidisciplinary applications. Chemistry 4(1):121–136. https://doi.org/10.3390/chemistry4010011

Ibrahim E, Zhang M, Zhang Y, Hossain A, Qiu W, Chen Y et al (2020) Green-Synthesization of silver nanoparticles using Endophytic bacteria isolated from garlic and its antifungal activity against wheat Fusarium head blight pathogen Fusarium graminearum. Nanomaterials (Basel) 10:219

Imani SM, Ladouceur L, Marshall T, Maclachlan R, Soleymani L, Didar TF (2020) Antimicrobial nanomaterials and coatings: current mechanisms and future perspectives to control the spread of viruses including SARS-CoV-2. ACS Nano 14:12341–12369

Imran M, Das KR, Naik MM (2019) Co-selection of multi-antibiotic resistance in bacterial pathogens in metal and microplastic contaminated environments: an emerging health threat. Chemosphere 215:846–857. https://doi.org/10.1016/j.chemosphere.2018.10.114

Jeevanandam J, Danquah MK, Pan S (2021) Plant-derived nanobiomaterials as a potential next generation dental implant surface modifier. Front Mater 8. https://doi.org/10.3389/fmats.2021.666202

Joung HJ, Kim JT, Park SH, Park HJ, Shin GH (2016) Development of food-grade curcumin nanoemulsion and its potential application to food beverage system: antioxidant property and in vitro digestion. J Food Sci 81:N745–N753. https://doi.org/10.1111/1750-3841.13224

Karnwal A, Kumar G, Pant G, Hossain K, Ahmad A, Alshammari MB (2023) Perspectives on usage of functional nanomaterials in antimicrobial therapy for antibiotic-resistant bacterial infections. ACS Omega 8:13492–13508. https://doi.org/10.1021/acsomega.3c00110

Karthik P, Jose PA, Chellakannu A, Gurusamy S, Ananthappan P, Karuppathevan R et al (2024) Green synthesis of MnO2 nanoparticles from Psidium guajava leaf extract: morphological characterization, photocatalytic and DNA/BSA interaction studies. Int J Biol Macromol 258:128869. https://doi.org/10.1016/j.ijbiomac.2023.128869

Katongole P, Nalubega F, Florence NC, Asiimwe B, Andia I (2020) Biofilm formation, antimicrobial susceptibility and virulence genes of Uropathogenic Escherichia coli isolated from clinical isolates in Uganda. BMC Infect Dis 20:453. https://doi.org/10.1186/s12879-020-05186-1

Kaur T, Putatunda C, Vyas A, Kumar G (2020) Zinc oxide nanoparticles inhibit bacterial biofilm formation via altering cell membrane permeability. Prep Biochem Biotechnol 51:309–319. https://doi.org/10.1080/10826068.2020.1815057

Khan SA, Shahid S, Lee CS (2020) Green synthesis of gold and silver nanoparticles using leaf extract of Clerodendrum inerme; characterization, antimicrobial, and antioxidant activities. Biomolecules 10:835. https://doi.org/10.3390/biom10060835

Lan J (2022) Overview of application of nanomaterials in medical domain. Contrast Media Mol Imaging 2022:3507383. https://doi.org/10.1155/2022/3507383

Liang Y, He J, Guo B (2021) Functional hydrogels as wound dressing to enhance wound healing. ACS Nano 15:12687–12722. https://doi.org/10.1021/acsnano.1c04206

Lin J, Tujios S (2023) Hidden dangers: herbal and dietary supplement induced hepatotoxicity. Liver 3:618–636. https://doi.org/10.3390/livers3040041

Ling L, Huang XY, Zhang WX (2018) Enrichment of precious metals from wastewater with Core-Shell nanoparticles of iron. Adv Mater 30:e1705703. https://doi.org/10.1002/adma.201705703

Liu F, Cheng Y, Tan J, Li J, Cheng H, Hu H et al (2021) Carbon nanomaterials with hollow structures: a mini-review. Front Chem 9. https://doi.org/10.3389/fchem.2021.668336

Makvandi P, Wang CY, Zare EN, Borzacchiello A, Niu LN, Tay FR (2020) Metal-based nanomaterials in biomedical applications: antimicrobial activity and cytotoxicity aspects. Adv Funct Mater 30:1910021

Maleki B, Jafari-Soghieh F, Alinezhad H, Ghani M, Jamshidi A (2022) Development of pamam dendrimer-modified magnetic polyoxometalate: a novel platform to reinforce mechanical and thermal properties of diglycidyl ether of bisphenol a/isophorone diamine hardener epoxy. High Perform Polym 34(7):768–777. https://doi.org/10.1177/09540083221089563

Mann GS, Singh L, Kumar P, Singh S (2018) Green composites: a review of processing technologies and recent applications. J Thermoplast Compos Mater 33(8):1145–1171. https://doi.org/10.1177/0892705718816354

Mehdipour M, Gloag L, Hagness D, Lian J, Alam M, Chen X et al (2022) Flow-based synthesis of gold-coated magnetic nanoparticles for magnetoplasmonic sensing applications. Part Part Syst Charact 39(8). https://doi.org/10.1002/ppsc.202200051

Meher R (2024) Nanomaterial-induced modifications in plant physiology and genetics for optimal crop production strategies. Int J Environ Clim Change 14(2):232–240. https://doi.org/10.9734/ijecc/2024/v14i23941

Mondal SK, Chakraborty S, Manna S, Mandal MS (2024) Antimicrobial nanoparticles: current landscape and future challenges. R Soc Chem Pharma 1:388. https://doi.org/10.1039/d4pm00032c

Morelli L, Cappelluti MA, Ricotti L, Lenardi C, Gerges I (2017) An injectable system for local and sustained release of antimicrobial agents in the periodontal pocket. Macromol Biosci 17(8). https://doi.org/10.1002/mabi.201700103

Mousaa I, Abdelreheim N, Elnaggar SA, Fathy RM (2022) Green synthesis and use of copper oxide nanoparticles to prepare a highly thermal, antimicrobial, and anticorrosion hybrid coating for mild steel under electron beam irradiation. Anti-Corros Methods Mater 69(2):160–170. https://doi.org/10.1108/acmm-11-2021-2566

Moussa Z, Hmadeh M, Abiad MG, Dib OH, Patra D (2016) Encapsulation of curcumin in cyclodextrin-metal organic frameworks: dissociation of loaded CD-MOFs enhances stability of curcumin. Food Chem 212:485–494. https://doi.org/10.1016/j.foodchem.2016.06.013

Nandhini SN, Sisubalan N, Vijayan A, Karthikeyan C, Gnanaraj M, Gideon DAM et al (2023) Recent advances in green synthesized nanoparticles for bactericidal and wound healing applications. Heliyon 9:e13128
Nasim I, Vishnupriya V, Jabin Z, Nathan S (2021) Effectiveness of silver nano particles on root canal disinfection. Bioinformation 17(1):218–222. https://doi.org/10.6026/97320630017218
Österberg M, Henn A, Farooq M, Valle-Delgado J (2023) Biobased nanomaterials-the role of interfacial interactions for advanced materials. Chem Rev 123(5):2200–2241. https://doi.org/10.1021/acs.chemrev.2c00492
Panneerselvam T, Arumugam S, Ali M, Kunjiappan S, Murugan I, Sivakumar A et al (2017) Design, network analysis, in silico modeling and synthesis of benzimidazoles nanocomposites as anticancer agent. ChemSelect 2(7):2341–2347. https://doi.org/10.1002/slct.201601763
Patil S (2023) Hygrophila schulli assisted green synthesis of silver and gold nanoparticles: a review. Adv Pharm J 8(5):112–118. https://doi.org/10.31024/apj.2023.8.5.1
Peressini D, Melchior S, Berlese M, Calligaris S (2020) Application of high-pressure homogenization to tailor the functionalities of native wheat starch. J Sci Food Agric 101(7):2668–2675. https://doi.org/10.1002/jsfa.10891
Pushpamalini T, Keerthana M, Sangavi R, Nagaraj A, Kamaraj P (2021) Comparative analysis of green synthesis of TiO2 nanoparticles using four different leaf extract. Mater Today Proc 40:180–184
Putri M, Chaerunisaa A, Abdassah M (2020) Cellulose nanocrystals preparation as pharmaceuticals excipient: a review. Indones J Pharm 2(2):42. https://doi.org/10.24198/idjp.v2i2.26422
Quichaba MB, Moreira TFM, de Oliveira A, de Carvalho AS, de Menezes JL, Gonçalves OH (2022) Biopreservatives against foodborne bacteria:combined effect of nisin and nanoncapsulated curcumin and co encapsulation of nisin and curcumin. J Food Sci Technol. https://doi.org/10.1007/s13197-022-05641-8
Rahman F, Patwary MA, Siddique M, Bashar M, Haque M, Akter B et al (2022) Green synthesis of zinc oxide nanoparticles using Cocos nucifera leaf extract: characterization, antimicrobial, antioxidant and photocatalytic activity. R Soc Open Sci 9:220858. https://doi.org/10.1098/rsos.220858
Raju A, Das M, Dash B, Dey P, Nair S, Naskar K (2022) Variation of air permeability in bromobutyl rubber/epoxidized natural rubber composites: influence of structure of filler particle. Polym Eng Sci 62(3):718–729. https://doi.org/10.1002/pen.25879
Rashidi L (2021) Different nano-delivery systems for delivery of nutraceuticals. Food Biosci 43:101258. https://doi.org/10.1016/j.fbio.2021.101258
Rizki IN, Klaypradit W (2023) Utilization of marine organisms for the green synthesis of silver and gold nanoparticles and their applications: a review. Sustain Chem Pharm 31:100888
Ruttkay-Nedecký B, Kryštofová O, Nejdl L, Adam V (2017) Nanoparticles based on essential metals and their phytotoxicity. J Nanobiotechnol 15:1. https://doi.org/10.1186/s12951-017-0268-3
Sahu S, Rathi P, Jakhar P, Kumar S, Rathore RS, Saran D (2022) Effect of educational session on antimicrobial consumption pattern in the medical intensive care unit of government tertiary care hospital, Jaipur, Rajasthan. Asian J Med Sci 13(7):171–175. https://doi.org/10.3126/ajms.v13i7.43968
Sharma S (2024) Revolutionizing agriculture with nanotechnology: advances, applications, and sustainability considerations. International Journal of Environment and Climate Change 14(7):1–9. https://doi.org/10.9734/ijecc/2024/v14i74246
Sharma S, Barkauskaite S, Jaiswal AK, Jaiswal S (2021) Essential oils as additives in active food packaging. Food Chem 343:128403. https://doi.org/10.1016/j.foodchem.2020.128403
Sharma D, Gautam S, Singh S, Srivastava N, Khan AM, Bisht D (2025) Unveiling the nanoworld of antimicrobial resistance: integrating nature and nanotechnology. Front Microbiol 15:1391345. https://doi.org/10.3389/fmicb.2024.1391345
Shen L, Li B, Qiao Y (2018) Fe_3O_4 nanoparticles in targeted drug/gene delivery systems. Materials 11(2):324. https://doi.org/10.3390/ma11020324

Sims KR, He B, Koo H, Benoit DSW (2020) Electrostatic interactions enable nanoparticle delivery of the flavonoid Myricetin. ACS Omega 5:12649–12659. https://doi.org/10.1021/acsomega.9b04101
Singh P, Mijakovic I (2022) Green synthesis and antibacterial applications of gold and silver nanoparticles from Ligustrum vulgare berries. Sci Rep 12:7902
Sirotkin N, Khlyustova A, Costerin D, Naumova I, Titov V, Agafonov A (2022) Applications of plasma synthesized ZnO, TiO2, and Zn/TiO x nanoparticles for making antimicrobial wound-healing viscose patches. Plasma Process Polym 19:2100093
Smagulova G, Kaidar B, Yesbolov N, Prikhod'ko N, Maxumzhanova N (2020) Synthesis of carbon nanotubes from benzene in a fluidised bed reactor. Eurasian Chem-Technol J 22(3):235. https://doi.org/10.18321/ectj982
Spirescu VA, Chircov C, Grumezescu AM, Vasile BS, Andronescu E (2021) Inorganic nanoparticles and composite films for antimicrobial therapies. Int J Mol Sci 22:4595
Taha RH (2022) Green synthesis of silver and gold nanoparticles and their potential applications as therapeutics in cancer therapy; a review. Inorg Chem Commun 143:109610
Taher MA, Khojah E, Darwish MS, Elsherbiny EA, Elawady AA, Dawood DH (2022) Biosynthesis of silver nanoparticles by polysaccharide of Leucaena leucocephala seeds and their anti-cancer, antifungal properties and as preservative of composite milk sample. J Nanomater 2022:7490221: 1–16. https://doi.org/10.1155/2022/7490221
Ternullo S, Schulte-Werning LV, Holsaeter AM, Skalko-Basnet N (2019) Curcumin-in-deformable liposomes-in-chitosan-hydrogel as a novel wound dressing. Pharmaceutics 12:0008. https://doi.org/10.3390/pharmaceutics1201000
Thakor AS, Jokerst J, Zavaleta C, Massoud TF, Gambhir SS (2011) Gold nanoparticles: a revival in precious metal administration to patients. Nano Lett 11:4029–4036. https://doi.org/10.1021/nl202559p
Tibolla H, Czaikoski A, Pelissari FM, Menegalli FC, Cunha RL (2020) Starch-based nanocomposites with cellulose nanofibers obtained from chemical and mechanical treatments. Int J Biol Macromol 161:132–146. https://doi.org/10.1016/j.ijbiomac.2020.05.194
Tilahun E, Adimasu Y, Dessie Y (2023) Biosynthesis and optimization of ZnO nanoparticles using Ocimum lamifolium leaf extract for electrochemical sensor and antibacterial activity. ACS Omega 8:27344–27354. https://doi.org/10.1021/acsomega.3c02709
Trabelsi M, Mamun A, Klöcker M, Moulefera I, Pljonkin A, Elleuch K et al (2021) Magnetic carbon nanofiber mats for prospective single photon avalanche diode (spad) sensing applications. Sensors 21(23):7873. https://doi.org/10.3390/s21237873
Wang J, Wang L, Zhao W, Cheng M, Su M, Hu J, et al (2020) Aged zinc oxide nanoparticles did not induce cytotoxicity through apoptosis signaling pathway as fresh NPs. https://doi.org/10.21203/rs.3.rs-132318/v1. Accessed 4 Sept 2024
Wang J, Zhao C, Zhang T, Yang L, Chen H, Yue F (2022) In-depth identification of phenolics fractionated from eucalyptus Kraft lignin. Adv Sustain Syst 6:5. https://doi.org/10.1002/adsu.202100406
Wong MH, Misra RP, Giraldo JP, Kwak S, Son Y, Landry MP et al (2016) Lipid exchange envelope penetration (leep) of nanoparticles for plant engineering: a universal localization mechanism. Nano Lett 16(2):1161–1172. https://doi.org/10.1021/acs.nanolett.5b04467
World Health Organization (WHO). Antimicrobial resistance. Available: https://www.who.int/news-room/fact-sheets/detail/antimicrobial-resistance. Accessed 21 Mar 2023
Yadeta GL, Lealem BA (2024) Green synthesis of ZnO, CuO and NiO nanoparticles using neem leaf extract and comparing their photocatalytic activity under solar irradiation. Green Chem Lett Rev 17:2293841. https://doi.org/10.1080/17518253.2023.2293841
Yan Y, Zhu X, Yu Y, Li C, Zhang Z, Wang F (2022) Nanotechnology strategies for plant genetic engineering. Adv Mater 34:7. https://doi.org/10.1002/adma.202106945
Yılmaz GE, Göktürk I, Ovezova M, Yılmaz F, Kılıç S, Denizli A (2023) Antimicrobial nanomaterials: a review. Hygiene 3:269–290. https://doi.org/10.3390/hygiene3030020

Yuan Y, Gu Z, Yao C, Luo D, Yang D (2019) Nucleic acid–based functional nanomaterials as advanced cancer therapeutics. Small 15:26. https://doi.org/10.1002/smll.201900172

Zhang H, Feng H, Ling J, Ouyang XK, Song X (2021) Enhancing the stability of zein/fucoidan composite nanoparticles with calcium ions for quercetin delivery. Int J Biol Macromol 193:2070–2078. https://doi.org/10.1016/j.ijbiomac.2021.11.039

Zhao Y, Zhu Y (2015) Graphene-based hybrid films for plasmonic sensing. Nanoscale 7(35):14561–14576. https://doi.org/10.1039/c5nr03458b

Zhi H, Zhou S, Pan W, Shang Y, Zeng Z, Zhang H (2022) The promising nanovectors for gene delivery in plant genome engineering. Int J Mol Sci 23(15):8501. https://doi.org/10.3390/ijms23158501

Zielińska A, Carreiró F, Oliveira A, Neves A, Pires B, Venkatesh D et al (2020) Polymeric nanoparticles: production, characterization, toxicology and ecotoxicology. Molecules 25(16):3731. https://doi.org/10.3390/molecules25163731

Chapter 12
Controlled Release of Antioxidant Biomolecules by Plant-Based Nanomaterials

Nileema S. Gore, Vedika M. Kulkarni, and Subodh R. Joshi

12.1 Introduction

Materials that originate from plants and are employed for biological functions are known as plant-derived biomaterials. Natural substances generated from plants are employed in medicine and are readily produced for use in biological applications based on tissue engineering (Venugopal et al. 2024). Plant-based biomaterials entail a wide range of biodegradable, bio-derived materials drawn from various forms of plant matter, especially cellulose, lignin, starch, and protein-based materials like pineapple leaf-like fibers (Gaba et al. 2021). Such properties characterize these materials as biodegradability, biocompatibility, and the ability to be functionalized for specific applications (Kurowiak et al. 2023).

Shifting emphasis toward eco-friendly and sustainable technologies has motivated the development and use of applications of these plant-based materials apart from those based on non-renewable petroleum sources (Kurowiak et al. 2023). Plant-based biomaterials are indispensable due to their potential to overcome many limitations imposed by conventional synthetic materials. They are eco-friendly as they can be obtained from sustainable sources and more often than not, compostable waste dumpsters possess less environmental threat by biodegradation of such materials. In addition, these substances may have desirable physical and mechanical properties including high rigidity, flexible and barrier properties, etc. which widen the scope of their usage from food wraps to orthopedic implants.

The potential of plant-based nanomaterials strategies for controlled antioxidant biomolecule release. Antioxidants are important in numerous areas, such as protection (Liu et al. 2018). These substances, found mostly in fruits, vegetables, and

N. S. Gore (✉) · V. M. Kulkarni · S. R. Joshi
Institute of Biosciences and Technology, Mahatma Gandhi Mission University, Chhatrapati Sambhajinagar, Maharashtra, India

V. Ghosh et al. (eds.), *Applications of Plant Bioactive Compounds-Based Nanoformulations*, Nanotechnology in the Life Sciences,
https://doi.org/10.1007/978-3-032-13086-0_12

whole grains, are known to alleviate oxidative damage by scavenging free radicals, which cause a number of illnesses linked to ageing (Vaiserman et al. 2020). However, it is observed that the bioavailability and gastrointestinal stability of several such therapeutic antioxidants are minimal, hence rendering them less effective. To solve these issues, the researchers have turned to nanotechnology for assistance and the corresponding techniques for demonstrating improvement in the delivery and controlled release of antioxidant biomolecules.

Oxidative stress, a common occurrence in every person's body, can become chronic and lead to various diseases, such as cardiovascular diseases and cancer. Antioxidants are the very compounds that inhibit this oxidative stress and, consequently, prevent and cure diseases. They act on unstable free radicals, which have the potential to cause harm, injury, or illness in humans or other organisms and can even destroy the essential cellular components. Traditional antioxidant delivery has found difficulties, such as poor bioavailability and stability, limiting the roles they were to play. For example, quercetin and curcumin have many antioxidant properties, but poor solubilization and absorption rates make them ineffective when used in standard formats (Matencio et al. 2022). Additional developments in nanotechnology for antioxidant encapsulation in plant materials may offer promising opportunities for novel antioxidant delivery methods. These techniques make use of chitosan nanoparticles, which enhance bioavailability and provide sustained drug release profiles, leading to improved health outcomes and disease prevention (Mady 2021).

A significant limitation lies in the traditional means of delivering antioxidants. Very quick degradation of many antioxidants takes place on exposure to environmental factors like light and oxygen, leading to reduced bioavailability. Furthermore, conventional formulations typically lack targeted distribution, which implies that the targeted tissues or other biological structures are frequently unable to absorb them sufficiently. This inefficacy is primarily in topical applications, where the ingredients do not usually penetrate the skin very well. For instance, skin aging necessitates novel approaches, as noted in the discussion on topical formulations for improved delivery of active ingredients (Souto et al. 2020). Synthetic carriers could even lead to undesired responses, necessitating a better alternative delivery system. In contrast, plant-based nanomaterials offer promising progress toward the flaws explained, a combination of ensuring better stability and biocompatibility as well as improving targeted delivery for more efficacious therapeutic results.

Traditional antioxidant compositions struggle with bioavailability and stability when trying to achieve therapeutic use. Poor water solubility and low permeability characterize these antioxidant formulations, limiting their ability to become absorbed in the gastrointestinal tract digestively (Matencio et al. 2022). Moreover, previous chemical instability would lead to active compound degradation before therapeutic efficacy is achieved. The other obstacle is that which presents itself in the format of the blood-brain barrier (BBB) which seems to hinder the action of adaptive antioxidants over neurodegenerative disorders, considering that most classical formulations fail to permeate the effective barrier (Cascione et al. 2022). These elements so highlight the necessity of developing new delivery methods that would increase their efficacy through stability and bioavailability, resulting in enhanced clinical applications. Facing the above challenges, using modern technologies will take specific steps toward the advancement that would bring to the application of antioxidants in health and disease management.

Plant-based nanomaterials thus can be potential candidates to develop biocompatible and sustainable antioxidant delivery systems, but studies to find more effective ways are still ongoing. These nanomaterials utilize the naturally derived properties of plant compounds for encapsulating antioxidants and their targeted release, facilitating the effective therapeutic use thereof. The example of silver and gold (Ag and Au) nanoparticles synthesized using green synthesis methods provides an innovative approach to the meeting of environmental and surgical biomedical applications. Natural resources for producing nanoparticles are said to have reduced amounts of toxic byproducts and environmental effects as discussed by Ahmad et al. (2024). The biogenic materials are also expected to convey significantly improved bioavailability and stability of antioxidants, which can overcome many inherent restrictions of conventional delivery mechanisms (Cascione et al. 2022). This assists in efficacy and sustainable methodology in pharmaceutical applications.

The integration of plant-based nanomaterials into antioxidant delivery systems offers several compelling characteristics and advantages that address limitations found in conventional approaches. Plant-derived materials exhibit inherent biocompatibility and biodegradability, making them perfect for cosmetic and medicinal applications without the concerns associated with synthetic additives. These nanomaterials can encapsulate active antioxidant compounds effectively, ensuring a sustained release that enhances their stability and bioavailability over time. Moreover, the diversity of phytochemicals available allows for the customization of formulations tailored to specific therapeutic needs. For instance, lignin, a major component of lignocellulosic biomass, has shown potential in creating nanomaterials with antioxidant, UV-absorbing, and antimicrobial properties, which can be utilized in targeted drug delivery systems. Additionally, the incorporation of these biodegradable nanocarriers can minimize environmental impact, aligning with the increasing interest in sustainable practices in both pharmaceuticals and cosmetics (Fig. 12.1).

12.2 Types of Plant-Based Nanomaterials (PBN)

Nanomaterials are bringing about mega changes to various sectors, mostly agriculture, medicine, and the environment. In agriculture, they have high efficacy in nutrient delivery and pest control, much more so than the old ways. For example, it can improve food production and at the same time control the unwanted release of harmful substances, thus offering a more sustainable way of fighting against soil contamination, which can be conjoined by other agricultural management practices for healthier agriculture (Bachini et al. 2013). In a similar vein, medicines are also utilized for the delivery of drugs targeted at specific points in the body, and their effects are used in diagnostic imaging techniques that promise more effective

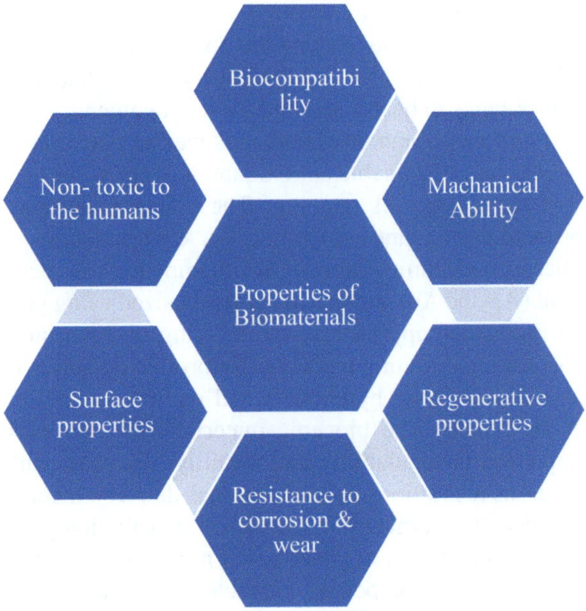

Fig. 12.1 Properties of biomolecules

treatment regimens associated with reduced side effects. The significance of nanomaterials in the environment can be highlighted in terms of soil remediation and pollutant detection, which deal with soil contaminants highlighted by SEPA (2001), thus posing dangers to ecosystem integrity. Their flexibility and utility thus accentuate the importance of nanomaterials in improving various fields while providing scope for environmental protection and sustainable development (Fig. 12.2).

The vast spectrum of plant-based nanomaterials reveals a range of possible application sites in multiple disciplines and those explicitly targeted toward agriculture and food. Such plant-sourced nanomaterials are cellulose nanocrystals, chitosan nanoparticles, and bioactive compounds from plant extracts. These materials have immense, flexible, and sustainable value and add great effect, enhancing soil health in addition to dynamic nutrient flow but serving as efficacious factors in pest control and disease management. For instance, recent studies have demonstrated that nanobiosensors from natural sources can be of great help in significantly improving the monitoring of soil and crop health for better resource management and increased agricultural productivity (Thakur et al. 2020). Besides, understanding their ecotoxicological impacts, especially on microplastics in horticulture, would be useful in trying to alleviate risks to human health and the environment (Gayathri et al. 2024). Such a holistic view strongly advocates the application of plant-based nanomaterials in sustainable agricultural practice.

Nanomaterials originating from plant sources are known as potentially an exciting area at the interface of biology and advanced materials science. This can be classified based on the origins and the synthesis methods, and they fall largely under naturally occurring nanoparticles and engineered nanomaterials. Naturally

Fig. 12.2 Types of PBN

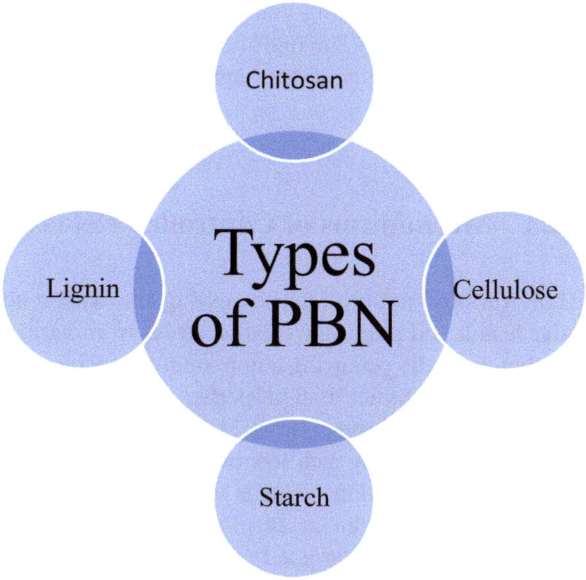

occurring ones are nanoparticles synthesized through various biological processes, while engineered nanomaterials are intentionally designed and produced using biological organisms, generally with a specific application. The main advantages of plant-based nanomaterials would be that they are very environmentally green and biocompatible, which is increasingly required in various industries. The health concerns and environmental impacts of using nanomaterials should not be taken lightly. Research indicates that, despite engineered nanomaterials coming with risks of exposure in the work environment, the plant-based ones are considered safer (Adamec et al. 2019). The dynamic scene of Nanotechnology certainly lays the premise for more investigations and regulations regarding these new materials in terms of classification and regulation.

A new term in the land of nanotechnology is plant-based nanomaterials. These have great prospects in horticulture since they are mostly anticipated to boost food production and ecology in very significant ways. Such nanomaterials could derive from living sources and act as modulators for a more refined delivery of nutrients and management of plants from pests for forming much healthier crops, and improve devices for bio-tracking the condition of soil and monitoring crop diseases to act promptly to increase yield outcomes, but using fewer chemicals. For example, it could detect soil health and monitor crop diseases, allowing farmers to adopt a high-yield outcome but with reduced chemical applications. Absorbing pollutants from the soil and water, and thus rebalancing ecological construct, is also propitious with plant-based nanomaterials, specially designed for contaminant remediation (Bachini et al. 2013). Besides, these materials are normally toxin-free and environmentally friendly, thus supporting the conversion into sustainable agricultural practices (Thakur et al. 2020). The multifunctional aspects of plant-nano-materials point to a bright future in terms of both agricultural productivity and environmental hazard reduction.

12.3 Mechanisms of Controlled Release

Many various biomolecules possessing antioxidant qualities are responsible for cellular health and purging oxidative stress from various diseases. From polyphenols to other naturally occurring compounds, these biomolecules reveal exceptional antioxidants and antimicrobial properties, enabling them to counteract harmful reactive oxygen species (Costa et al. 2021). However, the wide utility of these antioxidants is always hindered by their low bioavailability, which becomes quite a challenge for therapeutic applications. This is where nanomaterial-based delivery systems come in; antioxidants can be encapsulated into them to increase their solubility and permeability and allow targeted release at sites suffering from oxidative stress. For example, how antioxidant nanomaterials can be set to interact with biological entities fundamentally can influence redox mechanisms and provide a double benefit because it does not only serve as a delivery resource for the antioxidants but also modulate biological responses to oxidative damage (Hanna et al. 2017). Thus, it is vital to know the properties and delivery mechanisms for antioxidant biomolecules to develop strong therapeutic solutions (Table 12.1).

An important step towards designing an efficient biomedical application for the antioxidant biomolecule is understanding the mechanisms controlling the release rate of the antioxidant biomolecules into significant efficacy-enhancing applications. Interaction with the delivery systems considerably affects the release profile of the biomolecules. Such an attribute of the antioxidant release would often rely on the size and surface charge of the nanomaterials involved. Thus, smaller nanoparticles are generally construed to have quicker release due to greater surface area; surface modifications thus can tune the kinetics of adsorption/desorption hence release characteristics (Murphy 2015). Furthermore, changes in environmental conditions like pH and temperature lead to alterations in the release kinetics, thereby affecting the bioavailability and stability of antioxidants. Thus, incorporating such knowledge in developing those nanomaterial platforms would eventually optimize the therapeutic potential of antioxidant biomolecules and open avenues for their advancement in practical applications within medicine (Jampilek and Placha 2019).

Nanomaterials' physical properties significantly affect the rates of release for antioxidant biomolecules, like in applications for drug delivery and agricultural enhancement. Size, shape, and surface chemistry are important considerations in the modulation of these release rates. Smaller nanoparticles have been found to improve the bioavailability of attached biomolecules and present a more controlled and sustained release scheme since they penetrate biological barriers more efficiently. So because of surface modifications, these can even adapt their contact with

Table 12.1 Mechanisms and factors influencing controlled release of antioxidant biomolecules via plant-based nanomaterials (2020–2025)

Sr. no.	Release mechanism	Influencing factors	Type of nanomaterial	Observed outcome	References
1	Diffusion-controlled release	Particle size, concentration gradient	Zein nanoparticles	Sustained release of lutein and folic acid	Chuacharoen (2015)
2	pH-responsive release	pH sensitivity, matrix degradation	MOFs	Targeted delivery at acidic pH	Mhettar et al. (2024)
3	Erosion-controlled release	Polymer matrix degradation	Pullulan hydrogels	Prolonged release of essential oils	Teixeira et al. (2023)
4	Surface adsorption-desorption	Surface chemistry, environmental pH	Chitosan nanoparticles	Tunable release based on interactions	Mady (2021)
5	Ionic interaction	Nanoparticle charge, membrane affinity	Silver nanoparticles	Cellular uptake via strong membrane interaction	Ahmad et al. (2024)
6	Inclusion complexation	Hydrophobic interactions	Cyclodextrins	Increased solubility of antioxidants like piperine	Fitrani et al. (2021)
7	Redox-sensitive release	ROS levels in target environment	Green-synthesized nanoparticles	Triggered release under oxidative stress	Hanna et al. (2017)

the environment, affecting adsorption and desorption rates (Balusamy et al. 2023). Illustrates that the physicochemical properties of nanoparticles would affect the soil health as well as overall crop performance, and thus it should be judiciously selected based on the need. Thus (Díez-Pascual 2022), shows how functionalized nanomaterials differ regarding cytotoxicity and therapeutic efficiencies and, therefore, require an optimization of the properties to have specific release profiles for target applications.

The properties of Nanomaterials are essential for determining their dynamics of release, particularly in controlled delivery systems for antioxidant biomolecules. These include size, surface charge, and surface chemistry, all of which can greatly affect the interaction of such nanomaterials with their environment, thereby influencing the rate and extent of biomolecular release. One example is that small nanoparticles tend to have much higher surface areas to volumes, increasing reactivity and making them likely to release biomolecules more quickly under certain conditions. Surface modification may also affect how they interact with environmental components, thereby influencing stability and bioavailability. This also goes in line with the findings that the physicochemical properties of nanoparticles affect nanoparticle efficacy in agricultural applications, as controlled release may enhance crop yield with minimal nanotoxicity (Balusamy et al. 2023). Yet again, bio-based nanomaterials, especially iron-based, call into play the greenest of synthesis to achieve the least disturbance to the environment while maximizing the remediation of pollutants (Quintero and Palencia 2021). Of course, understanding these dynamics is crucial for furthering the application of medical applications of nanotechnology and environmental sciences.

The size and surface charge of nanomaterials will greatly affect the controlled release mechanisms of the antioxidant biomolecules, which could prove beneficial for their therapeutic efficacy. Smaller nanoparticles tend to have a greater surface area-to-volume ratio, thus releasing faster when interacting with the biological environment than larger particles. For example, nanoparticles synthesized in nanometric dimensions can carry bioactive agents and assist in their target action while minimizing the toxicity associated with their short-lived action (Balusamy et al. 2023). Surface charge also plays an important role: positively charged nanoparticles tend to interact more robustly with negatively charged cellular membranes, facilitating their reuptake and consequent release into the cells of the incorporated antioxidants. In contrast, slower-release kinetics are expected from neutral or negatively charged particles because of their limited interactions with biological matrices. However, size and surface charge should always be optimized for prolonged and effective release of therapeutic antioxidant biomolecules.

12.4 Methods of Synthesis of PBN for Antioxidant Delivery

12.4.1 Green Synthesis Techniques

Green synthesis approaches have become a significant subject of inquiry over the last decade and a half because they are becoming more important as sustainable alternatives to traditional methods of synthesis, which normally use very toxic materials and by-products, leaving behind substantial waste (Cheng et al. 2018). One of the major advantages attributed to green synthesis methods is the fact that they use plant extracts for reducing and stabilizing agents in nanoparticle fabrication (Kiwumulo et al. 2022). The mechanism behind this is usually attributed to phytoremediation, in which plants can take and accumulate some metals from the environment. Plant extracts reduce the reaction time, keeping the cell structure more intact, thus minimizing the chances of other contamination.

The use of plant extracts in biosynthesis has come to remain the most affordable approach and one of the striking breakthroughs in nanotechnology. This approach utilizes the naturally reducing nature of plant biomolecules to produce biosynthesized nanoparticles without generating toxic byproducts. For example, silver ions were reduced into silver nanoparticles with a mean size of around 20 nm, which brought into health antioxidant activity through effective radical scavenging by Atropa acuminata roots. Such properties demonstrate the potential of plant-based nanoparticles for the antioxidant delivery route and also help improve bioavailability and therapeutic efficiency. Moreover, Nanoparticles of zinc oxide have been created from natural sources for applications ranging from drug delivery to antibacterial properties and have indicated diverse benefits of green synthetic methods applications in bio-medicinal applications (Younas et al. 2023).

The traditional methods of synthesizing nanoparticles are sometimes involve the use of hazardous substances and by using a lot of energy, escalating the environmental and health issues. In contrast, green synthesis techniques are based on using natural resources, such as plant extracts or bacteria, and fungi, making them eco-friendly alternatives to harmful synthetic processes. Recent developments have made plant extracts active constituents, such as terpenoids and flavonoids, which have both reduction and stabilization functions for nanoscale size production without any harmful byproducts (Mohammadi Dargah et al. 2023). This biogenic approach would thus reduce environmental threats along with safety and efficacy benefits for nanoparticles to be used in medicine. Furthermore, the inclusion of such approaches into nanomedicine augurs well for wider public acceptance as they allay growing regulatory anxieties about how nanoparticles are produced and used. Green synthesis would thus provide a way toward sustainable development of advanced nanomaterials (Fig. 12.3).

12.4.2 Surface Modification for Enhanced Stability and Release of the Control

On account of surface modification techniques, antioxidants significantly enhance their efficacy for food packaging and therapeutic applications. Its alteration leads to modified surface properties for stabilizing and/or adjusting the release kinetics for

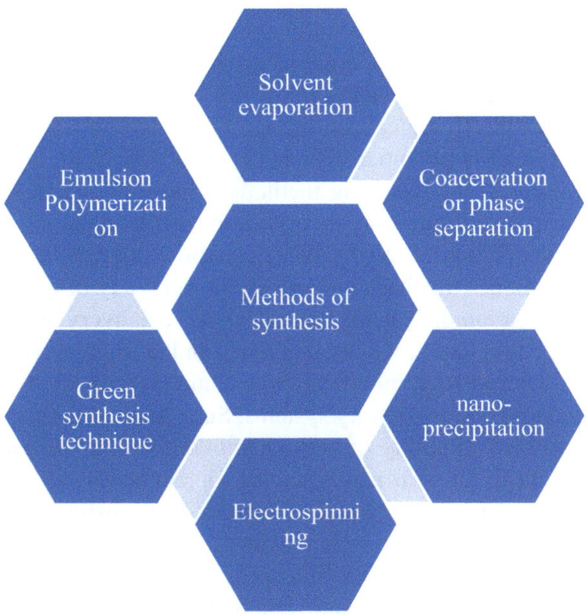

Fig. 12.3 Methods of synthesis of PBN

antioxidative agents to ensure long-term applications. For example: Plant extracts and nano-sized materials incorporated into polymer blends improved not only the mechanical strength of the blend but also antimicrobial and antioxidant properties (Maurizzi et al. 2022). The help release the antioxidants within the target, thus avoiding degradation and helping the antioxidant out of maximal protection against oxidative stress. Furthermore, as sustainability gains more ground, the specialized adaption of these surface modifications with biopolymers portends for eco-friendly packaging solutions to these issues: environmental and food safety concerns. All the same, the multifarious advantages of surface modification demonstrate its importance in the optimization of various antioxidant applications.

Nanocarrier efficacy for antioxidant delivery is dependent on effective surface modification techniques for stability and controlled release. An important method for surface modification involves surfactants because, apart from stabilizing nanoparticles, they also improve significantly the chemical stability of entrapped bioactive substances. For example, zein nanoparticles stabilized by surfactants have been reported to effectively shield antioxidants like lutein from degradation thereby offering a slow release under physiological conditions (Chuacharoen 2015). Further, the use of stimuli-responsive materials tailors drug release for local, site, and therapeutic effect-specific delivery. For example, metal-organic frameworks (MOFs) are representative of customizable surface chemistry which enables targeting of drug release through environmental triggers (Mhettar et al. 2024). These types of innovative surface modifications will not only improve in loading of antioxidants but also optimal release kinetics, resulting in more efficient delivery systems in the food and pharmaceutical applications.

Chemical techniques are perhaps among the best ways to modify surfaces in order to maximise antioxidant loading and release kinetics in a variety of applications. This can significantly enhance the stability and efficacy of bioactive by using polymer blending and nanoencapsulation techniques. For example, zein nanoparticles enhance the chemical stability of lutein and provide a sustained release profile under physiological conditions. These demonstrated their food and pharmaceutical application potential (Chuacharoen 2015). Drug-loaded microfibers created using improved wet-spinning methods exhibit unique surface morphologies, which will further facilitate drug incorporation while maintaining favorable mechanical properties (Wu et al. 2016). All these show impressive aspects of surface modification techniques that have underlined the necessity of particular chemical methods in the application of systems that allow value delivery and good retention of antioxidants for better health-related outcomes and product performance.

The surface modification techniques impact the effectiveness of antioxidant loading within delivery systems. Such improvements result in the enhanced stability and bioavailability of antioxidant compounds. Research has shown the successful synthesis of zein nanoparticles for alterations in surface properties of materials, such as nanoparticles, for improved encapsulation of hydrophobic antioxidants. These zein nanoparticle carriers also provided GI-conditioned stability and controlled release profiles, thus providing a protective environment for very sensitive compounds, such as lutein and β-carotene, which tend to degrade during processing and digestion (Chuacharoen 2015). MOFs have also emerged as a versatile class of carriers owing to their tunable surface modifications, allowing them to provide better loading capacities and a site-specific delivery of drugs and antioxidants (Mhettar et al. 2024). Tailor-made surface modifications, therefore, can improve antioxidant loading and target the release of those bioactive compounds, substantially evidencing their importance in innovating food applications and pharmaceutical uses.

Surface characteristics are crucial to the retention and action of the antioxidants within different delivery systems. The chemical composition, porosity, and texture of a surface can affect the adsorption and stability of antioxidant molecules significantly. Improving antioxidant activity by surface modifications for solubility and enhanced bioavailability is one such important criterion. Recent research suggests that cyclodextrins (CDs) can highly improve the solubility of poorly soluble antioxidants through their inclusion complexation leading to an enhancement of dissolution rates and stability. In contrast, surface modification, such as coating or grafting of specific functional groups, allows the establishment of conditions that reduce oxidative degradation and strengthen the antioxidant-molecule interaction (Fitrani et al. 2021). Hence, these surface functionalization strategies will not only extend the retention of antioxidants but also allow a controlled release of them for a better therapeutic response.

12.5 Mechanism of Controlled Releases of Antioxidant Biomolecules

12.5.1 Impact on Release Rate Impact of Nanomaterial Type, Size, Surface Charge and Environmental Conditions

They are found in a wide variety of antioxidant biomolecules that support cell health and the defence against oxidative stress, which is directly linked to numerous illnesses. These biomolecules, which include polyphenols and certain other naturally occurring chemicals, have been demonstrated to possess exceptional antioxidant and antibacterial qualities. In essence, they can neutralise harmful reactive oxygen species (ROS) (Costa et al. 2021). Efficacy for most antioxidants is often decreased due to their low bioavailability, a tough barrier in therapeutic applications, as it is at this point that nanomaterial-based delivery systems become very vital: encapsulating antioxidants in nanomaterials can enhance solubility and permeability while allowing targeted delivery at oxidative stress sites. In such ways, antioxidant nanomaterials interacting with biological entities can have fundamental effects on redox mechanisms so that antioxidant delivery occurs along with biological response modulation to oxidative damage (Hanna et al. 2017). Hence, learning the properties and mechanisms by which antioxidant biomolecules deliver will be crucial for effective therapeutic strategies.

An understanding of the mechanisms that can be needed to control the rate of release of antioxidant biomolecules is quite an important factor in improving the use of these materials in biomedical applications. The interaction between these biomolecules and delivery systems will strongly influence their release profiles. For example, some of the physicochemical features of the nanomaterials, size and surface charge for instance, determine the rate at which antioxidants are released into their targeted environment. Smaller nanoparticles may exhibit a release rate faster compared to bigger particles due to their comparatively larger surface area that can be released, while surface modifications are capable of manipulating adsorption/desorption dynamics to effectively tune the release characteristics (Murphy 2015). Moreover, the kinetic release changes with environmental conditions, such as pH and temperature, which may additionally affect the bioavailability and stability of antioxidants. These phenomena could ultimately be incorporated into the design of nanomaterial platforms that would fully exploit the therapeutic potential of antioxidant biomolecules, thereby advancing their real applications in medicine (Jampilek and Placha 2019).

The properties of nanomaterials have direct impacts on the rates of release of antioxidant bio-molecules from nanocarriers, governing efficiency in drug delivery in addition to agricultural applications. Surface chemistry, size, and form are important elements that modulate release rates. For example, smaller nanoparticles were proven to be more bioavailable, making the coupled biomolecules even more controlled and sustained, because they easily penetrate biological barriers. Also, surface modifications can be made to tailor- make the interaction between the nanomaterials and their surroundings for changes in adsorption and desorption rates (Balusamy et al. 2023). Have studied the physicochemical properties of nanoparticles vis-a-vis their effects on soil health and crop performance, emphasizing the need for correct application of selection according to what is required of the end results. Similarly, property optimization becomes important with (Díez-Pascual 2022) illustrate the functionalized nanotubes being different in cytotoxicity and therapeutic efficiency, thus highlighting the need for property optimization for achieving specific release profiles in targeted applications.

Nanomaterials exhibit characteristics that allow spreading kinetics, especially in regards to controlled delivery systems that carry antioxidant biomolecules. Size, charge, and surface chemistry are eminent characteristics influencing the biomolecule release dynamics. Their interaction with the environment will determine the biomolecule release dynamics or slow down the process. As an example, decreasing the size corresponds with increased surface area-to-volume ratios that increases reactivity and result in faster release rates under certain conditions. Surface modification also alters interaction with environment components for stability and bioavailability because this corresponds with the findings that show physicochemical properties of nanoparticles influence their effectiveness during application in agriculture where controlled release improves crop performance and reduces nanotoxicity (Balusamy et al. 2023). Such approaches include developments of biobased nanoparticles especially of iron kinds. In dealing with synthesis methods, these methods of synthesis must be environmentally friendly for minimal damage to the environment while they optimize pollutant remediation processes (Quintero and Palencia 2021). Understanding these dynamics serves as a stepping stone towards some uses for nanotechnology in medicine and environmental sciences.

The controlled release mechanisms related to antioxidant biomolecules are significantly influenced by the size and surface charge of nanomaterials, which is crucial for the better therapeutic potential of these biomolecules. Smaller nanoparticles, often having a higher surface area-to-volume ratio, will have faster release rates, as they can communicate with biological environments more easily than larger ones. Engineered at the nanoscale, the nanoparticles can effectively deliver bioactive compounds to ensure site-specific action with reduced toxicity (Balusamy et al. 2023). The surface charge has a significant effect; positively charged nanoparticles usually interact with the surface of negatively charged cellular membranes, allowing for better uptake of incorporated antioxidants and thus a higher release after cellular entry. Conversely, neutral or particles that are negatively charged may have slow-release profiles due to weak bio-analytical interactions. Thus, size and surface charge must both be optimized to ensure that a prolonged release of the biomolecule is attained and that its entire therapeutic potential is harnessed.

12.5.2 *Kinetic Models of Release*

Antioxidant biomolecules are important for health since they should shield cells from oxidative stress, which causes a number of chronic illnesses, such as cancer and heart disease. The antioxidants have many special existing compounds, like vitamins C and E, flavonoids, and polyphenols; these act by neutralizing free radicals to prevent cellular damage. The delicate balance achieved through antioxidants is essential for a successful immunological response and general well-being. The addition of antioxidant biomolecules in controlled release systems would provide similar extended antioxidant activity, thereby maximizing therapeutic effects in health improvement and disease prevention. Hydrogen sulfide, for instance, is a signaling molecule implicated in several physiological processes, which supports evidence as to why modulation of such molecules would be useful in therapeutic intervention (Magli et al. 2021). The controlled release mechanism is thus one of the crucial areas of any future study, which could go on to inform understanding into optimization of health benefit brought about by antioxidant biomolecules, as extrapolated through kinetic models such as the Higuchi model.

Different mechanisms are behind the controlling delivery of antioxidant biomolecules, contributing to their unique therapeutic effects. At the molecular level, diffusion is the most critical mode; diffusion is the process by whereby molecules are liberated from the matrix through concentration gradients for continued bioavailability. For instance, research has demonstrated that models such as the Higuchi model, which captures the impact of the drug's solubility in the delivery system on the release rate, may frequently be used to depict the kinetics of release from such systems. As Bagas stated, a system with biodegradable polymers can develop a particular release profile: it will encapsulate and retain the stability and bioactivity of compounds like salicylamide and a broad range of nutraceuticals (Pudjiastuti et al. 2020). Therefore, this is important since some matrices can be designed to respond to conditions of physiology by effectively controlling bioactive release via swelling or erosion. Such system understanding can reveal avenues for optimizing antioxidant delivery systems to ensure maximum efficacy and health benefits.

Controlled release systems will herald a new era of improved drug delivery in pharmaceutical and biomedical applications, both in effectiveness and adherence to drugs by the patients themselves. Such systems release therapeutic agents at a pre-stabilized rate, preserving lowering the high and low peaks in medication plasma concentrations by maintaining the therapeutic doses for extended periods, an inherent characteristic of classical therapy regimens. The importance of these methodologies concerns the vastness of public enthusiasm, with the promise of having a reduced range of adverse effects while demonstrating increased therapeutic effects example being diseases such as diabetes, where stable concentrations of blood glucose are highly required. For example, studies show that nanoparticles can enhance the export of herbal compounds, and sweetening turns on the hopeful note in its antidiabetic activity (Haritha and Bhagyalakshi 2014). The advancement in novel materials this case, zein-for drug-eluting stents, typifies the prospect of controlled release technologies into improving clinical biocompatibility with minimized adverse effects. Such sophistication in guaranteeing controlled release technologies is of utmost priority for improved therapeutic modalities for the benefit of patients.

Understanding the kinetics models of release processes between antioxidant biomolecules and their delivery systems is necessary for the prediction of behaviour. The Higuchi, Zero-order, and First-order kinetics programmed models serve as important paradigms that describe these mechanisms of release. Thus, through an example wherein salicylamide's release from carrageenan-based hard-shell capsules showed zero order kinetics because the rate of release remained steady over time, particularly at a pH of 6.8. Such a level of constant rates is important in reliable nutraceutical systems (Pudjiastuti et al. 2020). Likewise, thereby showcasing the principles of kinetic models in predicting release profiles from large biomolecules such as vitamin B12, this would be an application of organic-inorganic hybrid systems. The research elaborates on protease-controlled release successes that controlled boosting bioavailability and efficacy. The Higuchi model is one of the numerous kinetic models used in controlled-release systems. However, it is unique in its widest application in describing the release of a drug or biomolecules from solid matrices, where the drug's solubility in the specified matrix and the release time determine the release rate. This makes it possible to achieve a predictable release profile. It is significantly useful in nutraceutical fields wherein the controlled release of biomolecules like antioxidants can improve bioavailability. Recent findings have illustrated the effectiveness of the Higuchi model in different hydrogels taking its modelling foundation toward optimizing the release kinetics of some compounds, such as vitamin B2 and essential oils. Hydrogels based on biodegradable polymers like pullulan also demonstrate the versatility of the model in biomedical engineering concerning medication delivery methods that focus on various pathologies (Teixeira et al. 2023). With that, the Higuchi model has an essential role in the advancement of controlled release strategies for pharmaceutical applications.

The bottom of the spectrum is that the Higuchi model is among the many kinetic models used in controlled release systems; however, it finds a particular place in being widely applicable to explaining the publication of a drug or biomolecules from solid matrices. According to this approach, the drug's solubility in the matrix and release time determines the release rate, with the promise of offering a predictable release profile. How it is mainly useful for the nutraceuticals field, whereby the antioxidant biomolecules are delivered to the body in a controlled manner to improve bioavailability, can be seen. Newer studies have demonstrated the applicability of the Higuchi model to various hydrogels that take the model basis toward the optimization of release kinetics for certain kinds of compounds, including vitamin B2 and essential oils. Hydrogels made from biodegradable polymers like pullulan really demonstrate the versatility of the model with regard to biomedical engineering for drug delivery systems aimed at treatment of various pathologies (Teixeira et al. 2023). Therefore, the Higuchi model has a lot of promise in the consideration of making it a very important possibility furthering controlled release strategies involving and for pharmaceuticals. These models, therefore, lay the foundation for designing more efficacious antioxidant delivery systems in nutraceutical applications.

12.6 In Vitro and in Vivo Release Studies-Studying Release Profiles and Antioxidant Activity in Various Biological Models

Many different types of biomolecules can act as antioxidants, and they are valuable for cellular well-being against oxidative stresses related to numerous disorders. Such biomolecules include, among others, polyphenols, which have also demonstrated their considerable nutraceutical and medicinal antioxidant and antimicrobial effectiveness by neutralizing very damaging reactive oxygen species (ROS) (Costa et al. 2021). However, such low bioavailability often diminishes the effectiveness indicated in these antioxidants, which poses a huge challenge for therapeutic applications. This is where nanomaterial-based delivery systems become important; they encapsulate antioxidants, improving their solubility and permeability for targeted release at the site of oxidative stress. Antioxidant nanomaterials interact with biological entities, usually influencing redox mechanisms or, in other words, providing the application of antioxidant delivery forms plus modulation of biological responses to oxidative damage (Hanna et al. 2017). Therefore, it is necessary to comprehend the properties and delivery pathways of these antioxidant biomolecules for creating efficient therapy treatments for the management of oxidative stress disorder.

Grasping mechanisms controlling the release rate of antioxidant biomolecules is vital for further increasing their medicinal uses. The interaction of biomolecules with their delivery vehicles could largely influence the release profiles. For instance, the size and surface charges of the nanomaterials can affect the release velocities of the antioxidants into the target environment. Smaller nanoparticles may have a faster release because they have a larger surface area, whereas surface modifications can be used to modify the adsorption/desorption dynamics, tuning the release behaviours (Murphy 2015). Additionally, factors like pH and temperature change release kinetics and the bioavailability and stability of an antioxidant over time. Finally, the incorporation of the above factors in the nanomaterial platforms' designs can be used for optimizing the therapeutic potentials of antioxidant biomolecules, thus taking practical applications in medicine a step further (Jampilek and Placha 2019).

Release rates of biomolecules possessing antioxidant qualities happen to be influenced mostly by the properties of nanomaterials; such rates determine the effectiveness of nanosystems for drug delivery and agricultural enhancement applications. The release rates are modulated by size, shape, and surface chemistry. Smaller nanoparticles have shown the ability to increase the bioavailability of coupled biomolecules for a more controlled and sustained release because they penetrate biological barriers easily. In addition, surface modification can tailor the interaction of nanomaterials with their environment according to the adsorption-desorption rates (Balusamy et al. 2023). Beautifully reveals how soil health and crop performance can be affected through different lenses of physicochemical properties, implying the importance of selecting the right one depending on desired outcomes, i.e. antibiotic or fungicidal effects. On the other hand, (Díez-Pascual 2022) prove that functionalized nanomaterials present differences in terms of cytotoxicity and therapeutic efficiency, which drive the importance of property optimization for achieving precise release profiles for applications targeted.

What defines the behaviour of these nanomaterials is the nature of the released characteristics These are particularly controlled biomolecular delivery systems for antioxidants. Size, surface charge, and surface chemistry can determine the interactions of these nanomaterials with their respective environment and thus the release rates and efficiencies of biomolecules. For example, the small-sized nanoparticles typically have a greater surface area-to-volume ratio, which translates to higher reactivity and further potential fast release under certain conditions, while manipulating the surfaces can change interfaces with environmental components and affect stability and bioavailability as such. This is due to findings that physicochemical traits of nanoparticles might also influence their efficiencies in agricultural applications, whereby controlled release can improve performance while commensurately minimizing nanotoxicity (Balusamy et al. 2023). There is more concern on developing environmental-friendly ways as more emphasis is given to the development of biobased nanomaterials such as iron-based ones. Such environmentally friendly techniques ensure no or little disruption to the environment as it seek to enhance pollutant remediation processes (Quintero and Palencia 2021). Understanding these dynamics is key to progress in all possible nano-dimensions in the biomedical and environmental spheres.

The size and surface charge of nanomaterials considerably interact and thereby control the release mechanisms of antioxidant biomolecules, which is important for improving their therapeutic efficiency. Smaller nanoparticles usually have much higher surface area-to-volume ratios; thus, they have relatively higher release rates as the accessibility of smaller nanoparticles to a biological environment enhances the release compared to larger ones. For instance, when engineered at the nanoscale, nanoparticles release bioactive compounds most efficiently, but it is directed action and possibly decreased toxicity at the same level (Balusamy et al. 2023); Naked animals may also essential to this factor; the positively charged nanoparticles tend to interact strongly with located cellular membranes as negatively charged rather than facilitate increased uptake and further release of protected antioxidants. However, neutral or negatively charged particles can provide a slower release profile as weaker relationships with biological buffers are responsible. Therefore, the place and surface charge must be optimized in a sustained-releasing system to make the most out of antioxidant biomolecules.

12.7 Applications in Pharmaceuticals, Food, Agriculture, and Others

Antioxidant biomolecules are very important to cellular health because they are involved in negating free radicals, which can lead to oxidative stress and a variety of diseases. Vitamins C and E, polyphenols, flavonoids, and bioactive peptides-roots of up to many plant-based biomaterials with therapeutic properties. These antioxidants are becoming more used in nanotechnology delivery systems to improve their bioavailability and thus their effectiveness and accessibility for therapeutic applications (Durazzo et al. 2020). In addition, by controlled release of antioxidant biomolecules through nanomaterials, a sustained delivery could be warranted, thereby maximizing therapeutic uses in pharmaceutical, food, and agricultural domains. In addition, their 'biochemical properties', such as size and surface charge, will determine their interaction with biological systems, and this underlines that it shall research on this to grasp fully their potential and safety (Puttasiddaiah et al. 2022). Thus, by integrating antioxidant biomolecules with nanotechnology, many avenues are created because of its advantages in health and the sustainable future it will signify in the areas it will touch: health-beneficial and revolutionizing holistic practices.

During the last decade, national transformation in the delivery of drugs using nanotechnology as a part of its setup has shown several controlled methods. The use of were then increased and utilized more on plant-based nanomaterials such as embankment by-products, more preferably because of their biocompatibility and low toxicity into the encapsulation of antioxidant biomolecules. This approach would on the one hand help in maximizing activities and increasing stability and bioavailability while on the other providing a targeting delivery system, thus enhancing therapeutic efficacy. The incorporation of nanomaterials loaded with cosmeceuticals like vitamins and growth factors reportedly addresses several skin problems associated with ageing through sustained release of such active components (Souto et al. 2020). New advances in nanofabricated delivery systems glaringly show that these technologies are set to improve absorption and efficacy of pharmaceutical compounds while lessening side effects (Puttasiddaiah et al. 2022). This trajectory of plant-based nanomaterials alongside controlled release technology stands for a significant forward leap in pharmaceutical delivery mechanisms towards safer and thus more efficacious alternatives for treatment.

Very recently plant-based nanomaterials integrated into drug delivery systems have shown excellent biocompatibility, and at the same time, have been quite sustainable. Such materials from natural sources can be processed to encapsulate bioactive compounds, enhancing their safety and bioavailability. This is most important in nutraceuticals, wherein the activity of certain polyphenolics and flavonoids can be greatly enhanced through nanoencapsulation, thereby enabling controlled release and improved therapeutic results (Puttasiddaiah et al. 2022). Further, the unique physicochemical features such as size and charge of such nanomaterials play an important part in the targeted delivery of drugs and thus enable precise interactions with biological systems. Adding plant-based nanomaterials may, therefore, not only increase the effectiveness of antioxidant biomolecules in pharmaceuticals but also become one of the growing needs of environmentally friendly healthcare solutions, further strengthening their essential role in future drug delivery systems.

The food sector is actively promoting the adoption of highly advanced technologies with high-speed systems on plant-based nanomaterials for food safety and preservation purposes. These techniques assist in the complete encapsulation and proper targeted delivery of antioxidant biomofticles that help lengthen the shelf life of food items through their oxidation degradation prevention mechanism. For instance, engineered nanomaterials (ENMs) are available for smart delivery preservatives such that antioxidants are released in response to a given stimulus, thus maximizing the effectiveness of the active ingredients (Patil et al. 2024). Besides, cerium oxide (CeO_2) nanostructures are well on track in continuing as key players in this area because they are unique in their ability to begin displaying antioxidant properties in different settings, making them useful in multiple applications in food (Rozhin et al. 2021). This interfacing holds promise, as changing trends are set to revolutionize how one thinks about preservation and quality management by adopting such technologies in future safe and sustainable food systems.

The implementation of controlled-release mechanisms in food preservation technologies opens new horizons for product life optimization and nutritional improvement. Encapsulation of bioactive ingredients with selective release through the use of plant-derived nanomaterials can result in the sustained delivery of antioxidants, countering oxidative stress in foods. Rather than just ensuring protection against spoilage, this futuristic thrust is also ensuring that bioactive nutrients remain bioavailable for consumption. Beyond bioavailability, studies have also shown that stabilizing encapsulation of these compounds, ensuring their efficacy. In addition, appropriate release control of these antioxidants could significantly reduce the dependence on artificial preservatives, thereby addressing consumers' demands for cleaner labels and healthier products. Thus, controlled-release technologies represent a remarkable convergence between scientific advancement and public health to drive smarter food preservation technologies. In general integration of controlled release mechanisms into food preservation technologies promises bright approaches for the improvement of both storage and nutritional quality of products.

12.8 Challenges and Future Perspectives

Antioxidant biomolecules are very important for the protection of cells against the assault of oxidative stress linked to some chronic diseases and aging processes. These are the natural components, mostly derived from plant sources, required to neutralize free radicals and inhibit outcomes of damage at the cellular level. Recently, increasing efforts have been geared towards the application of such biomolecules in plant-based nanomaterials functioning as delivery systems, which seem to represent an exciting leap towards improving their bioavailability and efficacy. Of great importance, technological advancement development in nanotechnology applying fungal types has been a new way to foster the sustainable production of antioxidant nanostructures. Well, the prosperity of such approaches provides better solutions to predicaments and derails many limitations suffered using conventional antioxidant applications, thus creating an opportunity for biogenic factories to form new materials as by-products from waste. Natural processes coupled with nanotechnology present further ground for research in improved delivery systems, ultimately aiming at enhancing the therapeutic applications of antioxidant biomolecules in medicine and agriculture (Griffin et al. 2017).

Antioxidant biomolecules are efficacious; however, there are problems with their stability and bioavailability. Most antioxidants are easily degraded due to light, temperature, and oxygen, resulting in poor therapeutic efficacy. The hydrophobic characteristics of these antioxidants, which require newer delivery systems for bioavailability enhancement, impeded their absorption at the cellular level. The application of plant-derived nanomaterial is the best option for providing a protective covering for these sensitive biomolecules to stabilize them from environmental stress. Research on green synthesis of nanoparticles has shown in recent studies extravagance towards improving stability. The strategies prescribed in the literature, including using smaller-sized metal nanoparticles with tailored surface coatings, maximise efforts to support the intended development of delivery systems that seek to reduce ecological toxicity while enhancing antioxidant stability and bioavailability (Balusamy et al. 2023; Ahmad et al. 2024).

12.9 Conclusion

Biomaterials and nanomaterials of plant origin have emerged as promising new avenues for advancing antioxidant delivery systems beyond the drawbacks associated with conventional means such as poor bioavailability and instability. These inherently biocompatible and biodegradable materials with excellent functional versatility make eco-friendly and sustainable alternatives for biomedical and agricultural applications. They are plant-list nanomaterials into controlled release systems and enhance the stability and bioavailability of antioxidants for targeted delivery and sustained release, thus improving therapeutic outcomes.

Besides, more recent developments of green syntheses and surface modification techniques have further ameliorated such delivery systems, thus making them much less harmful to the environment while much more functional. Size-dependent release, modulation of surface charge, and environmental responsiveness are the mechanisms used to provide efficient and controlled delivery of antioxidants to their specific sites of oxidative stress. The application ranges from medicine to food preservation and agriculture and encompasses the improvement of health as well as its management and promotion of sustainable practices.

While very much progress has been made, there are still lots of challenges remaining especially on scaling production, its long-term environmental impacts as well as regulatory considerations. Future investigations should pay much attention to improvement on synthesis methods, identifying new plant materials, and incorporating kinetic modeling for better developed release profiles. Bringing these innovations to bear can potentially open doors to greater use for antioxidants and much sustainable innovation in all fields.

References

Adamec V, Pavlikova S, Prochazkova D, Zeman T (2019) Current approaches of occupational and safety health management in work environments containing nanoparticles. Saf Eng Ser 14(2):25–37

Ahmad S, Ahmad S, Ali S, Esa M, Khan A, Yan H (2024) Recent advancements and unexplored biomedical applications of green synthesized Ag & AU nanoparticles; a review. Int J Nanomedicine 19:3187–3215. https://doi.org/10.2147/IJN.S453775

Bachini F, Mura S, Sedddaiu G, Roggero PP, Greppi GF (2013) Advances of nanotechnology in agro-environmental studies. Ital J Agron 3(8). https://doi.org/10.4081/ija.2013.e18

Balusamy SR, Joshi AS, Perumalsamy H, Mijakovic I, Singh P (2023) Advancing sustainable agriculture: a critical review of smart and eco-friendly nanomaterial applications. J Nanobiotechnol 21:372. https://doi.org/10.1186/s12951-023-02135-3

Cascione M, Martano S, De matteis M, Rinald R (2022) Nanomaterials versus polymer-based nanoparticles for overoing neurodegeneration. Nanomaterials 12:2337. https://doi.org/10.3390/nano12142337

Cheng HN, Richard AG, Patrick BS (2018) Green polymer chemistry: pipelines toward new products and processes. In: Am Chem Soc, vol 1310, pp 1–11. https://doi.org/10.1021/bk-2018-1310.ch001

Chuacharoen T (2015) Release, stability and functionality of nanodelivered hydrophobic (lutein and beta-carotene) and hydrophilic (folic acid) antioxidants using zein nanoparticles. LSU Doctoral Dissertations, p 297. https://repository.lsu.edu/gradschool_dissertations/297

Costa M, Sezgin-Bayindir Z, Losada-Barreiro S, Paiva-Martins F, Saso L, Bravo-Díaz C (2021) Polyphenols as antioxidants for extending food shelf-life and in the prevention of health diseases: encapsulation and interfacial phenomena. Biomedicine 9:1909. https://doi.org/10.3390/biomedicines9121909

Díez-Pascual AM (2022) Surface engineering of nanomaterials with polymers, biomolecules, and small ligands for Nanomedicine. Materials 15(9):3251. https://doi.org/10.3390/ma15093251

Durazzo A, Nazhand A, Lucarini M, Atanasov AG, Souto EB, Novellino E, Capasso R, Santini A (2020) An updated overview on nanonutraceuticals: focus on nanoprebiotics and nanoprobiotics. Int J Mol Sci 21(7):2285. https://doi.org/10.3390/ijms21072285

Fitrani L, Saisabila H, Zaini E (2021) Recent strategies for improving solubility and oral bioavailability of piperine. Int J Appl Pharm 13(4). https://doi.org/10.22159/ijap.2021v13i4.41596

Gaba E, Asimeng B, Kaufmann E, Katu S, Foster E, Tiburu (2021) Mechanical and structural characterization of pineapple leaf fiber. Fibers 9(8):51. https://doi.org/10.3390/fib9080051

Gayathri NP, Prasad G, Prabhakaran V, Priya V (2024) Understanding the impact of microplastic contamination on soil quality and eco-toxicological risks in horticulture: a comprehensive review. Case Stud Chem Environ Eng 9:article 100633. https://doi.org/10.1016/j.cscee.2024.100633

Griffin S, Masood MI, Nasim MJ, Sarfraz M, Ebokaiwe AP, Schäfer KH, Keck CM, Jacob C (2017) Natural nanoparticles: a particular matter inspired by nature. Antioxidants (Basel) 29;7(1):3. https://doi.org/10.3390/antiox7010003. PMID: 29286304; PMCID: PMC5789313

Hanna SK, Sims C, Heller DA, Christopher P, Horoszko Johnson ME, Bustos ARM, Reipa V, Kathryn RR, Nelson BC (2017) Redox-active nanomaterials for nanomedicine applications. Nanoscale 9(40):15226–15251. https://doi.org/10.1039/c7nr05429g

Haritha H, Bhagyalakshi J (2014) Green synthesis and characterization of silver nanoparticles using Pterocarpus marsupium and assessment of its in vitro antidiabetic activity. Am J Adv Drug Deliv 5(03):118–130. http://www.imedpub.com/advanced-drug-delivery/

Jampilek J, Placha G (2019) Graphenic materials for biomedical applications. Nanomaterials 9(12):1758. https://doi.org/10.3390/nano9121758

Kiwumulo HF, Muwonge H, Ibingira C, Lubwama M, Kirabira JB, Seekitoleko RH (2022) Green synthesis and characterization of iron-oxide nanoparticles using *Moringa oleifera*: a potential protocol for use in low and middle-income countries. BMC Res Notes 15:149

Kurowiak J, Klekiel T, Bedzinski R (2023) Biodegradable polymers in biomedical applications: a review -developments, perspectives and future challenges. Int J Mol Sci 42:16952. https://doi.org/10.3390/ijms242316952

Liu Z, Ren Z, Zhang J, Chaung CC, Kandaswamy E, Zhou T, Zuo L (2018) Role of ROS and nutritional antioxidants in human diseases. Front Bioeng Biotechnol 9:477. https://doi.org/10.3389/fphys.2018.00477

Mady M (2021) The antioxidant & antimicrobial activities of cornsilk extract encapsulated in polysaccharide nanoparticles. American University in cairo AUC knowledge Fountain. https://fount.aucegypt.edu/etds/1526

Magli E, Perissutti E, Santagada V, Caliendo G, Corvino A, Esposito G, Esposito G, Fiorino F, Migliaccio M, Scognamiglio A, Severino B, Sparaco R, Frecentese F (2021) H_2S donors and their use in medicinal chemistry. Biomolecules 11(12):1899. https://doi.org/10.3390/biom11121899

Matencio A, Hoti G, Pedrazzo AR, Cecone C, Appleton SL, Monfared YK, Caldera F, Trotta F (2022) Nutraceutical concepts and dextrin-based delivery systems. Int J Mol Sci 23(8):4102. https://doi.org/10.3390/ijms23084102

Maurizzi E, Bigi F, Quartieri A, De Leo R, Volpelli LA, Pulvirenti A (2022) The green era of food packaging: general considerations and new trends. Polymers 14(20):4257. https://doi.org/10.3390/polym14204257

Mhettar P, Kale N, Pantwalawalkar J, Nangare S, Jadhav N (2024) Metal-organic frameworks: drug delivery applications and future prospects. ADMET DMPK 17;12(1):27–62. https://doi.org/10.5599/admet.2057. PMID: 38560715; PMCID: PMC10974818

Mohammadi Dargah M, Pedram P, Cabrera-Barjas G, Delattre C, Nesic A, Santagata G, Cerruti P, Moeini A (2023) Biomimetic synthesis of nanoparticles: a comprehensive review on green synthesis of nanoparticles with a focus on Prosopis farcta plant extracts and biomedical applications. Adv Colloid Interf Sci 332. https://doi.org/10.1016/j.cis.2024.103277

Murphy A (2015) An investigation into the biological responses induced following oral exposure to silver nanoparticles. Doctoral thesis, Technological University Dublin. https://doi.org/10.21427/D73K5W

Patil VW, Suryawanshi NS, Bokhad MN, Parsodkar VJ, Narhe AS (2024) Engineered nanomaterials in plant protection: their controlled, site-directed delivery and phytotoxicity. Curr Bot 15:7–24. https://doi.org/10.25081/cb.2024.v15.7906

Pudjiastuti P, Wafiroh S, Hendradi E, Darmokoesoemo H, Harsini M, Fauzi M, Nahar L, Sarker S (2020) Disintegration, *in vitro* dissolution, and drug release kinetics profiles of $_{k}$-carrageenan-based nutraceutical hard-shell capsules containing Salicylamide. Open Chem 18(1):226–231. https://doi.org/10.1515/chem-2020-0028

Puttasiddaiah R, Lakshminarayana R, Somashekar NL, Gupta VK, Inbaraj BS, Usmani Z, Raghavendra VB, Sridhar K, Sharma M (2022) Advances in nanofabrication Technology for Nutraceuticals: new insights and future trends. Bioengineering 9(9):478. https://doi.org/10.3390/bioengineering9090478

Quintero G, Palencia M (2021) "Green" nanomaterials: synthesis, assessment, and their potential application in soil remediation. Research Group in Science with Technological Applications (GI-CAT), Chemistry Department, Faculty of Natural and Exact Sciences, Universidad del Valle, Cali

Rozhin P, Melchionna M, Fornasiero P, Marchesan S (2021) Nanostructured ceria: biomolecular templates and (bio)applications. Nanomaterials 11(9):2259. https://doi.org/10.3390/nano11092259

Souto EB, Fernandes AR, Martins-Gomes C, Coutinho TE, Durazzo A, Lucarini M, Souto SB, Silva AM, Santini A (2020) Nanomaterials for skin delivery of cosmeceuticals and pharmaceuticals. Appl Sci 10(5):1594. https://doi.org/10.3390/app10051594

Teixeira MO, Marinho E, Silva C, Antunes JC, Felgueiras HP (2023) Pullulan hydrogels as drug release platforms in biomedicine. J Drug Deliv Sci Technol 89:105066. https://doi.org/10.1016/j.jddst.2023.105066

Thakur M, Wang B, Verma ML (2020) Development and applications of nanobiosensors for sustainable agricultural and food industries: recent developments, challenges and perspectives. Environ Technol Innovation:102371. www.elsevier.com/locate/eti

Vaiserman A, Loiada A, Zayachkivska A, Lushchak O (2020) Nanodelivery of natural antioxidants: an anti-aging perspective. Front Bioeng Biotechnol 7:447. https://doi.org/10.3389/fbioe.2019.00447

Venugopal J, Sridhar S, Ramakrishna S (2024) Electrospun plant-derived natural biomolecules for tissue engineering. Plant Sci Today 1(3):151–154. https://doi.org/10.14719/pst.2014.1.3.65

Wu H-l, Bremner DH, Li H-y, Shi Q-q, Wu J-z, Xiao R-q, Zhu L-m (2016) A novel multifunctional biomedical material based on polyacrylonitrile: preparation and characterization. Mater Sci Eng 62:702–709

Younas Z, Mashwani ZU, Ahmad R, Khan I, Zaman M, Sawati S, Sohail L (2023) Mechanistic approaches to the application of Nano-zinc in the poultry and biomedical industries: a comprehensive review of future perspectives and challenges. Molecules *28*(3):1064

Chapter 13
Nano-scale Physics of Controlled Release Pesticide Diffusion in Agriculture

Sourodeep De, Anindita Seth, Prasanta Murmu, and Sabyasachi Ghosh

13.1 Introduction

Pesticides play a ubiquitous role in modern agricultural practices to ensure global food security (Cooper and Dobson 2007; Sharma et al. 2019; Li et al. 2021). There are above 1000 of pesticides used worldwide (WHO 2022) to increase yields of crops by preventing or controlling weeds, pests, diseases and also plant pathogens (Damalas and Eleftherohorinos 2011), which cause significant damage of crops and foods (Oerke 2005). Pesticides are a broad group of chemicals including herbicides, insecticides, fungicides and others used in agricultural and forestry sector to manage or prevent pests, the vectors of plants, animal and humans (Smalling et al. 2013; Solomon et al. 2014). Globally, among all pesticides usage, herbicides account for the highest *i.e.* 40%, followed by insecticides (17%), and fungicides (10%) while the other pesticides comprises 33% on average (Ali et al. 2021). As per the available data reported by the Ministry of Agriculture and Farmers Welfare, Government of India with the help of the State Governments and Union Territories, it is noticed that consumption of pesticides has been varied respective to years and also regions depending on the need for plant protection of growing crops while in the state of Maharashtra the pesticides consumption has been reported as the maximum among all states (Fig. 13.1).

Pesticides are broadly classified into five different groups i.e. carbamates, organochlorines, organophosphates, pyrethroids and phenylpyrazoles (IRAC 2017). Again, each pesticide exhibits unique properties and toxicological effects. The toxicity of a pesticide can be explained by two key factors viz., dose as well as time which refer the existing concentration of pesticide or its residue and the exposure

S. De · A. Seth · P. Murmu · S. Ghosh (✉)
Department of Physics, Indian Institute of Technology Bhilai, Bhilai, Chhattisgarh, India
e-mail: sabya@iitbhilai.ac.in

V. Ghosh et al. (eds.), *Applications of Plant Bioactive Compounds-Based Nanoformulations*, Nanotechnology in the Life Sciences,
https://doi.org/10.1007/978-3-032-13086-0_13

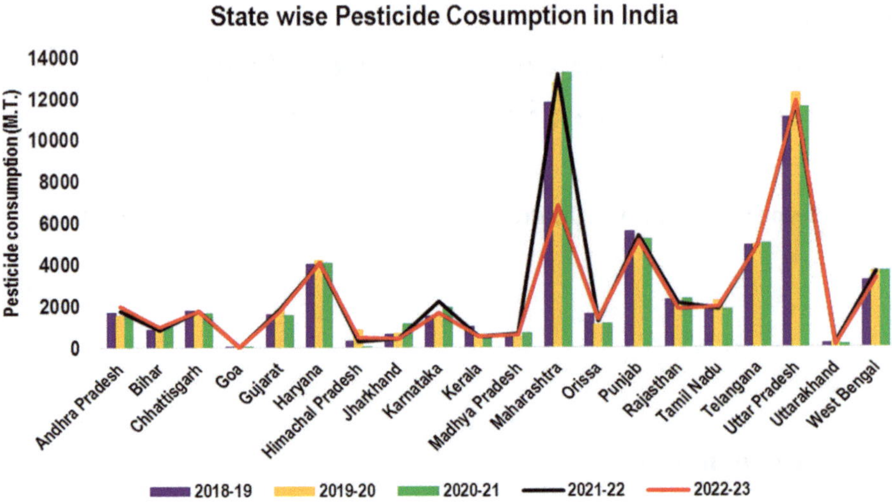

Fig. 13.1 Pesticide consumption in India. (Source: https://ppqs.gov.in/statistical-database)

effect of the same. These two variables are the cause of two main types of toxicity viz., acute toxicity and chronic toxicity. Acute toxicity describes the harmful effects of a pesticide due to short-term exposure. Pesticides with high acute toxicity can be lethal even while it is used in a very small quantity. Acute toxicity is usually measured in three ways like (i) acute oral toxicity, (ii) acute dermal toxicity, and (iii) acute inhalation toxicity. Chronic toxicity can be caused by the delayed poisonous effects resulting from continuous or long-term exposure to a pesticide over a certain or extended period of time. The symptoms may not appear immediately due to chronic toxicity unlike acute toxicity causes. Chronic toxicity is a major concern for human and those who work directly with pesticides, as exposure can occur through contaminated soil, water, air or by food. The harmful effect of pesticide to human lead to severe hazards by causing several chronic diseases like heart problems, respiratory problem, neurological disorder, paralysis, and even cancer (Garud et al. 2024). Hence, the indiscriminate use of pesticides causes environmental contamination due to a large distribution of pesticide residues in soil, water and air (FAO 1990), which is a menace for future sustainability. Such residues are prevailed as destructive as these can persist in the environment for a longer time resulting a direct or indirect effect on non-target organisms which in turns creates an imbalance in the ecosystem (Rosell et al. 2008).

13.1.1 Environmental Pollution Due to Pesticide

Pesticides currently threaten our ecosystem and hampers its function as well as structure leaving negative impacts on its components (Chin 2010). The persistent organic pollutants (POPs) can spread and contaminate worldwide because of their physico-chemical properties, causing a potential threat to humans and wildlife both. There are reported 16 POPs only as pesticides among officially recognized 24 POPs as per Stockholm convention. These types of pollutants are carbon based organic chemicals exposing as severely hazardous which can persist in the environment for long-term, even their bio-accumulation is possible in living organisms. Pesticides categorized as POPs, for example aldrin, DDT, dieldrin, endrin, chlordane, chloropyriphos, hexachlorobenzene, mirex, heptachlor, toxaphene, and etc., can be airborne and can travel a long distances, posing a greater and prolonged risks to human health and ecosystems. They are long-lived, and resistant to degradation, leading to their widespread distribution and long-term presence in the environment.

The abiotic (soil, water and air) and biotic (humans, plants and animals) components of the environment are under a major concern due to pesticide pollution (Rosell et al. 2008; Bhunia 2017) and its lingering effect in contrary to rapid economic and technological development with increased population (Ali et al. 2021; Rasool et al. 2022). Pesticides are distributed and retained in all components of the environment and even affect the non-target organisms in addition to causing severe health hazard to humans (Pastor et al. 2003). Pesticide residues, persist in soils, can cause detrimental effects on soil biota causing a negative effect on soil quality and soil health (Mu et al. 2023). Soil microorganisms as well as soil fauna together termed as soil biota, crucially helps to perform essential soil functions such as nutrient cycling, soil quality, and manages overall biodiversity; all are key indicators of soil health (Bhandari et al. 2021). However, persistence of pesticide residues in the soil can cause significant threats to soil organisms by disrupting their gene expression and disturbing enzymatic activity, which can lead to reduced fecundity, stunted growth, and lower survival rates resulting ecological risks for soil biota and soil health ultimately (Mu et al. 2023).

Again polarity of pesticides governs or assess the tendency of interaction with environment and health components. Polar pesticides tend to exhibit a greater risk to aquatic environment by contaminating aquifers while nonpolar pesticides are more hazardous in terms of long-term persistence in environment causing health risks due to bioaccumulation in living organisms cells (Lainsbury 2018). Pesticides can contaminate organic matter, the major food source of aquatic lives; it can be further adsorbed and accumulated in aquatic environments particularly on the sedimentary layers, leading to degradation of ecosystem causing water pollution (AbuQamar et al. 2024). The water pollution also caused when rain and irrigation water transports pesticides into surface waters and aquatic ecosystem (De Vlaming et al. 2000). Eventually, uncontrolled use of pesticide can enter into food chain and further can be stored into animal or human tissues (Liu et al. 2016) during the unknowing intake the contaminated foods (Gullick et al. 2016) and drinking water (Mekonen et al. 2016) resulting a higher risk to non-target objects. The efficacy of pesticide depends on the fate of the same, as influenced by several possible physical and chemical process like evaporation, surface runoff, leaching, degradation, photolysis, and others (Hassaan and Nemr 2020; Pathak et al. 2022). It is essential to monitor the production, application and distribution of pesticides with necessary regulation to control the undesired fate of pesticides in order to manage intrinsic toxicity and deliberate spreading in the environment. Controlled-release (CR) formulations can be an alternative to regulate the release of active ingredients at optimum level at a desired concentration for the target-object for a required time-period (Li et al. 2021; Huang and Jiang 2024). This can minimize dissipation or degradation or other losses of pesticides and enhance the efficacy potentially with a higher precision.

13.2 Diffusion Physics of Pesticides

It is rational to monitor the transport phenomena of pesticides during applications to the targeted objects to avoid undesirable environmental pollution aiming to safety and sustainability. Diffusion plays a pivotal role to predict the movement and distribution of pesticides which influences its efficiency or undesirable fate predicting benefit or hazards respectively (Green and Obien 1969; Bhandari et al. 2020; Tudi et al. 2021). Pesticide can transport through both of the vapor phase and solution phase in porous soil as well as can be even adsorbed onto soil solid which in turns makes possible the movement of the pesticide either along the solid surfaces or through the solid phase. Furthermore, once the pesticide gets adsorbed onto the soil solid, it tends to diffuses in a significant slowest manner there compared to does in the soil pores which in turns makes it unavailable for diffusion and even accounts for the negligible contribution to the total diffusion flux. Eventually, a significant amount of the applied concentration of pesticide remains strongly adsorbed onto the soil (Graham-Bryce 1969). When pesticide molecules are applied and heterogeneously distributed in soil in macroscopic scale, they tend to diffuse from a higher concentration zone towards lower concentration one due to the difference in chemical potential in the presence of concentration gradient. Diffusion phenomena of pesticide can be well explained by macroscopic physics of diffusion, and also from a microscopic physics point of view, considering a simple relaxation time approximation based kinetic theory or standard Einstein's theory of molecular diffusion. The rate or speed of diffusion of pesticides is governed by the physical and chemical processes in the soil like adsorption, desorption or repulsion and again physico-chemical properties of the pesticide, viz. solubility and volatility, influence the diffusion coefficient of the pesticide in the vapor phase and solution phase in soil. The diffusion phenomena becomes more complex in soil carrying variable charges i.e. both positive and negative surface charges as there occurs further interaction between pesticide ions/ molecules and surface charges. Under static conditions, if the ion concentration at initial point is higher than at final point, comparatively more ions will migrate from initial to final point than that in the opposite direction because of random thermal motion. The macroscopic framework is governed by Fick's law, where diffusion coefficient is entered as an important input and in the microscopic framework of diffusion coefficient, it is governed by Einstein's relation (Fig. 13.2). But for the concept of estimating diffusion coefficient for real soil medium instead of water there also needs to be incorporate some soil factors like tortuosity.

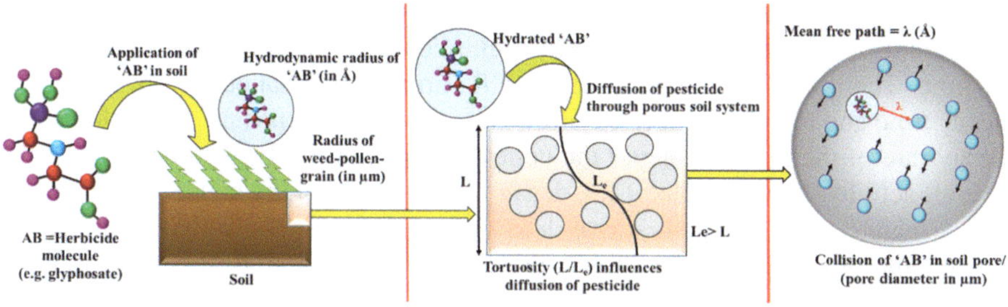

Fig. 13.2 Schematic of pesticide diffusion in soil from macroscopic and microscopic point of view

The net diffusion flux (J) per unit time along a unit area can be expressed by Fick's first law of diffusion: $J = -D\frac{\partial C}{\partial x}$, where $\frac{\partial C}{\partial x}$ is called the concentration gradient and D is termed as the diffusion coefficient (Yaron 1989). The negative sign in the equation indicates the direction of the flux from higher concentration to lower one *i.e.* opposite to that of the concentration gradient. Fick's first diffusion law can be applied in homogeneous and heterogeneous medium both such as soil. Considering the law of continuity, $\frac{\partial J}{\partial x} = \frac{\partial C}{\partial x}$, Fick's first law can be extended as a function of time (t) by developing Fick's second law as $\frac{\partial C}{\partial t} = D\frac{\partial^2 C}{\partial x^2}$. The Fick's second law solutions is included soil geometry factor, initial condition and the boundary of soil (Graham-Bryce 1969).

13.2.1 Physics of Control Release Formulation of Pesticides

Traditional agricultural practices often involve the use of pesticides in large quantities or in their pure form, typically without the aid of adjuvants which are the substances that enhance the effectiveness or stability of pesticides. One of the major drawbacks of this conventional approach is the significant loss of pesticides during application. These losses are influenced by factors such as the application method and prevailing weather conditions, leading to reduced efficiency, environmental contamination and increased costs. The unregulated use or application of pesticides involve inefficient diffusion causing pollution in environment and food systems (Rasool et al. 2022; Singh et al. 2020). To avoid environmental contamination aiming for safety and sustainability, it is desirable to optimize the spread of pesticide beyond the intended area. The controlled release (CR) formulation of a pesticide regulates the diffusion process in sustained manner through the porous soil matrix or cell membrane by controlling the release rate of active ingredients. The release rate of CR formulations influence the time constant over which the activity of the material is produced.

The challenge of CR formulations is to control the concentration of the active ingredients which is both influenced by their release and decay over a certain period of time. The changes in activity of the CR formulations happens with respect to decay time and toxicity with release time upon pesticide exposure. The release and decay rate of the material as per the controlled release mechanism determines the overall activity. For example, if the decay rate is low, then the concentration of the active material follows the first rate decay principle where the activity depends on the amount of active material already present. However, if the decay rate is comparable to the time scales involved, a more intricate interplay of the two competing process will occur resulting in the rate of release to be dependent on the release time which is a characteristic of the controlled release system and the time at which any given measurement is taken. The active substance, for instance, a herbicide first gets dispersed through a granule and when it is applied in an agricultural field, it gets released into the soil via diffusion. The solution for this process is largely determined by the error function solution of Fick's law governed by the boundary conditions like the slab and the sphere (Collins and Doglia 1973). Let us explore the theoretical framework for the changes in activity of the CRFs with respect to decay time and toxicity with release time.

13.3 Mathematical Formalism of Control Release of Pesticides

Let us consider granules with radius R, then its spherical volume will be

$$V = \frac{4}{3}\pi R^3$$

After time t, the radius is r. The rate of change of volume is

$$\frac{dV}{dt} = 4\pi r^2 \frac{dr}{dt}$$

Assuming that $\frac{dV}{dt}$ is proportional to the area of the granule $4\pi r^2$, implies that $\frac{dr}{dt}$ is a constant. We have τ as the release time for the complete degradation.

$$\tau = \frac{\frac{-R}{dr}}{dt} = \frac{-Rdt}{dr}$$

The negative sign implies that the radius decreases with time. Thus for r we have

$$r = R\left(1 - \frac{t}{\tau}\right)$$

And we get the rate of release

$$\frac{dV}{dt} = \frac{3V}{\tau}\left(1 - \frac{t}{\tau}\right)^2$$

We see that the release rate decreases with time as $t \to \tau$. The total amount A released in time t is

$$A = \int_0^t \frac{dV}{dt'} dt'$$

The herbicides have their unique characteristic decay times over which they decay and it actually depends on the chemical composition of the same and also on the release medium with the environmental factors. Thus, to study the effects of controlled release, we need to consider the degradation to comprehend the mechanics of controlled release of herbicides in the environment. We consider first order rate law for the process of studying the degradation, where the total amount of degraded material depends on the amount present at any time. Thus,

$$\frac{dA}{dt} = -\lambda A$$

With λ being the rate constant and A being the amount present at time t.

$$A = A_0 e^{-\lambda t}$$

A_0 is the amount initially present. The half life comes as

$$T_{\frac{1}{2}} = \frac{\ln 2}{\lambda}$$

Considering the inflow of herbicides and its controlled release via outflow and degradation, the net concentration at any time t is

$$C = \frac{3V}{\tau}\int_0^t \left(1 - \frac{t'}{\tau}\right)^2 e^{-\lambda(t'-\tau)} dt'$$

The analytic solution is

$$C = \frac{3V}{\lambda^3 \tau^2}\left(-e^{-\lambda t}\left[(\lambda t + 1)^2 + 1\right] + \left[(\lambda(t - \tau) - 1)^2 + 1\right]\right) \quad \text{when } t \le \tau$$

and

$$C = C(\tau)e^{-\lambda(t-\tau)} \quad \text{when } t > \tau$$

The surface concentration has the units of kg/ha, g/m^2 etc. In the above expressions, we can see that before and after the release time or characteristic time τ, control release concentration expressions are different.

13.3.1 Fick's Law of Diffusion

In the one-dimensional diffusion in isotropic substance, the rate of transfer of the diffusing substance per unit area is proportional to the concentration gradient normal to that area. This is Fick's first law

$$F = -D\frac{\partial C}{\partial x}$$

F is the rate of transfer per unit area, C is concentration of diffusing substance and D is the diffusion coefficient. We assume the diffusion occurs in the x direction. To consider the change in gradient with position x, we have Fick's second law (in 3D picture)

$$\frac{\partial C}{\partial t} = D\left(\frac{\partial^2 C}{\partial x^2} + \frac{\partial^2 C}{\partial y^2} + \frac{\partial^2 C}{\partial z^2}\right) = D\nabla^2 C,$$

Where ∇ is the divergence gradient operator in Cartesian coordinates. We focus on the active substance (in our case the herbicide), which is dispersed initially through a granule and gets released by outward diffusion into the soil. The usage of Fick's law involves the boundary conditions along with the shape of the granule. We consider the cases of the slab which involves one dimensional diffusion and the sphere which includes radial diffusion. The boundary conditions are at $t = 0$, $C = (C_0) =$ constant throughout the pellet and is zero outside. For $t \geq 0$, C is sensibly zero outside. This (boundary) condition is applicable for materials which migrate from the bounding matrix such that the concentration is maintained at a low level (Damalas and Eleftherohorinos 2011).

13.3.2 Slab

Accurate treatments of the Fick's law involve the error function. The solution for a one-dimensional Fick's law for the slab is

$$C = \frac{A}{t^{\frac{1}{2}}} e^{\frac{-x^2}{4Dt}}$$

with the general solution being linear combinations of solutions meeting the boundary conditions.Here, C is the concentration of pesticides, t is the time elapsed during spreading, D is the diffusion coefficient of the pesticides, x is the distance from the source where the pesticides are applied, and A is a constant.

At t = 0, the concentration C exhibits an infinitely high value at x = 0 and is zero elsewhere, a behavior characterized by the Dirac delta function. Accordingly, the initial condition is expressed as-

$$C(x,0) = C_0 \delta(x)$$

With increasing time (t), the concentration diffuses symmetrically about x = 0, forming a Gaussian distribution profile but total amount of the concentration will be constant all over time, Mathematically shows as like this-

$$\int_{-\infty}^{+\infty} C(x,t)\, dx = C_0 \int_{-\infty}^{+\infty} \delta(x) dx = C_0$$

If we plot the concentration (C) of pesticides versus position (x) at a particular time (t = 0.5 s) and temperature (T = 25 °C), we obtain a graph like this-

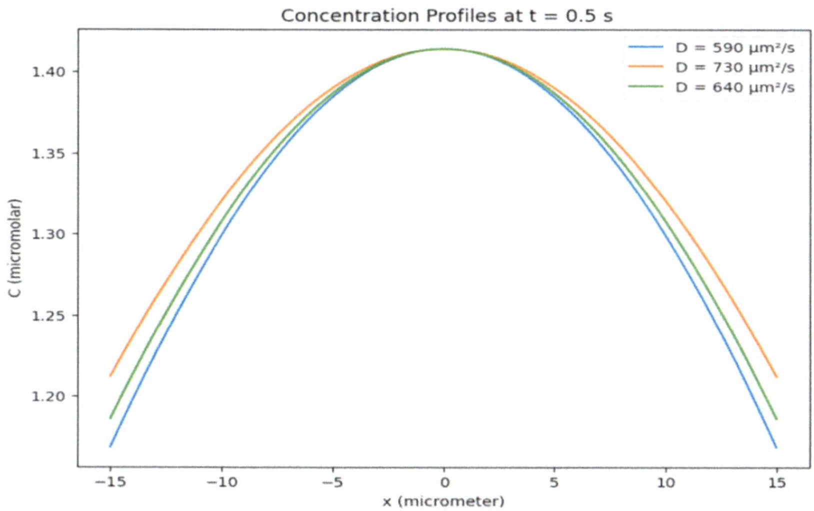

Three different types of pesticides are used, such as Cyromazine, Chlorotoluron, Pirimicarb, with diffusion coefficients 0.73 × $10^{-9} m^2/s$, 0.64 × $10^{-9} m^2/s$, 0.59 × $10^{-9} m^2/s$, respectively. In the graph, Pirimicarb, Cyromazine, and Chlorotoluron are represented by blue, orange, and green colours, respectively. The data table presents the concentration (C) versus distance (x) values for these three pesticides.

Position (x, μm)	Concentration of Pesticides (C, μM)		
	Pirimicarb	Chlorotoluron	Cyromazine
−15	1.1687	1.1862	1.2122
−14	1.1978	1.2134	1.2366
−13	1.2255	1.2393	1.2596
−12	1.2517	1.2637	1.2814
−11	1.2764	1.2867	1.3017
−10	1.2993	1.3079	1.3206
−9	1.3204	1.3275	1.3379
−8	1.3396	1.3452	1.3536
−7	1.3567	1.3611	1.3675
−6	1.3717	1.3750	1.3798
−5	1.3846	1.3869	1.3902
−4	1.3952	1.3966	1.3988
−3	1.4035	1.4043	1.4055
−2	1.4094	1.4098	1.4103
−1	1.4130	1.4131	1.4132
0	1.4142	1.4142	1.4142
1	1.4130	1.4131	1.4132
2	1.4094	1.4098	1.4103
3	1.4035	1.4043	1.4055
4	1.3952	1.3966	1.3988
5	1.3846	1.3869	1.3902
6	1.3717	1.3750	1.3798
7	1.3567	1.3611	1.3675
8	1.3396	1.3452	1.3536
9	1.3204	1.3275	1.3379
10	1.2993	1.3079	1.3206
11	1.2764	1.2867	1.3017
12	1.2517	1.2637	1.2814
13	1.2255	1.2393	1.2596
14	1.1978	1.2134	1.2366
15	1.1687	1.1862	1.2122

An alternative approach involves assuming the solution to be of the form

$$C = X(x)T(t)$$

The general solution is of the form

$$C = \sum_{m=1}^{\infty} \{A_m \sin(\lambda_m x) + B_m \cos(\lambda_m x)\} e^{-\lambda_m^2 Dt}$$

The solution converges for large t. Next considering the slab of thickness a, the solutions are

$$\frac{C_0-C}{C_0}=\sum_0^{\infty}(-1)^n\, erfc\frac{\frac{(2n+1)a}{2}-x}{2\sqrt{Dt}}+\sum_{n=0}^{\infty}(-1)^n\, erfc\frac{\frac{(2n+1)a}{2}+x}{2\sqrt{Dt}}$$

$$\Rightarrow\frac{C_0-C}{C_0}=1-\frac{4}{\pi}\sum_{n=0}^{\infty}\frac{(-1)^n}{2n+1}e^{\frac{-D(2n+1)^2\pi^2 t}{a^2}}\cos\frac{(2n+1)\pi x}{a}$$

The amount of material left in the slab at time t can be obtained by integrating $C(x)dx$ for x from 0 to a. Considering the amount of diffused material at time t as M_t we have

$$\frac{M_t}{M_\infty}=2\sqrt{\frac{4Dt}{a^2}}\left\{\frac{1}{\sqrt{\pi}}+2\sum_{n=1}^{\infty}(-1)^n\, ierfc\frac{na}{2\sqrt{Dt}}\right\}$$

With M_∞ being the total amount of substance to be eventually released and $ierfcx=\int_x^{\infty}erfc\zeta\, d\zeta=\frac{1}{\sqrt{\pi}}e^{-x^2}-xerfcx.$ Thus we have

$$\frac{M_t}{M_\infty}=1-\sum_{n=0}^{\infty}\frac{8}{(2n+1)^2\pi^2}e^{-\frac{D(2n+1)^2\pi^2 t}{a^2}}$$

13.3.3 Sphere

For the spherical case we have the Laplacian as $\nabla^2=\frac{\partial^2}{\partial r^2}+\frac{2}{r}\frac{\partial}{\partial r}$. The angular terms are discarded considering the radial symmetry. Putting $u = rC$, we have

$$\frac{\partial C}{\partial t}=D\left(\frac{\partial^2 C}{\partial r^2}+\frac{2}{r}\frac{\partial C}{\partial r}\right)$$

which for the one-dimensional case becomes

$$\frac{\partial u}{\partial t}=D\frac{\partial^2 u}{\partial r^2}$$

The solutions are

$$\frac{C_0-C}{C_0}=\frac{a}{r}\sum_{n=0}^{\infty}\left[erfc\frac{(2n+1)a-r}{2\sqrt{Dt}}-erfc\frac{(2n+1)a+r}{2\sqrt{Dt}}\right]$$

$$\Rightarrow \frac{C_0 - C}{C_0} = 1 + \frac{2a}{\pi r}\sum_{n=1}^{\infty}\frac{(-1)^2}{n}\sin\left\{\frac{n\pi r}{a}\right\}e^{\frac{-Dn^2\pi^2 t}{a^2}}$$

Where a is the radius. The diffused amount depends on the integration of the concentration through the sphere

$$\frac{M_t}{M_\infty} = 6\sqrt{\frac{Dt}{a^2}}\left[\frac{1}{\sqrt{\pi}} + 2\sum_{n=1}^{\infty} ierfc\frac{na}{\sqrt{Dt}}\right] - \frac{3Dt}{a^2}$$

$$\frac{M_t}{M_\infty} = 1 - \frac{6}{\pi^2}\sum_{n=1}^{\infty}\frac{1}{n^2}e^{\frac{-Dn^2\pi^2 t}{a^2}}$$

13.3.4 Active Material Concentration Due to Release

In absence of controlled release, we have $C = C_1 e^{-\lambda t}$ with λ as the decay constant. Thus,

$$\lambda = \frac{\ln 2}{T_{\frac{1}{2}}}$$

where the half-life indicates the persistence of the substance.

The current concentration of the active substance released at a rate $M_t^{\cdot}$, given by Collins et al. (1973);

$$M(t) = \int_0^t M_{t'}^{\cdot} e^{-\lambda(t-t')}dt'$$

The first class (*erf*) solutions lead to non-integrable integrals and hence the approximation of $t^{\frac{1}{2}}$ is valid. The second class solutions are

$$M_{t'}^{\cdot} = M_\infty \frac{8D}{a^2}\sum_{n=0}^{\infty} e^{\frac{-D(2n+1)^2\pi^2 t}{a^2}}$$

$$M(t) = M_\infty \frac{8D}{a^2}\int_0^t \sum_{n=0}^{\infty} e^{\frac{-D(2n+1)^2\pi^2 t}{a^2}} e^{-\lambda(t-t')}dt'$$

$$= M_\infty 8D\sum_{n=0}^{\infty}\left\{\frac{e^{-\frac{D(2n+1)^2\pi^2 t}{a^2}} - e^{-(\lambda t)}}{\lambda a^2 - D(2n+1)^2\pi^2}\right\}$$

Substituting $\lambda = \frac{\ln 2}{\tau_0}$, with τ_0 as the released material's half life and $\tau = \frac{a^2 \ln 2}{D\pi^2}$ as the effective persistence of the material undergoing controlled release, for the slab we have

$$M(t) = \frac{8}{\pi^2} M_\infty \sum_{n=0}^{\infty} \frac{e^{-\frac{(2n+1)^2 t \ln 2}{\tau}} - e^{-\frac{t \ln 2}{\tau_0}}}{\frac{\tau}{\tau_0} - (2n+1)^2}$$

For the spherical case, we have the *erf* solutions which are intractable equations. The Fourier solution leads to,

$$M_t(t) = \frac{6DM_\infty}{a^2} \sum_{n=1}^{\infty} e^{\frac{-Dn^2\pi^2 t}{a^2}}$$

$$M(t) = \frac{6DM_\infty}{a^2} \int_0^t \sum_{n=1}^{\infty} e^{\frac{-Dn^2\pi^2 t'}{a^2}} e^{-\lambda(t-t')} dt'$$

$$M(t) = \frac{6DM_\infty}{a^2} \sum_{n=1}^{\infty} \frac{e^{\frac{-Dn^2\pi^2 t}{a^2}} - e^{-\lambda t}}{\lambda - \frac{Dn^2\pi^2}{a^2}}$$

Considering the τ and τ_0, we have for the sphere

$$M(t) = \frac{6DM_\infty}{a^2} \sum_{n=1}^{\infty} \frac{e^{\frac{-n^2 t(\ln 2)}{\tau}} - e^{\frac{-t(\ln 2)}{\tau_0}}}{\frac{\ln 2}{\tau_0} - \frac{n^2 \ln 2}{\tau}}$$

$$M(t) = \frac{6M_\infty}{\pi^2} \sum_{n=1}^{\infty} \frac{e^{-n^2 t\frac{(\ln 2)}{\tau}} - e^{-t\frac{(\ln 2)}{\tau_0}}}{\frac{\tau}{\tau_0} - n^2}$$

τ_0 is chosen equal to one. For $\tau_0 \neq 1$, time is read as $\frac{t}{\tau_0}$ and τ as $\frac{\tau}{\tau_0}$. The natural unit of time is half life τ_0 of the active material. We seen the long term effect of diffusion for materials dissipating after release via a first order rate.

13.3.5 Application of Pesticides for Optimum Efficiency

For optimum efficiency in the use of a pesticide, we need to consider the circumstances in which the application is to be done. For example, if the pest is already present, a large but transient concentration is desirable. However, if the problem is pre-emergent, maintenance of an effective level is required.

We are interested in the maintenance of active concentration M' for time t'. Long term diffusion results are equivalent to an increased persistence of the active material. We need to obtain the value of τ which gives the maximum efficiency, followed by the determination of the efficiency.

Considering the diffusion controlled release as instant release material with $\tau > \tau_0$. The graph with the logarithm of concentration *vs* time is obtained as a straight line passing over M', t'. To obtain an initial concentration M_0,

$$M_\infty = M_0 \frac{\tau}{\tau_0}$$

Where M_0 is the active initial concentration. $M(t)$ falls exponentially, we have

$$M(t) = M_0 e^{-t\frac{(\ln 2)}{\tau}}$$

$$\frac{M_\infty \tau_0}{\tau} e^{-t\frac{(\ln 2)}{\tau}}$$

and at time t', we have $M' = \frac{M_\infty \tau_0}{\tau} e^{-t'\frac{(\ln 2)}{\tau}}$

Solving for M_∞, representing the actual amount of active material applied, we have

$$M_\infty = \frac{M\tau}{\tau_0} e^{t'\frac{(\ln 2)}{\tau}}$$

Optimization for M_∞, implying its derivative with respect to τ is 0, we have

$$\frac{dM_\infty}{d\tau} = \frac{M'}{\tau_0}\left[1 - \frac{t' \ln 2}{\tau}\right] e^{t'\frac{(\ln 2)}{\tau}} = 0$$

$$\Rightarrow \tau = t' \ln 2$$

Hence, the amount of required material is

$$M_\infty = \frac{eM' t' \ln 2}{\tau_0}$$

Thus, we find that the amount of required material depends on the time of protection. This indicates constant efficiency for extended missions in this case, rather than a continual decline of efficiency. In fact, we can get a measure of efficiency. The ideal efficiency from Collins et al. (1973).

$$M_{\infty} = M'\left(1 + \frac{t' \ln 2}{\tau_0}\right)$$

Neglecting the constant term, we have ideal efficiency to be

$$Eff_{\text{ideal}} = \frac{M't'\dfrac{(\ln 2)}{\tau_0}}{eM't'\dfrac{(\ln 2)}{\tau_0}} = \frac{1}{e} = 38\%$$

Realistic configurations include

$$Eff_{\text{slab}} = \frac{8}{\pi^2}\frac{1}{e} = 30\%$$

$$Eff_{\text{sphere}} = \frac{6}{\pi^2}\frac{1}{e} = 22\%$$

13.4 Factors Influencing Diffusion of Pesticides Including Herbicides in Soil

From the Sect. 13.3, we get a mathematical framework theory which we can quantify the diffusion of pesticides and its control release picture, there are many factor influencing diffusion of pesticide which is discussed in present section. The soil being a dynamic and heterogeneous system generates several complexities to study diffusion process. It is the primarily soil properties viz., soil organic matter, porosity, bulk density, soil water content, soil temperature and mineral composition of soil influence the variation of the diffusion behavior of a substance and ultimately its diffusion coefficient. Temperature and pressure gradients along with external forces, as contributing or influencing factors, attribute to diffusive flux. Again diffusion of pesticide is influenced by important factors like: (i) the diffusion coefficient depends on concentration, and (ii) the diffusion of a molecule becomes restricted to particular areas in soil system due to the inability of the molecule to diffuse through a soil solid particle, (iii) the possibility of sorption of the diffusing pesticides on soil particles. Furthermore, the gaseous diffusion coefficients in soil air change proportionally with temperature and inversely with pressure as well as it is almost independent on composition of soil air, whereas in soil water, diffusion coefficients increase with temperature as well as it is dependent on solute concentration and types (Kutilek and Nielsen 1992; Kutilek and Nielsen 1994). Thus, the behavior or fate of pesticides including herbicides is influenced or regulated by, the physio-chemical properties the organic pesticide molecule, climate factors, types of soil and its properties which determine in a combined way about the consequent response of the applied pesticide to targeted pest or the environment (Upchurch 1966). Additionally, there are several variables viz., formulation, application rate, solubility, volatility, sorption-coefficient and degradation pathway of herbicide along with rate of soil water flow and evapotranspiration as well affect desperately the diffusion and adsorption of the same.

Diffusion induced transport of herbicide occurs in soil at slower rate compared to the transport by mass flow, the first process solely is surpassingly significant for a considerable short distance to be covered by the herbicide under the static water-flow conditions (Bode et al. 1973). The molecular diffusion of organic pesticides through a porous and heterogeneous soil system may be further complicated due to soil consists of soil solid (minerals and organic matter), soil water (solution) and soil air (void). The applied pesticide might be adsorbed, dissolved or suspended, and existed on soil solids including organic matter or minerals, in soil solution, and in the vapor phase as well respectively. Pesticide can transport through both of the vapor phase and solution phase in porous soil as well as can be even adsorbed onto soil solid which in turns makes possible the movement of the pesticide either along the solid surfaces or through the solid phase. The vapor phase shows a significant higher diffusion coefficient of a pesticide compared to the liquid phase, and again than the adsorbed or solid phase having at the lowest (Yaron 1989). Furthermore, once the pesticide gets adsorbed onto the soil solid, it tends to diffuses in a significant slowest manner there compared to does in the soil pores; this phenomena makes the applied pesticide molecule unavailable for diffusion and even accounts for the negligible contribution to the total diffusion flux. Eventually, a significant amount of the applied concentration of pesticide remains strongly adsorbed onto the soil solid and the sorption isotherm usually appeared as linear (Graham-Bryce 1969). Moreover, the value of diffusion coefficient of a molecule in soil porous system depends on the behaviour of the existing counter-diffusing solute or gaseous species (Kutilek and Nielsen 1994). When the available pathway for diffusion of the pesticide in soil is desired from larger pores to smaller ones sequentially, their absorption by soil solids is the phenomena which restricts the movements of pesticides (Bode et al. 1973).

Moreover, the specific geometry or the tortuosity of the diffusion pathway of pesticide depends on the soil structure i.e. arrangement of soil particle, porosity and water content in a particular soil. Once a pesticide applied, the pathway of diffusion through the soil pores gets hindered and becomes tortuous due to sorption of a part of the pesticides on the solids resulting the retardation reasonably. A numerous studies explained the effective diffusion coefficient and how it is influenced by factors such as porosity, tortuosity, and water content (Millington 1959; Van Brakel and Heertjes 1974; Fu et al. 2021). A proper and thorough knowledge about the influencing factors will expedite to know how the transport of pesticides occur in the soil and hence robust theoretical and experimental studies on interrelations of soil, climate, herbicide and crops would conclude to specific predictions about the of effectiveness of herbicide and also of chemical pesticides.

It is important to comprehend the speed of diffusion of a pesticide through soil to predict how its distribution occurs in soil so that the actual concentration of the pesticide which moves to plant roots, to become available to them, can be measured. Estimation of speed of diffusion is important to explain probable interactions of pesticides with soil solids. To avoid environmental contamination aiming to safety and sustainability, it is desirable to monitor and optimize the spread of pesticide beyond the intended area. The controlled release (CR) formulation of a pesticide regulates the diffusion process in sustained manner through the porous soil matrix or cell membrane by controlling the release rate of pesticide active ingredients. The concept of CR technology has been fruitfully developed to ensure judicious and safe use of pesticides to provide maximum efficiency by securing them against unwanted fate and environmental losses (Hussain 1989).

13.5 Advantages and Limitations of Controlled Release Formulation of Pesticides

The amendment made by European Union was reported (Singh et al. 2020) for regulations on registration of pesticide by commanding a significant reduction in availability of permissible active ingredients up to 15%. In such a certified way, pesticides should be formulated with using minimum active ingredients for fruitful but hazardless application than earlier for minimal risks for environment. However, such efficient formulations based on confined active ingredients need to be included under existing pesticide regulatory policies. The commercialization of controlled-release pesticide formulations address several challenges due to cost of production and ideal performance both at the same time. The introduction of carrier materials increases production costs, making it challenging to balance ideal performance with cost-effectiveness. Although CR formulations ensures sustained release but it may not be controllable sometimes due to the variation of the environmental factors like light, humidity, temperature and also because of limitations of the carrier materials (Li et al. 2021).

CR pesticides can be formulated in particular way as per the occurrence of weeds with their characteristics, diseases of plants, pathogens, insect or pests, and also based on environmental or soil conditions in which they manifest. By employing various physical and chemical mechanisms, these formulations ensure continuous and stable release and target specific delivery of active ingredients at precise doses over designated periods without premature dissipation (Bahadir and Pfister 1990; Huang et al. 2018). The active persistence period after applying CR pesticides demands to be regulated based on the requirement of different targets to prevent undesired fate or loss of the pesticide in the environment which can balance the persistence time and the efficacy of CR formulations. It can be optimized like if prevailing high pest densities, CR formulations need to act rapidly to ensure efficiency and timely control or management of pests. Conversely, in case of involving pests or pathogens with longer life cycles, CR formulations should sustain their activity over an extended period. Additionally, CR formulations should possess lower toxicity towards non-targeted objects or surface than that by their conventional counterparts (Kumar et al. 2019).

Some characteristically differences in structure, size, components, and materials used for slow released formulations of pesticides lead to varying release behaviors of CR pesticides. These properties significantly impact field efficacy, necessitating standardized use and evaluation protocols as a large-scale production of CR formulations requires robust technologies than conventional pesticide formulations. However, advancements in nano-agriculture have significantly enhanced the capabilities of CR formulations of pesticides as required. These technologies enable the permeation driven transfer of active ingredients from the source to a target surface in desired level, maintaining a desired and targeted dose for a specified duration based on application requirements. Such innovations not only optimize pesticide application but also play a pivotal role in advancing the commercialization of controlled-release technologies.

However, the initial CR formulation of some pesticide only allows release of active ingredients in sustained manner which is prone to change by external environmental factors as required for growth of crops. For instance, abamectin, a broad-spectrum insecticide, is susceptible to photodegrade and so the release rate cannot be optimized as per the predetermined design which puts a limitation on the benefits of the CR formulations as its efficacy (Li et al. 2021). If the composition and structure of CR pesticides can be modified to adapt alternations in environmental conditions and align with crop growth characteristics, these formulations could achieve better performance with limited or no toxicity to non-targeted surfaces, posing to minimum losses to environment and along with improved pest-control efficiency. However, use and applications of CR formulated pesticides are discrete still (Singh et al. 2020).

13.6 Controlled Release Nanoparticles of Pesticides

Nano-pesticides have a wide range of applications, particularly in agriculture and the food sector. In agriculture, they are effectively used to manage pest-related diseases and control weed growth, offering a more efficient and environmentally-friendly alternative to traditional pesticide methods. A lot of observations reported that, <2% of conventionally used pesticides can efficiently reach to the target object while <1% can terminate pest manifestation. Conversely, nano-pesticides have gained priority to be adopted worldwide (Arora et al. 2022) as it can overcome the limitations of conventional pesticides. Nano-pesticides, which are alternate and effective weapon used for efficient plant protection, comprises active ingredients or the carrier molecules are designed by use of nanotechnology. Similar to conventional pesticide, nano-pesticide formulations include the active ingredient, a carrier molecule, and surfactants. The primary aim in formulating nano-pesticides is to reduce the environmental toxicity, caused by active ingredients of pesticides by enhancing their efficacy (Nair et al. 2010).

Nano-pesticide formulations offer diverse properties that enhance pesticide efficiency during their application by several advanced formulations viz., micro-emulsions, nano-dispersions, nano-emulsions, and nano-encapsulations; These formulations help to improve the solubility of hydrophobic pesticides, protect active ingredients from early degradation, and enhance target specificity against pests. By modification of the chemical nature of carrier molecules, different formulation are designed by using organic polymer, lipids, nano-sized metals or metal oxides, clay, porous inorganic, and nanocomposites (Kumar et al. 2019). CR pesticides often are classified based on their structure as particles, capsules, films, emulsions, hydrogels, porous beads, fibers, and layered systems (Li et al. 2021). Solid nanoparticles can also be formulated as nano-pesticides. The inert silica, alumina, and clays cause a mechanical damage on the insect cuticle by both sorption and abrasion resulting in dehydration of the insect/ pest cells. Nano-silica particle is a preferable candidate as a solid nanoparticle due to its exceptional benefit of higher tolerance to abiotic and biotic stresses, showed by the plant.

The extremely small particle size and unique shape is a key advantage of using nanoparticles where it typically ranges from 1 to 100 nm (Prasad 2014) and in this nanoscale particles exhibit a higher ratio of surface area and volume which allows a greater volume of pesticide to reach the targeted pests in contact. The forces prevails at the nanoscale are Van der Waals and electrostatic interactions; these acting forces dominates to directly influence some physical phenomena like colloid stability, adsorption, friction, and surface adhesion, and but their role on diffusion of nanoparticle is not yet clear (Giorgi et al. 2019). The shape of nanoparticles along with zeta potential influence or facilitates their penetration into negatively charged cell membrane of target cells. Both organic and inorganic nanocarrier improve stability of such nano-pesticides promoting smart diffusion in low concentration also with a sustained release over time; again, it minimizes non-targeted toxicity and prevents degradation, volatilization or leaching loss during application in field.

Several bioactive components in plant extract like polyphenols, lipids, proteins, flavonoids, carbohydrates reduce complex metal ions to nano forms at different pH values. Increase in the electrostatic repulsion between nanoparticles at suitable pH influences not only the reaction kinetics but also allows to maximize the efficiency of reduction of metal ions resulting the fabrication of nano-pesticides in improved shape and size which influences the diffusion dynamics. The Stokes-Einstein equation alone is not adequate to explain the complexity of the diffusion of nanoparticles in non-Newtonian fluids as it excludes the potential effects of electrochemical forces but depends only on dimension of nanoparticles, viscosity of the medium and temperature. The target-specific, lower toxicity, intrinsic charge, wettability, improved pesticide droplet resilience and larger surface area of nano-pesticide formulation increase their superiority over conventional pesticides (Huang et al. 2018).

13.7 Advantages and Limitations of Usage of Nano-pesticides

It is simultaneously essential to recognize the profound impact of nano-pesticides which can have on environment and human health, leading to both positive and negative outcomes. The unique features of nanomaterials is that their behavior at the nanoscale often differs significantly from their behavior at the micro or macro scale. This remarkable properties make them functionally valuable in a wide range of applications. Nanoparticles, typically being at least one dimension smaller than 100 nanometers, exhibits novel physical, chemical, and biological characteristics which are not seen in their larger-scale counterparts.

The innovative approach to develop eco-friendly nano-pesticide formulations includes the advantage of being highly target-specific (Prasad et al. 2014). The controlled release mechanisms and targeted delivery of nano-pesticides enhance the efficacy of pesticide utilization by reducing their application frequency (Zhang and Schwendeman 2012) but decreasing the persistence of residue and further to minimize environmental pollution. Again, nano-microcapsule are formulated for slow release and controlled performance by using the active ingredients sensitive to environmental condition like light, temperature, humidity, soil pH, soil enzyme to optimize delivery of pesticides (Huang et al. 2018). Formulations of nano-pesticides also enhance adhesion of droplets to plant surfaces, reducing drift losses. This improved adhesion enhances the dispersion and bioactivity of the active ingredient, making the formulations more effective. For instance, use of nano-encapsulated pesticides can reduce phyto-toxicity (Kumar et al. 2012) and furthermore, nanocapsules imparts improved diffusion by controlled release of active ingredients on the targeted objects (Yearla and Padmasree 2016). Nano-capsulation of pesticides with biocompatible polymers can help in targeted delivery and sustained release of pesticides (Rani et al. 2015).

Sandhya et al. (2017) formulated CR nano-carbendazim-chitosan and CR nano-carbendazim pectin showed no further toxicity to the nontargeted bacterial strain; when phytotoxicity tests performed on *Zea mays*, *Lycopersicum esculentum* and *Cucumis sativa* plants, they exhibited reduced toxicity by nano-formulated carbendazim compared to conventional carbendazim when applied to seeds; furthermore pure carbendazim reduced germination up to 60% rather than the nano-CR formulation of the same. Singh et al. (2015) and Atta et al. (2015) formulated nano-CR formulated photo-responsive 2,4-D using organic photo-sensitive fluorescent nanomaterials and using coumarin copolymer respectively, which ensured effective transmission of herbicides in plants which proved efficient and fluorescent herbicidal activity of nano-CR 2,4-D. Therefore, nano-pesticides having higher efficiency in comparison with the conventional pesticide formulations, due to their unique properties, which ensures a means of an eco-friendly, safe and sustainable agricultural practices.

The use of nanoparticle-mediated DNA and gene editing in plants (Priyanka et al. 2019) is emerging as a promising strategy to enhance insect pest resistance diversely and also agricultural productivity in improved manner. By utilizing encapsulated nanoparticles, crops can be provided with a steady and controlled release of nutrients, leading to improved growth and resilience. In contrast, conventional pest management techniques which often rely on synthetic pesticides such as DDT, chloropyriphos, atrazin, malathion, metachlor, endosufan and etc. have adverse effects on both human and animal health leads to a decline in soil fertility and soil health. These limitations highlight the need for more sustainable and precise alternatives, such as those offered by nanotechnology.

Nano-pesticides may possess some different type of contamination or toxicity for the environment since their persistency are apparently higher compared to their conventional counterparts. While research into nano-pesticide development is progressing rapidly, the subject has yet to gain significant attention from the public or regulatory authorities (Kookana et al. 2014). Like many innovations, nano-pesticides present both opportunities and challenges for environmental as well as for human health. However, the present scenario regarding environmental fate of nano-pesticides remains insufficient to allow a comprehensive assessment of the potential advantages and disadvantages associated with their use. Therefore a better insightful research on the fate and behavior of nano-pesticides is required in physics and chemistry perspective after their application.

13.8 Summary

Let us summarized our content. We first introduce about pesticides and then control-release techniques in agriculture to protect plants from weeds, fungi, insects, etc. We have also briefly introduce the impact of pesticides on the environment and addressed about the classification of pesticides. In comparison with normal pesticides sprays, the control-release technique of pesticides is introduced as fruitful method for optimizing environmental contamination. Addressing the brief mechanism of control-release pesticides, we have gone through its detailed mathematical formulation, where we have considered slab- and sphere-shaped soil geometry for the diffusion process. We can see that, before and after the release time or characteristic time, control-release concentration expressions are different. Former is slow concentration decay, which get sufficient time to act on targeted area and later is rapid concentration decay, which is also good for optimizing environmental contamination. So, from mathematical understanding, reader can realize that control release technique is better method than normal pesticide spray technique.

Next, we have also discussed different factors which can influence the diffusion of pesticides in soil. Also, we have compiled different advanced limitations of control-release pesticides. There is another advanced technique by using nano-pesticides, which is also briefly discussed here.

Acknowledgement The authors acknowledge funding support from the IIT Bhilai Innovation and Technology Foundation (IBITF) and express their gratitude to the Agri-Sci based project titled

"Novel Strategies to Tackle Water and Soil Contamination by Health Hazardous Pollutants using Computational Tools: A Small Step Towards Sustainable Agricultural Practice." They thank the Principal Investigator (PI), Dr. Hemant Kumar Kishan, Assistant Professor at IIT Bhilai, and the Co-Principal Investigator (Co-PI), Dr. Sabyasachi Ghosh, Associate Professor at IIT Bhilai, along with the funding body IIT Bhilai Innovation and Technology Foundation (IBITF) under the Tribal Area Sub Plan (TSP) scheme. They also extend their thanks to other members of the Agri-Sci team: Jayanta Dey, Deepika Sharma, Biswajit Sahariah, Anand Rai, Sanju Singh, and Kiran Devi Tulsiyan.

References

AbuQamar SF, El-Saadony MT, Alkafaas SS, Elsalahaty MI et al (2024) Ecological impacts and management strategies of pesticide pollution on aquatic life and human beings. Mar Pollut Bull 206:116613. https://doi.org/10.1016/j.marpolbul.2024.116613

Ali S, Ullah MI, Sajjad A, Shakeel Q, Hussain A (2021) Environmental and health effects of pesticide residues. In: Inamuddin, Ahamed MI, Lichtfouse E (eds) Sustainable agriculture reviews, vol 48. Springer, Cham. https://doi.org/10.1007/978-3-030-54719-6_8

Arora S, Murmu G, Mukherjee K, Saha S, Maity D (2022) A comprehensive overview of nanotechnology in sustainable agriculture. J Biotechnol 355:21–41. https://doi.org/10.1016/j.jbiotec.2022.06.007

Atta S, Paul A, Banerjee R, Bera M, Ikbal M, Dhara D, Singh NDP (2015) Photoresponsive polymers based on a coumarin moiety for the controlled release of pesticide 2,4-D. RSC Adv 5:99968–99975

Bahadir M, Pfister G (1990) Controlled release formulations of pesticides. In: Haug G, Hoffmann H (eds) Controlled release, biochemical effects of pesticides, inhibition of plant pathogenic fungi, Chemistry of Plant Protection, vol 6. Springer, Berlin/Heidelberg, pp 1–12. https://doi.org/10.1007/978-3-642-46674-8_1

Bhandari G, Atreya K, Scheepers PTJ, Geissen V (2020) Concentration and distribution of pesticide residues in soil: non-dietary human health risk assessment. Chemosphere 253:126594

Bhandari G, Atreya K, Vasickova J, Yang X, Geissen V (2021) Ecological risk assessment of pesticide residues in soils from vegetable production areas: a case study in S-Nepal. Sci Total Environ 788:147921

Bhunia P (2017) Environmental toxicants and hazardous contaminants: Recent advances in technologies for sustainable development. J Hazard Toxic Radioact Waste 21:02017001

Bode LE, Day CL, Gebhardt MR, Goering CE (1973) Mechanism of trifluralin diffusion in silt loam soil. Weed Sci 21(5):480–484

Chin NP (2010) Environmental toxins: physical, social, and emotional. Breastfeed Med 5:223–224

Collins RL, Doglia S (1973) Concentration of pesticides slowly released by diffusion. Weed Sci. Cambridge University Press on behalf of the Weed Science Society of America 21(4):343–349. https://www.jstor.org/stable/4042008

Collins RL, Doglia S, Mazak RA, Samulski ET (1973) Controlled release of herbicides—theory. Weed Sci 21(1):1–5

Cooper J, Dobson H (2007) The benefits of pesticides to mankind and the environment. Crop Prot 26:1337–1348. https://doi.org/10.1016/j.cropro.2007.03.022

Damalas CA, Eleftherohorinos IG (2011) Pesticide exposure, safety issues, and risk assessment indicators. Int J Environ Res Public Health 8:1402–1419

De Vlaming V et al (2000) Application of whole effluent toxicity test procedures to ambient water quality assessment. Environ Toxicol Chem 19(1):42–62. https://doi.org/10.1002/etc.5620190106

FAO (1990) http://www.fao.org/3/w2598e/w2598e07.htm

Fu Y, Horton R, Ren T, Heitman JL (2021) A general form of Archie's model for estimating bulk soil electrical conductivity. J Hydrol 597:126160. https://doi.org/10.1016/j.jhydrol.2021.126160
Garud A, Pawar S, Patil MS, Kale SR, Patil S (2024) A scientific review of pesticides: classification, toxicity, health effects, sustainability, and environmental impact. Cureus 16(8):e67945. https://doi.org/10.7759/cureus.67945
Giorgi F, Coglitore D, Curran JM et al (2019) The influence of inter-particle forces on diffusion at the nanoscale. Sci Rep 9:12689. https://doi.org/10.1038/s41598-019-48754-5
Graham-Bryce J (1969) Diffusion of organophosphorus insecticides in soils. Agr Food Chem J 20:489–494
Green RE, Obien SR (1969) Herbicide equilibrium in soils in relation to soil water content. Weed Sci 17(4):514–519
Gullick DR, Mott KB, Bartlett MG (2016) Chromatographic methods for the bioanalysis of pyrethroid pesticides. Biomed Chromatogr 30:772–789
Hassaan MA, Nemr AE (2020) Pesticides pollution: classifications, human health impact, extraction and treatment techniques. Egypt J Aquat Res 46(3):207–220
Huang L, Jiang G (2024) Photothermal controlled-release microcapsule pesticide delivery systems constructed with sodium lignosulfonate and transition metal ions: construction, efficacy and on-demand pesticide delivery. Pest Manag Sci 80(6):2827–2838. https://doi.org/10.1002/ps.7991
Huang B, Chen F, Shen Y, Qian K, Wang Y, Sun C, Zhao X, Cui B, Gao F, Zeng Z et al (2018) Advances in targeted pesticides with environmentally responsive controlled release by nanotechnology. Nanomaterials 8:102. https://doi.org/10.3390/nano8020102
Hussain M (1989) Atoms in agriculture: nuclear techniques in "controlled-release" pesticide research. IAEA Bull (2):36–40
IRAC (Insecticide Resistance Action Committee) (2017) Classify a new insecticide. https://irac--online.org/documents/classify-a-new-insecticide
Kookana R, Boxall A, Reeves P, Ashauer R, Beulke S, Chaudhry Q, Cornelis G et al (2014) Nanopesticides: guiding principles for regulatory evaluation of environmental risks. J Agric Food Chem 62(19):4227–4240
Kumar RSS, Shiny PJ, Anjali CH, Jerobin J, Goshen KM, Magdassi S, Mukherjee A, Chandrasekaran N (2012) Distinctive effects of nano-sized permethrin in the environment. Environ Sci Pollut Res 20:2593–2602. https://doi.org/10.1007/s11356-012-1161-0
Kumar S, Nehra M, Dilbaghi N, Marrazza G, Hassan AA, Kim KH (2019) Nano-based smart pesticide formulations: emerging opportunities for agriculture. J Control Release 294:131–153
Kutilek M, Nielsen DR (1992) Soil hydrology. Schweizerbart Science Publishers, Stuttgart. http://www.schweizerbart.de//publications/detail/isbn/9783510653874/Kutilek_Nielsen_Soil_Hydrology_GeoEcol
Kutilek M, Nielsen DR (1994) Soil hydrology: geo-ecology textbook: [Textbook for students of soil science, agriculture, forestry, geo-ecology, hydrology, geomorphology or other related disciplines]. Catena Verlag, Cremlingen-Destedt
Lainsbury A (2018) The UK Pesticide Guide 2018. BCPC, London
Li N, Sun C, Jiang J, Wang A, Wang C, Shen Y, Huang B et al (2021) Advances in controlled-release pesticide formulations with improved efficacy and targetability. J Agric Food Chem 69(43):12579–12597. https://doi.org/10.1021/acs.jafc.0c05431
Liu Y, Li S, Ni Z, Qu M, Zhong D, Ye C, Tang F (2016) Pesticides in persimmons, jujubes and soil from China: residue levels, risk assessment and relationship between fruits and soils. Sci Total Environ 542:620–628
Mekonen S, Argaw R, Simanesew A, Houbraken M, Senaeve D, Ambelu A et al (2016) Pesticide residues in drinking water and associated risk to consumers in Ethiopia. Chemosphere 162:252–260
Millington RJ (1959) Gas diffusion in porous media. Science 130(3367):100–102

Mu H, Yang X, Wang K, Tang D, Xu W, Liu X, Ritsema CJ, Geissen V (2023) Ecological risk assessment of pesticides on soil biota: an integrated field-modelling approach. Chemosphere 326:138428. https://doi.org/10.1016/j.chemosphere.2023.138428

Nair R, Varghese SH, Nair BG, Maekawa T, Yoshida Y, Kumar DS (2010) Nanoparticulate material delivery to plants. Plant Sci 179:154–163

Oerke EC (2005) Crop losses to pests. J Agric Sci 144:31–43

Pastor S, Creus A, Parrón T, Cebulska-Wasilewska A, Siffel C, Piperakis S, Marcos R (2003) Biomonitoring of four European populations occupationally exposed to pesticides: use of micronuclei as biomarkers. Mutagenesis 18:249–258

Pathak VM, Verma VK, Rawat BS, Kaur B, Babu N, Sharma A, Dewali S et al (2022) Current status of pesticide effects on environment, human health and its eco-friendly management as bioremediation: a comprehensive review. Front Microbiol 13:962619. https://doi.org/10.3389/fmicb.2022.962619

Prasad R (2014) Synthesis of silver nanoparticles in photosynthetic plants. J Nanopart 2014:963961. https://doi.org/10.1155/2014/963961

Prasad R, Kumar V, Prasad KS (2014) Nanotechnology in sustainable agriculture: present concerns and future aspects. Afr J Biotechnol 13:705–713. https://doi.org/10.5897/AJBX2013.13554

Priyanka P, Kumar D, Yadav K, Yadav A (2019) Nanopesticides: synthesis, formulation and application in agriculture. In: Abd-Elsalam K, Prasad R (eds) Nanobiotechnology applications in plant protection, Nanotechnology in the life sciences. Springer, Cham, pp 107–123. https://doi.org/10.1007/978-3-030-13296-5_7

Rani R, Dilbaghi N, Dhingra D, Kumar S (2015) Optimization and evaluation of bioactive drug-loaded polymeric nanoparticles for drug delivery. Int J Biol Macromol 78:173–179. https://doi.org/10.1016/j.ijbiomac.2015.03.070

Rasool S, Rasool T, Gani KM (2022) A review of interactions of pesticides within various interfaces of intrinsic and organic residue amended soil environment. Chem Eng J Adv 11(2666-8211). https://doi.org/10.1016/j.ceja.2022.100301

Rosell G, Quero C, Coll J, Guerrero A (2008) Biorational insecticides in pest management. J Pestic Sci 33:103–121

Sandhya, Kumar S, Kumar D, Dilbaghi N (2017) Preparation, characterization, and bio-efficacy evaluation of controlled release carbendazim-loaded polymeric nanoparticles. Environ Sci Pollut Res 24(1):926–937

Sharma A, Kumar V, Shahzad B et al (2019) Worldwide pesticide usage and its impacts on ecosystem. SN Appl Sci 1:1446. https://doi.org/10.1007/s42452-019-1485-1

Singh PND, Atta S, Bera M, Chattopadhyay T, Paul A, Ikbal M, Maiti MK (2015) Nano-pesticide formulation based on fluorescent organic photoresponsive nanoparticles: for controlled release of 2,4-D and real-time monitoring of morphological changes induced by 2,4-D in plant systems. RSC Adv 5:86990–86996

Singh A, Dhiman N, Kar AK, Singh D, Purohit MP, Ghosh D, Patnaik S (2020) Advances in controlled release pesticide formulations: prospects to safer integrated pest management and sustainable agriculture. J Hazard Mater 385:121525. https://doi.org/10.1016/j.jhazmat.2019.121525

Smalling KL, Kuivila KM, Orlando JL, Phillips BM, Anderson BS, Siegler K et al (2013) Environmental fate of fungicides and other current-use pesticides in a central California estuary. Mar Pollut Bull 73:144–153

Solomon KR, Williams WM, Mackay D, Purdy J, Giddings JM, Giesy JP (2014) Properties and uses of Chlorpyrifos in the United States. In: Ecological risk assessment for Chlorpyrifos in terrestrial and aquatic systems in the United States, Reviews of environmental contamination and toxicology, vol 231. Springer, Cham, pp 13–34

Tudi M, Daniel RH, Wang L, Lyu J, Sadler R, Connell D, Chu C, Phung DT (2021) Agriculture development, pesticide application and its impact on the environment. Int J Environ Res Public Health 18(3):1112

Upchurch RP (1966) Behavior of herbicides in soils. Residue Rev 16:46–85

Van Brakel J, Heertjes PM (1974) Analysis of diffusion in macroporous media in terms of a porosity, a tortuosity and a constrictivity factor. Int J Heat Mass Transf 17(9):1093–1103. https://doi.org/10.1016/0017-9310(74)90190-2

WHO (2022) https://www.who.int/news-room/fact-sheets/detail/pesticide-residues-in-food

Yaron B (1989) General principles of pesticide movement to groundwater. Agric Ecosyst Environ 26(3–4):275–297

Yearla SR, Padmasree K (2016) Exploitation of subabul stem lignin as a matrix in controlled release agrochemical nanoformulations: a case study with herbicide diuron. Environ Sci Pollut Res:6983–6988. https://doi.org/10.1007/s11356-016-6983-8

Zhang Y, Schwendeman SP (2012) Minimizing acylation of peptides in PLGA microspheres. J Control Release 162(1):119–126

Chapter 14
Applications of Plant-Based Nanomaterials to Increase the Shelf Life of Food Products

Uroosa Noor, Nidhi Singh, Shashi Soni, Mona Kumari, and Ena Gupta

14.1 Introduction

Globally, the food wastage has become the major concern as certain quantity of food is wasted throughout the food manufacturing process and due to an inadequate supply network. As per the reports, the total food waste generation includes 60% consumer waste, 20% production waste, and 1% processing waste. Whereas the major reasons for food wastage are high consumer demand, climatic changes, and manufacturing issues (Sridhar et al. 2021). As per the 2021 Food Waste Index Report by the United Nations Environment Programme (UNEP), over 931 million tonnes of food waste were produced in 2019, accounting for 17% of the world's total food production. In today's scenario, the utilization of nanotechnology for the development of nanoscale products is growing in the field of research. Apart from this, nanotechnology is considered to be developing technology in food sectors, as approximately 400 companies globallyhave advanced applications by utilizing nanomaterials within the food sector (Chadha et al. 2022). As food safety is considered to be a wide concern due to the growing demands of healthy, fresh, and nutritious food (Saravanan et al. 2021). The idea of food preservation techniques gained popularity as a way to prevent food from spoiling by limiting the activity of harmful microorganisms and blocking metabolic processes. As a result, these methods enable food materials to have their shelf lives extended (Sridhar et al. 2021). Nanotechnology constitutes an emerging field that encompasses various disciplinary fields offering numerous applications in science and technology. It also deals with certain characteristics, exploitation, creation, and synthesis associated with nanotechnology. The fast-expanding multidisciplinary field comprises physics,

U. Noor · N. Singh · S. Soni · M. Kumari · E. Gupta (✉)
Department of Family and Community Sciences, University of Allahabad, Allahabad, Uttar Pradesh, India

V. Ghosh et al. (eds.), *Applications of Plant Bioactive Compounds-Based Nanoformulations*, Nanotechnology in the Life Sciences,
https://doi.org/10.1007/978-3-032-13086-0_14

chemistry, and biology with engineering that shows concern towards the development of nanomaterials via various novel methods (Gul et al. 2021; Nadeem et al. 2018; Saleem et al. 2019). The production of nanomaterials using scientific chemical and physical procedures is hazardous and does not provide a sustainable environment. Green plant-derived nanomaterials, on the other hand, offer a variety of applications in specialized biological sectors such as biotechnology, imaging, electronics, optics, food packaging, agriculture, medicine, catalysis, and sensors. The word "Nano" is obtained from the Greek word i.e., nanos, which means "dwarf or tiny". The size dimension of nanoparticles ranges from 1 to 100 nm. Each nanomaterial are unique components due to their distribution, size, and morphology. Nanotechnologies are basically categorized into dry, wet, and computational methods, in which dry nanotechnology is linked with the inorganic components like carbon and silicon components, wet nanotechnology is based on the tissues, cellular components, enzymes, and membranes as well as computational nanotechnology is related to the simulations of nanostructures (Chung et al. 2016; Sridhar et al. 2021). According to the previous literature, the green nanotechnology is focusing on the deeper exploration of plant-based nanoparticles that are eco-friendly forms of nanofactories. It also encourages meeting the demands for the sustainable nano products in various aspects (Singh et al. 2016). Comparatively, the biological system comprises fungi, bacteria, yeast, and plants that are often involved in green synthesis for the creation of nanomaterials (Jadoun et al. 2021). Additionally, the plant-derived nanomaterials incorporate the non-toxic solvents with mild response conditions, adaptability, and feasibility in the utilization for various applications such as surgical and pharmaceutical applications (Hano and Abbasi 2021). Additionally, different types of metals such as silver nitrate (Ag), copper (Cu), zinc (Zn), titanium dioxide (TiO_2), gold (Au), and later, palladium (Pd), platinum (Pt), cerium (Ce), nickel (Ni), and manganese (Mn) were used for the development of plant-based nanomaterial. Few reports have suggested that the environmentally sustainable production of silver nanoparticles (AgNPs) through the utilization of biological agents leading to a non-toxic alternative to chemically synthesized nanomaterials (Hano and Abbasi 2021).

Perhaps the future trend of nanotechnology focuses on the important role in food packaging methods, which is considered to be the vital benefit for the food industry. It emphasizes thermal, mechanical, barrier activity, and physicochemical qualities that prolong shelf life and maximize food quality (Ash et al. 2021). Generally, nanotechnology is utilized for the production of films incorporated with nanomaterial that are further responsible for inhibiting undesirable carbon dioxide gases and balancing the transportation of the gases from plastic films. Additionally, it acts as a protective barrier that retards the food deterioration by the gases like ethylene and oxygen (Duncan 2011; Silvestre et al. 2011; Ash et al. 2021). Thereby, recent evaluations have encouraged the food industry sectors day by day. This chapter presents an introduction of nanotechnology, several ways for producing plant-based nanomaterials, and their importance in extending food shelf life, along with a comprehensive examination of nanoparticles for future applications in food preservation.

14.2 Overview of Various Types of Plant-Based Nanomaterials

Nowadays, plant-derived nanoparticles are being employed as novel therapies for a variety of diseases. Natural products, primarily plant extracts, also have anti-inflammatory, insulin-sensitizing, and antioxidant qualities. In addition, they are regarded as a possible choice due to their low risk of negative consequences. This revolutionary green nanotechnology is now accessible and may be utilized to improve plant-mediated green synthesis of metallic nanoparticles, signifying a significant development in the green synthesis of metal nanoparticles (Okeke et al. 2023). Silver nanomaterials (AgNMs) generated from plants are the simplest to make. The green production of silver nanomaterials involves the use of reducing biological agents in silver metal ion solution. In order to produce AgNMs, a variety of biomolecules, including amino acids, alkaloids, polysaccharides, phenolics, vitamins, and saponins, are used to stabilize and reduce Ag ions (Khan et al. 2020). The low toxicity and excellent biocompatibility of gold nanoparticles (AuNMs) make them easy to produce and have acceptable characteristics. A number of biogenic complexes serve as reducing agents in the nanomaterial manufacturing process. Various studies have depicted that bio-molecules such as proteins, flavonoids, and polyphenolic compounds effectively reduce metallic ions in gold nanomaterials derived from plant extracts (Jeong et al. 2011). Zinc oxide nanomaterials (ZnONMs) are considered to be the second most effective metal oxide. ZnONMs is low in toxicity, ecofriendly, inexpensive, has a high UV absorption capacity, and can be easily prepared. Various synthesis procedures involve the addition of different precursors alter the physio-chemical structure of ZnONPs. Different parts of plants, such as seeds, roots, leaves, and flowers, are used for the synthesis of ZnONMs. It is reported that zinc oxide nanomaterial is an inorganic form of an antibacterial agent that shows the ability to resist microorganisms. These nanomaterials consist of a high exciton binding energy of around 60 meV, which provides the properties of semiconductors (Kalpana and Devi Rajeswari 2018; Supraja et al. 2016; Supraja et al. 2016; Bala et al. 2015). Copper nanomaterials (CuNMs) are recognized to be an exclusive form of biomaterials that range from 1 to 100 nm in diameter. It is considered the most stable, low-cost metal and exhibits high electric conductivity with a high melting point. CuNMs are generally synthesized by reducing Cu ion concentration with the various plant extracts. As per the studies, the CuNMs exhibits potential antibacterial and antifungal properties as it is broadly utilized in the field of biomedical sciences (Tata et al. 2023). Murthy et al. (2020), demonstrated the copper nanomaterials synthesis derived from plant extract of *H. abyssinica*. The extract consists of essential phenolic compounds, tannins, and anthraquinone glycosides, which are known to be a good source of high antioxidant levels. Therefore, these components get bound with Cu2+ ions precursors in the presence of salt solution that results in reduced Cu2+ ions to form CuO. The presence of biomolecules acts as a stabilizing and enveloping agent for the nanomaterials. Studies have revealed that the concentration of salt solutions and their precursors are responsible for the shape and size of the Copper nanomaterial prepared from leaf extract of A. *Indica* (Nguyen et al. 2022). However, various metals like manganese (Mn), nickel (Ni), cerium (Ce), titanium (Ti), and palladium (Pd) are plant-derived nanomaterials that are also used in different industrial applications (Hano and Abbasi 2021).

14.3 Potential Methods for Extracting Plant-Based Nanomaterials

There are various techniques to synthesize nanomaterials, including chemical, biological, and physical methods. These techniques are found to be eco-friendly and cost-effective in nature. The chemical method involves means of precipitation/reduction reactions, whereas biological synthesis involves utilization of plant extracts or microbial agents as reducing agents (Kumari et al. 2023). Physical methods focus on the techniques that utilize physical processes like condensation, mechanical forces and evaporation for the production of nanomaterials. As per the studies, these methods are contributing towards developing the eco-friendly biogenetic procedure for nanomaterial synthesis (Popescu et al. 2010). In addition to this, the most common form of method used for nanomaterial synthesis is the physical and chemical process. It consists of less application due to the presence of toxic components and yields (Fig. 14.1).

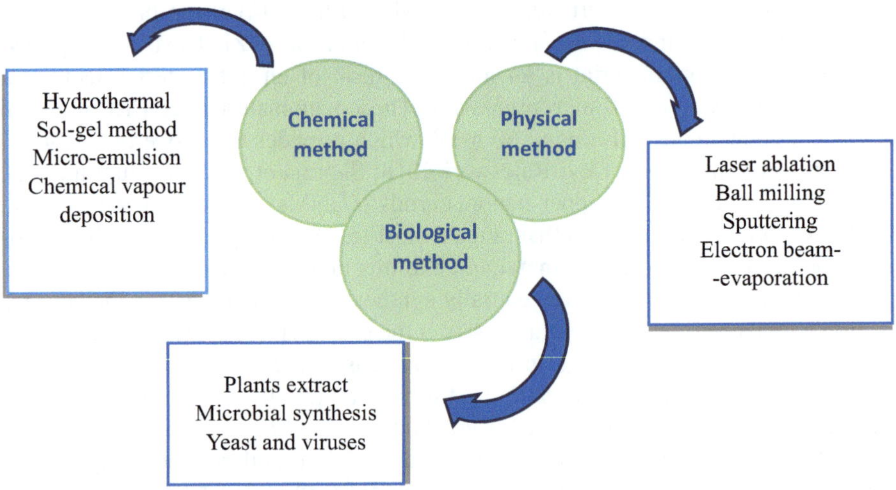

Fig. 14.1 Various method utilized in the nanomaterial synthesis

14.3.1 Chemical Method

In the chemical method, the organic and inorganic forms of reducing agents are utilized for the synthesis of nanomaterial resulting in metal formulation and oligomeric clusters which produces metallic colloidal particles. Basically, the common techniques used in the chemical methods are sol-gel or sono-chemical, hydrothermal, pyrolysis, microemulsion, intercalation, co-precipitation, vapour deposition, ion exchanges, and solvothermal technique (Kumari et al. 2023). Some of these are described below:

14.3.1.1 Sol-Gel Method and Sono-Chemical Method

The sol-gel is a form of wet chemical method used for the synthesis of various nanomaterials. In this process, the precursor of components (usually metal alkoxide) is mixed with alcohol or water concentration, which further gets converted into the gel substrate with the help of heating or stirring through the hydrolysis process. The gel obtained after the process of hydrolysis and alcoholysis is present in the form of wet and damp gel. According to the desired characteristics of the substrate, these wet gels are further dried by the use of appropriate methods. For instance, if the process is done in the alcoholic condition, then the drying process of gel is done through burning alcohol (Bokov et al. 2021). Apart from it, the sono-chemical process involves high ultrasound radiations of around 20 kHz to 10 kHz for molecules that promote different chemical reactions. The physical phenomenon that triggers sonochemical reactions is called acoustic cavitation. This method is utilized for the synthesis of various metallic nanomaterials, especially for iron nanomaterials. It involves the suitable operating system with ambient conditions; it is also responsible for controlling the size of nanomaterials with the help of specific precursors at different concentrations in the solvent. The mechanism of sonolysis involves transferring the sound wave with fixed frequency by a solution of certain metal precursors. Due to the presence of vapour pressure of a specific threshold in the solvent, it results in expansion and compression in waves that form cavities to grow and implode. As per the studies, sono-electrochemical synthesis involves both ultrasonic pulses and electrolytes for the synthesis of nanomaterials (Kumari et al. 2023).

14.3.1.2 Hydrothermal Synthesis

This method is used for the hydrophilic and hydrophobic magnetic nanomaterials having controlled shape and size. The hydrothermal synthesis is a quick method for the development of nanomaterials such as Iron (II) tungstate ($FeWO_4$), Cobalt Ferrite ($CoFe_2O_4$), Zinc Oxide (ZnO), Cadmium Sulfide (CdS), etc. In hydrothermal synthesis, the heated water utilized to result in the diffusion of particles. Typically, this process is carried out in a large closed vessel with high pressure known as an autoclave. The first step includes a general reaction, which involves mixing the precursor with the solvent in a specific ratio. Secondly, it includes a crystallization process that takes place at a supersaturated state and promotes the nucleation and growth phenomenon of particles that helps to achieve the final product (Kumar et al. 2021).

14.3.2 Biological Method

Biosynthesis is a technique that involves the green synthesis of nanomaterials through a bottom-up approach method. Bottom-up approach is referred to as self-assembly, that involves constructing nanomaterials through assembling the tiny components such as atoms, molecules and other small components. It is highly expensive, environmentally toxic, and involves long period of time for production of by-product (Pandit et al. 2022). However, biological methods utilize the microbial strains as a bio-reducing agent, such as bacteria, algae, fungus, protozoa, and macromolecules, including enzymes, carbohydrates, proteins, nucleic acids, and plant residues, for the nanomaterial synthesis. These biological agents secrete certain enzymes that act as stabilizing and reducing agents, whereas these enzymes bind with the metal ions to form nanomaterials. Comparatively, the phytotoxicity and cytotoxicity of plant-derived nanomaterials were less harmful than the wet chemical nanomaterials that are utilized in several applications (Abinaya et al. 2021; Nguyen et al. 2022).

14.3.2.1 By Bacteria, Algae, Yeast and Plant Extract

Bacterial species are utilized for the production of metals and other nanomaterials. Actinomycetes and *prokaryotes are widely employed in the production of metal oxide nanoparticles* (Gobinath et al. 2021). Species such as *B. cecem, B. amyloliquefaciens, B. indicus, Klebsiella pneumoniae, B. cecembensis*, *Geobacter* sp., *Morganella psychrotolerans, and B. licheniformis, as well as SH10 Phaeocystis antarctica,* were used for the synthesis of Ag nanomaterials. Additionally, the species of *Lactobacillus* and *Bacillus subtilis* are utilized for the production of Titanium dioxide nanomaterials. Bacterial strains have the ability to produce intracellular and extracellular nanomaterials during the time of incubation period. As a by-product, the biosynthesis of nanomaterials through bacterial species is considered to be cheap, versatile, and feasible for larger productivity. A diverse group of algae is investigated to have a vital role in the field of nanotechnology. The microalgae are considered to be single-celled, colony forming, filamented, photosynthetic microorganisms. Algae is recognized to be an eco-friendly and convenient form of biological substrate that is responsible for the nanomaterial synthesis at ambient pressure and temperature with normal pH value. Mukherjee et al. (2021), depicted that the cells of filamentous *Cyanobacteria* (*Plectonemaboryanum)* promote the production of silver (Ag), gold (Au), and platinum (Pt) nanomaterials. The strains of *Chlorella vulgaris* and *Lyngbya majuscule* are found to be suitable, cost-effective methods for the synthesis of Ag nanomaterial. Additionally, dried edible algae *(Spirulina platensis)* were utilized for the extracellular synthesis of bimetallic nanoparticles, i.e., Au/Ag and Au, Ag. Another study demonstrated that the incorporation of *S. muticum*-was used for the synthesis of ZnONMs, whereas the brown algae were utilized for the synthesis of ZnONMs and TiO_2NMs (Kumari et al. 2023). The red algae, i.e., *Hypneamusciformis,* and green algae, *Ulva latica,* are used for Ag-NP synthesis, which further hinder the growth of fungal species such as *Candida albicans* and *Candida parasitosis.* Yeast is a single-celled eukaryote that acts as a template and is used in the primary mechanism for the production of nanomaterials (NMs). For example, the Au NMs are formed through the utilization of yeast, i.e., *Thermomonosporachromogena* and *Thermomonosporacurvata.* Likewise, the yeasts, namely, *Rhodosporidiumdiobovatum* and *Candida glabrata,* are used for producing Cd and Pb5 nanoparticles. The extracts of yeast act as reducing agents for metal precursors and stabilize the outer surface of the nanomaterial, like AgNMs, and also provide a well-uniform spherical shape (Kumari et al. 2023).

Apart from this, the synthesis of nanomaterials through incorporation of plant extract is recognized under the green process synthesis. Plant extract is investigated to be an eco-friendly, stable, less toxic, and low-cost method for the biosynthesis process. Additionally, it is a source of various biochemical agents such as polysaccharides, terpenoids, organic acids, flavonoids, alkaloids, and polyphenols, which act as stabilizers and reductants of metal ion formulationsThe nanoparticle synthesis using plant extract was identified to be more advantageous over the other biological processes as it eliminates the longer process that maintains the cell cultures. Green synthesis of nanomaterials shows high catalytic action and a large surface area between the metal ions and reducing enzymes (Borase et al. 2014). Studies have investigated that the synthesis of gold nanomaterial is done by the leaves of geranium herbs, i.e., *Pelargonium graveolens.* Additionally, alfalfa plant leaves have been utilized for the production of gold and silver extracellular nanomaterials (Baraka et al. 2017) (Fig. 14.2).

14.3.3 Physical Method

Physical methods focus on the working principle for the development of nanomaterial through the use of mechanical pressure, radiation, and thermal energy, which are additionally responsive to the processes of abrasion, condensation, and dissolution of nanomaterials. The physical methods produce nanomaterials through fragmentation of bulky material and follow the top-down approaches for synthesis. Top-down approach refers to breaking down of bulk block materials which is considered as a base substrate into tiny particles or even micron-sized components. It involves physical process such as milling, crushing and grinding for reducing the size of large particle. Top-down approach is regarded as simple, bulk fabrication

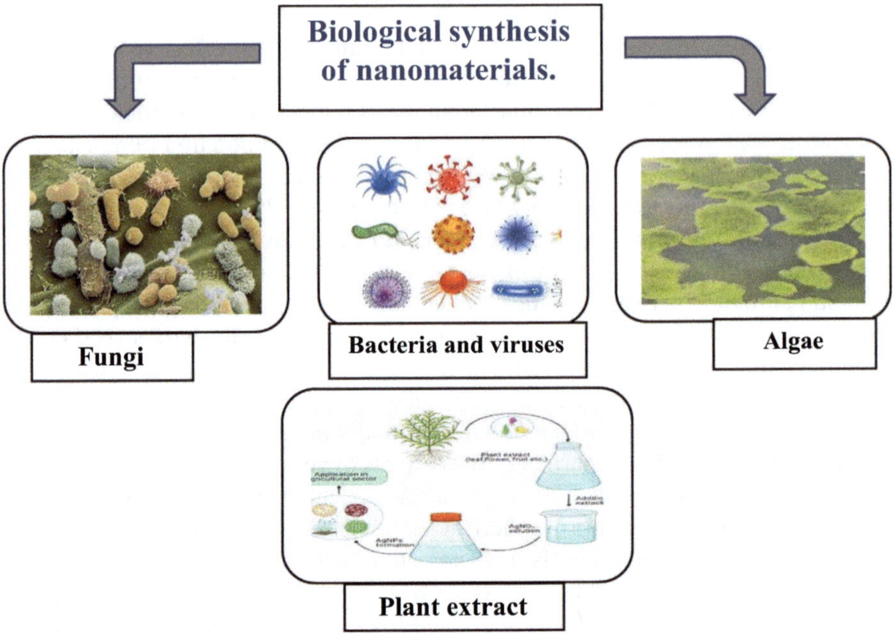

Fig. 14.2 Various biological methods of nanomaterial synthesis (Nguyen et al. 2022)

procedure for obtaining appropriate structures having desired properties. According to the studies, the physical methods are investigated to be different from the chemical methods as they generate a homogenous form of nanomaterials that are environmentally friendly in nature, whereas they restrict the use of chemical reagents in this process. Various physical techniques involved in the production of nanomaterials are inert gas condensation (IGC), mechanical ball milling, physical vapour deposition (PVD), laser pyrolysis, ion sputtering, laser ablation, pulsed wire discharge, and arc discharge process. These different procedures for nanomaterial synthesis depend upon the physio-chemical characteristics with the quantity and the type of nanomaterials (Baig et al. 2021; Kumari et al. 2023).

14.3.3.1 Ball Milling and Inert Gas Condensation (IGC)

This method works on the principle based on the utilization of ball milling by reducing the particle size with high energy. It was discovered by. The reduction in the particle size results in the modification of the surface area of the material (Rajput 2015). Ball milling is an alternative approach for the production of nanoparticles. Various types of mills are vibratory, planetary, tumbler, and rod. Moreover, the container comprises hard balls that are prepared from steel and carbide. The synthesis of nanocrystalline materials, such as Cr, W, Co, and Ag-Fe, is done by a ball milling procedure. It is reported that the ratio of balls to material is 2:1. The container gets filled with air and inert gas concentrations that are centrifuged at high speed and velocity around the central region. These materials are squeezed between the solid balls and the wall present inside the container. Apart from this, the specific duration and speed of the milling process are responsible for the synthesis of nanomaterial with optimum size (Patil et al. 2021). The inert gas condensation (IGC) technique is involved in the production of a wide range of nanomaterials, intermetallic components, ceramics, alloys, semiconductors, composites, and oxide nanoparticles. This method depends upon the squeezing of the metal particles with the incorporation of inert gas components such as helium, argon, and neon under the high pressure and temperature present inside the metallic chamber that regulates the size of nanomaterials ranging from 2 to 100 nm and evaporates the metal particle. As the metal vapor gets cooled off with the help of liquid nitrogen, it loses its energy. For example, the inverse gas chromatography method was utilized for controlling the size of the copper nanomaterial. Along with this, the production of copper nanomaterial was done through the evaporation of particles in the presence of inert gas. The inert gas condensation process is also utilized to create ultrafine metallic forms of dielectric multishell nanomaterials, with high-quality platinum and silver nanomaterials (Zheng and Branicio 2020).

14.3.3.2 Laser Pyrolysis

Laser pyrolysis is considered to be an effective vapor-phase synthesis technique to create magnetic nanoparticles. It involves a high-power carbon dioxide laser, which is used to excite gaseous and liquid components and quench them. In this reaction, a salt solution is sprayed in a series of reactions, where the reactants get evaporated, and later it gets dried, which converts it into a concentrated form of particle (Jing et al. 2014). Large-scale production scalability is maintained by laser pyrolysis; however, it is being researched as an expensive piece of equipment. The process involves the synthesis of nanoparticles such as silicon carbide, silicon, and molybdenum disulfide. Previous literature has depicted that metals, oxides, composites, and carbide nanomaterials are created by using this method. The titanium isopropoxide, tetraethyl orthosilicate, and silane as precursor materials were used for the synthesis of ceramic nanomaterials, titanium dioxide and silicon dioxide nanoparticles (D'Amato et al. 2022; Jing et al. 2014).

14.4 Mechanism of Plant-Based Nanomaterial Synthesis

The main phases involved in the synthesis of the plant-based nanomaterials include the following mechanism:

- The activation phase: The substrate of metal salt is converted into the tiny metal ions at specific pH and temperature levels. This involves the reduction of the metal ions substrate that further leads to the development of metal atoms, followed by the nucleation process of reduced metal ions.

- The growth phase: Due to the process of growth and nucleation, the nanomaterials coalesce into the larger nanoparticles. Furthermore, it reduces the metal ions, which results in an increase in the thermodynamic stability of the nanoparticles. The overall process is termed to be Ostwald ripening.
- The terminal phase: This step involves the final determination of the shape of the nanomaterial. As the overall stability of nanomaterial is dependent on the original shape of the particle that is found after the process of capping and stabilization. However, it is investigated that the capping and stabilizing agents present in biological entities that are responsible for inhibiting the mechanism of agglomeration act as a growth terminator for maintaining the persistency and stability of nanomaterials (Si and Mandal 2007).

14.5 Antimicrobial Action of Plant-Derived Nanomaterials

In recent year, plant-based nanomaterials have gaining popularity due to their antimicrobial properties. Previous researches investigated that plant-based nanomaterialsis an alternative antimicrobial optionwith high efficacy to resist pathogenic bacteria. The antimicrobial action of plant-derived nanomaterials involves the aggregation of nanoparticles (NPs) on bacterial surfaces, hindering cellular transport and disrupting signal induction processes. Additionally, NPs penetrate the cytoplasm of microbes, leading to alterations in their metabolic functions. Studies suggest that the charge on NPs plays a critical role in their impact on bacterial activity. For instance, green-synthesized silver nanomaterials (AgNMs) possess positively charged ions, which have demonstrated effectiveness against negatively charged bacterial strains (Banasiuk et al. 2020). Previous researches also revealed that the plant extract of copper oxide NPs were derived from species *Caesalpinia bonducella*, which further showed an inhibitory action with the membrane disrupting ability against gram-positive and gram-negative strains of bacteria *i.e.*, *Staphylococcus aureus* and *Aeromonas* respectively (Sukumar et al. 2020; Okeke et al. 2023). Similarly, the leaf extract of *Carissa carandas*was used for the preparation of green AgNMs that further showed the potent inhibitory activity against pathogenic bacteria. It is investigated that the presence of synergistic effects of biological active phytoconstituents obtained from the plant extract determines the antimicrobial effects through reactive oxygen species (ROS). In addition, with this, the antibiofilm effect of gold nanomaterials (AuNMs) was observed against the bacterial species i.e., *Serratia marcescens* and *Listeria monocytogenes* (Okeke et al. 2023). Certain evidences have showed that zinc oxide nanomaterials (ZnO-NMs) synthesized from bark of *Cinnamomum auriculata* has ability in disrupting the bacterial cell integrity. Another nanomaterial namely, NiO-NMs prepared from stevia leaf extract which shows positive effect against gram-negative bacteria (Prasad et al.

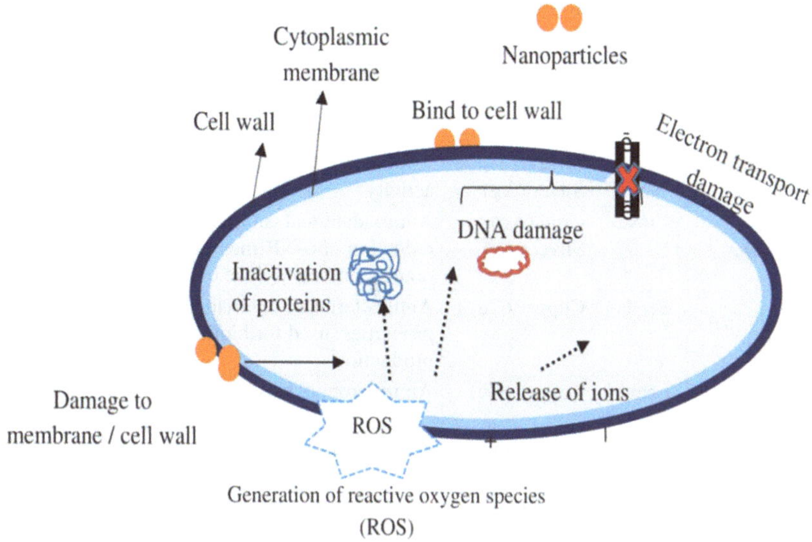

Fig. 14.3 Action of nanomaterials on the bacterial cell (Huang et al. 2018)

2020; Srihasam et al. 2020). The antimicrobial mechanism of NPs is depicted bellow in Fig. 14.3 and Table 14.1.

14.6 Synergistic Antimicrobial Action with Nanomaterials

Research has indicated that different types of nanomaterials *i.e.* copper, zinc, chitosan, iron, platinum, silver, gold, and carbon nanotubes when combined with natural derivatives like essential oils, provide a synergistic antibacterial function. It was also analyzed that utilization of essential oil droplets with the combination of gold nanomaterial to encapsulate peppermints and cinnamon aldehyde provide antimicrobial activity. In contrast to this, essential oil of *Lippiasidoides* as a natural substrate of thymol was nano encapsulated in the chitosan gum nanomaterials, which is widely used in the fields of food, chemicals, and pharmaceuticals. The combination of iron oxide nanomaterials and essential oils was used in the preparation of magnetic nanofluid. One of the studies discovered that incorporating zein-sodium caseinate nanomaterial with thymol increases its antibacterial activity (Li et al. 2012). The antifungal and antibacterial activity of essential oils is increased through the incorporation of various nanomaterials. It has also observed that the combination of essential oil, namely *Origanum dictamnus,* and liposome-based nanomaterial showed high antimicrobial potential against gram-negative and gram-positive bacteria. Additionally, *Santolina insularis* essential oil derivative, such as p-cymene, γ-terpinene, thymol, and carvacrol, incorporated with phosphatidylcholine

Table 14.1 Various biological activities of plant-based nanomaterials

Plant species	Plant part	Nanomaterial	Biological activities	References
Catharanthus roseus Linn	Flower	Gold (Au), Silver (Ag)	Antibacterial and anticandidal activity	Singh et al. (2018)
Gloriosa superb	Flower	Gold (Au), Silver (Ag)	Antibacterial and antibiofilm activity	Salem and Fouda (2021)
Syzygium aromaticum	Clove	Gold (Au), Silver (Ag)	Antioxidant and catalytic reduction of p-NP, methyl orange, and methylene blue	
Punica granatum	Seeds	Copper (Cu)	Antibacterial and antiviral properties, used for biofilm production	Krkobabić et al. (2023)
Azadirachta indica	Leaves	Copper (Cu)	Antimicrobial activities	Ikhioya et al. (2024)
Pinus resinosa	Peel	Palladium (Pd)	Biocatalyst	Arora et al. (2024)
Cinnamomum camphora	Leaves	Palladium (Pd)	Biocatalyst activity	Devarajan (2023)
Brassica oleracea	Leaves	Silver (Ag)	Antibacterial activity against foodborne pathogens	Ansar et al. (2020)
Artocarpus integer	Leaves	Silver (Ag)	Antibaterial activity against *E. coli, Staphylococcus aureus, Bacillus cereus*	Majeed et al. (2019)
Coffee	Seeds	Silver (Ag)	Completely retards the growth of *E. coli*	Baghaienezhad et al. (2020)
Passiflora	Fruit	Silver (Ag)	Antifungal effect against the *fusarium* species	Elangovan et al. (2022)
Justicia adhatoda	leaf	Zinc Oxide (ZnO)	Antifungal effect against *Aspergillus flavus, Aspergillus niger*	Pachaiappan et al. (2021)

liposome nanomaterials, showed potential bacteriostatic property (Nile et al. 2020) (Fig. 14.4).

14.7 Active and Smart Food Packaging

Nowadays, consumers pay attention to new packaging methods known as active packaging. Additionally, the creative approach used for active food packaging lengthens shelf life and ensures food safety. It involves the inhibition of the microbial growth, prevents moisture loss, and its degradation process that further shows appropriate results than other forms of traditional packaging. It is reported that active packaging can be utilized in the absorption of chemicals present either in the food or the environment. Headspace in the packaging allows flavorings, preservatives, and antioxidants to be released into the food product. Active components are

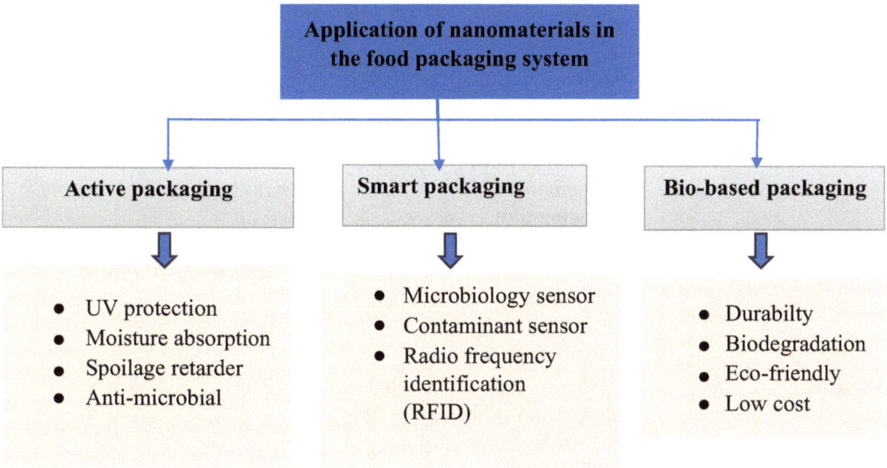

Fig. 14.4 Utilization of nanomaterials in a variety of food packaging methods (Jafarzadeh et al. 2024)

directly incorporated into the polymer matrix that results in antioxidant and antimicrobial forms of packaging. Therefore, due to various characteristics, active packaging is considered to be a promising application in food preservation (Dobrucka and Przekop 2019; Janani et al. 2020; Jafarzadeh et al. 2024).

Smart packaging has a significant role in improving food quality and its traceability. This emerging technology focuses on various sets of action to monitor the external and internal environments of the food packaging system. Smart packaging systems also determine the food's freshness and deliver relevant status to the retailer, manufacturers, and customers (Garavand et al. 2022). Generally, smart packaging involves multiple functions such as sensing, tracking, communicating, monitoring, and recording and utilizes scientific logics. In contrast to this, intelligent packaging comprises essential elements like communicating, detection, recognition, and recording. It also includes screening/sensors, freshness identification, time-temperature indicators, recorders, etc., which enhance the efficacy of managing the entire product distribution chain (Jafarzadeh et al. 2024) (Table 14.2).

14.8 Plant Derived Nanomaterials and Their Applications in the Food Preservation

14.8.1 Copper Oxide Nanomaterial (CuO-NM)

Copper oxide nanomaterial shows the tendency in the formation of antimicrobial nanohybrids. According to previous investigations, the incorporation of CuO-NMs has the potential of increasing the activity of components present in the

Table 14.2 Applications of nanomaterials in shelf-life extension of food items

Nanomaterial	sizes	Matrix	Features	Food products	References
AgNMs	NA	AgNMs added with polylactic acid used as Active nanocomposite packaging film	Preserve the Vitamin C content.	Strawberries	Zhang et al. (2018)
AgNMs	7–40	Cellulose acetate with AgNMs used as active food packaging having antioxidant	Antibacterial activity against Escherichia coli strain	Fruits and vegetables	Dairi et al. (2019)
AgNMs	200	AgNMs mixed chitosan-based film are used as antimicrobial packaging	Antioxidant properties	Pork	Wu et al. (2019)
AgNMs	NA	Nanocomposite combined with ginger extract and Ag NMs	Prolong the shelf- life of the food item	Chicken sausage	Mathew and Radhakrishnan (2021)
ZnONMs	200	Composite film	Extends the shelf life of grapes for 21 days	Grapes	Basumatary et al. (2022)
ZnONMs	NA	Chitosan film with cellulose acetate phthalate and ZnO NMs	Extended the shelf life of fruit for 9 days	Black grapes	Basumatary et al. (2022)
ZnONMs	NA	Banana powder-based nanocomposite reinforced with ZnONMs and cellulose nanofibre	Extend the shelf life of fruit for 6 days	Pomegranate arils	Leta et al. (2024)
ZnO NMs	NA	ZnO NMs with chitosan film	Prolong shelf life upto 14 days	Tomatoes	Iqbal et al. (2022)
AgNMs	NA	Silver nanomaterial based low-density polyethylene (LDPE)	Maintains the freshness of the food product	Barberry	Ghosh et al. (2019)
TiO_2NMs	10–30 nm	Polyvinyl matrix	Enhance water vapour barrier properties, protect against UV radiation	Garlic	Lieu et al. (2024)

(continued)

Table 14.2 (continued)

Nanomaterial	sizes	Matrix	Features	Food products	References
TiO_2NMs	30 nm	Titanium oxide nanomaterial incorporated with rosemary oil for cellulose nanofiber formation and protein matrix	Prevent lipid oxidation, reduces microbial load and improves shelf life from 6 to 18 days	Lamb meats	Alizadeh-Sani et al. (2020)

nanocomposites. However, using the chemical precipitation process, new nanocomposites are incorporated with the CuO-NMs to create the bacterial cellulose and chitosan nanofibers (Almasi et al. 2018). In order to measure the amount of coliform in ultrafiltered (UF) cheese, Beigmohammadi et al. (2016) looked into the antibacterial capability of low-density polyethylene (LDPE) packaging sheets that include CuO-NPs, Ag-NPs, and ZnO-NPs. Additionally, these nanoparticles can improve the mechanical and barrier qualities of packaging materials. Similarly, one of the prior studies determined the antimicrobial action of copper oxide nanomaterials (CuO-NMs) against the Gram-negative and Gram-positive bacterial species after getting attached to the bacterial cellulosic nanofibers. Additionally, the synergistic action was observed between CuO-NMs and chitosan nanofibers (CHNF) (Almasi et al. 2018).

Gvozdenko et al. investigated that the technique utilized for the synthesis of nanomaterial is stabilized with gelatin that shows high potential for the purpose of food packaging in the form of independent nanofilm. Furthermore, it also contains antifungal properties that result in food nanopackaging and improve the longevity of the food product. As per the report, the nanopackaging containing CuO-NM with stabilized gelatin has shown a positive effect in shelf-life improvement of tomato and strawberry food items. Due to the presence of high stability of CuO nanomaterial, it can be utilized in the production of active food packaging material. It was also demonstrated that CuO nanomaterial incorporated with methylcellulose films showed a high inhibitory effect against the growth of yeast, coliform, and mold in cheese samples. Therefore, methylcellulose film containing CuO-NM is used as a packaging material for shelf-life improvement of hard cheese "Holland" (Gvozdenko et al. 2022).

14.8.2 Gold Nanomaterial (Au-NM)

Gold nanomaterials are identified to be significant forms of heterogeneous oxidative catalysts. The DPPH (2,2-diphenyl-1-picrylhydrazyl) is considered to be a scavenging model and is broadly used method for identifying the efficacy of the compounds to inhibit free radicals. According to the previous studies, the bio-nanocomposites prepared by gold nanomaterial (Au-NMs), mixed with unbleached kraft softwood fiber (AuNM-UBK), are determined to be reliable component for antioxidant-based active packaging methods that are used to preserve food (Khan et al. 2019).

14.8.3 Titanium Oxide Nanomaterials (TiO_2NMs)

Titanium oxide nanomaterials (TiO_2 NMs) are broadly utilized in various application due their high photocatalytic effects, chemical stability and low toxicity. The plant-mediated TiO_2NMs have received promising attention towards having excellent anti-inflammatory, anti-fungal, anti-bacterial and anti-microbial properties, including several biological functions. As per the previous research, several bio-mediated extracts, such as *Annona squamosa, cinnamon powder, Azadirachtaindica leaf, Cynodondactylon, Mangifera indica, Curcuma longa, Citrus reticulate, fungus-mediated, Morindacitrifolia, Jatropha curcas, and Psidium guajava* are utilized forthe production of TiO_2NMs in various biological applications (Aravind et al. 2021). TiO_2NMs exhibits anti-bacterial potential against different pathogenic microbes such as *Escherichia coli, Campylobacter*, *Salmonella*, *Staphylococcus aureus*, and *Listeria monocytogenes.* Additionally, it has the tendency to scavenge the reactive oxygen species (ROS) and deactivates the microbial load that ensuresthe safety of food products. TiO_2 NMs are mostly used in condiments, powdered foods, and candies for shelf-life enhancement and also used as a food packaging material, such as milk cartons, foils, and bottles. Along with this, TiO_2-based biosensors are prepared for detecting the sensitivity and identifying the contaminants in food (Mao and Hao 2024).

14.8.4 Silver Nanomaterial (Ag-NM)

Silver is found to be a widely explored form of nanomaterial that has antimicrobial potential against several fungi, viruses, pathogens, and commensal bacteria. It shows bacteriostatic effects by targeting the metabolic action through binding with proteins, enzymes, and DNA. It destabilizes the cytoplasmic membrane of the bacterial cell (Kumar and Münstedt 2005). They also have the capability to scavenge reactive oxygen species that prevents certain enzymes involved in the respiratory chain mechanism. According to studies, nanoparticles implanted in the porous structure of zeolite (a substance used to make low-density polyethylene), has the capacity to increase the durability and shelf life of beverages, such as orange juices (Emamifar et al. 2010; Vieira et al. 2020). In another experiment it was reported that the nanocomposites are highly active in conferring heat treatment capacity and antibacterial activities which have been carried out for the development of polymeric matrixes by incorporating silver nanomaterial (AgNM). A study conducted by Singh and Sahareen (2017), reported that the incorporation of 15.3 mg/mL of

Ag-NMs into the cellulosic form of food packaging showed effective results for extending the shelf life of the cabbage and tomatoes. Another study investigated the efficacy of silver nanomaterials AgNMs on the physiochemical characteristics of different films such as film-forming solution (FFS), nanocomposite films (NCF), and hydroxypropyl methylcellulose (HPMC). Additionally, the effect of these films on the production of anthracnose which is a plant disease caused by genus namely, *C. gloeosporioides*present in papaya. However, FSS was incorporated with AgNMs with the concentration of 0.25 wt% for the coating papayawhich modified the morphology, chemical bonds, thickness, moisture content and thermal stability of the film. It has also inhibited the effect of *C. Gloeosporioides* on papaya which reduces the weight of the fruit during storage. Silver-based HPMC coating enhanced the shelf life of the papaya fruit (Vieira et al. 2020).

The polyvinyl polyethylene coating materials are mixed with 72 and 98 mg/mlAg-NMs that show inhibitory action against *Bacillus cereus* and *E. coli* and therefore prevent the formation of molds in pepper chilies for 15 days (Adeyemi and Fawole 2023). Thus, due to the notable antimicrobial characteristics of Ag-NMs, it can utilize for packaging materials for food preservation purposes. The bacterial cellulose embedded with AgNMs shows the effective role in reducing the exposure of corrosive vapor content (Vieira et al. 2020). According to research by Tavakoli et al. (2017), nuts including hazelnuts, pistachios, almonds, and walnuts can have their shelf lives extended by up to 18–20 months when packaged with silver nanomaterials. Another study found that adding AgNMs to edible films improved the food quality of turkey meat products for around 12 days when they were stored in a vacuum-sealed system (Deus et al. 2017).

14.8.5 Zinc Oxide Nanomaterial (ZnO-NMs)

Zinc oxide nanomaterial (ZnO-NMs) is found to be the most inexpensive category of nanomaterial having high antibacterial properties. The US Food and Drug Administration (FDA) lists zinc oxide as a GRAS (generally regarded as safe) metal oxide (Agarwal et al. 2017). It also exhibits high semiconducting properties due to the presence of high exciton binding energy, a large band gap of up to 3.37 eV, high catalytic activity, UV filtering features, and wound healing capacity. Previous studies have shown that phytochemicals released by plant extracts are used as a stabilizing and capping agent. The application of ZnO-NMs in food packaging systems focuses on the antibacterial impact and shelf-life enhancement of fresh food products. Li et al. (2017), examined the role of ZnO-NMs mixed with Polylactic acid films improves the freshness and quality of apples. Another study demonstrated that ZnO-NMs-blended films are responsible for maintaining the color, total phenolic content, firmness, and sensory quality of the food and also contain high inhibitory activity against the microorganisms. Beak et al. (2017) determined the potential of olive flounder bone gelatin which is a temperate marine species blended with ZnO-NMs-based film on spinach quality, which demonstrated antibacterial activity against *L. monocytogenes* without compromising the spinach's color or vitamin C content. The development of carboxymethyl cellulose/chitosan and ZnO-bionanocomposite film was done to prolong the shelf life of Egyptian soft white cheese. The study by Suo et al. (2017) evaluated that the ZnO-NMs-coated packaging films are accountable for the increase in microbial injury that preserves the pork meat items even in cold storage. Al-Shabib et al. (2016), did the preparation of biofilm blended with Zn-NMs that are extracted from seeds of *Nigella sativa*, whereas, it shows the inhibitory action against pathogenic strains such as *L. monocytogenes*, *E. coli*, *PAO*1, *C. violaceum* 12472.

14.8.6 Lignin-Based Nanomaterial

Lignin is considered to be a polymer made up of three monolignols, such as guaiacyl (G) units, syringyl (S) units, and *p*-hydroxyphenyl (H) units. It also includes phenolic groups like coniferyl, *p*-coumaryl, and sinapyl. Lignin is a plant-based material that exhibits good physiochemical characteristics and acts as an antibacterial, antiviral, antioxidative agent, prevents UV radiation, and improves the shelf life of the food item. Therefore, it makes them suitable substitutes for the food packaging applications (Lu et al. 2022). Research has determined that the natural form of lignin polymer in combination with collagen-based gelatin polymer elevates its mechanical activity, prevents microbial growth, and also ensures the strongest free radical scavenging activity. Despite of these benefits, it also shows poor separating and mechanical characteristics in the case of lignin combined with agar, whereas it enhances thermal durability, including barrier and transparency ability (Zubair et al. 2024). One of the researchformed composite film by incorporating hydroxyethyl cellulose with lignin and gellan gum at different percentages of 0, 1, 5, and 10%. However, the incorporation of 10% of lignin causes protection against UVA and UVB at 90% and 100%, respectively. It also resulted in a 59.2% improvement in tensile strength. Another lignin-based compound, known as alkali lignin-lignosulfonate-soy protein, demonstrates good mechanical and thermal stability (Zubair et al. 2024).

14.8.7 Pectin-Based Nanomaterial

Pectin is one of the essential polysaccharides present in cell wall constituents which is α1,4-linked polygalacturonic acids with colloidal carbohydrates. According to the Food and Drug Administration, pectin is considered to be generally recognized as safe (GRAS). Due to its excellent structural and mechanical characteristics, the polysaccharides are obtained as a base material for edible film development (Kalia and Parshad 2015). Studies have demonstrated that pectin is utilised in the formation of nanostructures that increase adsorption, exhibit protection and provide availability of various bioactive components, which make it suitable for the production of coating nanomaterial. Essential hydrophobic components are incorporated into functional coatings to slow down the moisture barrier properties in the coating films. The edible coating material exhibits the potential of changing the environmental condition of food items such as fruits and vegetables. Additionally, it also inhibits the production of ethylene and slows down the oxygen level within the fruits; therefore, it alters the physiological deterioration of the food item. The process of ripening involves deterioration of food items over the period of time that causes loss of texture and essential bioactive components. However, the selective permeability present in edible coating is used to prevent spoilage in fruits and vegetables (Rohasmizah and Azizah 2022).

14.9 Future Perspectives

The increasing demand for nanotechnology and green chemistry has focused on the growth of green synthesis methods for obtaining various nanomaterials. Researchers have been focusing on the utilization of plant-derived nanomaterials through environmentally favorable techniques. Plant-mediated nanomaterials are non-toxic, eco-friendly, and cost-effective in nature. Future prospects of plant-derived nanomaterial synthesis and its various applications involve the elucidation of phytoconstituents and the extension of industrial-scale functioning. The bioinformatic tools used in nanomaterial production drive the process for suppressing harmful bacterial strains (Rani et al. 2023). Plant-based nanoparticles were claimed to be widely employed in the food industry, cosmetics, and pharmaceuticals. Plant-derived nanomaterials and edible coatings incorporated with nanomaterials are considered to be ideal packaging materials for food preservation purposes. Nanomaterial has the potential of maintaining the mechanical and physical characteristics of packaging polymer that further balance the flexibility, strength, barrier, durability, and reusable capabilities (Gupta et al. 2024). In today's scenario, plant-based nanomaterials are depicted to be more valuable for fabricating active and smart packaging, which results in extending the shelf life of the food items. Despite all these, the nanomaterial-based food packaging is promoted and may dominate the fields of food packaging systems in the upcoming future. The development of green synthesis methods for producing different nanomaterials has been fulfilled by the growing need for green chemistry and nanotechnology. Researchers have been concentrating on using ecologically friendly methods to use nanomaterials generated from plants. Future prospects for the synthesis of plant-derived nanomaterials and their diverse applications include the extension of industrial scale functioning and the clarification of phytoconstituents. The technique for suppressing pathogenic bacterial strains is driven by bioinformatic tools used in the fabrication of nanomaterials (Ettadili et al. 2022).

14.10 Conclusion

The innovative approaches have emerged towards the wide application of food packaging systems that include storage, food preservation, and its distribution. However, the nanotechnology provides numerous methods that are cost-effective, degradable, and eco-friendly for the formation of renewable packaging materials. Various applications of nanotechnology in food sectors involve food production, packaging, processing, and preservation. It has contributed towards solving the problem of food crises by ensuring the food safety. In this way, plant-based synthesis of nanomaterials is utilized for obtaining various nanomaterials that are beneficial for food preservation purposes, such as protecting against UV lights, moisture, and oxidative stress. Different plants-derived nanomaterials, such as AgNMs, AuNMs, and ZnO NMs, show antibacterial properties against a wide range of bacteria, including the foodborne pathogens. Several plants synthesized metal nanomaterials that show antimicrobial effects against the broad categories of microbes, especially the foodborne pathogens. It is an innovative method for the widespread use of food packaging systems covers distribution, storage, and preservation of food items. Nonetheless, nanotechnology offers a variety of economical, biodegradable, and environment friendly techniques for creating sustainable packaging materials. By guaranteeing food safety, it has helped to address the issue of food crises. Thus, research has made it possible to incorporate necessary plant-derived nanoparticles into food packaging systems.

Acknowledgement None declared.

Disclosure Statement The authors have declared no conflict of interest.

References

Abinaya S, Kavitha HP, Prakash M, Muthukrishnaraj AJ (2021) Green synthesis of magnesium oxide nanoparticles and its applications: a review. Sustain Chem Pharm 19:100368

Adeyemi JO, Fawole OA (2023) Metal-based nanoparticles in food packaging and coating technologies: a review. Biomolecules. 13(7):1092

Agarwal H, Kumar SV, Rajeshkumar S (2017) A review on green synthesis of zinc oxide nanoparticles–an eco-friendly approach. Resour-Eff Technol 3(4):406–413

Alizadeh-Sani M, Mohammadian E, McClements DJ (2020) Eco-friendly active packaging consisting of nanostructured biopolymer matrix reinforced with TiO2 and essential oil: application for preservation of refrigerated meat. Food Chem 322:126782

Almasi H, Jafarzadeh P, Mehryar L (2018) Fabrication of novel nanohybrids by impregnation of CuO nanoparticles into bacterial cellulose and chitosan nanofibers: characterization, antimicrobial and release properties. Carbohydr Polym 186:273–281

Al-Shabib NA, Husain FM, Ahmed F, Khan RA, Ahmad I, Alsharaeh E, Khan MS, Hussain A, Rehman MT, Yusuf M, Hassan I (2016) Biogenic synthesis of Zinc oxide nanostructures from Nigella sativa seed: prospective role as food packaging material inhibiting broad-spectrum quorum sensing and biofilm. Sci Rep 6(1):36761

Ansar S, Tabassum H, Aladwan NS, Naiman Ali M, Almaarik B, AlMahrouqi S, Abudawood M, Banu N, Alsubki R (2020) Eco friendly silver nanoparticles synthesis by Brassica oleracea and its antibacterial, anticancer and antioxidant properties. Sci Rep 10(1):18564

Aravind M, Amalanathan M, Mary MS (2021) Synthesis of TiO2 nanoparticles by chemical and green synthesis methods and their multifaceted properties. SN Appl Sci 3:1–0

Arora G, Srivastava A, Srivastava M, Dwivedi J (2024) Current trends in biogenic synthesis and applications of palladium nanoparticles: a sustainable approach to environmental remediation. In: Green solutions for degradation of pollutants. Bentham Science Publishers, pp 229–248

Ash M, Raza MB, Roy A, Golui D (2021) Role of green synthesized nanoparticles in food packaging. In: Microbial synthesis of nanoparticles. Nova Science Publishers, pp 180–199

Baghaienezhad M, Boroghani M, Anabestani R (2020) Silver nanoparticles synthesis by coffee residues extract and their antibacterial activity. Nanomed Res J 5(1):29–34

Baig N, Kammakakam I, Falath W (2021) Nanomaterials: a review of synthesis methods, properties, recent progress, and challenges. Mater Adv 2(6):1821–1871

Bala N, Saha S, Chakraborty M, Maiti M, Das S, Basu R, Nandy P (2015) Green synthesis of zinc oxide nanoparticles using Hibiscus subdariffa leaf extract: effect of temperature on synthesis, anti-bacterial activity and anti-diabetic activity. RSC Adv 5(7):4993–5003

Banasiuk R, Krychowiak M, Swigon D, Tomaszewicz W, Michalak A, Chylewska A, Ziabka M, Lapinski M, Koscielska B, Narajczyk M, Krolicka A (2020) Carnivorous plants used for green synthesis of silver nanoparticles with broad-spectrum antimicrobial activity. Arab J Chem 13(1):1415–1428

Baraka A, Dickson S, Gobara M, El-Sayyad GS, Zorainy M, Awaad MI, Hatem H, Kotb MM, Tawfic AF (2017) Synthesis of silver nanoparticles using natural pigments extracted from alfalfa leaves and its use for antimicrobial activity. Chem Pap 71:2271–2281

Basumatary IB, Mukherjee A, Katiyar V, Kumar S (2022) Biopolymer-based nanocomposite films and coatings: recent advances in shelf-life improvement of fruits and vegetables. Crit Rev Food Sci Nutr 62(7):1912–1935

Beak S, Kim H, Song KB (2017) Characterization of an olive flounder bone gelatin-Zinc oxide nanocomposite film and evaluation of its potential application in spinach packaging. J Food Sci 82(11):2643–2649

Beigmohammadi F, Peighambardoust SH, Hesari J, Azadmard-Damirchi S, Peighambardoust SJ, Khosrowshahi NK (2016) Antibacterial properties of LDPE nanocomposite films in packaging of UF cheese. LWT 65:106–111

Bokov D, Turki Jalil A, Chupradit S, Suksatan W, Javed Ansari M, Shewael IH, Valiev GH, Kianfar E (2021) Nanomaterial by sol-gel method: synthesis and application. Adv Mater Sci Eng 2021(1):5102014

Borase HP, Salunke BK, Salunkhe RB, Patil CD, Hallsworth JE, Kim BS, Patil SV (2014) Plant extract: a promising biomatrix for ecofriendly, controlled synthesis of silver nanoparticles. Appl Biochem Biotechnol 173:1–29

Chadha U, Bhardwaj P, Selvaraj SK, Arasu K, Praveena S, Pavan A, Khanna M, Singh P, Singh S, Chakravorty A, Badoni B (2022) Current trends and future perspectives of nanomaterials in food packaging application. J Nanomater 2022(1):2745416

Chung IM, Park I, Seung-Hyun K, Thiruvengadam M, Rajakumar G (2016) Plant-mediated synthesis of silver nanoparticles: their characteristic properties and therapeutic applications. Nanoscale Res Lett 11:1–4

D'Amato R, Iacono SD, Santoni A, Reale P, Limosani F, Terranova G, Botti S, Bonfigli F, Migliori A, Scalbi S (2022) Silicon nanoparticles as active materials for LED: laser synthesis and functionalization for tuning luminescence. J Phys Conf Ser 2315(1):012002. IOP Publishing

Dairi N, Ferfera-Harrar H, Ramos M, Garrigós MC (2019) Cellulose acetate/AgNPs-organoclay and/or thymol nano-biocomposite films with combined antimicrobial/antioxidant properties for active food packaging use. Int J Biol Macromol 121:508–523

Deus D, Kehrenberg C, Schaudien D, Klein G, Krischek C (2017) Effect of a nano-silver coating on the quality of fresh Turkey meat during storage after modified atmosphere or vacuum packaging. Poult Sci 96(2):449–457

Devarajan Y (2023) Catalytic activities of oxidation on hydrazine and azo-dye of nano palladium particles synthesis from bio-waste Cinnamomum camphora leaf extract. Biomass Conv Biorefin 13(16):15135–15148

Dobrucka R, Przekop R (2019) New perspectives in active and intelligent food packaging. J Food Process Preserv 43(11):e14194

Duncan TV (2011) Applications of nanotechnology in food packaging and food safety: barrier materials, antimicrobials and sensors. J Colloid Interface Sci 363(1):1–24

Elangovan D, Rahman HB, Dhandapani R, Palanivel V, Thangavelu S, Paramasivam R, Muthupandian S (2022) Coating of wallpaper with green synthesized silver nanoparticles from Passiflora foetida fruit and its illustrated antifungal mechanism. Process Biochem 112:177–182

Emamifar A, Kadivar M, Shahedi M, Soleimanian-Zad S (2010) Evaluation of nanocomposite packaging containing Ag and ZnO on shelf life of fresh orange juice. Innovative Food Sci Emerg Technol 11(4):742–748

Ettadili FE, Aghris S, Laghrib F, Farahi A, Saqrane S, Bakasse M, Lahrich S, El Mhammedi MA (2022) Recent advances in the nanoparticles synthesis using plant extract: applications and future recommendations. J Mol Struct 1248:131538

Garavand F, Cacciotti I, Vahedikia N, Rehman A, Tarhan Ö, Akbari-Alavijeh S, Shaddel R, Rashidinejad A, Nejatian M, Jafarzadeh S, Azizi-Lalabadi M (2022) A comprehensive review on the nanocomposites loaded with chitosan nanoparticles for food packaging. Crit Rev Food Sci Nutr 62(5):1383–1416

Ghosh C, Bera D, Roy L (2019) Role of nanomaterials in food preservation. In: Microbial Nanobionics: volume 2, basic research and applications. Springer International Publishing, Cham, pp 181–211

Gobinath R, Bandeppa, Manasa V, Rajendiran S, Kumar K, Paul R, Basavaraj K (2021) Nanoparticle-mediated adsorption of pollutants: a way forward to mitigation of environmental pollution. In: Microbial rejuvenation of polluted environment: volume 2, pp 317–348

Gul A, Gallus I, Tegginamath A, Maryska J, Yalcinkaya F (2021) Electrospun antibacterial nanomaterials for wound dressings applications. Membranes 11(12):908

Gupta D, Lall A, Kumar S, Patil TD, Gaikwad KK (2024) Plant based edible films and coatings for food packaging applications: recent advances, applications, and trends. Sustain Food Technol 2:1428–1455

Gvozdenko AA, Siddiqui SA, Blinov AV, Golik AB, Nagdalian AA, Maglakelidze DG, Statsenko EN, Pirogov MA, Blinova AA, Sizonenko MN, Simonov AN (2022) Synthesis of CuO nanoparticles stabilized with gelatin for potential use in food packaging applications. Sci Rep 12(1):12843

Hano C, Abbasi BH (2021) Plant-based green synthesis of nanoparticles: production, characterization and applications. Biomolecules 12(1):31

Huang Y, Mei L, Chen X, Wang Q (2018) Recent developments in food packaging based on nanomaterials. Nanomaterials 8(10):830

Ikhioya IL, Nkele AC, Ochai-Ejeh FU (2024) Green synthesis of copper oxide nanoparticles using neem leaf extract (Azadirachta indica) for energy storage applications. Mater Res Innov 28(3):214–220

Iqbal T, Raza A, Zafar M, Afsheen S, Kebaili I, Alrobei H (2022) Plant-mediated green synthesis of zinc oxide nanoparticles for novel application to enhance the shelf life of tomatoes. Appl Nanosci 12:1–3

Jadoun S, Arif R, Jangid NK, Meena RK (2021) Green synthesis of nanoparticles using plant extracts: a review. Environ Chem Lett 19(1):355–374

Jafarzadeh S, Nooshkam M, Zargar M, Garavand F, Ghosh S, Hadidi M, Forough M (2024) Green synthesis of nanomaterials for smart biopolymer packaging: challenges and outlooks. J Nanostructure Chem 14(2):113–136

Janani N, Zare EN, Salimi F, Makvandi P (2020) Antibacterial tragacanth gum-based nanocomposite films carrying ascorbic acid antioxidant for bioactive food packaging. Carbohydr Polym 247:116678
Jeong S, Choi SY, Park J, Seo JH, Park J, Cho K, Joo SW, Lee SY (2011) Low-toxicity chitosan gold nanoparticles for small hairpin RNA delivery in human lung adenocarcinoma cells. J Mater Chem 21(36):13853–13859
Jing Y, Wang H, Chen X, Wang X, Wei H, Guo Z (2014) Pulsed laser deposited Ag nanoparticles on nickel hydroxide nanosheet arrays for highly sensitive surface-enhanced Raman scattering spectroscopy. Appl Surf Sci 316:66–71
Kalia A, Parshad VR (2015) Novel trends to revolutionize preservation and packaging of fruits/fruit products: microbiological and nanotechnological perspectives. Crit Rev Food Sci Nutr 55(2):159–182
Kalpana VN, Devi Rajeswari V (2018) A review on green synthesis, biomedical applications, and toxicity studies of ZnO NPs. Bioinorg Chem Appl 2018(1):3569758
Khan T, Ullah N, Khan MA, Mashwani ZU, Nadhman A (2019) Plant-based gold nanoparticles; a comprehensive review of the decade-long research on synthesis, mechanistic aspects and diverse applications. Adv Colloid Interface Sci 272:102017
Khan M, Al-Hamoud K, Liaqat Z, Shaik MR, Adil SF, Kuniyil M, Alkhathlan HZ, Al-Warthan A, Siddiqui MR, Mondeshki M, Tremel W (2020) Synthesis of au, ag, and au–ag bimetallic nanoparticles using pulicaria undulata extract and their catalytic activity for the reduction of 4-nitrophenol. Nanomaterials 10(9):1885
Krkobabić A, Radetić M, Tseng HH, Nunney TS, Tadić V, Ilic-Tomic T, Marković D (2023) Green in situ synthesis of Ag-and Cu-based nanoparticles on viscose fabric using a Punica granatum peel extract. Appl Surf Sci 611:155612
Kumar R, Münstedt H (2005) Silver ion release from antimicrobial polyamide/silver composites. Biomaterials 26(14):2081–2088
Kumar V, Gupta R, Bansal A (2021) Hydrothermal growth of ZnO nanorods for use in dye-sensitized solar cells. ACS Appl Nano Mater 4(6):6212–6222
Kumari S, Raturi S, Kulshrestha S, Chauhan K, Dhingra S, András K, Thu K, Khargotra R, Singh T (2023) A comprehensive review on various techniques used for synthesizing nanoparticles. J Mater Res Technol 27:1739–1763
Leta TB, Adeyemi JO, Fawole OA (2024) Shelf-life extension of pomegranate arils using banana powder-based nanocomposite coatings functionalized with cellulose nanofiber and zinc-oxide nanoparticles. CyTA-J Food 22(1):2405633
Li KK, Yin SW, Yang XQ, Tang CH, Wei ZH (2012) Fabrication and characterization of novel antimicrobial films derived from thymol-loaded zein–sodium caseinate (SC) nanoparticles. J Agric Food Chem 60(46):11592–11600
Li W, Li L, Cao Y, Lan T, Chen H, Qin Y (2017) Effects of PLA film incorporated with ZnO nanoparticle on the quality attributes of fresh-cut apple. Nanomaterials 7(8):207
Lieu DM, Dang TK, Nguyen TH (2024) Green synthesized silver nanoparticles, a sustainable approach for fruit and vegetable preservation: an overview. Food Chemistry: X 23:101664
Lu H, Yadav V, Bilal M, Iqbal HM (2022) Bioprospecting microbial hosts to valorize lignocellulose biomass–environmental perspectives and value-added bioproducts. Chemosphere 288:132574
Majeed S, Bakhtiar NF, Danish M, Ibrahim MM, Hashim R (2019) Green approach for the biosynthesis of silver nanoparticles and its antibacterial and antitumor effect against osteoblast MG-63 and breast MCF-7 cancer cell lines. Sustain Chem Pharm 12:100138
Mao X, Hao C (2024) Recent advances in the use of composite titanium dioxide nanomaterials in the food industry. J Food Sci 89(3):1310–1323
Mathew S, Radhakrishnan EK (2021) Silver-based nanomaterials for food packaging applications. In: Silver nanomaterials for agri-food applications. Elsevier, pp 125–146
Mukherjee A, Sarkar D, Sasmal S (2021) A review of green synthesis of metal nanoparticles using algae. Front Microbiol 12:693899

Murthy HA, Desalegn T, Kassa M, Abebe B, Assefa T (2020) Synthesis of green copper nanoparticles using medicinal plant Hagenia abyssinica (Brace) JF. Gmel. leaf extract: Antimicrobial properties. J Nanomater 2020(1):3924081

Nadeem M, Tungmunnithum D, Hano C, Abbasi BH, Hashmi SS, Ahmad W, Zahir A (2018) The current trends in the green syntheses of titanium oxide nanoparticles and their applications. Green Chem Lett Rev 11(4):492–502

Nguyen TH, Nguyen VC, Phan TN, Le VT, Vasseghian Y, Trubitsyn MA, Nguyen AT, Chau TP, Doan VD (2022) Novel biogenic silver and gold nanoparticles for multifunctional applications: green synthesis, catalytic and antibacterial activity, and colorimetric detection of Fe (III) ions. Chemosphere 287:132271

Nile SH, Baskar V, Selvaraj D, Nile A, Xiao J, Kai G (2020) Nanotechnologies in food science: applications, recent trends, and future perspectives. Nano-Micro Lett 12:1–34

Okeke ES, Nweze EJ, Anaduaka EG, Okoye CO, Anosike CA, Joshua PE, Ezeorba TP (2023) Plant-derived nanomaterials (PDNM): a review on pharmacological potentials against pathogenic microbes, antimicrobial resistance (AMR) and some metabolic diseases. 3 Biotech. 13(9):291

Pachaiappan R, Rajendran S, Ramalingam G, Vo DV, Priya PM, Soto-Moscoso M (2021) Green synthesis of zinc oxide nanoparticles by Justicia adhatoda leaves and their antimicrobial activity. Chem Eng Technol 44(3):551–558

Pandit C, Roy A, Ghotekar S, Khusro A, Islam MN, Emran TB, Lam SE, Khandaker MU, Bradley DA (2022) Biological agents for synthesis of nanoparticles and their applications. J King Saud Univ Sci 34(3):101869

Patil N, Bhaskar R, Vyavhare V, Dhadge R, Khaire V, Patil Y (2021) Overview on methods of synthesis of nanoparticles. Int J Curr Pharm Res 13(2):11–16

Popescu M, Velea A, Lőrinczi A (2010) Biogenic production of nanoparticles. Digest J Nanomater Biostruct (DJNB) 5(4):187

Prasad KS, Prasad SK, Ansari MA, Alzohairy MA, Alomary MN, AlYahya S, Srinivasa C, Murali M, Ankegowda VM, Shivamallu C (2020) Tumoricidal and bactericidal properties of ZnONPs synthesized using Cassia auriculata leaf extract. Biomolecules 10(7):982

Rajput N (2015) Methods of preparation of nanoparticles-a review. Int J Adv Eng Technol 7(6):1806

Rani N, Singh P, Kumar S, Kumar P, Bhankar V, Kumar K (2023) Plant-mediated synthesis of nanoparticles and their applications: a review. Mater Res Bull 163:112233

Rohasmizah H, Azizah M (2022) Pectin-based edible coatings and nanoemulsion for the preservation of fruits and vegetables: a review. Appl Food Res 2(2):100221

Saleem K, Khursheed Z, Hano C, Anjum I, Anjum S (2019) Applications of nanomaterials in leishmaniasis: a focus on recent advances and challenges. Nanomaterials 9(12):1749

Salem SS, Fouda A (2021) Green synthesis of metallic nanoparticles and their prospective biotechnological applications: an overview. Biol Trace Elem Res 199(1):344–370

Saravanan A, Kumar PS, Hemavathy RV, Jeevanantham S, Kamalesh R, Sneha S, Yaashikaa PR (2021) Methods of detection of food-borne pathogens: a review. Environ Chem Lett 19:189–207

Si S, Mandal TK (2007) Tryptophan-based peptides to synthesize gold and silver nanoparticles: a mechanistic and kinetic study. Chem Eur J 13(11):3160–3168

Silvestre C, Duraccio D, Cimmino S (2011) Food packaging based on polymer nanomaterials. Prog Polym Sci 36(12):1766–1782

Singh M, Sahareen T (2017) Investigation of cellulosic packets impregnated with silver nanoparticles for enhancing shelf-life of vegetables. LWT 86:116–122

Singh P, Kim YJ, Zhang D, Yang DC (2016) Biological synthesis of nanoparticles from plants and microorganisms. Trends Biotechnol 34(7):588–599

Singh J, Dutta T, Kim KH, Rawat M, Samddar P, Kumar P (2018) 'Green' synthesis of metals and their oxide nanoparticles: applications for environmental remediation. J Nanobiotechnol 16:1–24

Sridhar A, Ponnuchamy M, Kumar PS, Kapoor A (2021) Food preservation techniques and nanotechnology for increased shelf life of fruits, vegetables, beverages and spices: a review. Environ Chem Lett 19:1715–1735

Srihasam S, Thyagarajan K, Korivi M, Lebaka VR, Mallem SP (2020) Phytogenic generation of NiO nanoparticles using Stevia leaf extract and evaluation of their in-vitro antioxidant and antimicrobial properties. Biomolecules. 10(1):89

Sukumar S, Rudrasenan A, Padmanabhan Nambiar D (2020) Green-synthesized rice-shaped copper oxide nanoparticles using Caesalpinia bonducella seed extract and their applications. ACS Omega 5(2):1040–1051

Suo B, Li H, Wang Y, Li Z, Pan Z, Ai Z (2017) Effects of ZnO nanoparticle-coated packaging film on pork meat quality during cold storage. J Sci Food Agric 97(7):2023–2029

Supraja N, Prasad TN, Krishna TG, David E (2016) Synthesis, characterization, and evaluation of the antimicrobial efficacy of Boswellia ovalifoliolata stem bark-extract-mediated zinc oxide nanoparticles. Appl Nanosci 6:581–590

Tata P, Bhavya KS, Ganesan R, Dutta JR (2023) Nanoparticle-based treatment of bacterial biofilms. In: Viral, parasitic, bacterial, and fungal infections. Academic, pp 563–573

Tavakoli H, Rastegar H, Taherian M, Samadi M, Rostami H (2017) The effect of nano-silver packaging in increasing the shelf life of nuts: an in vitro model. Ital J Food Saf 6(4):6874

Vieira AC, de Matos Fonseca J, Menezes NM, Monteiro AR, Valencia GA (2020) Active coatings based on hydroxypropyl methylcellulose and silver nanoparticles to extend the papaya (Carica papaya L.) shelf life. Int J Biol Macromol 164:489–498

Wu Z, Zhou W, Pang C, Deng W, Xu C, Wang X (2019) Multifunctional chitosan-based coating with liposomes containing laurel essential oils and nanosilver for pork preservation. Food Chem 295:16–25

Zhang W, Xiao B, Fang T (2018) Chemical transformation of silver nanoparticles in aquatic environments: mechanism, morphology and toxicity. Chemosphere 191:324–334

Zheng K, Branicio PS (2020) Synthesis of metallic glass nanoparticles by inert gas condensation. Phys Rev Mater 4(7):076001

Zubair M, Rauf Z, Fatima S, Ullah A (2024) Lignin-derived bionanocomposites as functional food packaging materials. Sustain Food Technol 2:945–966

Chapter 15
Plant Bioactive-Based Nanomaterials for Wound Dressing Applications

A. Ch. Pradyutha and P. Sakuntala

15.1 Introduction

15.1.1 Overview of Wound Healing

The process of wound healing is dynamic and intricate, restoring the integrity of injured tissues. It involves multiple stages and cell types, working together to repair the wound site. The process is vital for maintaining skin and tissue integrity after injury and ensuring infection protection (Strodtbeck 2001; Litvinova et al. 2025).

15.1.2 Stages of Wound Healing

Hemostasis: Hemostasis is the body's initial response to injury, with the main goal of stopping bleeding. Whea cut occurs, blood vessels constrict to reduce blood flow, and platelets form a clump together to seal the injured area, preventing excessive blood loss and laying a foundation for healing to begin (Hosgood 2006; Libretti and Puckett 2023). Inflammation follows within a few hours and lasts several days. This phase is marked by redness, heat, swelling, and pain, as white blood cells combat infection and clear away debris, laying the groundwork for tissue repair.

A. Ch. Pradyutha (✉)
Department of Microbiology, R.B.V.R.R Women's College,
Narayanguda, Hyderabad, Telangana, India

P. Sakuntala
Department of Physics, Geethanjali College of Engineering and Technology,
Hyderabad, Telangana, India

V. Ghosh et al. (eds.), *Applications of Plant Bioactive Compounds-Based Nanoformulations*, Nanotechnology in the Life Sciences,
https://doi.org/10.1007/978-3-032-13086-0_15

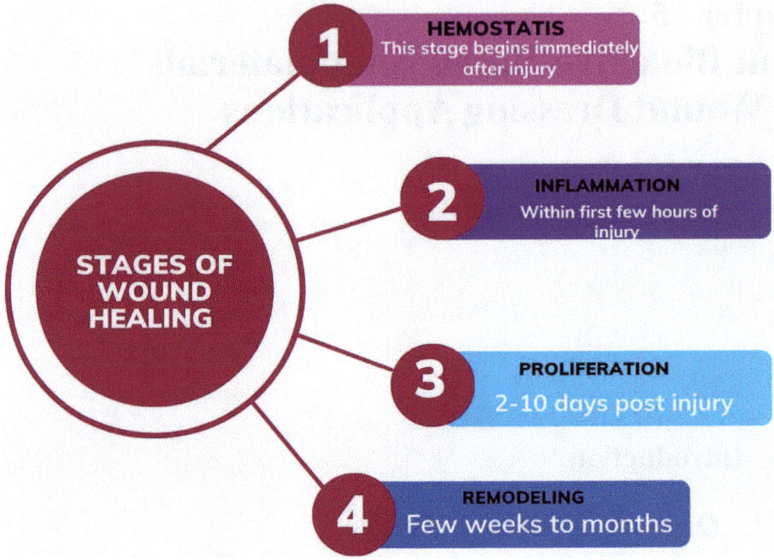

Fig. 15.1 Stages of wound healing

Active tissue reconstruction takes place during the proliferation phase, which begins a few days after injury and lasts for several weeks. Collagen and other extracellular matrix constituents, which serve as the building blocks of new tissue, are produced by fibroblasts. Keratinocytes proliferate to repair the skin's surface, while blood vessels regenerate to deliver nutrients and oxygen (Enoch and Leaper 2008).

Remodeling is the stage where newly formed tissue is refined and strengthened. Collagen is reorganized to improve tensile strength, and this process can last months or even years. It results in scar tissue that may not fully resemble the original yet demonstrates the body's impressive regenerative capacity (Billman 2020) (Fig. 15.1).

15.1.3 Factors Affecting Wound Healing

15.1.3.1 Age

Older individuals typically heal more slowly for the following reasons.

The healing process tends to slow down as people age because of a number of physiological changes. One important factor is delayed cellular turnover, which causes the skin's capacity to regenerate to deteriorate over time. Fibroblasts, which are essential for the production of collagen and the contraction of wounds, become less active. Furthermore, as people age, their skin becomes thinner and less elastic, which increases their vulnerability to injury and slows their recovery. Together, these aging-related changes cause older adults' wounds to take longer to heal (Earle 2024).

15.1.3.2 Nutrition

Proteins, vitamins and Minerals are essential for the healing of wounds. Since proteins provide the amino acids required for collagen synthesis, angiogenesis, and fibroblast proliferation, they are crucial for tissue repair. Among the vitamins, vitamin C is essential for the synthesis of collagen and for boosting immunity to ward off infections. While vitamin E is a potent antioxidant that shields cells from oxidative damage during the healing process, vitamin A promotes epithelialization and strengthens the immune system. Minerals like zinc are essential for the enzymatic processes that produce collagen and DNA; a zinc deficiency can cause major delays in the healing of wounds. Additionally, iron is necessary for the transportation of oxygen through hemoglobin, which is essential for cellular energy (Hajj et al. 2024).

15.1.3.3 Infection

Through a number of mechanisms, infection can seriously impede the healing process of wounds. Prolonged inflammation brought on by bacterial colonization frequently causes the body to release toxins and enzymes that can harm the surrounding tissues. Bacteria can occasionally form structured communities called biofilms, which erect a physical barrier that prevents immune cells and antibiotics from getting to the pathogens and killing them. Furthermore, infection prevents the formation of granulation tissue by slowing down angiogenesis and obstructing the extracellular matrix's deposition, both of which are necessary for efficient tissue repair. As a result, the overall healing process is considerably delayed (Uberoi et al. 2024).

15.1.3.4 Chronic Conditions

Wound healing can be severely hampered by diabetes mellitus and its related complications. Diabetic vascular problems impair blood flow, which limits the amount of oxygen and vital nutrients that reach the wound site and impedes the healing process. Because diabetic neuropathy reduces pain perception, people are less conscious of wounds, increasing the possibility of secondary infections and unseen wounds. Because hyperglycemia inhibits white blood cell function and decreases collagen production, it makes healing even more difficult. Furthermore, arterial blood flow is restricted by peripheral vascular disease (PVD), which is frequently observed in diabetic patients. This deprives tissues of the oxygen and nutrients required for efficient repair. Another frequent consequence, chronic kidney disease, slows down the healing process by interfering with tissue regeneration and immune function because of underlying metabolic imbalances (Yachmaneni et al. 2023).

15.1.3.5 Medications

Chemotherapy and corticosteroids can inhibit wound healing in a number of ways. Because of their anti-inflammatory properties, corticosteroids inhibit the natural inflammatory response, which is essential in the early stages of healing. By preventing fibroblast proliferation, which impacts the extracellular matrix's deposition, they also lessen the synthesis of collagen. Corticosteroids are immunosuppressants that impair immunity, raising the risk of infection and delaying recovery even more. Chemotherapy drugs also target cells that divide quickly, such as cancer cells, keratinocytes, fibroblasts, and endothelial cells that are necessary for wound healing. Recovery is delayed and tissue regeneration is severely hampered by this disruption (Bennett et al. 2024).

15.1.3.6 Additional Factors

Numerous environmental and lifestyle factors can affect how quickly wounds heal. For example, smoking reduces blood flow and oxygen delivery to the wound site because of nicotine's effects, which also prevent collagen synthesis and angiogenesis. Maintaining adequate hydration is equally crucial because dehydrated skin loses its elasticity and becomes more prone to cracking, which can postpone re-epithelialization. The environment surrounding the wound is also very important; moist wounds heal more quickly than dry ones because moisture promotes cell migration and inhibits the development of crusts, which can obstruct healing Knowing these elements enables more individualized and successful interventions, especially for high-risk groups like the elderly or people with long-term medical conditions (Lee et al. 2024; Dobson et al. 2024).

15.1.4 Need for Advanced Wound Dressings

As traditional wound care methods have become less effective over the years, the need for advanced wound dressings has grown. Chronic wounds, including diabetic ulcers, pressure sores, venous leg ulcers, burns, and post-surgical wounds, present significant challenges that conventional dressings such as gauze, cotton pads, or bandages cannot adequately resolve. These traditional methods often don't work well for modern wound care, which includes controlling moisture, preventing infection, and speeding up tissue regeneration. As a result, advanced wound dressings have become a sophisticated solution that meets these needs by creating the best environment for healing.

15.1.4.1 Limitations of Traditional Dressings

- *Inadequate Moisture Management*: Traditional wound dressings often fail to maintain the necessary moisture balance in the wound bed. A dry wound environment can lead to desiccation and delayed healing, while excessive moisture can result in the maceration of surrounding tissues, increasing the risk of infection.
- *Limited Infection Control*: Simple gauze or cotton dressings provide minimal protection against microbial contamination. They are often permeable to bacteria and lack the antimicrobial properties required to prevent infection in high-risk wounds, such as chronic ulcers or burns.
- *Lack of Bioactive Properties*: Conventional dressings do not actively participate in the healing process. They merely cover the wound and provide physical protection but cannot deliver therapeutic agents, such as drugs, growth factors, or antimicrobial agents, directly to the wound site.
- *Pain and Trauma During Dressing Changes*: Changing traditional dressings can be painful and traumatic for patients, as they often adhere to the wound bed, causing further damage to healing tissues and increasing patient discomfort.
- *Inconsistent Gas Exchange*: Effective wound healing requires an optimal balance of oxygen and carbon dioxide exchange. Traditional dressings do not provide the controlled permeability required for adequate gas exchange, which can impede the healing process.(Jiang et al. 2023)

15.1.4.2 Advantages of Advanced Wound Dressings

Advanced wound dressings are made to get around the problems with traditional options by providing customized solutions for different types of wounds and conditions (Alberts et al. 2025). Some of the most important features and benefits of advanced wound dressings are:

- *Moisture Management*: Hydrocolloids, hydrogels, and foams are examples of advanced dressings that are made to keep the wound bed at the right level of moisture. They soak up extra exudate while keeping enough moisture to keep cells active and help tissues heal. This controlled setting stops drying out and breaking down, which speeds up healing.
- *Infection Prevention*: Numerous cutting-edge dressings are impregnated with antimicrobial substances, like iodine, chlorhexidine, or silver nanoparticles, which prevent bacterial growth and lower the risk of wound infections. For instance, antimicrobial dressings made of nanomaterials offer persistent defense against a variety of infections.
- *Delivery of Therapeutic Agents*: Advanced dressings can act as carriers for drugs, growth factors, or stem cells, enabling localized and sustained delivery of these bioactive compounds to the wound site. This feature enhances wound healing by

promoting angiogenesis, reducing inflammation, and accelerating tissue regeneration.

- *Minimized Pain and Trauma*: Non-adherent advanced dressings, such as silicone-coated foam dressings, reduce pain and trauma during dressing changes. They do not adhere to the wound bed, allowing for easy removal without damaging newly formed tissues.
- *Improved Gas Exchange*: Advanced dressings are designed to facilitate controlled gas exchange, ensuring an adequate supply of oxygen to the wound while preventing excessive water vapor loss. This balance promotes cellular respiration and supports the wound-healing process.
- *Enhanced Mechanical Properties*: Many advanced dressings are flexible, conformable, and durable, making them suitable for wounds in challenging anatomical locations. They provide mechanical support and protection against external factors while allowing for patient mobility.
- *Reduction of Scarring*: By creating a conducive healing environment, advanced wound dressings help minimize scarring and promote the regeneration of healthy tissue. This is particularly important for burn wounds and surgical incisions, where aesthetic outcomes are a concern.

15.1.4.3 Types of Advanced Wound Dressings

Advanced wound dressings are specifically made to encourage quicker and more efficient healing, particularly in complicated or chronic wounds. These are frequently antimicrobial, moisture-retentive, or bioactive, in contrast to conventional dressings (Farahani and Shafiee 2021).

The primary kinds consist of

- *Hydrocolloids*: Hydrocolloid dressings are composed of hydrophilic particles, such as gelatin or pectin, embedded in an adhesive matrix. They form a gel when they come into contact with wound exudate, maintaining a moist environment that supports autolytic debridement and promotes granulation tissue formation. These dressings are particularly effective for low-to-moderate exuding wounds, such as pressure ulcers and shallow burns.
- *Hydrogels*: Hydrogels are three-dimensional polymeric networks that contain a high water content, providing hydration to dry wounds. They are ideal for wounds with minimal exudate, such as necrotic or sloughy wounds, as they facilitate autolytic debridement and soothe pain. Additionally, hydrogels can be infused with antimicrobial agents or growth factors for enhanced therapeutic effects.
- *Foam Dressings*: Foam dressings are composed of semipermeable polyurethane foam that absorbs excess exudate while maintaining a moist wound environment. They are suitable for moderate-to-high exuding wounds and provide cushioning to protect the wound from external pressure and trauma. Foam dressings are often used for venous leg ulcers, diabetic foot ulcers, and pressure sores.

- *Nanomaterial-Based Dressings*: Nanotechnology has revolutionized wound care with the development of nanomaterial-based dressings. These dressings incorporate nanoparticles, such as silver, gold, or zinc oxide, to provide antimicrobial properties and promote tissue regeneration. Additionally, plant-based nanomaterials, such as cellulose nanocrystals, offer biodegradable and biocompatible options for advanced wound care.
- *Alginate Dressings*: Made from natural alginic acid derived from seaweed, alginate dressings are highly absorbent and form a gel-like layer when in contact with wound exudate. They are effective for managing heavily exuding wounds, such as venous ulcers or surgical wounds, and support a moist environment for faster healing.
- *Collagen Dressings*: Collagen-based dressings provide a structural scaffold for cell migration and tissue regeneration. They are particularly beneficial for chronic wounds that fail to progress through the normal stages of healing. Collagen dressings can also be combined with growth factors or antimicrobial agents to enhance their efficacy.
- *Bioactive Dressings*: Bioactive dressings incorporate biologically active compounds, such as growth factors, peptides, or cytokines, to actively stimulate the wound-healing process. These dressings target specific aspects of wound healing, such as angiogenesis or fibroblast proliferation, to accelerate tissue repair.

15.1.4.4 Applications and Benefits of Advanced Wound Dressings

The versatility of advanced wound dressings makes them suitable for a wide range of applications, including:

- Chronic Wounds: Diabetic ulcers, venous leg ulcers, and pressure sores benefit significantly from advanced dressings due to their ability to manage exudate, prevent infection, and promote granulation tissue formation.
- Burns: Advanced dressings, such as hydrogels and nanomaterial-based options, provide a soothing effect, reduce pain, and prevent microbial colonization in burn wounds. They also facilitate faster re-epithelialization and reduce scarring.
- Surgical Wounds: Post-operative wounds require dressings that protect against infection and promote tissue regeneration. Foam dressings and collagen-based dressings are commonly used in this context.
- Traumatic Wounds: Advanced dressings provide mechanical protection and support for traumatic wounds, reducing the risk of infection and accelerating healing.
- Pediatric and Geriatric Wounds: These populations benefit from advanced dressings due to their non-adherent properties, which minimize pain and discomfort during dressing changes.

By overcoming the drawbacks of conventional techniques, the introduction of sophisticated wound dressings has revolutionized wound care. By controlling moisture levels, avoiding infections, and delivering bioactive agents straight to the wound site, these cutting-edge dressings create the ideal healing environment. Advanced wound dressings are essential to contemporary wound care management because of their capacity to minimize scarring, lessen pain, and speed healing. Ongoing research and development in this field continue to expand the possibilities for tailored and effective wound care solutions, ultimately improving patient outcomes and quality of life (Su et al. 2023).

15.1.5 Recent Advances in Wound Healing

The creation of sophisticated wound dressings is being revolutionized by nanotechnology, especially with the application of plant-based nanomaterials. These materials' inherent antioxidant and antimicrobial qualities aid in lowering oxidative stress and infection at the wound site, hastening the healing process. To administer therapeutic agents, like medications, growth factors, or antioxidants, directly and gradually at the site of injury, controlled release systems are being developed concurrently. By sustaining ideal concentrations of bioactive substances over time, this focused strategy improves tissue repair and raises the general efficacy and efficiency of the healing process.

Wound healing is an essential area of research in medical and biological sciences, and advancements in biomaterials, drug delivery systems, and regenerative medicine continually improve outcomes.

15.1.6 Role of Nanomaterials in Wound Healing

Based on their special qualities and capacity to improve several aspects of the healing process, nanomaterials are essential to wound healing. Because of their nanoscale size, they can interact with biological systems more effectively, facilitating quicker and more efficient tissue repair. While gold and silica nanoparticles can be designed to deliver medications, growth factors, or antioxidants directly to the wound site, guaranteeing targeted and sustained release, silver nanoparticles and other nanomaterials have antimicrobial qualities that help prevent infections in wounds. Furthermore, scaffolds based on nanofibers and nanoparticles can replicate the extracellular matrix, offering structural support for the development of new tissue. These substances also improve collagen synthesis and encourage angiogenesis, or the growth of new blood vessels, which speeds up the healing process. All things considered, nanomaterials are revolutionizing wound care by providing sophisticated (Wang et al. 2021).

15.2 Plant Bioactives and Their Therapeutic Potential

15.2.1 Types of Plant Bioactives

Naturally occurring substances known as plant bioactives affect living things biologically and frequently support a plant's defense or signaling systems. Although these substances are not necessary for the plant's basic metabolism, they are important for interactions with the environment, including defense against stress, disease, and pests. Antioxidant, antimicrobial, anti-inflammatory, and anticancer properties are just a few of the therapeutic advantages that plant bioactives have been acknowledged for in human health (Stephane et al. 2022; Ramawat et al. 2009).

15.2.1.1 Phenolics

Important categories of phenolic compounds include flavonoids like quercetin and catechin, phenolic acids like gallic acid and caffeic acid, and tannins like ellagitannins and condensed tannins.

Phenolics are one of the most abundant classes of bioactives, characterized by their aromatic structure with hydroxyl groups. They possess strong antioxidant activity, effectively neutralizing free radicals that cause oxidative stress. Their ability to reduce oxidative damage makes them crucial in preventing chronic diseases like cancer, cardiovascular disorders, and neurodegenerative conditions. Additionally, phenolics exhibit significant anti-inflammatory and antimicrobial properties. Tannins, for example, inhibit bacterial growth by precipitating microbial proteins, while flavonoids modulate immune responses.

15.2.1.2 Alkaloids

A class of naturally occurring substances called alkaloids is well-known for having potent physiological effects on both people and animals. Caffeine, a stimulant of the central nervous system that is frequently present in coffee and tea, is one example. Because of its potent analgesic effects, morphine—which is derived from the opium poppy—is used extensively. Quinine has long been used to treat malaria after being extracted from the bark of cinchona trees. The stimulant and addictive qualities of nicotine, which is extracted from tobacco plants, are well-known. These alkaloids are significant in both conventional and contemporary medicine because of their wide range of biological activities.

Alkaloids are nitrogen-containing organic compounds that are well-known for their pharmacological activities. These compounds have been utilized for centuries in traditional medicine, and many of them serve as the basis for modern drugs. Below is an in-depth exploration of their characteristics and applications:

- Pain Relief: Morphine is among the most potent analgesics derived from plants. It acts on the central nervous system (CNS) by binding to opioid receptors, thus blocking the transmission of pain signals. While morphine is indispensable in pain management, especially in terminal illnesses and post-surgical care, it is also associated with a risk of dependency and side effects such as respiratory depression.
- Stimulation of the Central Nervous System: Caffeine, a widely consumed alkaloid, is a natural stimulant that temporarily wards off fatigue and enhances alertness. It functions by antagonizing adenosine receptors in the brain, which leads to increased release of neurotransmitters like dopamine and norepinephrine. Apart from its stimulating effects, caffeine has shown promise in neuroprotective roles, potentially reducing the risk of Alzheimer's and Parkinson's diseases.
- Antimalarial Activity: Quinine, derived from cinchona bark, was historically a breakthrough in malaria treatment. Quinine works by interfering with the parasite's ability to digest haemoglobin, thus inhibiting its growth. Despite being largely replaced by synthetic antimalarials, quinine remains a vital treatment option in certain regions.
- Addiction and Dependence: Nicotine is another alkaloid with profound effects on the CNS. It stimulates the release of neurotransmitters like dopamine, providing a sense of pleasure and relaxation. However, its addictive potential has made nicotine one of the most controversial bioactives, primarily due to its association with tobacco use and related health risks.
- Antimicrobial and Anticancer Effects: Many alkaloids, such as berberine and camptothecin, exhibit antimicrobial and anticancer properties. Berberine disrupts bacterial cell membranes, while camptothecin inhibits topoisomerase enzymes essential for DNA replication in cancer cells. These properties make alkaloids valuable in combating multidrug-resistant pathogens and certain types of cancer.

15.2.1.3 Terpenoids

One of the biggest and most varied classes of plant bioactive substances are terpenoids, which are derived from isoprene units. They are well-known for a variety of biological activities, such as antibacterial, anti-inflammatory, and antioxidant properties. Eugenol from clove oil, carotenoids like beta-carotene and lycopene, and menthol from peppermint are examples of terpenoids. Because of its analgesic and cooling effects, menthol is frequently used topically. Carotenoids protect cells from oxidative stress by acting as antioxidants in addition to being precursors to vitamin A. Because of its potent antibacterial properties and capacity to lessen pain and inflammation, eugenol is especially prized in the field of dentistry.

15.2.1.4 Saponins

Saponins are amphipathic glycosides that exhibit a range of health benefits. Examples of Saponins include Ginsenosides (from ginseng), Glycyrrhizin (from licorice). They possess immunomodulatory properties, enhancing the body's defense mechanisms. Ginsenosides, for instance, improve cognitive function and reduce inflammation. Saponins also lower cholesterol levels by forming insoluble complexes with bile acids, thereby increasing their excretion. Additionally, they exhibit anticancer activity by inducing apoptosis in cancer cells.

15.2.1.5 Glucosinolates

Sulfur-containing substances called glucosinolates are well-known for their significant contribution to the prevention of cancer. They contain bioactive substances like sulforaphane and indole-3-carbinol and are present in cruciferous vegetables like broccoli. Glucosinolates are hydrolyzed into active compounds by the enzyme myrosinase when plant tissues are damaged, as happens when they are chopped or chewed. These bioactive products shield cells from oxidative stress and aid in the activation of detoxification enzymes. A useful substance in the field of chemoprevention, sulforaphane in particular has shown the capacity to suppress the growth of cancer cells by affecting gene expression and triggering phase II detoxification enzymes.

15.2.1.6 Sterols

Beta-sitosterol and campesterol are examples of phytosterols, which are substances derived from plants that share structural similarities with cholesterol. They aid in lowering blood cholesterol levels by competing with dietary cholesterol for intestinal absorption. Phytosterols have anti-inflammatory qualities and may help lower the risk of cardiovascular diseases in addition to lowering cholesterol. Phytosterols are important for maintaining general health because beta-sitosterol, in particular, has been extensively researched for its possible advantages in supporting prostate health and boosting immune function.

15.2.1.7 Essential Oils

Essential oils, including eucalyptus, lavender, and tea tree oils, are volatile aromatic compounds that are frequently used in pharmaceutical, cosmetic, and aromatherapy applications. Strong antibacterial, antifungal, and anti-inflammatory qualities are well-known for these natural extracts. For example, tea tree oil works well against a variety of pathogens, including bacteria that are resistant to antibiotics. Lavender oil is widely used in relaxation and stress-reduction treatments because of its calming and relaxing properties. Conversely, eucalyptus oil is well known for its anti-inflammatory and decongestant properties and is frequently used to treat respiratory conditions. When combined, these essential oils provide a variety of therapeutic advantages.

Plant bioactives, with their diverse chemical structures and bioactivities, are invaluable in both traditional and modern medicine (Sargin 2021; Dar et al. 2023). Alkaloids, in particular, stand out for their profound pharmacological effects, ranging from pain relief to antimicrobial and anticancer properties. As research continues to uncover the potential of these compounds, their role in preventing and treating diseases is expected to expand, paving the way for innovative therapeutic applications.

15.2.2 Mechanisms of Action of Biomaterials in Wound Healing

Biomaterials facilitate wound healing through several key mechanisms designed to support, protect, and accelerate the body's natural repair processes. Their action primarily revolves around providing structural support, enhancing cellular interactions, modulating immune responses, and facilitating tissue regeneration (Agrawal et al. 2014).

- Structural Support and ECM Mimicry: Biomaterials often serve as scaffolds, mimicking the extracellular matrix (ECM) to provide a framework for cell adhesion, migration, and proliferation. This scaffold-like structure ensures that new cells can easily populate the wound area, forming a foundation for tissue regeneration. Materials like collagen and fibrin closely resemble the natural ECM, promoting fibroblast activity and the formation of new tissue.
- Antimicrobial Properties: Infected wounds are challenging to heal, making antimicrobial biomaterials particularly beneficial. Some biomaterials are designed to either inherently resist bacterial colonization or are infused with antimicrobial agents like silver nanoparticles. These properties help maintain a sterile environment, reducing infection risks and preventing biofilm formation, which can delay wound healing.
- Immune Modulation and Anti-inflammatory Effects: Biomaterials can modulate the immune response by interacting with macrophages and other immune cells. By promoting an anti-inflammatory or "pro-healing" phenotype in these cells, they minimize prolonged inflammation and support a smoother transition into tissue formation. This is especially important for chronic wounds, where inflammation is often dysregulated.
- Angiogenesis Stimulation: Effective wound healing requires a robust blood supply to deliver nutrients and oxygen to the wound site. Certain biomaterials are designed to promote angiogenesis through the release of vascular endothelial growth factor (VEGF) or by creating a scaffold that encourages endothelial cell

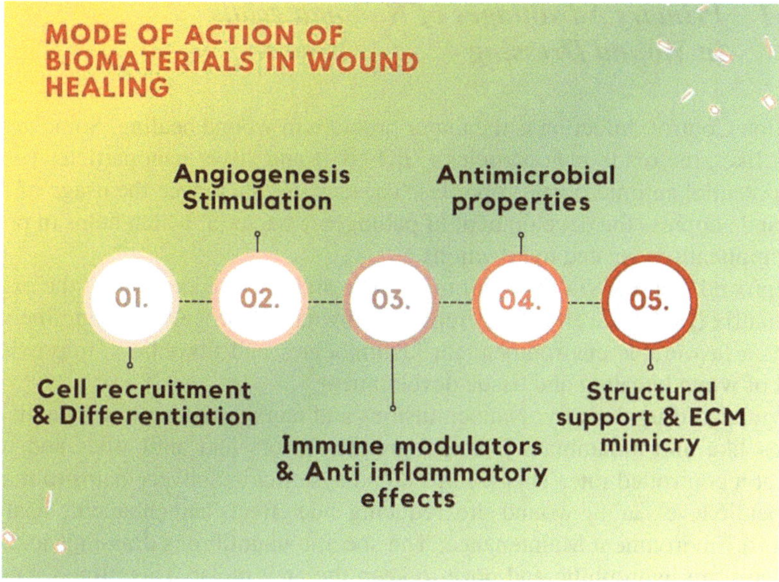

Fig. 15.2 Mode of action of biomaterials in wound healing

growth. Enhanced blood flow aids in faster tissue repair and reduces necrosis risk in the healing area.

- Cell Recruitment and Differentiation: Advanced biomaterials can be engineered to attract specific types of cells, such as stem cells, to the wound site. They can also create a microenvironment that encourages these cells to differentiate into the types necessary for tissue regeneration, such as fibroblasts or keratinocytes in skin wounds. This targeted approach ensures that the right cells populate the wound, leading to improved structural and functional recovery.

These diverse actions highlight the importance of biomaterials in wound healing. By combining structural support, antimicrobial action, immune modulation, and bioactive molecule release, biomaterials effectively accelerate wound repair and improve tissue outcomes, particularly in complex or chronic wounds (Fig. 15.2).

15.3 Nanomaterials in Wound Dressing

Nanomaterials have revolutionized wound care by offering advanced functionalities that traditional dressings cannot provide. Their unique nanoscale properties enable enhanced antimicrobial action, controlled drug delivery, and faster tissue regeneration, making them ideal candidates for wound dressing applications. Here's an overview of nanomaterials' roles in wound dressings and how they support the healing process (Stoica et al. 2020; Zeng et al. 2022).

15.3.1 Primary Advantages of Nanomaterials in Wound Dressing

Infection Control: Infection is the major problem in wound healing. Some nanomaterials like zinc oxide nanoparticles (ZnO NPs) and silver nanoparticles (AgNPs) have potential antimicrobial properties. These materials reduce the usage of antibiotics and suppress the development of pathogenic bacteria, which helps in preventing complications related to infections.

Improved Healing via ECM Mimicry: The architecture and role of the extracellular matrix (ECM) are effectively replicated by nanohydrogels and nanofibers. This creates a favourable environment for keratinocytes and fibroblasts, improving the speed of wound healing and tissue development.

Regulated drug delivery: Nanoemulsions and nanoparticles carry bioactive substances like anti-inflammatory drugs, growth factors and antibiotics and release them at a controlled rate. This localized and systematic delivery helps to maintain therapeutic levels at the wound site, reducing side effects and enhancing healing.

Moist Environment Maintenance: The specific nanofibrous dressings and nanohydrogels are hydrophilic and ooze to keep the area moist. This provides cellular processes required for wound healing. This facilitates a moist environment to maintain optimal conditions to minimize scarring and cell function.

Reduced Scarring and Fibrosis: Nanomaterials can influence cellular responses and Inflammation, resulting in well structured tissue regeneration and reduced fibrosis. They can help in reducing scar formation and enhanced aesthetic results by regulating prolonged inflammation and improving cell signalling for organized tissue growth.

15.3.2 Types of Nanomaterials in Wound Dressings

Nanoparticles: Metal-based and polymeric nanoparticles such as silver, zinc oxide and gold etc. are used to provide active cellular responses and antimicrobial effects.

Nanofibers: These are made from collagen, polyactic acid and chitosan. These materials replicate the extracellular matrix (ECM) to support proliferation, interactions and cellular mitigation resulting a s structure suitable for tissue regeneration.

Nanohydrogels: These materials absorb large volumes of water. This maintains the wound wet and can be used for controlled drug release, providing hydration to the wound site.

Nanoemulsions: These substances serve as carriers for lipophilic drugs and improve the penetration and bioavailability, while maintaining a moist and protective layer on the wound surface.

Factors influencing safety and biocompatibility: Nanoparticles have many advantages in the medical field, especially for wound healing. The biocompatibility and efficacy are crucial challenges that needd to be evaluated. The unique physical and chemical properties allow them for many biomedical applications and also pose significant risks to biological structures. These nanoparticles interact with the cells in many ways due to their large surface area, small size and reactivity. This section deals with the key safety concerns, methods to improve biological compatibility to assure safe applications of nanoparticles in wound healing.

15.3.3 Potential Risks Associated with Nanomaterials

- Oxidative Stress and Cytotoxicity: The potential for oxidative stress inflicted by nanomaterials, especially metallic nanoparticles (such as titanium dioxide, gold, and silver), is one of the main issues with their utilization. When nanoparticles interact with cells, they produce reactive oxygen species (ROS), which can cause apoptosis, DNA alterations, and cellular damage. In addition to interfering with regular cellular processes, oxidative stress can lead to chronic inflammation, which is detrimental to wound healing.
  - For example, research has indicated that while silver nanoparticles are excellent antibacterial agents, at greater quantities they may be harmful to human cells. One of the main challenges in creating safe wound dressings based on nanoparticles is striking a balance between cytotoxicity and antibacterial activity.
- Bioaccumulation and Persistence: Certain metallic or ceramic nanoparticles are examples of non-biodegradable nanomaterials that can cause tissue bioaccumulation. In critical organs like the liver, kidneys, and lungs, prolonged retention of these particles can cause fibrosis, chronic inflammation, or even toxicity. This emphasizes the necessity of biodegradable substitutes that the body can safely process and eliminate.
- Immune Response and Inflammation: Unintentionally triggering immune responses by nanoparticles can result in inflammation and possibly allergic reactions. When the immune system perceives nanoparticles as alien substances, macrophages may phagocytose them and trigger inflammatory reactions. Immune system activation over time can hinder wound healing and have negative consequences.
- Interaction with Cellular Structures: Nanoparticles can readily interact with intracellular structures like mitochondria and nuclei and pass through cellular membranes because of their nanoscale size. Unintended interactions can cause toxicity by upsetting cellular homeostasis, even though this property is beneficial for targeted drug delivery.
- Environmental Impact: The effects of nanoparticles on the environment must be taken into account in addition to issues with human safety. When nanomaterials are improperly disposed of, they can contaminate water and soil systems, which could have an impact on ecosystems and make their way into the food chain.

15.3.4 Strategies to Enhance Safety and Biocompatibility

- Surface Modifications: Surface engineering of nanoparticles is a common approach to reduce their toxicity and enhance biocompatibility. By coating nanoparticles with biocompatible polymers such as polyethylene glycol (PEG), chitosan, or dextran, their interaction with biological systems can be controlled. For example, PEGylation reduces protein adsorption and prevents rapid clearance by the immune system, enhancing both safety and efficacy. Functionalizing the surface with specific ligands can also target nanoparticles to specific cells or tissues, minimizing off-target effects and reducing systemic toxicity.
- Use of Biodegradable Nanomaterials:
  - For wound care applications, biodegradable nanoparticles—such as those derived from synthetic or natural polymers—like collagen and alginate—are preferred. By breaking down into non-toxic byproducts that are safe to metabolize and eliminate, these substances lower the chance of bioaccumulation and chronic inflammation.
  - Because plant-based nanomaterials are naturally biocompatible and biodegradable, recent developments in this field have shown promise. For instance, preclinical research has shown that cellulose nanocrystals and starch-based nanoparticles have exceptional safety profiles.
- Dose Optimization: The dose and concentration of nanoparticles play a crucial role in determining their safety. Lower concentrations are generally less toxic while still being effective. Careful dose optimization through in vitro and in vivo studies is essential to strike a balance between therapeutic benefits and potential side effects.
- Incorporation of Antioxidants: To mitigate oxidative stress caused by nanoparticles, incorporating antioxidants into the nanoparticle formulation is an effective strategy. Antioxidants such as vitamin E, quercetin, or ascorbic acid can scavenge reactive oxygen species, reducing the potential for oxidative damage to cells.
- Tailored Size and Shape: The size and shape of nanoparticles significantly influence their interactions with biological systems. Smaller nanoparticles may penetrate deeper into tissues but are more likely to induce toxicity due to their high reactivity. Larger nanoparticles, on the other hand, may have reduced toxicity but limited tissue penetration. Tailoring the size and shape based on the specific application can enhance safety and efficacy.
- Controlled Release Systems: Incorporating nanoparticles into controlled release systems, such as hydrogels or polymeric matrices, can prevent their uncontrolled release into the wound environment. This approach ensures a sustained and localized release of therapeutic agents, minimizing systemic exposure and potential side effects.

15.3.5 Regulatory Oversight and Preclinical Testing

- Toxicological Studies: Before being used in clinical settings, thorough toxicological research is necessary to assess the safety of nanomaterials. In these investigations, genotoxicity, acute and chronic toxicity, and possible carcinogenicity are evaluated. To comprehend how nanoparticles interact with cells, tissues, and organ systems, both in vitro and in vivo models are employed. Cutting-edge methods like omics and high-content imaging can shed more light on the molecular processes behind the toxicity caused by nanoparticles.
- Biocompatibility Testing: The ability of nanomaterials to interact with biological systems without causing negative effects is evaluated through biocompatibility testing. Cytotoxicity, hemocompatibility, immunogenicity, and irritation potential are important factors. These assessments are guided by standards like ISO 10993 to guarantee dependability and consistency.
- Regulatory Frameworks: Guidelines for the assessment of nanomaterials in medical applications have been established by regulatory bodies like the European Medicines Agency (EMA) and the US Food and Drug Administration (FDA). The necessity of thorough characterization, safety testing, and risk assessment is emphasized in these guidelines.
  - Regulatory approval in wound care necessitates proving the product's safety, efficacy, and superiority over current alternatives. In real-world situations, post-market surveillance is also essential for tracking long-term safety.
- Considerations: Ethical considerations play a vital role in the development and testing of nanomaterials. Ensuring informed consent, minimizing the use of animal models through alternative testing methods, and addressing potential environmental impacts are critical aspects of ethical compliance.

15.3.6 Emerging Trends and Future Directions

- Green Synthesis: Green synthesis methods use natural extracts or biological systems to produce nanoparticles, reducing the need for hazardous chemicals. This approach not only enhances the biocompatibility of nanoparticles but also aligns with sustainability goals. For example, plant-mediated synthesis of silver and gold nanoparticles has shown reduced toxicity compared to chemically synthesized counterparts.
- Nanomaterials with Dual Functions: Designing nanomaterials with dual functions, such as antimicrobial and anti-inflammatory properties, can address multiple aspects of wound care while minimizing the need for additional therapeutic agents. For instance, zinc oxide nanoparticles exhibit both antibacterial activity and the ability to promote tissue regeneration.

- Integration with Advanced Technologies: The integration of nanomaterials with advanced technologies, such as 3D printing and wearable sensors, offers exciting possibilities for wound care. For example, 3D-printed wound dressings embedded with nanoparticles can be customized to fit specific wound shapes, while Nano sensors can monitor wound healing in real-time.
- Personalized Medicine: Advances in personalized medicine are paving the way for customized nanoparticle formulations tailored to individual patient needs. Factors such as genetic predisposition, wound type, and patient-specific risks can be considered to optimize safety and efficacy.

Nanomaterials are playing a key role in wound care with their novel approaches for infection control, tissue regeneration and quick healing. However their safety and biocompatibility continue to be major challenges, which require systematic study and advancement. The threats associated with the nanoparticles can be minimised by using methods such as use of biodegradable materials, surface modifications, dose refinement and exact regulatory oversight. The researchers, regulatory bodies and physicians must coordinate and work together to use nanomaterials safely.

Nanomaterials have the ability to produce adaptive and sophisticated wound dressings with respect to changes in the wound environment. Research is in the progress to provide systematic and dynamic support for the doable recovery and to create dressings according to stage of wound healing. A crucial area of wound care is illustrated by nanomaterials, where precision, control and enhanced healing is achieved through nanoscale engineering. These approaches may become more accessible, safer and adaptable with further advancements, reinforcing the wound care strategies for patient needs.

15.4 Plant Bioactive-Based Nanomaterials

15.4.1 Incorporation of Plant Bioactives into Nanomaterials

Numerous biomolecules found in plant extracts and essential oils are important in a variety of industries, including medicine, cosmetics, and the food industry. Significant biological activities, including bactericidal, antitumor, anti-inflammatory, and antifungal properties, are possessed by these secondary metabolites (Mahady 2005; Antunes Filho et al. 2022). Because of their potential biological activities, these natural products are essential to the health of both humans and animals. To properly identify and use each class of compounds, however, a variety of extraction, purification, and characterization methods are needed due to their complex compositions. The ability of these plant extracts to synthesize nanomaterials has also drawn attention. With the bioactive qualities of the extracts themselves, they provide a more environmentally friendly option to conventional synthesis techniques, which frequently use hazardous chemical reagents, and may even improve the properties of nanomaterials (Hao et al. 2015; Süntar 2020; Alasmari 2020; Horstmanshoff 1999).

More investigation is still required into the precise molecules in charge of synthesis and stabilization within these processes, despite the fact that plant extracts and essential oils are known to be chemically complex and have been used in the synthesis of nanomaterials. In a similar vein, not much research has thoroughly examined whether the molecules that cover nanomaterials made from plant extracts alter, improve, or lessen their biological activity. Since a synergistic effect between these natural compounds and nanomaterials may open up new therapeutic or industrial applications, this presents an exciting area for future research. The primary molecular groups found in essential oils and plant extracts concentrate on characterization and extraction procedures. The use of plant-based products in the synthesis of metal nanomaterials, investigation of their antimicrobial potential, and evaluation of the potential synergy between their natural components and antimicrobial properties.

15.4.2 Methods of Fabrication (e.g., Electrospinning, Green Synthesis)

Nanofibers are incredibly versatile, and how we produce them often depends on the specific goal of the project. In simple terms, we have two main approaches: 'top-down' and 'bottom-up.' The top-down approach involves breaking down larger materials into extremely small particles. This can be useful for applications where the initial material is robust but needs to be refined into a finer form.

On the other hand, bottom-up methods start with tiny components—often on the scale of just a few nanometers—and build up. This approach is suitable for scenarios where the submicroscopic precision is essential. Chemical, biological and physical methods come under these approaches. Physical techniques generate nanofibers using energy in the form of mechanical or radiation based without any chemicals. Chemical methods use regulated reactions to cluster atoms into predefined forms. Finally, biological techniques use enzymes or microorganisms to produce nanofibers. These are attaining popularity due to their sustainability and eco friendly approach. Each method has its own advantages and uses, the choice of method depends on the end use of nanofibers (Bahrami et al. 2019; Barhoum et al. 2019; Spizzirri et al. 2021; Zhao et al. 2016).

15.4.2.1 Electrospinning

This method is widely used to produce nanofibers because of its versatility and low cost (Nune et al. 2017). The nanofibers created through electrospinning have several benefits, including a large surface area, very fine diameters ranging from nano to microscale, and a porous structure, making them ideal for various applications (Figen 2020).

15.4.2.2 Working of Electrospinning

Figure 15.3 shows the experimental setup for the production of nanofibers using the electrospinning method. It consists of four major components: 1. High voltage power supply

An electrospinning setup typically includes four key components:

- High Voltage Power Supply: Provides the electric field that drives the process.
- Syringe Pump: Controls the flow of the polymer solution.
- Spinneret (Needle): Dispenses the polymer solution into the electric field.
- Collector: Collects the nanofibers as they are formed.

This diagram shows electrospinning components work together to produce a continuous flow of nanofibers.

The high voltage at the spinneret produces an electric field that affects the polymer droplet when a polymer solution—often combined with additives like medications, peptides, or nanoparticles—is loaded into the syringe. The droplet is shaped by this force into a cone, or "Taylor cone." A jet of polymer is released in the direction of the collector once the electric field overcomes the surface tension. The

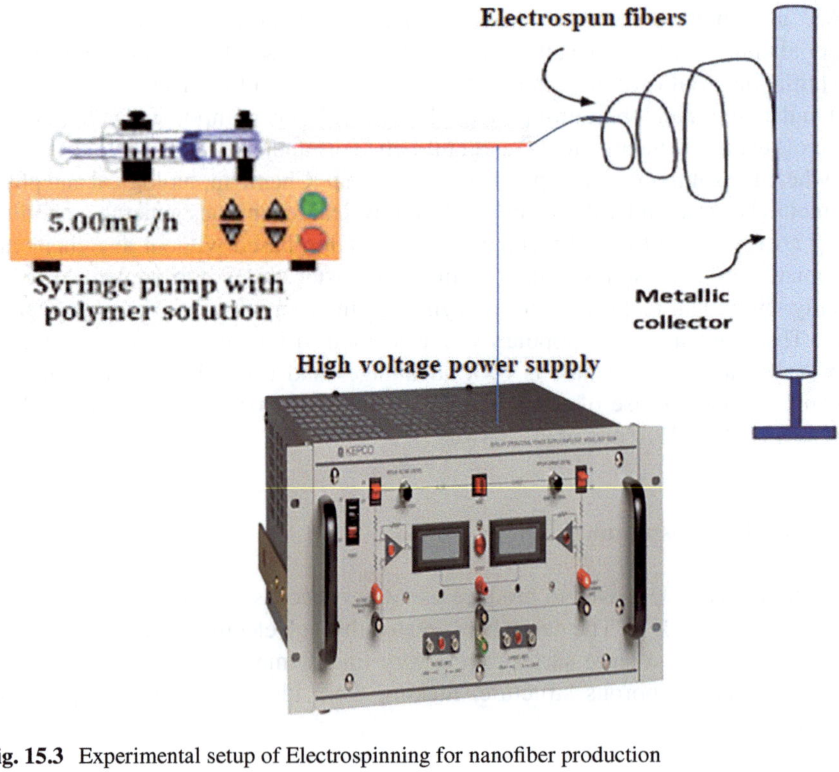

Fig. 15.3 Experimental setup of Electrospinning for nanofiber production

solvent evaporates as this jet passes through the atmosphere, leaving behind a thin, solid fiber that gathers in a nonwoven mat on the grounded collector.

15.4.2.3 Important Parameters

The success of electrospinning depends on a few essential factors, including:

- Polymer properties: Its concentration, conductivity, and viscosity.
- Flow rate: Speed at which the polymer solution is dispensed.
- Distance between the needle and collector: This affects fibre formation and structure.

Electrospinning's Drawbacks
Despite its widespread use, electrospinning has certain drawbacks, such as the need for a high voltage, which may be dangerous. Furthermore, scaling up the method can be costly and has a relatively low production yield (Zhang and Lu 2014).

Examples of Nanomaterials Derived from Plants for Wound Dressing
Because of their inherent bioactivity, biocompatibility, and biodegradability, plant-based nanomaterials have attracted a lot of attention for use in wound care applications. Here are a few instances of these materials that are frequently found in wound dressings:

The high mechanical strength, flexibility, and biocompatibility of cellulose nanofibers—which are extracted from plant cell walls—make them valuable. These fibers are perfect for wound dressings because they can produce a moist environment that aids in healing and acts as a barrier against bacteria. Because of its purity and capacity to retain water, bacterial nanocellulose—which is produced by the fermentation of plant-based sugars by bacteria—is also extensively utilized.

Mushroom-Based Chitosan: Although chitosan is typically made from the shells of crustaceans, plant-based sources such as mushrooms are being investigated as potential vegan substitutes. Chitosan has antimicrobial qualities, encourages cell renewal, and can be incorporated into wound dressings as a hydrogel or nanofiber.

Aloe Vera Nanogels: Known for its antibacterial, anti-inflammatory, and skin-repairing qualities, aloe vera can be successfully added to nanogel formulations. By improving the stability and bioavailability of aloe vera's active ingredients, these nanogels enable deeper skin penetration and controlled release. Because they provide a calming and restorative effect with little irritation, aloe vera nanogels are being investigated extensively for use in burn treatment, wound healing, and cosmetic applications.

Curcumin Nanoparticles: Curcumin, which is extracted from turmeric, possesses potent antibacterial, anti-inflammatory, and antioxidant qualities. Curcumin can be applied to a wound site in controlled dosages when it is in nanoparticle form, which speeds up wound closure and lowers the risk of infection.

Citrus Fruit Pectin: Fruits like apples and citrus contain pectin, a naturally occurring polysaccharide with superior gelling qualities. Pectin, when applied in a

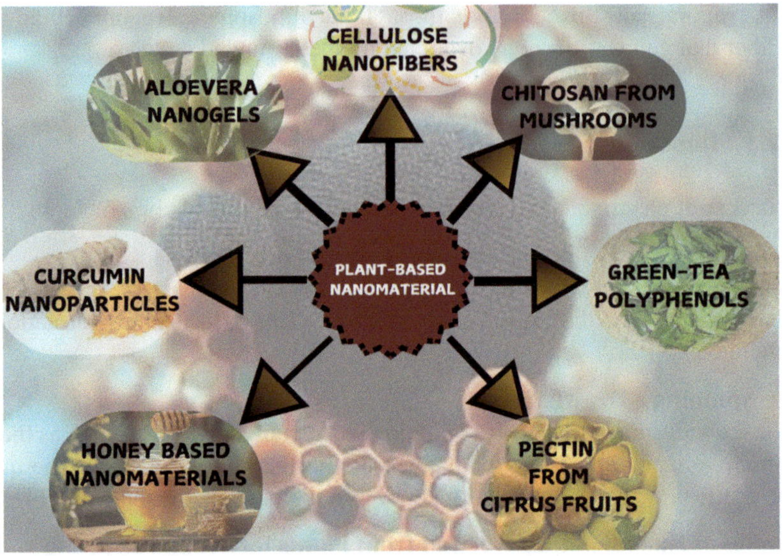

Fig. 15.4 Plant based nanomaterials

nanostructured form, can aid in preserving the moisture content of the wound environment, which is crucial for the best possible healing. To improve wound care qualities, it can also be mixed with other bioactive substances.

Green Tea Polyphenols: Antioxidant polyphenols found in green tea aid in lowering microbial growth and inflammation. Dressings can incorporate polyphenol coatings or nanoparticles to provide these bioactive benifits enhancing both antimicrobial protection and anti-inflammatory activity (Fig. 15.4).

15.5 In Vitro and In Vivo Studies

The experiment conducted within a living organism, such as humans or animals is referred to as in vivo. To ensure the well-being of the subjects involved, these studies often follow preliminary *in vitro* safety evaluations. Additional caution is required before progressing to *in vivo* testing if a chemical exhibits significant toxicity during *in vitro* experiments on cells, tissues, or organs.

In vivo, studies in animals help determine safe and effective dosages of medications for humans and animals. These dosages are identified as "no adverse effect" levels. Additionally, in vivo research is utilized to assess the potential harmful effects of low-dose exposure to both natural and synthetic substances.

In vitro studies are conducted outside a living organism, typically in a laboratory setup. These studies focus on tissues, chemicals, cells, or isolated organs. This is useful in understanding environmental factors and evaluating the effects of medications or physical agents like light, heat and radiation on biological systems.

15.5.1 Evaluation of Antibacterial Properties

Antibacterial evaluation focuses on combating drug-resistant bacteria in a clinical setting. This specifically refers to the ability of a substance to kill bacterial cells or to inhibit the growth of bacteria. Various methods are used to evaluate the antibacterial properties including broth dilution, disk-diffusion assays and other methods.

15.5.2 Testing of Biocompatibility and Cytotoxicity

Biocompatibility is the capability of a material to perform the desired application without having any side effects on the body. The material should be nontoxic and have an interface with the human body to carry out specific physiological functions. These materials integrate with medical devices and are embedded into the organs and tissues. Thus, It is necessary to know the basic principles related to the structure of healthy and diseased cells, organs and tissues. The methods used to examine their function and structure as well as the basic mechanisms involved in disease processes.

15.5.3 Animal Model Studies on Wound Healing Efficacy

Animal model studies are essential for assessing how well new treatments, such as plant-based nanomaterials, heal wounds. These animal models—rats, mice, and rabbits—offer important insights into the biological mechanisms underlying wound healing, including angiogenesis, re-epithelialization, inflammation, and collagen deposition. Researchers can evaluate variables like wound contraction rate, histological alterations, and microbial clearance by administering experimental treatments to standardized wound sites. When compared to traditional treatments, studies using plant-based nanogels or dressings have shown improved tissue regeneration, decreased infection, and accelerated healing. These results lay the groundwork for clinical translation in the treatment of human wounds in addition to validating the therapeutic potential of such nanomaterials.

15.6 Commercial Applications and Market Potential

The commercial applications of plant-based nanomaterials in wound care span acute wound management, chronic wound treatment, and surgical applications. These dressings meet the growing demand for advanced wound care products that enhance healing and reduce infection risks. With their biocompatibility, antimicrobial properties, and ability to deliver therapeutic agents, they hold significant promise in addressing conditions like diabetic ulcers, burns, and pressure sores. The global wound care market, projected to grow steadily, creates opportunities for the introduction of eco-friendly and innovative products based on plant bioactives. Their alignment with sustainability goals also appeals to environmentally conscious consumers and healthcare providers, enhancing market potential.

15.6.1 Current Commercial Wound Dressings Incorporating Plant Bio Actives

Several commercial wound dressings have incorporated plant bio actives due to their natural healing properties:

- Alginate Dressings: Derived from seaweed, these dressings promote moisture balance and are widely used for exudative wounds.
- Chitosan-Based Products: Leveraging chitosan's antimicrobial and hemostatic properties, these dressings are utilized for both acute and chronic wounds.
- Essential Oil-Infused Dressings: Dressings incorporating tea tree oil, lavender oil, or eucalyptus oil offer antimicrobial and soothing effects.
- Hydrogels with Plant Extracts: These dressings are enriched with compounds like Aloe vera or curcumin, known for their anti-inflammatory and antioxidant benefits.

These products demonstrate the feasibility of integrating plant-derived components into advanced wound care solutions, setting the stage for nanomaterial-based innovations.

15.6.2 Challenges in Commercialization and Large-Scale Production

- Consistency in Raw Material Quality: Variability in plant-derived bio actives due to environmental and cultivation conditions can affect product efficacy and scalability.
- High Production Costs: Advanced fabrication techniques like electrospinning and 3D printing require significant investment, increasing the cost of production.
- Regulatory Hurdles: Gaining approval from regulatory bodies for new materials, particularly nanomaterials, involves stringent testing and validation, delaying market entry.
- Stability and Shelf Life: Ensuring the long-term stability of bio actives and nanomaterials in dressings poses technical challenges.

- Consumer Awareness and Adoption: Educating healthcare providers and patients about the benefits of plant-based nanomaterials is essential to drive market penetration.
- Scalability: Transitioning from laboratory-scale production to commercial-scale manufacturing while maintaining quality and efficacy is a critical challenge.

Addressing these challenges through advancements in processing technologies, cost optimization strategies, and supportive regulatory frameworks will be essential for unlocking the full potential of plant-based nanomaterials in the commercial wound care market.

15.7 Challenges and Future Prospects

15.7.1 Stability and Release Kinetics of Plant Bio Actives

The stability of plant bio actives is a critical challenge, as these compounds can degrade when exposed to environmental factors such as light, heat, or oxygen, reducing their therapeutic efficacy. Optimizing encapsulation techniques, such as incorporating bio actives into Nano carriers, can protect them from degradation and ensure prolonged activity. Controlling the release kinetics is equally vital; plant bio-actives must be delivered in a sustained and localized manner to maximize their healing potential without causing cytotoxic effects. Advances in nanotechnology, such as the development of responsive nanoparticles, can address these issues by enabling on-demand release based on the wound microenvironment.

15.7.2 Regulatory Challenges in Nanomaterial-Based Wound Dressings

Nanotechnology-based wound dressings face stringent regulatory scrutiny due to their novel nature. Regulatory bodies require extensive preclinical and clinical evidence to ensure their safety, efficacy, and biocompatibility. Challenges include standardizing testing protocols for nanomaterials, addressing concerns about long-term toxicity, and navigating complex approval pathways. Additionally, variations in global regulations can complicate the commercialization process, necessitating region-specific strategies for market entry. Collaborative efforts between researchers, manufacturers, and regulatory agencies are essential to establish clear guidelines and accelerate approval timelines.

15.7.3 Future Directions in Plant-Based Nanotechnology for Wound Care

Plant-based nanotechnology in wound care has a bright future thanks to the combination of cutting-edge advancements and environmentally friendly methods. Real-time pH, temperature, and infection marker monitoring is made possible by smart wound dressings with biosensors, allowing for individualized and adaptable wound care. The complex nature of wound healing is addressed by multifunctional nanomaterials that combine antimicrobial, anti-inflammatory, and regenerative qualities. Stability and therapeutic efficacy are improved by hybrid systems that combine synthetic polymers with plant bioactives. Eco-friendly development of nanomaterials is ensured by putting an emphasis on sustainable production using green chemistry and renewable resources. Precision medicine techniques also make it possible to customize wound care plans according to the unique characteristics of each patient, such as the type of wound and the presence of microorganisms. When taken as a whole, these developments show that plant based nanotechnology holds great promise for transforming wound care by offering safer, more effective and biologically compatible strategies for enhancing healing and preventing infections.

15.8 Conclusion

The application of plant-based nanomaterials in wound care signifies a groundbreaking advancement in medical science, combining natural biocompatibility with cutting-edge nanotechnology. These materials offer unparalleled benefits such as antimicrobial efficacy, controlled drug delivery, structural mimicry of the extracellular matrix, and the ability to create a moist, healing-friendly environment. Fabrication techniques like electrospinning and 3D printing further enhance their customizability for diverse clinical needs. Looking ahead, research into multifunctional nanomaterials, smart biosensors, and novel bio actives, coupled with scalability and regulatory validations, can bridge the gap between innovation and clinical application. By addressing critical challenges in wound management, particularly for chronic and complex wounds, plant-based nanomaterials pave the way for more effective, sustainable, and patient-centric healthcare solutions.

References

Agrawal P, Soni S, Mittal G, Bhatnagar A (2014) Role of polymeric biomaterials as wound healing agents. Int J Low Extrem Wounds 13(3):180–190

Alasmari A (2020) Phytomedicinal potential characterization of medical plants (Rumex nervosus and Dodonaea viscose). J Biochem Technol 11:113–121

Alberts A, Tudorache DI, Niculescu AG, Grumezescu AM (2025) Advancements in wound dressing materials: highlighting recent Progress in hydrogels, foams, and antimicrobial dressings. Gels 11(2):123

Antunes Filho S, Dos Santos MS, Dos Santos OA, Backx BP, Soran M, Opriş O, Lung I, Stegarescu A, Bououdina M (2022) Biosynthesis of nanoparticles using plant extracts and essential oils. Molecules 28(7):3060. https://doi.org/10.3390/molecules28073060

Bahrami, S., Adel, M., Esmaeili, F., Rezayat, S.M., .Mehravi, B., and Zahmatkeshan, M. (2019). Carbohydrate-based nanofibers: applications and potentials. Handbook of nanofibers, 263–285. https://doi.org/10.1007/978-3-319-53655-2_28, Springer International Publishing

Barhoum A, Pal K, Rahier H, Uludag H, Kim IS, Bechelany M (2019) Nanofibers as new-generation materials: from spinning and nano-spinning fabrication techniques to emerging applications. Appl Mater Today 17. https://doi.org/10.1016/j.apmt.2019.06.015

Bennett G, Abbott J, Sussman G (2024) The negative impact of medications on wound healing. Wound Pract Res 32(1):17–24

Billman GE (2020) Homeostasis: the underappreciated and far too often ignored central organizing principle of physiology. Front Physiol 11:200

Dar RA, Shahnawaz M, Ahanger MA, Majid IU (2023) Exploring the diverse bioactive compounds from medicinal plants: a review. J Phytopharmacol 12(3):189–195

Dobson GP, Morris JL, Letson HL (2024) Pathophysiology of severe burn injuries: new therapeutic opportunities from a systems perspective. J Burn Care Res 45:irae049

Earle L (2024) A better second half: dial Back your age to live a longer, healthier, happier life. The number 1 sunday times bestseller 2024. Hachette UK

Enoch S, Leaper DJ (2008) Basic science of wound healing. Surgery (Oxford) 26(2):31–37

Farahani M, Shafiee A (2021) Wound healing: from passive to smart dressings. Adv Healthc Mater 10(16):2100477

Figen AK (2020) History, basics, and parameters of electrospinning technique. In: Electrospun materials and their allied applications, pp 53–69. https://doi.org/10.1002/9781119655039.ch2

Hajj J, Sizemore B, Singh K (2024) Impact of epigenetics, diet, and nutrition-related pathologies on wound healing. Int J Mol Sci 25(19):10474

Hao DC, Gu XJ, Xiao PG (2015) Medicinal plants: chemistry, biology and omics. Woodhead Publishing, Sawston

Horstmanshoff M (1999) Ancient medicine between hope and fear: medicament, magic and poison in the Roman Empire. Eur Rev 7:37–51. https://doi.org/10.1017/S1062798700003720

Hosgood G (2006) Stages of wound healing and their clinical relevance. Vet Clin Small Anim Pract 36(4):667–685

Jiang P, Li Q, Luo Y, Luo F, Che Q, Lu Z et al (2023) Current status and progress in research on dressing management for diabetic foot ulcer. Front Endocrinol 14:1221705

Lee SH, Kim SH, Kim KB, Kim HS, Lee YK (2024) Factors influencing wound healing in diabetic foot patients. Medicina (B Aires) 60(5):723

Libretti S, Puckett Y (2023) Physiology, homeostasis. In: StatPearls [Internet]. StatPearls Publishing

Litvinova O, Kumetchko M, Pavlov S, Babenko N, Kolisnyk I (2025) Problems of healing soft tissue injuries. East Ukr Med J 13:1–13

Mahady GB (2005) Medicinal plants for the prevention and treatment of bacterial infections. Curr Pharm Des 11:2405–2427. https://doi.org/10.2174/1381612054367481

Nune, S.K.,. Rama, K.S., .Dirisala, V.R., Chavali, M.Y. (.2017). Electrospinning of collagen nanofiber scaffolds for tissue repair and regeneration. In Nanostructures for novel therapy: synthesis, characterization and applications https://doi.org/10.1016/B978-0-323-46142-9.00011-6.

Ramawat KG, Dass S, Mathur M (2009) The chemical diversity of bioactive molecules and therapeutic potential of medicinal plants. In: Herbal drugs: ethnomedicine to modern medicine, pp 7–32

Sargin SA (2021) Potential anti-influenza effective plants used in Turkish folk medicine: a review. J Ethnopharmacol 265:113319. https://doi.org/10.1016/j.jep.2020.113319

Spizzirri UG, Aiello F, Carullo G, Facente A, Restuccia D (2021) Nanotechnologies: an innovative tool to release natural extracts with antimicrobial properties. Pharmaceutics 13(2):230. https://doi.org/10.3390/pharmaceutics13020230

Stéphane FFY, Jules BKJ, Batiha GE-S, Ali I, Bruno LN (2022) Extraction of bioactive compounds from medicinal plants and herbs. Nat Med Plants. https://doi.org/10.5772/intechopen.98602

Stoica AE, Chircov C, Grumezescu AM (2020) Nanomaterials for wound dressings: an up-to-date overview. Molecules 25(11):2699

Strodtbeck F (2001) Physiology of wound healing. Newborn Infant Nurs Rev 1(1):43–52

Su L, Jia Y, Fu L, Guo K, Xie S (2023) The emerging progress on wound dressings and their application in clinic wound management. Heliyon 9:e22520

Süntar I (2020) Importance of ethnopharmacological studies in drug discovery: role of medicinal plants. Phytochem Rev 19:1199–1209. https://doi.org/10.1007/s11101-019-09629-9

Uberoi A, McCready-Vangi A, Grice EA (2024) The wound microbiota: microbial mechanisms of impaired wound healing and infection. Nat Rev Microbiol 22(8):507–521

Wang M, Huang X, Zheng H, Tang Y, Zeng K, Shao L, Li L (2021) Nanomaterials applied in wound healing: mechanisms, limitations and perspectives. J Control Release 337:236–247

Yachmaneni JA, Jajoo S, Mahakalkar C, Kshirsagar S, Dhole S (2023) A comprehensive review of the vascular consequences of diabetes in the lower extremities: current approaches to management and evaluation of clinical outcomes. Cureus 15(10):e47525

Zeng Q, Qi X, Shi G, Zhang M, Haick H (2022) Wound dressing: from nanomaterials to diagnostic dressings and healing evaluations. ACS Nano 16(2):1708–1733

Zhang X, Lu Y (2014) Centrifugal spinning: an alternative approach to fabricate nanofibers at high speed and low cost. Polym Rev 54(4):677–701. https://doi.org/10.1080/15583724.2014.935858

Zhao Y, Qiu Y, Wang H, Chen Y, Jin S, Chen S (2016) Preparation of nanofibers with renewable polymers and their application in wound dressing. Int J Polym Sci. https://doi.org/10.1155/2016/4672839

Chapter 16
Limitations and Challenges of Medicinal Plant-Based Nanoformulations

Prangya Rath, Laxmi Kant Bhardwaj, Shweta Khanna, Yogesh Kumar Sharma, and Amit Kumar

16.1 Introduction

There has been rapid growth in nanotechnology (NT) in the medical industry (Tokas et al. 2024). NT performs an important role in the medical field and helps in the prevention of illness (Kong et al. 2023). Medicinal plants are used in the prevention and treatment of diseases and are used as a raw material in pharmaceutical sectors (Bhardwaj et al. 2023a). Nanoformulations made up of medicinal plants possess the pharmacological activities of medicinal plants as well as the benefits of NT. Examples of nanoformulations comprise liposomes, nanoparticles (NPs), nano-micelles, nano-emulsions, and various other nanostructures (Janik et al. 2023).

Different metal or metal oxide NPs loaded with herbal/medicinal plant extracts are conjugated with nanoformulations. They have been reported with better bioavailability, longer circulation time in blood, as well as accurate targeting of the diseased site. These NPs in plant-based formulations are intended to serve as carriers for drugs that cannot be delivered through conventional methods (Sharma et al.

P. Rath
Amity Institute of Environmental Sciences, Amity University, Noida, Uttar Pradesh, India

L. K. Bhardwaj (✉)
Amity Institute of Environmental Toxicology, Safety & Management, Amity University, Noida, Uttar Pradesh, India

S. Khanna
Department of Physics and Nanotechnology, SRM University, Chennai, Tamil Nadu, India

Y. K. Sharma
Department of Chemistry, Kalindi College, University of Delhi, New Delhi, India

A. Kumar
Molecular Biology and Biotechnology Division, CSIR-National Botanical Research Institute, Lucknow, Uttar Pradesh, India

V. Ghosh et al. (eds.), *Applications of Plant Bioactive Compounds-Based Nanoformulations*, Nanotechnology in the Life Sciences,
https://doi.org/10.1007/978-3-032-13086-0_16

2021; Bhardwaj et al. 2023b). Unfortunately, their unique characteristics and probable reactivity in biological systems can lead to toxicity and biocompatibility issues (Sivakumar et al. 2018). NPs possess some unique beneficial properties. These include a high ratio of surface area to volume, permeability for biological barriers, and quantum effects. However, studies have shown that upon adversely interacting with different cells/tissues/organs, they increase the risks associated with toxicity (Subramani and Ganapathyswamy 2020). Thus, research studies are still required to investigate toxicity and biocompatibility prior to the wide clinical usage of these medicinal plant-based nanoformulations.

Phytochemicals such as alkaloids, flavonoids, other metabolites, etc., have played a significant role in nanomaterials synthesis and have main implications/applications as therapeutic agents for curing different types of diseases (Mohammadzadeh et al. 2022). Secondary metabolites that are synthesized by medicinal plants are studied to be important biomarkers in the synthesis of NPs (Mozafari et al. 2020). These phytochemicals are highly water soluble. They are, however, less absorbed by the human body. The penetration of NPs through lipid membranes is affected because of their high molecular weight and low absorption rate. Thus, it leads to decreased bioavailability as well as efficacy (Noor et al. 2022). Luteolin, which is present in fruits, vegetables, and plants, is used in Chinese medicine to treat cancer and high blood pressure (Li et al. 2023). The majority of people in poor nations utilize herbal medicine according to a World Health Organisation survey (WHO 2019).

China has the richest resources with respect to medicinal plants, i.e., 11,146 medicinal plant species have been reported (Fan et al. 2023). Also, it is the largest producer and exporter of herbal medicines in the world. Combining medicinal plants and NT to produce nanoformulations has been beneficial for various opportunities and applications such as medication delivery, imaging and diagnostics systems, agricultural sectors, environmental fields, as well as targeted therapies (Chowdhury et al. 2024). There have been studies that have shown some successful results of nanoformulations made with plant extracts. For example, plant-based nanoformulations were used for *in-vitro* studies in mice models to show positive effects on hypoglycemic index, mechanism of action, antioxidant activity, toxicity, etc. (Wickramasinghe et al. 2022). It has been shown that the targeted approach of nanoformulations improved the outcomes in patients and also promised minimal adverse effects in different diseases.

Patients who were given nanoformulated treatments were observed to experience better outcomes. This was because of improved stability of drugs, targeted delivery, as well as reduced side effects. Such research observations highlighted the superiority of NT-enhanced therapies (Shree et al. 2025). This chapter is focused on scaling up the production, applications, toxicity mechanisms, limitations and challenges, benefits, as well as strategies used for mitigating toxicity associated with these plant-based nanoformulations. This chapter also focuses on the present scenario and future perspectives, and some recommendations. Different types of nanoformulations with the active ingredient and their impacts are shown in Table 16.1. The overview of this study is very well presented in Fig. 16.1.

Table 16.1 Nanoformulations with Active Ingredient and Their Impacts

S. No.	Nanoformulations	Active Ingredient	Impacts	References
1	Nanoemulsion	Curcumin	Curcumin nanolatex exhibits thixotropic rheological behavior with an efficient increase of permeability into the skin as compared with curcumin that was dispersed inside conventional hydrogel systems	Kumari et al. (2022)
		Brucine	Brucine nanolatex possesses high anti-inflammatory as well as antinociceptive activities. They reduce swelling of the hind paw as well as inhibit acetic acid-induced writhing of the abdomen as compared with brucine gels	Zhang et al. (2023)
		Quercetin	Quercetin bioavailability was highest in the delivery systems	Kandemir et al. (2022)
2	NPs	Puerarin	Poly-puerarin NPs possess significant anti-tumor effects as compared to free puerarin	Elmowafy et al. (2023)
		Naringenin	Naringenin NPs were more effective in rat experimental models as compared to free naringenin	Motallebi et al. (2022)
		Quercetin	Quercetin with ZnO NPs was more effective in Michigan Cancer	Homayoonfal et al. (2024)
3	Nano micelles	Naringenin	Naringenin-containing bovine β-casein micelles showed a lower concentration of critical micelle as well as more aggregation as compared to pure bovine β-casein micelles, which significantly led to a high amount of naringenin concentration in aqueous solution	Motallebi et al. (2022)
		Podophyllotoxin	Podophyllotoxin micelles had the highest activity as compared to other cytotoxic agent micelles	Kalinova et al. (2024)
		Thymoquinone	Polymer micelles of thymoquinone have been reported to exhibit better effects of wound healing as compared with natural thymoquinone	Ma et al. (2022)

(continued)

Table 16.1 (continued)

S. No.	Nanoformulations	Active Ingredient	Impacts	References
4	Nanoliposomes	*Sargassum boveanum*	*Sargassum boveanum* nanoliposomes retained a higher amount of phenolic compounds with antioxidant properties	Feng et al. (2023)
		Capsaicin	Nanoliposomes containing capsaicin have been reported to possess improved pharmacokinetic properties, increased activities of anticancerous activities, as well as selectivity as compared to capsaicin	Faghihi et al. (2023)
		Lutin	Lutin nanoliposomes exhibit high levels of vitality protection of compounds with rutin bioactivity	Li et al. (2024)
5	Nanocrystals	Resveratrol	Nanocrystals of resveratrol showed better anti-tumorous effects as compared with resveratrol	Ancic et al. (2022)
		Isoliquiritigenin	Isoliquiritigenin nanocrystals have high solubility as well as low toxicity for cells, as compared to the free form	Liu et al. (2024)
		Breviscapine	The nanocrystals of Breviscapine have relatively more stable drug concentration levels in plasma for a longer duration as compared to free Brevisanthin	Zhang et al. (2022)
		Ginkgolide B	Nanocrystals of Ginkgolide B showed higher levels of drug plasma as compared to free ginkgolide B	Liu et al. (2024)

16.2 Scale Up Production of Plant-Based Nanoformulation

Manufacturing efficiency will be improved. Nanoformulations based on plants have enormous potential for long-term disease management, such as diabetes, due to the improvement in stability, targeted distribution, and bioavailability of bioactive compounds from plant origin (Dewanjee et al. 2020). Currently, nanoformulations with plant-based extracts are being researched and developed for enhancing their bioavailability and treatment against different cancers (lung, breast, and pancreatic, etc.). Some of the common herbal-based nanoformulated drugs have been derived from *Curcuma longa, Panax ginseng, Withania somnifera, Silybum marianum, Salvia miltiorrhiza*, and *Gymnema sylvestre*. Nanoformulations made from extracts

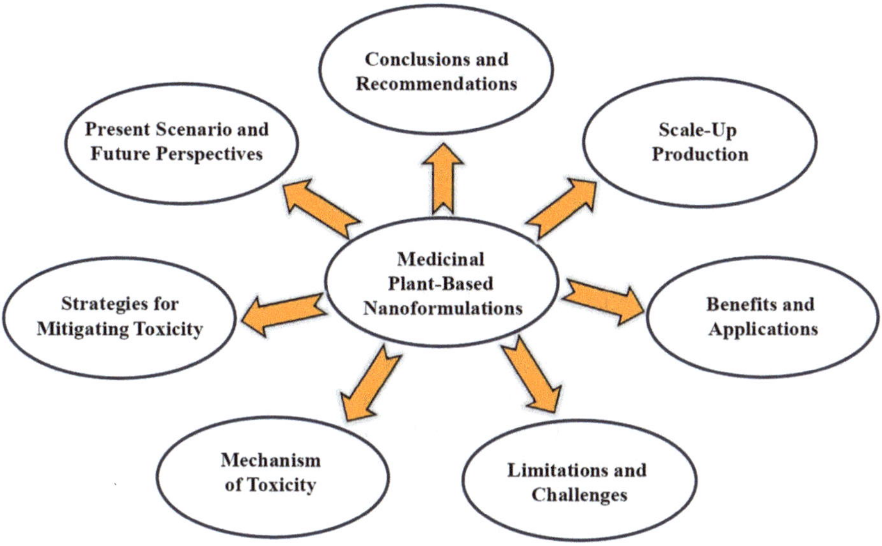

Fig. 16.1 Overview of Medicinal Plant-Based Nanoformulations

of *Azadirachta indica* (neem) and *Cinnamomum verum* (cinnamon) have been produced for treating antimicrobial infections (Azam et al. 2024).

There are several hurdles in transitioning complex formulations from lab scale to commercial production, including regulatory compliance, formulation stability, etc. The regulatory environment is one of the many factors affecting scale-up and must be complied with when marketing a product. Additionally, formulation technology must remain within its parameters during scaling; it needs to be optimized accordingly. Increase in the production of complex goods, which are made up of plant-based nanoformulations while maintaining their quality, stability, and effectiveness, a methodical strategy is needed. Phytochemicals such as curcumin, quercetin, and resveratrol have strong antidiabetic effects through mechanisms including increased insulin sensitivity and enzyme inhibition (Scarpa et al. 2023). To overcome some of these challenges, researchers are working towards developing innovative nanoformulations in the form of nanoemulsions, microemulsions, nanoencapsulation, as well as microencapsulation. These have advanced approaches that facilitate controlled release, enhancing stability (Sarmah et al. 2025).

16.2.1 Formulation Optimization

This is the first step to creating plant-based nanoformulations. It focuses on choosing ingredients and making the nanoformulation process better. The ingredients are active parts that keep it stable and substances that help mix things. Plant chemicals that fight diabetes can stop the enzymes α-amylase and α-glucosidase from working (Tiji et al. 2021). These enzymes play a key role in how the body uses glucose. Components enhance formulation stability during scaling by choosing stabilizers and surfactants. Nanoformulations are the latest innovation and are the future of pharmaceutical administration.

Drug delivery using NT may target the body more accurately. Drug nanoformulations improve stability and blood solubility. Drug absorption is faster and more effective when administered to the target tissue. Drug particles penetrate tissues and cells better due to their smaller size, improving therapeutic effectiveness (Barua and Mitragotri 2014). The pharmaceutical industry hopes that nanoformulations will improve medicine delivery and patient outcomes. The main things to consider are the size and shape of the particles, the charge on their surface, and how well they can hold things inside. To work well as medicine, it needs to be stable, easy for the body to use, and able to release. Some good ways to make uniform NPs that can be done on a big scale are using emulsion solvent evaporation, making them form by themselves in a solution, and using super-hot, high-pressure fluids. These methods affect overall therapeutic performance.

16.2.2 Pilot-Scale Testing

It is a common practice in the industry to carry out tests at a pilot scale to obtain feedback with lower risk. By verifying formulation consistency and optimizing process parameters, pilot-scale testing serves as a bridge between lab-scale and full-scale production. Pilot-scale homogenizers and microfluidizers can reveal a lot of scaling problems (Operti 2023). When variables like temperature, pressure, and stirring speed are optimized, the characteristics of NPs should hold true even when working with larger batch sizes. It is recommended to produce many batches at the pilot scale to check bioactivity, repeatability, and stability to ensure the therapeutic effects identified at lab size are sustained.

16.2.3 Process Scale-Up

Scaling up the operation requires the standardization of methods and the establishment of quality control criteria that will guarantee product uniformity (Christiansen 2020). To ensure consistency across batches, pilot data must be used to create batch-specific Standard Operating Procedures (SOPs). Testing of the finished product and quality control of the raw components should be done. The most important factors to consider during formulation are active component concentration, encapsulation efficiency, zeta potential, and particle size (Culas et al. 2024).

16.2.4 Stabilization and Preservation

Stabilization and preservation have to be effective so that the nanoformulations would live longer and remain useful under shelf conditions (Hosseini and Jafari 2020). Encapsulation efficiency should be a parameter aimed at in order to minimize the degradation of the active components. Capsule production in large batches may lead, sometimes, to spilling or spoilage; hence, encapsulation must be carefully optimized. Preservation methods such as lyophilization or encapsulation should be employed with biodegradable coatings to enhance shelf life and stability (Degobert and Aydin 2021). Since nanoformulations are especially susceptible to environmental factors, the packaging materials should be chosen that can safeguard against oxidation, moisture, and UV radiation.

16.2.5 Regulatory Compliance and Sustainability

Expansion of nanoformulation has to be done keeping in mind the regulatory requirements, especially when it concerns consumer health products. It has to be verified if the composition complies with the requirements of regulatory authorities concerning its safety and effectiveness, especially in the case of drugs or consumer goods (Gaur et al. 2020). All the documents at each stage of formulation, scale, and testing need to be well preserved, particularly since these types of documents bear significant importance in regulatory filings and for future use. Fulfilling commercially viable and sustainable aspects would steer the long-term success of expanded nanoformulations toward a victory march. While sustainability in procurements to fulfill customer, as well as regulatory demands for eco-friendly products, is to be focused on, scalability and market attractiveness would also need enhancements.

16.2.6 Benefits and Applications of Medicinal Plant-Based Nanoformulations

In recent years, plant-derived nanomaterials have attracted more interest in the application of treating human viral infections, owing to their low toxicity levels, biocompatibility, improved therapeutic efficacy, and reduced adverse effects (Krishnan and Thirunavukkarasu 2022). NPs, including gold (Au), iron (Fe), nickel (Ni), copper (Cu), and silver (Ag) NPs, have been exploited in various sectors like medicine, health care, material sciences, and physics (Khatami et al. 2021). Khatami et al. (2021) reported therapeutic deployment of metallic NPs against different types of bacterial, viral, as well as fungal pathogens. Plant extracts are being utilized in many biological, therapeutic, and medical domains to synthesize NPs (Nasrollahzadeh et al. 2020). Table 16.2 provides a clear presentation of the benefits

Table 16.2 Comparative Advantages and Limitations of Various Nanocarriers

S. No.	Type	Advantage	Limitations	References
1	Nanoemulsion	Stronger capacity for drug-loading. The size of the particle is uniform and small. Higher bioavailability.	Due to the potential toxicity of surfactants, their storage conditions need attention.	Sneha and Kumar (2022)
2	Nano micelles	Higher loading capacity, better stability levels in the blood, longer time in blood circulation, fewer side effects, and internally protected drugs from degrading.	Stability requires improvement; complex behavior in *in-vivo*	Ma et al. (2022)
3	Nanoliposomes	Passively targeting drugs, high efficiency of cargo delivery, and reduced cargo toxicity.	Storage, as well as transportation, conditions, and stability issues, require attention	Smruthi et al. (2022)
4	Upconversion NPs	Unique optical properties, lower levels of toxicity, better biocompatibility, and easy functionalization of the surface.	The optical conversion efficiency requires improvement. Comparatively lower levels of drug load.	Ovais et al. (2020)
5	Metal-organic framework	High specific surface area, structure, levels of porosity, biodegradability, as well as performance levels could be adjusted.	Synthesis and preparation are complex. There are less number of studies related to drug loading and its release kinetics.	Chen et al. (2024a)
6	Nanometallic elemental	Lower cytotoxicity levels, controlled size, surface, easier synthesis pathways, higher levels of cell permeability, ability to bind to different molecules on their surfaces, and higher capacity for drug loading.	Problems related to elimination from the body and difficulty in achieving complex patterns of drug delivery.	Xie et al. (2023)
7	Carbon dots	Significantly higher elastic modulus as well as mechanical strength, higher electrical as well as thermal conductivity, longer duration of circulation, permeability of cell membrane, and high aspect ratio allow for higher drug loading.	Limited drug loading. Preparation is complex and costly.	Lai et al. (2023)

and drawbacks of different nanocarriers (Chen et al. 2024b). A study showed facts about antioxidative activity and toxicity of plant-based nanoformulations on *in vitro* LC-540 cell lines. It showed a reduction in hydrogen peroxide quantity in a concentration-dependent manner, thereby reducing oxidative stress. It also showed that cell viability significantly increased to 70% when treated with such nanoformulation (Wickramasinghe et al. 2022).

16.2.7 Improve Bioavailability and Controlled Release/ Target Delivery

Plant-active compounds like curcumin, resveratrol, and quercetin have poor bioavailability because they don't dissolve well in water, are unstable, and break down (Yang et al. 2020). NT enhances the bioavailability of specific bioactive elements and enables more effective therapeutic results (Teng et al. 2023). Studies indicate that NPs solubilize and stabilize more active medicinal plant compounds than conventional formulations. For example, natural extracts of anti-cancer drugs like paclitaxel and curcumin, and flavonoids such as luteolin and quercetin, when nanoformulated, are more effective in the human body (Qu et al. 2023).

Nanoformulations made up of plant(s) extracts have been shown to have better solubility, stability, and pharmacokinetic profile (Majee et al. 2025). For example, nanoliposomes have been studied to have benefits like increased solubility of drugs. It also increased the bioavailability in *in-vitro* and *in-vivo* stability, as well as increased the cell-specific targeting of drugs (Wickramasinghe et al. 2022). Nanoformulations made from extracts of *Aloe vera, Mauritia flexuosa, Elaeagnus angustifolia, Punica granatum, Daucus carota, and Cocos nucifera* have also been used for cosmetic products and sunscreens as they possess hypoallergenic, antioxidant, anti-inflammatory, and non-toxic effects (Lungu et al. 2025).

Nanoformulations made with combining epigallocatechin gallate from green tea; proanthocyanidins from grape seed; flavonoids like apigenin, quercetin, rutin, kaempferol; phenolic acids, such as ferulic acid and caffeic acid), carotenoids (β-carotene, lycopene, and lutein), seed oils (shea butter, cocoa butter, jojoba oil, rosehip oil, and grapeseed oil), volatile oils (chamomile, lavender), and alkaloids (sanguinarine, piperine, and caffeine) were used as anti-inflammatory products, neutralized reactive oxygen species (ROS) generated due to UV radiations, and simultaneously smoothens skin and reduce erythema (sunburn) (Verma et al. 2024).

Nanoformulations control the rate of release of active ingredients and also enhance the rate of their absorption (Chen et al. 2024b). The targeting and triggered release performance of medicinal plant extracts can be improved by combining them with nanoformulations, and this combination can achieve more effective therapeutic effects over a prolonged period (Shi et al. 2023). This NT approach reduces the frequency of dosing, enhances patient compliance, and ensures a consistent therapeutic effect. Specifically, because of their efficiency as a drug delivery system, nanomaterials have been reported with significant potential in applications of medicinal plants. The development of NPs in the form of nanocarrier-assembled drugs has been reported with greater success as compared to conventional systems of drug delivery. Nanocarriers can be altered to maximize therapeutic advantages, reduce side effects, and target particular organs (Shah et al. 2021).

16.2.8 Enhanced Stability and Reduced Toxicity

NT can prolong the shelf life of plant-active chemicals and shield them from environmental deterioration, including oxidation and UV light exposure (Sagar et al. 2022). Nanoformulations can increase the potency of medicinal plant products and can reduce the concentrations of toxic plant compounds by the action of therapeutic agents at the site of action (Wickramasinghe et al. 2022). Nanoformulations can decrease the side effects associated with high doses.

16.3 Limitations and Challenges in the Use of Medicinal Plant-Based Nanoformulations

Several problems are still there that need to be overcome in order to create and produce nanoformulations, despite the fact that they may have some potential advantages. Some obstacles that can be characterized are: (i) requirement for designing and synthesizing safe NMs; (ii) the detailed role and mechanism of bioactives used; (iii) lack of procedures that could be easily employed for extract processes at larger scale (iv) lack of multidisciplinary approaches which are required for designing and implementing applications of NT in plants (Islam et al. 2025). One of the most critical issues when it comes to the creation of nanoformulations is the matter of repeatability and scalability changes in the manufacturing process that are quite minor, might result in significant changes in the effectiveness of the product.

Additionally, several of the substances used in nanoformulations have toxicity issues that require careful consideration and testing. Limitations still exist in some areas, despite developments, including clinical scaling levels, long-term stability profiles, and nanoformulation toxicity. The filling up of these gaps will be very necessary in order to translate research from the laboratory into options that are commercially viable and therapeutically meaningful. The development of nanocarrier systems derived from plants has the potential to enhance diabetes treatment and provide healthcare solutions that are safer, more inexpensive, and more long-term. This organized process that permits the reliable production of medications based on NT has made it possible for nanoformulations to be created from plants, which has a great deal of promise as a tool for the effective treatment of diabetes and other chronic disorders. Even though plant-based nanoformulations are effective for therapy, they encounter several obstacles. These include issues related to formulation consistency, formulation quality control, solubilization of the components, controlling dosing profiles while ensuring safety, regulatory compliance, product development, and environmental sustainability (Saadh et al. 2024).

16.3.1 Standardization and Quality Control

The extracts from medicinal plants are made up of a variety of bioactive components (Dar et al. 2023). These components impact the quality and reproducibility of the nanoformulations made from the extracts and introduce variation in the standardization of plant extracts. Furthermore, plant extracts are often contaminated with impurities such as pesticides, heavy metals, or microorganisms, and therefore standardization becomes tougher (Rao et al. 2011). Soil composition, climate conditions, and the timing of a specific harvest can contribute to inter-batch differences. This, in turn, makes it quite challenging to ensure a consistent level of quality.

16.3.2 Extraction and Solubility Issues

It is very hard to separate the bioactive constituents of a plant without subjecting it to degradation, as some of the constituents are very sensitive to heat, light, and solvents, which may be used during the extraction (Fongang Fotsing Yannick et al. 2021). This extract is in powdered form, which is quite hard. It proves very difficult to synthesize it into the nanofibrils. It also describes their therapeutic activities.

16.3.3 Concerns for Stability

Various stability issues may arise for the stability of nanoformulations based on medicinal plants, which may in turn affect their efficiency and safety (Thakur et al. 2011). This calls for the design of auxiliary substances or enveloping shells to enhance stability. Many plant-active components are often beta-unstable and are affected by environmental factors such as light and temperature, and changes in pH values. Irrespective of the nature of the storage and the conditions, the particle size and accumulation patterns of the nanoformulation may change. This susceptibility can lead to degradation, consequently reducing the effectiveness of health nanomaterials. Additionally, the nanoformulations may also be photodegraded, hydrolyzed, or oxidized (Patel et al. 2020). Nanocarriers that are in lipid or polymer form may show signs of agglomeration, settling down, or the two forms separated, and may affect their uniformity and the bioavailability of the product, respectively, over time (Gali et al. 2023).

16.3.4 Controlled Release and Targeting

Particularly in complex nanosystems, plant chemicals' fluctuating alterations make it challenging to regulate and sustain release. Because certain ligands are difficult to obtain, targeted delivery of plant-derived nanomedicines may be difficult, potentially resulting in non-specific distribution and decreased efficacy (Gorain et al. 2023). Nanoformulations' stability and efficacy may be hampered by the biological barriers they must overcome.

16.3.5 Legal and Policy Issues

Uncodified regulatory systems that vary by country confront plant-based nanoformulations (Saini et al. 2023). Since most naturally occurring chemicals and their nanoscale formulations are not patentable, obtaining patent coverage is difficult. Nanomedicine is biosimilars and other similar non-biosimilar medical products. Any product that is to be classified as NT and sold in the market should pass quality checks with safety, efficacy, and bioequivalency before getting to the market (Astier et al. 2017).

16.3.6 Manufacturing and Scalability

As a result of intricate production methods, it is often challenging to upscale nanoformulations without compromising their quality, stability, and efficacy (Wais et al. 2016). Furthermore, the NT method and its concomitant materials may impede commercial utilization and availability of medicinal plants in regions where they are most readily domestic. Nanoformulations can be costly to produce and test since they need expensive, advanced facilities and equipment.

16.3.7 Environmental Impacts

Hazardous wastes or their byproducts are occasionally produced during the nanoformulation process, which could be dangerous for the environment (Gaur et al. 2020). The excessive and uncontrolled harvesting of plant resources, particularly those with commercial potential, such as traditional medicine, will result in the loss of ecological diversity, and, in balancing this excessive use over time, ecosystems will be at risk. The use of NPs in therapeutics and subsequently laundering them will be the main route for NPs entering the natural ecosystem, thus disturbing its equilibrium (Tabasum et al. 2022).

16.3.8 Interactions with the Immune System and Inflammatory Responses

Nanoformulations can necessarily elicit immune responses, while the presence of plant materials may contour immune risk that is yet to be completely quantified (Zolnik et al. 2010). The interaction of nanoformulations with the immune system helps to study the biocompatibility of medicinal plants. Although metal NPs containing plant-based formulations can provoke an inflammatory response and excite immune cells to produce hypersensitivity or chronic inflammation, the physicochemical characteristics of the NPs, such as their surface charge, shape, and composition, will greatly define the extent and character of the immune response (Gholami et al. 2020). Relevant modifications on the surface could also influence the nature of the immune response.

16.3.9 Size and Shape-Dependent Toxicity

The physical dimensions and morphology of the NPs employed have a profound influence on particle behavior in a biological system (Rath et al. 2024). NPs of small size, because of their high surface area and reactivity, have been known to be more cytotoxic than their larger counterparts (Pattanayak et al. 2021). Similarly, it may in reality be the fact that spherical NPs are more readily internalized by the cells than rods or other elongated morphologies, and the toxicity will be affected, as well. For instance, NPs with a diameter of less than 10 nm are sufficient enough to accumulate in nervous tissues and employ neurotoxicity by ingoing the barrier between blood and brain (Barani et al. 2021).

16.3.10 Cellular and Organ-Specific Toxicity

The nano-form of plant compounds may exhibit unknown toxicities because many plant extracts contain multiple bioactive compounds with synergistic or antagonistic effects (Abbas et al. 2020). When the NPs get distributed or accumulated in specific organs, they may generate organ-specific toxicity. The organs that are primarily affected are the liver, lungs, kidneys, and brain. For example, because NPs that have accumulated in the liver would impair its function, deposition in the kidneys can cause failure of renal filtration (Díaz-Fernández et al. 2021). Nanoformulations based on medicinal plants could experience a similar problem if they are not appropriately designed and use non-degradable metal components that deposit over time.

Plant-based nanoformulations have been used for selective immunotherapy. It has been reported to enhance the drug delivery of antibodies. It induces antitumor immune responses by activating the functioning of neutrophils, lymphocytes, natural killer cells, as well as macrophages, thereby resulting in induced apoptosis and phagocytosis of tumor cells. Such successful studies are widening the scope to improve the design of immunotherapeutic strategies for targeting a large spectrum of cancers. This approach is being observed to be better as compared to current cytotoxic anticancer drugs that are commonly used in clinics (Pandey and Karmous 2025).

16.4 Mechanisms of Toxicity in Medicinal Plant-Based Nanoformulations

Oxidative stress has been reported as one of the main mechanisms of toxicity that is associated with medicinal plant-based nanoformulations. Herbal medicines as well as their nanoformulations have shown some manifestations in the form of neuro-, hepato-, nephro, and immunotoxicity, and transplacental toxicity. This has increased oxidative stress, inflammation, and DNA damage. Animal studies, as well as *in vitro* cytotoxicity assays associated with the conventional safety and toxicity assays of nanoformulations, have failed to simulate patho-physiological models of the diseases in humans, and also characterize the cellular or subcellular interactions and phenomena that would be responsible for therapeutic or toxic outcomes (Majee et al. 2025).

Also, upon the encounter of NPs with cellular components, ROS could occur. These are responsible for inducing oxidative damage to cellular molecules such as lipids, proteins, as well as DNA (Naz et al. 2019). Metal NPs, like Ag and ZnO, that are conjugated with medicinal plant-based nanoformulations possessing antimicrobial properties, have been studied to generate oxidative stress. The levels of ROS that are produced by these metal-NPs generally surpass the normal antioxidant defense mechanisms of the cells. They eventually lead to apoptosis, necrosis, or autophagy of affected cells (Mishra et al. 2021). For example, it has been observed that extracts of *Curcuma longa* conjugated with Ag-NPs produce significant ROS in hepatic cells. This also leads to dysfunctional mitochondria and induces apoptosis (Karthikeyan et al. 2020). During ZnO-NPs green synthesis, it was observed that ROS overexpression and cytotoxicity were induced in a dose-dependent manner (Ahmad et al. 2019). These observations illustrated that by using plant-based nanoformulations, oxidative stress-related toxicity could be reduced by regulating ROS levels.

16.5 Strategies for Mitigating Toxicity in Plant-Based Nanoformulations

Other alternative approaches, including bio-degradable polymers, non-metallic NPs, etc., have recently been utilized in medicinal plant-based nanoformulations. Studies have shown that these biodegradable materials degrade into non-toxic metabolites (Shah et al. 2021). For example, chitosan NPs encapsulating *Phyllanthus emblica* extract showed reduced cytotoxicity in human cell lines as compared to the metallic NPs used traditionally, thus being biocompatible (Ibrahim et al. 2020). As a biodegradable agent, chitosan stands to offer a decent strategy by reducing contrary effects while enhancing the therapeutic aids of medicinal plant-based nanoformulations. The nanoformulations made from medicinal plants offer the possibility of drugs being delivered more locally to the target tissues, and it is also possible to incorporate targeting ligands or stimuli-responsive elements in order to reduce off-target effects and toxicity (Hajialyani et al. 2018). pH-responsive, enzyme-triggered release or other controlled-release mechanisms might avoid premature drug release in non-target tissues and ultimately, reduce toxicity from the system (Farooq et al. 2021).

16.6 Present Scenario and Future Perspective

Recently, NT has gained attention in the medicinal field. One of the important aspects of the continuous diabetes treatment process is the requirement for patient compliance. Patient compliance has been reported to be improved by nanoformulations. They offer different administration routes, regulated release, improved biological stability, improved specificity towards targets, and reduced toxicity levels. Therefore, there has been a growing interest in the last 20 years towards developing nanoformulations that would be successful against diabetes. Similarly, nanoformulations of natural products for antidiabetic properties as well as for other diseases have gained interest amongst researchers. Hence, there has been an increased possibility of the development of appropriate nanoformulations made with natural agents or crude herbal products against diabetes and its associated complications.

Many natural agents having significant anti-diabetic potentials have been used as nanoformulations to improve their pharmacokinetics as well as therapeutic efficacies for managing diabetes mellitus. However, many studies are restricted due to insufficient data and statistics, especially about prolonged stability profiling, extended therapeutic effectiveness, as well as toxicological characteristics of plant-derived antidiabetic nanoformulations. Therefore, many findings are confined to laboratory experiments. Hence, significant attention is required to resolve these challenges. However, there have been some studies that have shown that plant-derived nanoformulations possess antidiabetic effects. These compounds have been developed into nanoformulations and have been reported to be successfully effective against various diseases that are etiologically similar or different from diabetes (Lombardo et al. 2019). Such studies are further directed toward finding nanoformulations and dosage manipulations to improve diabetes therapy management. Currently, approximately 51 FDA-approved nanoformulations are used clinically as medications and have been reported to improve the physicochemical, pharmaceutical, as well as pharmacokinetic attributes of drug molecules (Bobo et al. 2016). Such studies are expected to offer significant prospects of naturally synthesized nanoformulations as hypoglycaemic agents. This would help in improving the compliance of patients, simultaneously ensuring its cost-effectiveness, as well as reducing the toxicity associated with diabetes therapy.

16.7 Conclusions and Recommendations

A major challenge in developing more effective, safer, and sustainable treatment methods is the increase in demand from the healthcare field worldwide. Since ancient times, the therapeutic potential of medicinal plants has been utilized by mankind. Medicinal plant-based nanoformulations thus have played a role in evolving areas of pharmacology as well as natural medicine. They combine the benefits of traditional plant-based remedies as well as the technologies of modern drug delivery. Plant-based nanoformulations are of emerging interest in the nanomedicine industry. The inclusion of these formulations might have therapeutic advantages of plant extracts in improving the functionality of the body. A stable distribution of NPs is essential for achieving reliable therapeutic outcomes. The quality assessment of final drugs should be conducted by regulatory authorities based on their outcomes before they can be sold in the market. Sophisticated technologies are necessary for the creation and analysis of nanoformulations.

Continuous research is crucial for developing more resilient formulations that can endure environmental influences and preserve the effectiveness of active ingredients over time. However, nanoformulations can raise potential toxicity and biocompatibility issues, and therefore, the design, testing, and standardization of these nanoformulations must be taken seriously to protect patient safety. Despite these challenges, nanoformulations derived from medicinal plants still hold huge potential for improving drug delivery and efficacy. Given their therapeutic use, it is critical to resolve these constraints to ensure the safety, efficacy, and accessibility of treatments. In addition to making plant-based nanotherapies safer and more effective, overcoming these obstacles will advance the understanding of nanotoxicity in medicine more broadly.

Acknowledgments The authors are deeply grateful to Amity University, Noida, Uttar Pradesh, India, for providing the platform to do this study. Amit Kumar is thankful to ICMR, New Delhi, India, for the award and financial assistance in the form of ICMR-Research Associate.

References

Abbas Q, Yousaf B, Ullah H, Ali MU, Ok YS, Rinklebe J (2020) Environmental transformation and nano-toxicity of engineered nano-particles (ENPs) in aquatic and terrestrial organisms. Crit Rev Environ Sci Technol 50(23):2523–2581

Ahmad I, Ahmed E, Ahmad M (2019) The excellent photocatalytic performances of silver doped ZnO nanoparticles for hydrogen evolution. SN Appl Sci 1:1–2

Ančić D, Oršolić N, Odeh D, Tomašević M, Pepić I, Ramić S (2022) Resveratrol and its nanocrystals: a promising approach for cancer therapy? Toxicol Appl Pharmacol 435:115851

Astier A, Barton Pai A, Bissig M, Crommelin DJ, Flühmann B, Hecq JD, Knoeff J, Lipp HP, Morell-Baladrón A, Mühlebach S (2017) How to select a nanosimilar. Ann N Y Acad Sci 1407(1):50–62

Azam A, Nirala HK, Pandey JK, Ahmed R, Mishra T, Manisha AK (2024) A review of plant-based green nanoparticle synthesis: mechanisms, metals, and applications in nanotechnology. REDVET-Rev Electron Vet 25(1)

Barani M, Sangiovanni E, Angarano M, Rajizadeh MA, Mehrabani M, Piazza S, Gangadharappa HV, Pardakhty A, Mehrbani M, Dell'Agli M, Nematollahi MH (2021) Phytosomes as innovative delivery systems for phytochemicals: a comprehensive review of literature. Int J Nanomedicine:6983–7022

Barua S, Mitragotri S (2014) Challenges associated with penetration of nanoparticles across cell and tissue barriers: a review of current status and future prospects. Nano Today 9(2):223–243

Bhardwaj LK, Kumar D, Kumar A (2023a) Phytoremediation potential of Ocimum sanctum: a sustainable approach for remediation of heavy metals. In: Phytoremediation potential of medicinal and aromatic plants. CRC Press, pp 46–57

Bhardwaj LK, Rath P, Choudhury M (2023b) A comprehensive review on the classification, uses, sources of nanoparticles (NPs) and their toxicity on health. Aerosol Sci Eng 7(1):69–86

Bobo D, Robinson KJ, Islam J, Thurecht KJ, Corrie SR (2016) Nanoparticle-based medicines: a review of FDA-approved materials and clinical trials to date. Pharm Res 33:2373–2387

Chen X, Hong L, Wu Y, Gu Y, Luo J, Kong L (2024a) A dual recognition-based strategy employing Ni-modified metal-organic framework for in situ screening of SIRT1 inhibitors from Chinese herbs. Talanta 274:125975

Chen Y, Tang Y, Li Y, Rui Y, Zhang P (2024b) Enhancing the efficacy of active pharmaceutical ingredients in medicinal plants through nanoformulations: a promising field. Nanomaterials 14(19):1598

Chowdhury SS, Ibnat N, Hasan M, Ghosh A (2024) Medicinal plant-based nanoparticle synthesis and their diverse applications. In: Ethnopharmacology and OMICS advances in medicinal plants volume 2: revealing the secrets of medicinal plants. Springer Nature Singapore, Singapore, pp 213–250

Christiansen GD (2020) Quality control and quality assurance issues in biopharmaceutical processing. In: Biotechnology. CRC Press, pp 33–71

Culas MS, Popovich DG, Rashidinejad A (2024) Recent advances in encapsulation techniques for cinnamon bioactive compounds: a review on stability, effectiveness, and potential applications. Food Biosci 57:103470

Dar RA, Shahnawaz M, Ahanger MA, Majid IU (2023) Exploring the diverse bioactive compounds from medicinal plants: a review. J Phytopharm 12(3):189–195

Degobert G, Aydin D (2021) Lyophilization of nanocapsules: instability sources, formulation and process parameters. Pharmaceutics. 13(8):1112

Dewanjee S, Chakraborty P, Mukherjee B, De Feo V (2020) Plant-based antidiabetic nanoformulations: the emerging paradigm for effective therapy. Int J Mol Sci 21(6):2217

Díaz-Fernández A, Miranda-Castro R, de-Los-Santos-Álvarez N, Lobo-Castañón MJ, Estrela P (2021) Impedimetric aptamer-based glycan PSA score for discrimination of prostate cancer from other prostate diseases. Biosens Bioelectron 175:112872

Elmowafy M, Shalaby K, Elkomy MH, Alsaidan OA, Gomaa HA, Abdelgawad MA, Mostafa EM (2023) Polymeric nanoparticles for delivery of natural bioactive agents: recent advances and challenges. Polymers 15(5):1123

Faghihi H, Mozafari MR, Bumrungpert A, Parsaei H, Taheri SV, Mardani P, Dehkharghani FM, Pudza MY, Alavi M (2023) Prospects and challenges of synergistic effect of fluorescent carbon dots, liposomes and nanoliposomes for theragnostic applications. Photodiagn Photodyn Ther 42:103614

Fan P, Wu L, Wang Q, Wang Y, Luo H, Song J, Yang M, Yao H, Chen S (2023) Physiological and molecular mechanisms of medicinal plants in response to cadmium stress: current status and future perspective. J Hazard Mater 450:131008

Farooq MA, Xinyu H, Jabeen A, Ahsan A, Seidu TA, Kutoka PT, Wang B (2021) Enhanced cellular uptake and cytotoxicity of vorinostat through encapsulation in TPGS-modified liposomes. Colloids Surf B Biointerfaces 199:111523

Feng Y, Zarei V, Mousavipour N (2023) Provision and assessment properties of nanoliposomes containing macroalgae extracts of Sargassum boveanume and Padina pavonica. LWT 175:114194

Fongang Fotsing Yannick S, Bankeu Kezetas Jean J, Gaber El-Saber B, Iftikhar A, Lenta Ndjakou B (2021) Extraction of bioactive compounds from medicinal plants and herbs. IntechOpen, Rijeka

Gali L, Pirozzi A, Donsì F (2023) Biopolymer-and lipid-based carriers for the delivery of plant-based ingredients. Pharmaceutics 15(3):927

Gaur N, Sharma N, Dahiya A, Yadav P, Ojha H, Goyal RK, Sharma RK (2020) Toxicity and regulatory concerns for nanoformulations in medicine. In: The ELSI handbook of nanotechnology: risk, safety, ELSI and commercialization. Wiley, pp 333–357

Gholami N, Cohan RA, Razavi A, Bigdeli R, Dashbolaghi A, Asgary V (2020) Cytotoxic and apoptotic properties of a novel nano-toxin formulation based on biologically synthesized silver nanoparticle loaded with recombinant truncated pseudomonas exotoxin A. J Cell Physiol 235(4):3711–3720

Gorain B, Karmakar V, Sarkar B, Dwivedi M, Leong JT, Toh JH, Seah E, Ling KY, Chen KY, Choudhury H, Pandey M (2023) Biomacromolecule-based nanocarrier strategies to deliver plant-derived bioactive components for cancer treatment: a recent review. Int J Biol Macromol 253:126623

Hajialyani M, Tewari D, Sobarzo-Sánchez E, Nabavi SM, Farzaei MH, Abdollahi M (2018) Natural product-based nanomedicines for wound healing purposes: therapeutic targets and drug delivery systems. Int J Nanomedicine 13:5023–5043

Homayoonfal M, Aminianfar A, Asemi Z, Yousefi B (2024) Application of nanoparticles for efficient delivery of quercetin in cancer cells. Curr Med Chem 31(9):1107–1141

Hosseini H, Jafari SM (2020) Introducing nano/microencapsulated bioactive ingredients for extending the shelf-life of food products. Adv Colloid Interface Sci 282:102210

Ibrahim E, Zhang M, Zhang Y, Hossain A, Qiu W, Chen Y, Wang Y, Wu W, Sun G, Li B (2020) Green-synthesization of silver nanoparticles using endophytic bacteria isolated from garlic and its antifungal activity against wheat Fusarium head blight pathogen Fusarium graminearum. Nanomaterials 10(2):219

Islam A, Rahat I, Rejeeth C, Sharma D, Sharma A (2025) Recent advcances on plant-based bioengineered nanoparticles using secondary metabolites and their potential in lung cancer management. J Future Foods 5(1):1–20

Janik M, Hanula M, Khachatryan K, Khachatryan G (2023) Nano–/microcapsules, liposomes, and micelles in polysaccharide carriers: applications in food technology. Appl Sci 13(21):11610

Kalinova RG, Dimitrov IV, Ivanova DI, Ilieva YE, Tashev AN, Zaharieva MM, Angelov G, Najdenski HM (2024) Polycarbonate-based copolymer micelles as biodegradable carriers of anticancer podophyllotoxin or juniper extracts. J Funct Biomater 15(3):53

Kandemir K, Tomas M, McClements DJ, Capanoglu E (2022) Recent advances on the improvement of quercetin bioavailability. Trends Food Sci Technol 119:192–200

Karthikeyan A, Senthil N, Min T (2020) Nanocurcumin: a promising candidate for therapeutic applications. Front Pharmacol 11:487

Khatami M, Mosazade F, Raeisi M, Ghasemi M, Fazli Z, Arefkia K, Varma RS, Borhani F, Khatami S (2021) Simplification of gold nanoparticle synthesis with low cytotoxicity using a greener approach: opening up new possibilities. RSC Adv 11(6):3288–3294

Kong B, Liu R, Guo J, Lu L, Zhou Q, Zhao Y (2023) Tailoring micro/nano-fibers for biomedical applications. Bioact Mater 19:328–347

Krishnan S, Thirunavukkarasu A (2022) Plant-based nanoformulations for human viral diseases: an emphasis on COVID-19. In: Coronavirus drug discovery. Elsevier, pp 333–355
Kumari A, Raina N, Wahi A, Goh KW, Sharma P, Nagpal R, Jain A, Ming LC, Gupta M (2022) Wound-healing effects of curcumin and its nanoformulations: a comprehensive review. Pharmaceutics 14(11):2288
Lai H, Ming P, Liu Y, Wang S, Zhou Q, Zhai H (2023) MWCNTs and ZnO-based Ce-MOF nanocomposites as enhanced sensing platform for electrochemical detection of carbendazim in Chinese traditional herbs samples. Microchim Acta 190(8):281
Li R, Zhou J, Zhang X, Wang Y, Wang J, Zhang M, He C, Zhuang P, Chen H (2023) Construction of the Gal-NH2/mulberry leaf polysaccharides-lysozyme/luteolin nanoparticles and the amelioration effects on lipid accumulation. Int J Biol Macromol 253:126780
Li Q, Lv L, Liang W, Chen Z, Deng Q, Sun L, Wang Y, Liu Y (2024) Screening, characterization and mechanism of a potential stabiliser for nisin nanoliposomes with high encapsulation efficiency. Food Chem 457:140185
Liu Y, Zhang C, Cheng L, Wang H, Lu M, Xu H (2024) Enhancing both oral bioavailability and anti-ischemic stroke efficacy of ginkgolide B by preparing nanocrystals self-stabilized Pickering nano-emulsion. Eur J Pharm Sci 192:106620
Lombardo D, Kiselev MA, Caccamo MT (2019) Smart nanoparticles for drug delivery application: development of versatile nanocarrier platforms in biotechnology and nanomedicine. J Nanomater 2019(1):3702518
Lungu C, Grădinaru AC, Ivănescu B (2025) Plant-based sunscreens: innovations and new formulations. IntechOpen
Ma Z, Gao X, Raza F, Zafar H, Huang G, Yang Y, Shi F, Wang D, He X (2022) Design of GSH-responsive curcumin nanomicelles for oesophageal cancer therapy. Pharmaceutics 14(9):1802
Majee SB, Bera R, Avlani D, Das U, Singh S (2025) Safety profiles and potential toxicological concerns of herbal nanomedicine. In: Herbal pharmacopeia. CRC Press, pp 412–439
Mishra RK, Ahmad A, Vyawahare A, Alam P, Khan TH, Khan R (2021) Biological effects of formation of protein corona onto nanoparticles. Int J Biol Macromol 175:1–8
Mohammadzadeh V, Barani M, Amiri MS, Yazdi ME, Hassanisaadi M, Rahdar A, Varma RS (2022) Applications of plant-based nanoparticles in nanomedicine: a review. Sustain Chem Pharm 25:100606
Motallebi M, Bhia M, Rajani HF, Bhia I, Tabarraei H, Mohammadkhani N, Pereira-Silva M, Kasaii MS, Nouri-Majd S, Mueller AL, Veiga FJ (2022) Naringenin: a potential flavonoid phytochemical for cancer therapy. Life Sci 305:120752
Mozafari H, Khatami S, Kiani A, Rahimi Z, Vaisi-Raygani A, Afsharnaderi A, Alaei MR (2020) Oxidative stress parameters, trace elements, and lipid profile in Iranian patients with Gaucher disease. Biol Trace Elem Res 193:130–137
Nasrollahzadeh M, Sajjadi M, Dadashi J, Ghafuri H (2020) Pd-based nanoparticles: plant-assisted biosynthesis, characterization, mechanism, stability, catalytic and antimicrobial activities. Adv Colloid Interface Sci 276:102103
Naz F, Dinda AK, Kumar A, Koul V (2019) Investigation of ultrafine gold nanoparticles (AuNPs) based nanoformulation as single conjugates target delivery for improved methotrexate chemotherapy in breast cancer. Int J Pharm 569:118561
Noor F, Tahir ul Qamar M, Ashfaq UA, Albutti A, Alwashmi AS, Aljasir MA (2022) Network pharmacology approach for medicinal plants: review and assessment. Pharmaceuticals 15(5):572
Operti MC (2023) Scale-up manufacturing of polymeric immunomodulatory nanomedicines: from bench to the bedside
Ovais M, Mukherjee S, Pramanik A, Das D, Mukherjee A, Raza A, Chen C (2020) Designing stimuli-responsive upconversion nanoparticles that exploit the tumor microenvironment. Adv Mater 32(22):2000055
Pandey A, Karmous I (2025) Exploring the potential of plant-based nanotechnology in cancer immunotherapy: benefits, limitations, and future perspectives. Biol Trace Elem Res 203(3):1746–1763

Patel KS, Kumar P, Kohli K (2020) Validation of forced degradation and stability indicating studies of tamoxifen in nanoformulation using spectroscopic technique. Mater Today Proc 26:3265–3270

Pattanayak DS, Pal D, Thakur C, Kumar S, Devnani GL (2021) Bio-synthesis of iron nanoparticles for environmental remediation: status till date. Mater Today Proc 44:3150–3155

Qu N, Wang C, Meng Y, Gao Y (2023) Superior anticancer potential of nano-paclitaxel combined bevacizumab treatment in ovarian cancer. Curr Pharm Biotechnol 24(9):1204–1212

Rao MM, KumarMeena A, Galib (2011) Detection of toxic heavy metals and pesticide residue in herbal plants which are commonly used in the herbal formulations. Environ Monit Assess 181:267–271

Rath P, Bhardwaj LK, Yadav P, Bhardwaj AK (2024) A synthesis of biogenic nanoparticles (NPs) for the treatment of wastewater and its application: a review. In: Biogenic wastes-enabled nanomaterial synthesis: applications in environmental sustainability, pp 127–148

Saadh MJ, Mustafa MA, Kumar S, Gupta P, Pramanik A, Rizaev JA, Shareef HK, Alubiady MH, Al-Abdeen SH, Shakier HG, Alaraj M (2024) Advancing therapeutic efficacy: nanovesicular delivery systems for medicinal plant-based therapeutics. Naunyn Schmiedebergs Arch Pharmacol 397(10):7229–7254

Sagar NA, Kumar N, Choudhary R, Bajpai VK, Cao H, Shukla S, Pareek S (2022) Prospecting the role of nanotechnology in extending the shelf-life of fresh produce and in developing advanced packaging. Food Packag Shelf Life 34:100955

Saini P, Ahmed M, Iqbal U, Yadav N (2023) Challenges in stability and safety evaluation of nutraceutical and nanonutraceutical formulations. In: Handbook of nutraceuticals: science, technology and engineering. Springer International Publishing, Cham, pp 1–24

Sarmah K, Anbalagan T, Marimuthu M, Mariappan P, Angappan S, Vaithiyanathan S (2025) Innovative formulation strategies for botanical-and essential oil-based insecticides. J Pest Sci 98(1):1–30

Scarpa ES, Giordani C, Antonelli A, Petrelli M, Balercia G, Silvetti F, Pieroni A, Sabbatinelli J, Rippo MR, Olivieri F, Matacchione G (2023) The combination of natural molecules naringenin, hesperetin, curcumin, polydatin and quercetin synergistically decreases SEMA3E expression levels and DPPIV activity in in vitro models of insulin resistance. Int J Mol Sci 24(9):8071

Shah A, Aftab S, Nisar J, Ashiq MN, Iftikhar FJ (2021) Nanocarriers for targeted drug delivery. J Drug Delivery Sci Technol 62:102426

Sharma M, Bhat R, Usmani Z, McClements DJ, Shukla P, Raghavendra VB, Gupta VK (2021) Bio-based formulations for sustainable applications in agri-food-pharma. Biomolecules 11(5):768

Shi P, Cheng Z, Zhao K, Chen Y, Zhang A, Gan W, Zhang Y (2023) Active targeting schemes for nano-drug delivery systems in osteosarcoma therapeutics. J Nanobiotechnol 21(1):103

Shree JN, Anuf AR, Ahmad S (2025) Comparative analysis of traditional and nano-formulated approaches in viability, targeted delivery, and patient outcomes. In: Herbal pharmacopeia. CRC Press, pp 375–392

Sivakumar M, Sundaram NS, Thasthagir MH (2018) Effect of aluminium oxide nanoparticles blended pongamia methyl ester on performance, combustion and emission characteristics of diesel engine. Renew Energy 116:518–526

Smruthi MR, Nallamuthu I, Anand T (2022) A comparative study of optimized naringenin nanoformulations using nano-carriers (PLA/PVA and zein/pectin) for improvement of bioavailability. Food Chem 369:130950

Sneha K, Kumar A (2022) Nanoemulsions: techniques for the preparation and the recent advances in their food applications. Innov Food Sci Emerg Technol 76:102914

Subramani T, Ganapathyswamy H (2020) An overview of liposomal nano-encapsulation techniques and its applications in food and nutraceutical. J Food Sci Technol 57(10):3545–3555

Tabasum H, Bhat BA, Sheikh BA, Mehta VN, Rohit JV (2022) Emerging perspectives of plant-derived nanoparticles as effective antimicrobial agents. Inorg Chem Commun 145:110015

Teng H, Zheng Y, Cao H, Huang Q, Xiao J, Chen L (2023) Enhancement of bioavailability and bioactivity of diet-derived flavonoids by application of nanotechnology: a review. Crit Rev Food Sci Nutr 63(3):378–393
Thakur L, Ghodasra U, Patel N, Dabhi M (2011) Novel approaches for stability improvement in natural medicines. Pharmacogn Rev 5(9):48–54
Tiji S, Bouhrim M, Addi M, Drouet S, Lorenzo JM, Hano C, Bnouham M, Mimouni M (2021) Linking the phytochemicals and the α-glucosidase and α-amylase enzyme inhibitory effects of Nigella sativa seed extracts. Foods 10(8):1818
Tokas R, Bhardwaj LK, Kumar N, Jindal T (2024) Nanotechnology for sustainable development and future: a review. In: Green and sustainable approaches using wastes for the production of multifunctional nanomaterials. Elsevier, pp 221–233
Verma A, Zanoletti A, Kareem KY, Adelodun B, Kumar P, Ajibade FO, Silva LF, Phillips AJ, Kartheeswaran T, Bontempi E, Dwivedi A (2024) Skin protection from solar ultraviolet radiation using natural compounds: a review. Environ Chem Lett 22(1):273–295
Wais U, Jackson AW, He T, Zhang H (2016) Nanoformulation and encapsulation approaches for poorly water-soluble drug nanoparticles. Nanoscale 8(4):1746–1769
WHO (2019) Selection and use of essential medicines: report of the WHO expert committee on selection and use of essential medicines, 2019 (including the 21st WHO model list of essential medicines and the 7th WHO model list of essential medicines for children). WHO, Geneva
Wickramasinghe AS, Kalansuriya P, Attanayake AP (2022) Nanoformulation of plant-based natural products for type 2 diabetes mellitus: from formulation design to therapeutic applications. Curr Ther Res 96:100672
Xie Y, Liu H, Xie X, Li Y, Peng F, Zhao Y, Xu H (2023) Active ingredients biosynthesis and genic response of traditional Chinese medicine Ligusticum sinense cv. Chuanxiong under different cadmium level. Ind Crop Prod 202:117074
Yang B, Dong Y, Wang F, Zhang Y (2020) Nanoformulations to enhance the bioavailability and physiological functions of polyphenols. Molecules 25(20):4613
Zhang T, Li X, Xu J, Shao J, Ding M, Shi S (2022) Preparation, characterization, and evaluation of breviscapine nanosuspension and its freeze-dried powder. Pharmaceutics 14(5):923
Zhang H, Deng L, Yang J, Yang G, Fan H, Yin Y, Luo S, Li S, Liu L, Yang M (2023) Preparation and evaluation of a nanoemulsion containing cordycepin and its protective effect on skin. J Dispers Sci Technol 44(6):1094–1102
Zolnik BS, González-Fernández Á, Sadrieh N, Dobrovolskaia MA (2010) Minireview: nanoparticles and the immune system. Endocrinology 151(2):458–465

Chapter 17
Nanoformulations of Plant Compounds for Metabolic Syndrome: A Multitarget Approach

Ayodeji Johnson Ajibare, Olabode Oluwadare Akintoye, Kafilat Temidayo Aransi-Ola, Miracle Oyaname Adejumo, Esther-Oluwatomisin Adeola Folarin, and Abraham Olufemi Asuku

17.1 Introduction

The term metabolic syndrome is used to describe a cluster of conditions that increase the risk for severe and interrelated health problems, including cardiovascular disease and diabetes. The syndrome is usually accompanied by abdominal obesity, hypertension, dyslipidemia (abnormal blood lipids), and insulin resistance (Vesa et al. 2024). Overall, they play a role in exacerbating an unhealthy, exacerbated state of metabolic dysfunction and long-term health consequences (Silveira Rossi et al. 2022). It is thus an urgent public health problem for which concerted approaches to prevention and management are required. Oxidative stress, chronic inflammation, and free fatty acid accumulation are the implicated underpinning mechanisms that further entrench the established metabolic derangement cycle (Masenga et al. 2023). This complicates the management of metabolic syndrome and suggests required targeted therapeutic interventions (Willis et al. 2021).

The epidemiology of this syndrome equally presents a staggering prevalence and a growing concern about lifestyle-related factors. Studies show that up to one-third of adults in many countries have metabolic syndrome, with its prevalence rising in parallel with obesity prevalence (Naghipour et al. 2021; Orsini et al. 2023). These trends place a considerable load on healthcare systems and have far-reaching

A. J. Ajibare (✉) · O. O. Akintoye · K. T. Aransi-Ola · M. O. Adejumo · E.-O. A. Folarin
Neuro-Reproductive and Metabolism Unit, Department of Physiology, Faculty of Basic Medical and Health Sciences, College of Medicine, Lead City University, Ibadan, Oyo State, Nigeria
e-mail: ajibare.ayodeji@lcu.edu.ng

A. O. Asuku
Bioresources Development Centre, National Biotechnology Research and Development Agency, Ogbomosho, Oyo State, Nigeria

V. Ghosh et al. (eds.), *Applications of Plant Bioactive Compounds-Based Nanoformulations*, Nanotechnology in the Life Sciences,
https://doi.org/10.1007/978-3-032-13086-0_17

implications for global health, owing to the comorbidities associated with these trends that can result in premature death and diminished quality of life (Orsini et al. 2023). Today's metabolic syndrome intervention consists of lifestyle modifications, pharmacological treatments, and surgical options. Nevertheless, they are often limited by poor adherence to lifestyle changes, drug side effects, and a failed response in specific populations. They point out a pressing need for novel therapeutic strategies (Lillich et al. 2021). In this context, nanoformulations of plant compounds represent a promising approach for developing multi-target therapies for metabolic syndrome (Adetunji et al. 2022). Through the exploitation of the synergistic effects of bioactive phytochemicals and modern nanotechnology, it is essentially possible to improve the bioavailability and efficacy of therapeutics on the pathological basis of metabolic syndrome (Adetunji et al. 2022; Han et al. 2022). Subsequently, this chapter will further elucidate specific areas where these innovative approaches may change our approach to managing metabolic syndrome and improving patient outcomes.

17.2 Ethnopharmacology and Metabolic Syndrome

Research has indicated that the currently available strategies for managing the risk factors associated with Metabolic Syndrome (MetS) include lifestyle modifications and pharmaceutical drugs that primarily target specific biochemical processes involved in food metabolism (Mohamed et al. 2023; Ambroselli et al. 2023). Pharmaceutical medications, however, are associated with high cost, poor patient compliance, and the emergence of several unwanted side effects as a result of their long-term use; these medications have also been proven to be monotherapeutic, typically focusing on a small number of health outcomes linked to metabolic dysregulation (Ambroselli et al. 2023). Thus, there has been a periodic increase in research toward creating alternative and supplementary methods for treating metabolic illnesses.

Several studies have suggested medicinal plants as part of the aforementioned alternative management options for MetS. Plants or plant preparations with health-promoting qualities, short-term alleviation, or advantageous therapeutic and preventative qualities have been considered medicinal plants (Ambroselli et al. 2023; Bangash et al. 2024). The importance of these plant preparations must be acknowledged. Hence, healthcare practitioners are beginning to understand their potential in treating and preventing metabolic illnesses (Ambroselli et al. 2023; Lippman et al. 2024; Jeyasri et al. 2024). Moreover, herbal medicine use has not been limited to developing nations but has expanded into an industry worth billions of dollars, catering to individuals from various demographic and socioeconomic backgrounds (Ambroselli et al. 2023; Jeyasri et al. 2024).

As demonstrated in the literature, the recognition mentioned above of medicinal plants is said to be dependent on their constituting pharmacodynamic bioactive chemicals posing synergistic and additive therapeutic effects that help regulate metabolic diseases (Ambroselli et al. 2023; Sakhale et al. 2023; Mazzoni et al. 2021). Hence, most novel pharmaceutical medications are made from therapeutic plants using local people's knowledge and then the ensuing separation of the main active ingredients (Sruthi et al. 2023).

Furthermore, studies have proven that patterns that could be used to produce pharmaceutical medications can be derived from the plant material utilized to prepare herbal medicines, thus lessening the cost of using overpriced synthetic pharmaceutical medications, and identifying advantageous use of phytochemical substances found in medicinal plants to treat metabolic syndrome (Theodoridis et al. 2023).

In addition, it has also been reported that the WHO reports showed that despite the availability of modern medications, the majority of Africans, both rural and urban dwellers, still receive their primary medical treatment from traditional healers (Theodoridis et al. 2023); this is thought to have stemmed from the belief that medicinal plants do not have acute toxicity or adverse side effects, in contrast to a few of the pharmacological medications currently used to treat metabolic diseases (Pedroso et al. 2021).

Riaz et al. (2023a) described phytonutrients, also known as secondary metabolites of plants, as substances derived from plants with unique structural and functional characteristics; these substances found in plants that offer protection against a variety of illnesses but are not essential for nutrition have also been said to be known as bioactive compounds. Studies have further shown that based on their biosynthetic processes, these by-products of plants are commonly classified into six large-molecule groups: polysaccharides, alkaloids, phenolics, terpenes, saponins, and glycosides. They are found in various fruits, cereals, nuts, seeds, and vegetables (Riaz et al. 2023a).

17.2.1 Polyphenols

According to research, polyphenols have been described as secondary metabolites, or naturally occurring non-nutritional compounds, of plants commonly present in fruits, seeds, and vegetables. It was also mentioned that most polyphenols are made from l-phenylalanine through the phenylpropanoid route (Zagoskina et al. 2023). Findings also showed the main structural types of polyphenols to be flavonoids, phenolic acids, lignans, stilbenes, tannins, and coumarins; these are said to be distinguished by the presence of one or more phenolic groups that exhibit a wide variety of structures (Zagoskina et al. 2023).

Studies have also shown that polyphenols are becoming more significant, mainly due to their positive health effects; their antioxidant importance in the prevention and treatment of various diseases is growing, particularly concerning age-related degenerative diseases, cancer, cardiovascular issues, inflammation, neurological disorders, and oxidative stress (Hano and Tungmunnithum 2020).

The significance of such antioxidants, as mentioned earlier, in preventing and treating a variety of diseases is expressed through their interaction with

RNA-dependent protein kinase-like endoplasmic reticulum kinase (PERK), phosphatidylinositol 3-kinase (PI3K), protein kinase C (PKC), and mitogen-activated protein kinases (MAPKs), which phosphorylate Nrf2. Therefore, polyphenols can neutralize excess free radicals (ROS) and mitigate oxidative stress; examples are EGCG, resveratrol, quercetin, curcumin, and caffeic acid (Rudrapal et al. 2022).

Furthermore, literature has also shown that polyphenols can reduce adoptive cellular responses due to their suppression of COX2 expression, interleukin (IL)-1β, and inducible nitric oxide synthase (iNOS) and regulatory activation of NF-κB (Rudrapal et al. 2022).

Quercetin, a flavonoid, according to studies, is one of the flavonols that are found in many fruits and vegetables; based on its promising antioxidant potential, quercetin, myricetin, and kaempferol have been suggested as the most prevalent flavonols present in onions, apples, green teas, and capers (Nguyen et al. 2025). Many in vitro and in vivo studies have demonstrated quercetin's anti-inflammatory and antioxidant qualities (Wang et al. 2022). As a result, its antioxidant activities have been connected to its ability to scavenge free radicals, chelate metal cations, and donate electrons or hydrogen atoms (Elhusseiny et al. 2021). Quercetin and its glycosidic derivatives impact the functioning of cells, including the reduction of oxidative deoxyribonucleic acid (DNA) damage and cell signaling pathways because of their antioxidant properties (Zhu et al. 2022).

Furthermore, it has been demonstrated that quercetin increases the expression of antioxidant enzymes such as superoxide dismutase (SOD), glutathione peroxidase (GPX), and γ-glutamylcysteine synthetase (γGCS); it has also been linked to the increase in the level of glutathione reductase (GSH), improvement of Nrf2 stabilization, and stimulation of Nrf2-mediated transcription (Riaz et al. 2023a). However, quercetin prevents NF-κB from activating (Riaz et al. 2023a). According to research, onion shells (*Allium cepa*) contain a lot of quercetin (3, 5, 7, 3′, 4′-pentahydroxy flavones) (Kumar et al. 2022a). In light of this, dietary quercetin has been proposed to possibly prevent oxidative breakdown of low-density lipoproteins (LDL) by chelating metal ions produced by oxidation of lipids and scavenging reactive oxygen species (Zhu et al. 2022). All discussed properties of quercetin, a polyphenol, make it a strong plant-based therapeutic option for metabolic syndrome.

Based on studies, resveratrol has been recognized as the most researched stilbene with cardioprotective and anticarcinogenic properties; peanuts, berries, grapes, and vines have been shown to contain an abundant level of resveratrol. Also, it is believed that the suppression of the growth of vascular smooth muscle cells is the mechanism of action by which resveratrol, as a stilbene, achieves its cardioprotective effect (Nardini 2022). Furthermore, resveratrol has been shown to promote Nrf2 transcriptional activity and suppress oxidative stress by upregulating hemoxygenase-1 (HO-1) and γGCS expression (Rudrapal et al. 2022; Su et al. 2022) making it a potent plant-based bioactive compound with metabolic properties for the treatment of metabolic syndrome.

It has also been demonstrated that kaempferol, quercetin, and resveratrol inhibit the proteins that cause oxidation in cardiac tissues, preventing cardiac oxidative stress (Iqbal et al. 2023). Furthermore, by lowering LDL oxidation and generating

NO, polyphenols generally alter the tone of the artery endothelium, thus preventing atherosclerosis; consequently, polyphenols in cardiomyocytes have also been shown to prevent the synthesis of inflammatory markers and reduce their expression (Iqbal et al. 2023). The aforementioned indicates the effectiveness of these bioactive substances derived from plants on the cardiovascular disease component of metabolic syndrome.

17.2.2 Terpenes

Based on the literature, it has been revealed that more than 55,000 terpenes, also known as terpenoids or isoprenoids, have been identified, thus making them the largest class of bioactive compounds; this class of bioactive compounds has also been shown to find extensive application as spices, flavors, perfumes, and cosmetics in the industrial sector (Bolouri et al. 2022). There is also increasing interest in the therapeutic application of alkaloids in these drugs, according to Riaz (2023b) two well-known terpene-based medications are antimalarial (artemisinin) and anticancer (paclitaxel).

Furthermore, Dell (2021) described terpenes' remarkable anti-inflammatory qualities; they were said to work by blocking inflammatory pathways linked to various illnesses. In addition, *in vitro* and *in vivo* tests were said to have revealed that terpenes extracted from the turmeric species (*Curcuma* sp.) specifically had anti-inflammatory qualities (Jaiswal and Naik 2021). It has been revealed that these drugs' anti-inflammatory qualities depend on their ability to inhibit the primary enzymes of inflammatory pathways, lipoxygenase, and proteinase, according to Irshad (Elhusseiny et al. 2021). Terpenes could also express their ability to prevent oxidative stress by forming metal-ion chelates and by blocking the enzyme's ability, as such, act to catalyze the production of reactive oxygen species (Agus 2021).

Thus, their hepatoprotective properties have been linked to the antioxidant action of terpenoids (Ielciu et al. 2021) in addition to multiple pathways such as apoptotic and anti-inflammatory pathways which have been claimed to take effect by the decrease in the ratio Bax/Bcl2, inhibition of I-κB, ERK, and JNK activation, and the blockade of release of cytochrome c from hepatocytes to limit liver cell death. They also manage obesity-related disorders: hyperlipidemia, insulin resistance, and type 2 diabetes (Ghasemi-Gojani et al. 2022). Panigrahy (2021) have proved that terpenoids can be an efficient and secure treatment for diabetes.

Cox-Georgian (Rani et al. 2023a), also corroborated that the terpenes from curcumin have been commonly attributed to curing other illnesses, such as metabolic disorders, due to their anti-inflammatory and antioxidant capacities.

Based on several *in vitro*, *in vivo,* and clinical data available, various plant-based compounds such as resveratrol (polyphenol), rutin, hesperidin, naringin, nobiletin (all flavonoids), β-cryptoxanthin (carotenoid), and others have been shown to exhibit anti-inflammatory and anti-metabolic-syndrome properties established through the various underlying pathophysiological pathways of the metabolic syndrome

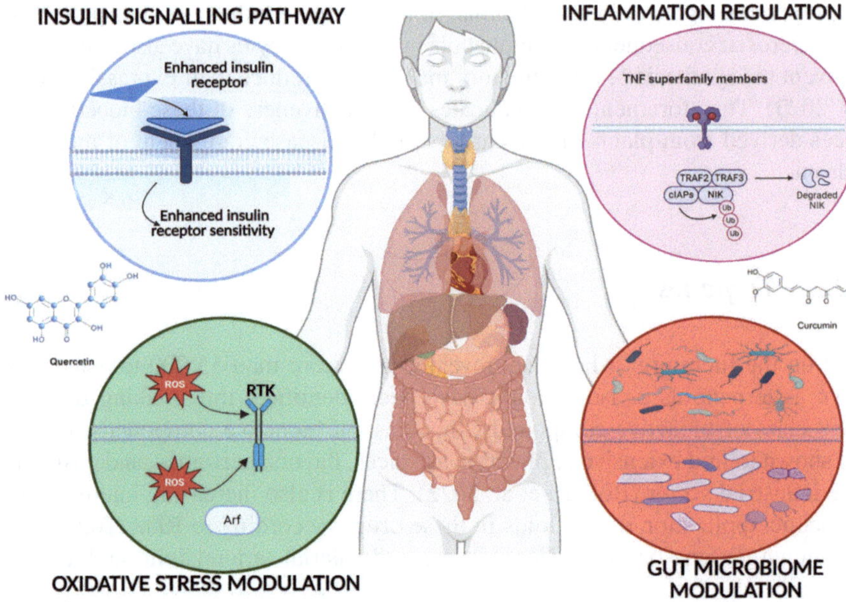

Fig. 17.1 Multitarget mechanisms of plant-derived nanoformulations addressing the complex pathophysiology of metabolic syndrome. (Created in https://BioRender.com)

(Rodríguez-Negrete et al. 2024). However, this chapter has focused more on the metabolic effects of polyphenols and terpenes, as well as classes of bioactive compounds with a specific discussion of quercetin, resveratrol, and curcumin (Fig. 17.1).

17.3 Limitations of Plant-Based Compounds

Although the therapeutic effects of those above plant-based bioactive compounds have been well elucidated, they are limited in bioavailability, solubility, rate of metabolism and efflux, stability, and degradation, thus their reduced efficiency as therapeutic models for metabolic syndrome. Such limited bioavailability of phenolic compounds has been revealed due to their low solubility in water and inability to remain stable at low pH values (Malode et al. 2021). Therefore, emerging studies have suggested that these limitations, such as the reduced availability of polyphenols in the body system, can be amplified by using nanoparticles (Adebayo-Tayo and Ogunrinade 2021), paving the path for the introduction of nanoscience and nanotechnology into biomedical research and the treatment of metabolic syndrome.

17.4 Nanotechnology and Nanoformulations

The world of medical science has come a long way from the traditional method of drug delivery to a more advanced phase of nanotechnology. Nanoformulation (the science of creating nanomedicines for biomedical applications) and nanotechnology are generally defined as the science and engineering of manipulating matter on a nanoscale (Patel et al. 2021). The advent of nanotechnology has greatly enhanced biomedical applications, mainly due to the opportunity of manipulating and controlling materials at the nanoscale, through the use of nanoparticles (NPs) and nanocarriers (NCs) (Mabrouk et al. 2021).

Current traditional drug delivery and treatment methods suffer from permeability, solubility, and bioavailability, rendering them inadequate with respect to the pharmacokinetic aspects of the administration routes (Stielow et al. 2023). With the advent of nanotechnology and nanoformulations, these challenges have been significantly overcome with the intent of creating a dosage form having a more optimal pharmacokinetic delivery system.

At the nanoscale, it includes a plethora of nanostructures (metallic and metal oxide nanoparticles, carbon nanotubes, liposomes, and nanopatterned surfaces), each with a unique feature for its use in medical applications (Kumar et al. 2022b). Drug delivery, diagnostics, bio-sensing, bio imaging, and implants are all good applications of nanoformulations (Petrovic et al. 2024). Drug nanoparticles not only serve as a delivery vehicle but can act as a drug itself or as a carrier to deliver a variety of therapeutic agents to specific areas of the body (Table 17.1).

In terms of specificity of delivery to affected tissues, bioavailability, sustained use, and reduced side effects, nanotechnology, compared with traditional drug delivery and treatment, has many advantages (Sultana et al. 2022). Even with such an array of advantages, nanoformulations still rely on nanoparticles' unique properties for the specificity of delivery to the affected tissues (Mabrouk et al. 2021). However, specific targeted delivery can be achieved by modifying the nanoparticle's surface with specific ligands such as antibodies or peptides, and then it can be delivered specifically to the diseased cell (Mitchell et al. 2021). All of the above reduce the impact on healthy cells in the area of the targeted cells and improve the rate of treatment. For example, one can conjugate antibodies to nanoparticles for them to home in on tumor cells (Mitchell et al. 2021).

Drugs can be controlled by modulating nanoformulations to only release drugs in response to a particular stimulus, like pH change or enzymatic activity, giving room for precise control of therapeutic effects (Abdelkawi et al. 2023). Nanoparticles also improve the solubility and stability of poorly soluble drugs, thus enhancing their bioavailability (Bhalani et al. 2022).

The development of the broad spectrum of nanoscale technologies has begun to change the scientific landscape regarding disease diagnosis, prevention, and treatment. Numerous studies have delved into discovering and developing new nanoformulations that can specifically target disease sites. Notwithstanding, there are still

Table 17.1 Key types of Nanoparticles and their applications

Key types of Nanoparticles	Applications
1. Iron Oxide Nanoparticles (IONPs)	IONPs are frequently used in medication delivery, magnetic resonance imaging (MRI), and cancer treatment via hyperthermia. Their magnetic features enable targeted delivery and regulated release of therapeutic agents when guided by external magnetic fields. They exhibit biocompatibility and non-toxicity (Salehirozveh et al. 2024).
2. Gold Nanoparticles	Due to their unique optical and electronic properties, they are utilized in bio-sensing, photothermal therapy, and targeted drug delivery. They can also be easily functionalized with biomolecules for specific targeting (Hammami and Alabdallah 2021).
3. Silver Nanoparticles	Silver nanoparticles are known for their antibacterial properties. They are used in wound dressings and coatings for medical devices. They also show potential in drug delivery and photothermal therapy (Bruna et al. 2021).
4. Cerium-Oxide Nanoparticles (Nanoceria)	These nanoparticles have antioxidant properties and are being explored as potential therapeutic agents against cancer and as biosensors (Lord et al. 2021).
5. Biogenic Nanoparticles	Biogenic nanoparticles have applications in drug delivery systems and bio-sensing technologies due to their natural origin and reduced toxicity (Harish et al. 2022). These nanoparticles are synthesized using biological materials such as plants, fungi, and bacteria. They are extensively explored due to their eco-friendly production methods and biocompatibility (Kaur 2021)

many gaps in knowledge about the effectiveness and efficacy of various nanoformulations, particularly in various metabolic syndromes.

17.5 Specific Plant Compounds Nanoformulations for Metabolic Syndrome

17.5.1 Curcumin Nanoformulations

Curcumin is the primary polyphenol extracted from the rhizomes of *Curcuma longa L.* (family Leguminosae), generally known as curry powder (turmeric) (Hafez Ghoran et al. 2022). Curcumin has many medicinal applications, including anti-inflammatory, antibacterial, antioxidant, antidiabetic, anticancer, antifungal, antiproliferative, and hepatoprotective properties (Fu et al. 2021). The hydrophobicity of curcumin makes it hard to pass cellular membranes because it binds to lipid fatty acyl chains membrane through hydrogen bonding and hydrophobic interactions. As a result, curcumin levels stay extremely low throughout the cytoplasm (Hafez Ghoran et al. 2022).

Curcumin exerts its effects through various mechanisms of action. It exhibits its anti-inflammatory effects through the modulation of several inflammatory pathways, which include inhibition of nuclear factor kappa B (NF-kB) and downregulation of pro-inflammatory cytokines such as TNF-alpha and Interleukin-6 (IL-6) (Zamanian et al. 2024). Insulin resistance, often induced by chronic inflammation, is mitigated by this modulation. Curcumin helps scavenge free radicals and boost the body's antioxidant defenses, thereby reducing oxidative stress and showing its antioxidant effect (Engwa et al. 2022). Furthermore, insulin sensitivity and glucose metabolism improved due to the influence of curcumin on the metabolic pathways, which helps manage conditions like type 2 diabetes (Balakumar et al. 2023).

Curcumin nanoformulations increase curcumin's nature as a medicinal agent by boosting bioavailability. Various nanoformulations have been developed to improve the solubility and bioavailability of curcumin, including polymeric nanoparticles, liposomes, micelles, cyclodextrin, solid lipid nanoparticles, magnetic nanoparticles, nano gel, and more (Hafez Ghoran et al. 2022; Gayathri et al. 2023).

Polymeric nanoparticles include poly (lactic-co-glycolic acid) (PLGA) nanoparticles that encapsulate curcumin, considerably boosting its oral bioavailability to up to nine times that of free curcumin (Pandey et al. 2025). Liposomes are globular vesicles with multiple or single phospholipid bilayers covering aqueous systems, resembling biological membranes (Liu et al. 2021). Liposomes are suitable delivery mechanisms for bioactive compounds in vivo and in vitro (Cardoso et al. 2022). Liposomes offer a range of benefits, which include increased biodegradability and biocompatibility, controlled release, high durability, enhanced dissolution rate, and delivery, and the capability to target individual cells, to mention a few (Dymek and Sikora 2022). Consequently, liposomes have become a potent tool for drug carrier mechanisms. Curcumin is shown to be soluble inside the phospholipidic bilayers of liposomes, which facilitates the transport of the curcumin into the aqueous phases and enhances its effect on the targeted cells. Curcumin is improved in cellular uptake and targeted delivery by the liposomal formulations, and the therapeutic efficacy is enhanced (Gholami et al. 2023; Obeid et al. 2023).

An amphiphilic surfactant molecule self-assembles to make a micelle, which is a spherical vesicle. This method is commonly used to deliver non-water-soluble medication such as curcumin (Silva et al. 2023a). Curcumin solubility in aqueous environments is increased by micelles, making it more available for absorption in the gastrointestinal tracts (Luo et al. 2022). Another formulation of curcumin is cyclodextrins. Solubilization of curcumin through cyclodextrin complexation has led to better solubility and stability than pure curcumin due to better anti-inflammatory properties (Mashaqbeh et al. 2021). A further drug formulation delivery system involves Solid Lipid Nanoparticles (SLNs), which help protect curcumin from degradation by providing a controlled release profile, enhancing its therapeutic potential against several diseases (Zhao et al. 2022). These magnetic nanostructures promote tailored drug delivery, enabling site-specific treatment while reducing associated systemic side effects (Nazir et al. 2022).

Previous research has been done on in vitro and in vivo curcumin nanoforms to study their quantitative effect on pure curcumin, as shown in (Urošević et al. 2022; Bolger et al. 2022). The lifespan of the curcumin nanoparticles was shown to be longer than curcumin alone. Additionally, they experienced increased bioavailability of curcumin due to increased absorption. Although much literature has primarily demonstrated the effects of various curcumin nanoformulations on cancerous cells, such results may well hint at its possible therapeutic effects on the metabolic syndrome by considering its effects on the cancerous cells through the inflammatory, oxidative stress, and apoptotic pathways (Bolger et al. 2022; Bateni et al. 2022).

Various researchers have looked into curcumin's safety, pharmacokinetic characteristics, and efficacy in treating various human illnesses. Clinical experiments have shown some promising outcomes of curcumin nanoformulations, with much research showing that curcumin can arrest or prevent the development of cancerous cells (Li et al. 2022; Zoi et al. 2021). Several clinical trials show that curcumin nanoformulations help treat metabolic syndrome patients, ankylosing spondylitis, amyotrophic lateral sclerosis, chronic kidney disease, multiple sclerosis, and malignancies (Bateni et al. 2022; Tagde et al. 2021). A review by Hafez (Bateni et al. 2022) showed that curcumin application in documented clinical trials is around 210, with about 92 of the clinical studies completed. The majority of these studies were focused on cancerous cells, leaving a gap in knowledge on its efficacy on other conditions like metabolic syndromes, chronic kidney disease, multiple sclerosis, and others, leaving a need for more clinical studies.

Using nanotechnology in curcumin formulations is a possible strategy for addressing the restrictions associated with its use in therapeutic settings. These nanoformulations boost curcumin's therapeutic efficacy by increasing solubility, bioavailability, and targeted administration. They also have the potential to help manage comorbidities like insulin resistance through anti-inflammatory mechanisms. Ongoing research investigates the efficacy of these formulations in various therapeutic applications, paving the way for novel treatment options.

17.5.2 Resveratrol Nanoformulations

This naturally occurring polyphenol and phytoalexin (resveratrol) has been of interest to the medical field (Annaji et al. 2021). It is mainly present in the grape and is anti-inflammatory, anti-cancer, and an antioxidant (Kaur et al. 2022).

At present, traditional dosage forms such as capsules, tablets, and powders are under investigation for the treatment with resveratrol. Broad literature describes its potential health benefits, including limited effects (Gal et al. 2021) to prevent or ameliorate aging and cancer, and the deleterious effects of dietary fat consumption that is more than the general recommendations (Sharifi-Rad et al. 2021a; Kumar et al. 2022c). Additionally, resveratrol can lower lipids and regulate insulin resistance in the liver as well as maintain glycemic homeostasis and even cure metabolic disease (Abou Assi et al. 2021).

However, poor pharmacokinetic properties have prevented resveratrol from becoming efficacious. However, bad aqueous solubility, bad photostability, and high first-pass metabolism have restricted its observed potential applications

(Jurczyk et al. 2022). However, it has continued to remain a compound of current interest due to its presence in grapes, red wine, and peanuts, among other things (Kaur et al. 2022).

The biological importance of resveratrol makes it the subject of intense research. It also has a known preventive effect against aging, cancer, and the pathological consequences of high-fat diet intake (Gal et al. 2021). Furthermore, lipid accumulation and fat in the liver are reduced, which improves insulin resistance in the liver, maintains glucose homeostasis, and solves metabolic disorders (Abou Assi et al. 2021).

However, poor pharmacokinetic properties such as poor photostability, limited aqueous solubility, and extensive first-pass metabolism restrict them to poor bioavailability (Jurczyk et al. 2022). Nanoformulations of resveratrol have been selected as an emerging massive bioavailability and therapeutic potency development using advanced delivery systems such as cyclodextrins, polymeric nanoparticles, micelles, and liposomes. Upon further increasing physicochemical instability, low aqueous solubility, and poor resveratrol bioavailability. Thus, resveratrol was deployed therapeutically in targeted cells (Yang et al. 2021). In fact, nanoformulations have also been shown to protect resveratrol from environmental degradation, increase the stability of the compound, and target it to a specific tissue. Conversely, this approach is likewise an ideal manner to treat different diseases like cancer and inflammatory conditions (Yang et al. 2021; Sharifi-Rad et al. 2021b). Leyane (Henein et al. 2022) describes many of these conditions as chronic inflammation, the leading cause of cardiovascular heart disease, diabetes, and arthritis, with elevated lipid peroxidation levels.

The mechanism of action was reported in several studies to be ameliorative, and with its action on lipid metabolism, the mechanism of action is targeted, and ameliorative mechanisms of action are applied to cardiovascular system diseases. In addition, it has a capacity that inhibits lipid peroxidation (Fenton Reaction), which also assists in lowering the oxidative chain complex and acts as a scavenger for ROS and other free radicals (Henein et al. 2022). Lipid peroxidases are estimated to measure free radical damage to cells. Ortho-diphenoxyl functionality of resveratrol resists oxidation caused by ROS in lipid peroxidation. At various stages of lipid metabolism, this inhibitor inhibits lipid peroxidation from initiation and propagation to the termination stages (Annaji et al. 2021; Briones-Valdivieso et al. 2024; Ju et al. 2023). Resveratrol blocks the formation of peroxy radicals at the start stage and lowers the production of conjugated alkenes at the termination stage by functioning as a competitive inhibitor of glutathione peroxidase activity (Zinatullina et al. 2021). However, when combined with nanoformulations, resveratrol's anti-inflammatory capabilities may mitigate inflammation and improve patient outcomes for affected patients. Resveratrol delivery by nanoformulations can achieve the therapeutic dose of resveratrol in patients suffering from many diseases, thereby increasing their efficacy by targeting a specific tissue of greater interest and allowing a simultaneous treatment of comorbiditie (Yang et al. 2021).

Resveratrol nanoformulations make a hopeful advance in the management of many comorbidities. As the nano-delivery system of resveratrol not only increases its therapeutic effects but also provides a new drug delivery system, it will make clinical applications in nanomedicine. Accurate nanotechnology formulation has been hampered by the difficulty in developing a nano delivery system and reliance upon extensive knowledge of that field. Thus, stable and efficient carriers for resveratrol that survive physiological conditions tend to be resource-intensive and complex to develop. A second nontrivial challenge is the tendency of nanoparticles to aggregate or disintegrate, which would mean losing effectiveness over time. To be practical, it is crucial to ensure that the drug will remain in an active form and also remain stable (Silva et al. 2023b; Chopra et al. 2022).

There is also a risk of adverse reactions if the nanocarriers provoke an immune response in the patient. Such situations warrant the careful assessment of the material biocompatibility used in the formulation (Boyton et al. 2024). Integrating nanomedicines into clinical use requires navigating stringent regulatory pathways; regulatory agencies need extensive safety and efficacy (Kardani 2024). Manufacturing scalability poses another significant challenge: processes must be refined to ascertain consistent quality and yield while strictly adhering to the regulatory standard to meet clinical or commercial demands (Rajnath et al. 2024).

Further studies must be conducted to see how these challenges can be combated or managed.

17.5.3 Quercetin Nanoformulations

Quercetin is a naturally occurring flavonoid. Quercetin is derived from quercetin (oak woodland). Quercetin is abundant in natural plants, such as berries, apples, brassica vegetables, spring onions, tea, grapes, onions, tomatoes, flowers, bark, seeds, nuts, and leaves (Vollmannová et al. 2024).

Its chemical structure includes four active groups: a dihydroxy group between the A ring, an o-dihydroxy group B, C ring C2, C3 double bond, and a 4-carbonyl. Quercetin is enriched with potent antioxidant properties due to a phenolic hydroxyl group and double bonds (Alizadeh et al. 2023). Furthermore, various in vivo and in vitro studies have shown that quercetin has antibacterial action and can efficiently inhibit biofilm formation by decreasing the expression of associated genes and anti-angiogenic and antitumor properties (Alizadeh and Ebrahimzadeh 2022; Hasan et al. 2022). It is documented that quercetin becomes important for deactivating the mycotoxin and preventing cell damage (Xia et al. 2022). These also have antioxidant and anti-inflammatory properties to combat cardiovascular disease, cancer, and metabolic syndrome (Xiong et al. 2023). Its use is quite effective but has limited bioavailability, solubility, permeability, and stability, limiting its effectiveness. Different quercetin nanoformulations yield advances in nanotechnology, which solve the problems mentioned above. This high potential for therapeutic applications is based on improvements in quercetin's bioavailability and efficacy using these innovative approaches (Manzoor et al. 2021). Multiple studies have shown that quercetin nanoformulation shows hypolipidemic and hypoglycemic effects, leading to weight management and improved metabolic profiles (Dinda and Dinda 2022). They can also lower the level of inflammatory markers and cytokines and help combat inflammation in obesity or other conditions, improving insulin sensitivity and protecting against metabolic diseases (Shivani et al. 2023).

To study various disease conditions like obesity, fat metabolism, adipogenesis, cardiovascular diseases, cancer, and others, several nanoformulation designs have been developed to test their effectiveness (Ahmed et al. 2021a) such as chitosan-PEG-PLGA nanoparticles.

Ahmed (2021b) study demonstrated that quercetin nanoformulation is effective over suspension, significantly improving the effectiveness of quercetin in treating dyslipidemia, abdominal obesity, elevated blood pressure, and insulin resistance.

A more effective drug delivery system for quercetin was studied as a carrier using fluorescent mesoporous silica nanoparticles (FMSNs) coated with polydopamine (PDA) as a carrier design in the adipogenesis suppressive agent by Kim (2024). Adipocyte overproduction of ROS that exceeds the antioxidant defense level leads to increased oxidative stress, increased secretion of leptin, MCP-1, IL-6, and TNF-α, and decreased adiponectin synthesis (Man et al. 2023). Inhibition of adipogenesis (Kim et al. 2024) was enhanced by delivery of quercetin using FMSNs that demonstrated enhanced cellular uptake and antioxidant efficacy of nanocarriers due to PDA coating of the silver nanowires (Dong et al. 2023).

Quercetin nanoformulation, through its bioavailability, antioxidant, and anti-inflammatory properties, provides an increased possibility in treating metabolic disorders, cardiovascular diseases, and many other conditions. However, further research is still required to fully understand the efficiency of various delivery mechanisms and their therapeutic potential (Fig. 17.2).

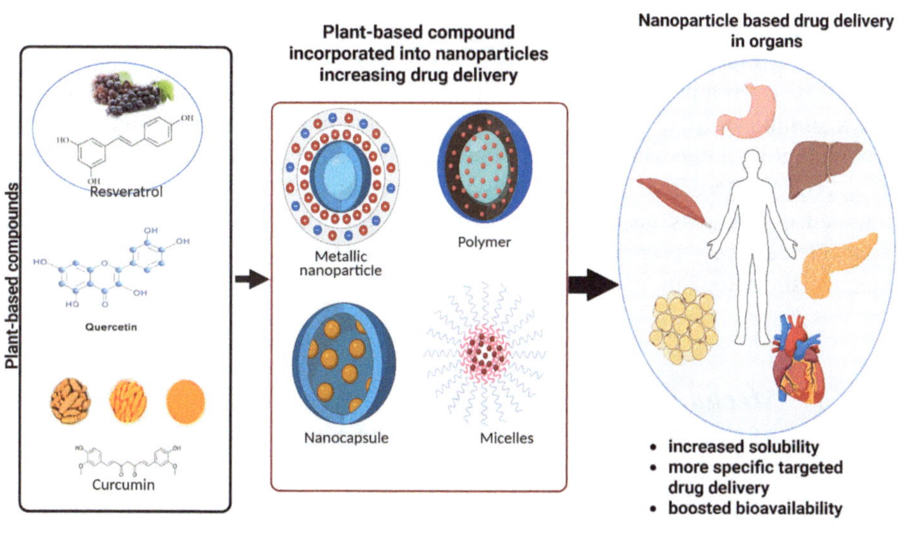

Fig. 17.2 Nanoformulation strategies enhance bioavailability and targeted delivery of plant bioactive compounds to metabolic syndrome-affected organs. (Created in https://BioRender.com)

17.6 Multitarget Approach in Metabolic Syndrome

Metabolic Syndrome (MetS) refers to a complex constellation of clustered physiological, biochemical, clinical, and metabolic factors associated with direct referral to cardiovascular disease, type 2 diabetes mellitus, and all-cause mortality (Masenga et al. 2023). This multifaceted condition requires a complete, multi-loop, therapeutic approach, spanning all aspects of the syndrome, with treatment customized to each patient.

Mets management starts with the primary therapeutic loop based on lifestyle modification (Ismail 2022). Based on this foundation, the secondary therapeutic loop includes pharmacological approaches to mitigate a subset of its components. Typically, blood pressure is controlled with ACE inhibitors, ARBs, and calcium channel blockers, and glycemic control is controlled with metformin, GLP1 receptor agonists, and SGLT2 inhibitors. Another vital aspect is lipid management, which includes statins, fibrates, and omega-3 fatty acids (Zeng et al. 2023). The sloop's operation of this loop depends on choosing medication and monitoring patient response.

Novel, tertiary therapeutic loop approaches involve nutraceutical interventions (Dama et al. 2024) and nanoformulation strategies (Abd El-Hacka et al. 2023). Its potential to improve the bioavailability of natural compounds and deliver targeted systems has lent itself to this unique loop, obtaining significant attention. Plant-based bioactive, special nutritional formulations, and multicomponent nanocarriers employed in MetS management are a promising frontier.

Convincing evidence is mounting for the personal use of plant compounds in nanoformulations in pharmaceutical development (Biganeh et al. 2023; Wahi et al. 2023; Bozzuto et al. 2024), thus representing an innovative approach to alleviate therapeutic benefits through the synergistic effect and the increase of deliverability of active compounds in body tissues. Justifications for synergistic combinations cannot be overemphasized: It allows multiple compounds to target multiple pathways simultaneously, reduces drug resistance potential and lowers effective doses when combined due to synergistic effects, and broadens therapeutic activity (Wahi et al. 2023). Alternative measures may include improved bioavailability of nanoformulated compounds based on improved solubility, membrane permeability, and controlled delivery properties, leading to improved tissue distributions and delivery mechanisms (Bozzuto et al. 2024).

17.6.1 Mechanisms of Action in the Delivery of the Multi-Targeted Effects of Nano-Therapy

1. Delivery mechanisms of action. Liposomes, Polymeric nanoparticles, Solid lipid nanoparticles and Nano-emulsions particles, Solid lipid nanoparticles and Nanoemulsions (Sazdova et al. 2023).

2. Cellular uptake enhancement includes better endocytosis, better penetration into biological barriers, and targeted delivery within tissues (Sazdova et al. 2023).
3. Comprising multiple pathway modulation, signal transduction pathways, metabolic processes, inflammatory responses, and oxidative stress regulation, with multi-targeted effects (Eity et al. 2024).

The above are possible for nanotherapeutic compounds' proper scientific, logical design considerations. Such compounds should be considered based on the compatibility of different plant compounds and their stability in combination with physicochemical properties such as particle size, surface charges, and solubility. In other words, formulation parameters are used to determine the optimum ratio of compounds, which informs the choice of the nanocarrier system. In addition, surface modifications add them the ability to be designed for improved absorption through biological membranes, better distribution in target tissues, and sustained release properties to provide these compounds with biopharmaceutics and pharmacokinetics (Biopharmaceutics and Pharmacokinetics or Biopharm) properties through their enhanced permeability and retention (EPR) effect (Sazdova et al. 2023).

The research on nanoformulation of existing compounds has been based on increasing the bioavailability and efficacy of natural metabolites. Anti-inflammatory, antioxidant, and lipid-modulating substances, such as curcumin, resveratrol, and green tea polyphenols. A polyphenolic compound derived from turmeric called curcumin-based nanoformulations has recently been demonstrated to increase insulin sensitivity and reduce inflammation. The formulation of curcumin in liposomes, polymeric, and solid lipid nanoparticles has improved bioavailability. Metabolic improvements in these models using animal models of MetS have been shown following curcumin solubility and stability improvements in these formulations (Sharma et al. 2024).

Similar reports on resveratrol (found in grapes and berries) were reported for cardiovascular protective effects (Salla et al. 2024). Its bioavailability and therapeutic potential have been enhanced using nanoencapsulation techniques. Research indicates that resveratrol nanoparticles can effectively reduce hyperglycemia, dyslipidemia, and oxidative stress, which are essential MetS components (Bejenaru et al. 2024). Green Tea Polyphenol Combinations are isolated from green tea, which is abundant in catechins, and formulated into nanoparticles to have metabolic effects. The potential modulation of lipid metabolism, improvement of glucose homeostasis, and weight management seen with these formulations in experimental models suggest they are worth investigating (Biganeh et al. 2023).

Action within nanoformulations encapsulates the bioactive compound, successfully delivering the bioavailable bioactive compound. However, nanoformulations increase the bioavailability of hydrophobic molecules, allow for targeted, specific delivery to particular tissues, and permit release properties. These characteristics allow them to achieve a wider therapeutic window for their target drugs by allowing a more efficient modulation of multiple metabolic pathways, including those involved in inflammation, oxidative stress, and lipid and glucose metabolism (Salla et al. 2024).

While there are promising findings, there remain gaps in the research of nanoformulations for MetS, which, if completed, will better position the effectiveness of nano-therapy in MetS management. Thus, future studies must elucidate the molecular and cellular mechanisms involved in the action of these formulations. Notably, it is significant to translate diverse studies from preclinical settings to clinical settings to validate their safer and more effective nanoformulations in clinical settings. This would help develop advanced formulation techniques to, for example, develop intelligent delivery systems that will respond to some physiological triggers to enhance the precision and effectiveness of the treatment.

Therefore, the above opens space for combining nanoformulations with other treatments to achieve synergistic effects and higher therapeutic outcomes through combination therapies. We must do more research to glean optimal combinations and dosing regimens. A personalized medicine approach could also be explored, including developing a personalized nanoformulation strategy that optimizes treatment efficacy and minimizes adverse effects based on individual patient profiles. In today's approach, it is necessary to identify biomarker profiles and devise patient-specific formulation designs. Regulatory and manufacturing factors concerned with large-scale production and regulatory approval issues will be crucial in successfully commercializing nanoformulations. Their assurance of these formulations' consistency, quality, and stability will be the greatest.

17.7 Safety Considerations

Nano-formulated plant compounds, as shown, have been perceived as a significant innovation in disease management (Chen et al. 2024). New therapeutics for diseases such as metabolic syndrome (Chen et al. 2024; Hu et al. 2022) may be opened by the characteristics that increase the medicinal effect of plant-based drugs with the help of the unique properties of nanoparticles. Specific guidelines and definitions regarding nanotechnology-enabled health products (NHPs) are lacking (Rodríguez-Gómez et al. 2025, 2023). Therefore, a new NHP classification system is proposed based on scientific and regulatory principles (Rodríguez-Gómez et al. 2023). For example, NHP quality and safety characterization has identified methodological gaps in surface properties, drug loading, and NHP interactions within the biological system (Halamoda-Kenzaoui et al. 2021). One of the advantages of particles is their ability to serve as nanocarriers for bioactive compounds. In addition, they can also be used to improve the delivery and stability of phytochemical compounds, which have traditionally been rejected as therapeutics. Additionally, these compounds are encapsulated, increasing plant extracts'bioavailability and therapeutic outcomes (Kumari et al. 2022).

Nano-formulated compounds are eco-friendly and biodegradable, and align with the trends to help minimize the potential for environmental accumulation and toxicity (Rani et al. 2023b). Given this, their potential use in metabolic syndrome is highly significant since the primary mechanisms of disease progression are determined by lifestyle and environmental circumstances. While some challenges must be overcome to enable the wide-scale use of nano-formulated plant compounds, their potential use in clinical settings is promising (Rani et al. 2023b; Lim et al. 2022). Nanotechnology healthcare product regulation should incorporate safety, efficacy, and ethical requirements as the basis of such regulation. Such studies are needed to fill our knowledge gaps regarding the mechanism of action, optimal dose, and potential side effects of nano-formulated plant compounds (Rodríguez-Gómez et al. 2023).

References

Abd El-Hacka ME, Kamalb M, Alazragic RS, Alreemic RM, Qadhid A, Ghafourid K, Azhard W, Shakoorie AM, Alsaffarf N, Naffadig HM, Tahah AE, Abdeln SA (2023) Impacts of chitosan and its nanoformulations on the metabolic syndromes: a review. Braz J Biol 83:e276530. https://doi.org/10.1590/1519-6984.276530

Abdelkawi A, Slim A, Zinoune Z, Pathak Y (2023) Surface modification of metallic nanoparticles for targeting drugs. Coatings 13(9):1660

Abou Assi R, Abdulbaqi IM, Siok Yee C (2021) The evaluation of drug delivery nanocarrier development and pharmacological briefing for metabolic-associated fatty liver disease (MAFLD): an update. Pharmaceuticals 14(3):215

Adebayo-Tayo BC, Ogunrinade JO (2021) Mycosynthesis and characterization of gold and silver nanoparticles using biomolecules of Pleurotus ostreatus as antibacterial agent against some selected pathogens. Acta Microbiol Bulg 69

Adetunji CO, Michael OS, Rathee S, Singh KR, Ajayi OO, Adetunji JB, Ojha A, Singh J, Singh RP (2022) Potentialities of nanomaterials for the management and treatment of metabolic syndrome: a new insight. Mater Today Adv 13:100198

Agus HH (2021) Terpene toxicity and oxidative stress. In: Toxicology. Academic Press, pp 33–42

Ahmed HH, Kotob SE, Abd-Rabou AA, Aglan HA, Elmegeed GA, Mohawed OA (2021a) Pre-clinical evidence for the anti-obesity potential of quercetin and curcumin loaded chitosan/PEG blended PLGA nanoparticles. Biomed Pharmacol J 14(4):1731–1759

Ahmed OA, Hassan NA, Azhar AS, El-Mas MM, El-Bassossy HM (2021b) A nano-pharmaceutical formula of quercetin protects from cardiovascular complications associated with metabolic syndrome. Front Pharmacol 12:696981

Alizadeh SR, Ebrahimzadeh MA (2022) Quercetin derivatives: drug design, development, and biological activities, a review. Eur J Med Chem 229:114068

Alizadeh SR, Savadkouhi N, Ebrahimzadeh MA (2023) Drug design strategies that aim to improve the low solubility and poor bioavailability conundrum in quercetin derivatives. Expert Opin Drug Discov 18(10):1117–1132

Ambroselli D, Masciulli F, Romano E, Catanzaro G, Besharat ZM, Massari MC, Ferretti E, Migliaccio S, Izzo L, Ritieni A, Grosso M (2023) New advances in metabolic syndrome, from prevention to treatment: the role of diet and food. Nutrients 15(3):640

Annaji M, Poudel I, Boddu SH, Arnold RD, Tiwari AK, Babu RJ (2021) Resveratrol-loaded nanomedicines for cancer applications. Cancer Rep 4(3):e1353

Balakumar P, Venkatesan K, Khan NA, Raghavendra NM, Venugopal V, Bharathi DR, Fuloria NK (2023) Mechanistic insights into the beneficial effects of curcumin on insulin resistance: opportunities and challenges. Drug Discov Today 28(7):103627

Bangash SA, Khan MN, Akhtar M, Khan S, Aziz R, Wahab S (2024) Medicinal plants and emerging trends in biotechnology. In: Plants as medicine and aromatics, pp 210–240

Bateni Z, Behrouz V, Rahimi HR, Hedayati M, Afsharian S, Sohrab G (2022) Effects of nano-curcumin supplementation on oxidative stress, systemic inflammation, adiponectin, and NF-κB in patients with metabolic syndrome: a randomized, double-blind clinical trial. J Herbal Med 31:100531

Bejenaru LE, Biţă A, Belu I, Segneanu AE, Radu A, Dumitru A, Ciocîlteu MV, Mogoşanu GD, Bejenaru C (2024) Resveratrol: a review on the biological activity and applications. Appl Sci 14(11):4534

Bhalani DV, Nutan B, Kumar A, Singh Chandel AK (2022) Bioavailability enhancement techniques for poorly aqueous soluble drugs and therapeutics. Biomedicine 10(9):2055

Biganeh H, Dizaji SM, Taghipour YD, Murtaza G, Rahimi R (2023) Nanoformulations of plant-derived compounds as emerging therapeutic approach for colorectal cancer. Curr Drug Deliv 20(8):1067–1094. https://doi.org/10.2174/1567201819666220823155526

Bolger GT, Pucaj K, Minta YO, Sordillo P (2022) Relationship between the in vitro efficacy, pharmacokinetics, and in vivo efficacy of curcumin. Biochem Pharmacol 205:115251

Bolouri P, Salami R, Kouhi S, Kordi M, Asgari Lajayer B, Hadian J, Astatkie T (2022) Applications of essential oils and plant extracts in different industries. Molecules 27(24):8999

Boyton I, Rennie C, Collins-Praino LE, Care A (2024) Nanoparticle formulations for the improvement of symptomatic treatments of neurodegenerative disorders. ChemNanoMat 10(12):e202400147

Bozzuto G, Calcabrini A, Colone M, Condello M, Dupuis ML, Pellegrini E, Stringaro A (2024) Phytocompounds and nanoformulations for anticancer therapy: a review. Molecules 29(16):3784

Briones-Valdivieso C, Briones F, Orellana-Urzúa S, Chichiarelli S, Saso L, Rodrigo R (2024) Novel multi-antioxidant approach for ischemic stroke therapy targeting the role of oxidative stress. Biomedicine 12(3):501

Bruna T, Maldonado-Bravo F, Jara P, Caro N (2021) Silver nanoparticles and their antibacterial applications. Int J Mol Sci 22(13):7202

Cardoso RV, Pereira PR, Freitas CS, Paschoalin VM (2022) Trends in drug delivery systems for natural bioactive molecules to treat health disorders: the importance of nano-liposomes. Pharmaceutics 14(12):2808

Chen Y, Tang Y, Li Y, Rui Y, Zhang P (2024) Enhancing the efficacy of active pharmaceutical ingredients in medicinal plants through nanoformulations: a promising field. Nanomaterials 14(19):1598

Chopra H, Bibi S, Islam F, Ahmad SU, Olawale OA, Alhumaydhi FA, Marzouki R, Baig AA, Emran TB (2022) Emerging trends in the delivery of resveratrol by nanostructures: applications of nanotechnology in life sciences. J Nanomater 2022(1):3083728

Dama A, Shpati K, Daliu P, Dumur S, Gorica EA, Santini A (2024) Targeting metabolic diseases: the role of nutraceuticals in modulating oxidative stress and inflammation. Nutrients 16:507. https://doi.org/10.3390/nu16040507

Del Prado-Audelo ML, Cortés H, Caballero-Florán IH, González-Torres M, Escutia-Guadarrama L, Bernal-Chávez SA, Giraldo-Gomez DM, Magaña JJ, Leyva-Gómez G (2021) Therapeutic applications of terpenes on inflammatory diseases. Front Pharmacol 12:704197

Dinda B, Dinda S (2022) Advances in nanoencapsulated phytomedicines (phytochemicals and their extracts) for the treatment of obesity, diabetes, and their associated complications. In: Natural products in obesity and diabetes: therapeutic potential and role in prevention and treatment. Springer International Publishing, Cham, pp 507–532

Dong W, Zhao Z, Liu F, Li P, Wang L, Zhou Y, Shen Y, Lang C, Deng B, Li H, Li D (2023) PVDF nanofiber modified with ZnO nanowires/polydopamine for the treatment of sewage containing heavy metals, organic dyes, and bacteria. ACS Appl Mater Interfaces 15(50):58994–59004

Dymek M, Sikora E (2022) Liposomes as biocompatible and smart delivery systems–the current state. Adv Colloid Interface Sci 309:102757

Eity TA, Bhuia MS, Chowdhury R, Ahmmed S, Sheikh S, Akter R, Islam MT (2024) Therapeutic efficacy of quercetin and its nanoformulation, both the mono- or combination therapies in the management of cancer: an update with molecular mechanisms. J Trop Med 2024(1):5594462

Elhusseiny SM, El-Mahdy TS, Awad MF, Elleboudy NS, Farag MM, Aboshanab KM, Yassien MA (2021) Antiviral, cytotoxic, and antioxidant activities of three edible agaricomycetes mushrooms: Pleurotus columbinus, Pleurotus sajor-caju, and Agaricus bisporus. J Fungi 7(8):645

Engwa GA, Nweke FN, Nkeh-Chungag BN (2022) Free radicals, oxidative stress-related diseases, and antioxidant supplementation. Altern Ther Health Med 28(1)

Fu YS, Chen TH, Weng L, Huang L, Lai D, Weng CF (2021) Pharmacological properties and underlying mechanisms of curcumin and prospects in medicinal potential. Biomed Pharmacother 141:111888

Gal R, Deres L, Toth K, Halmosi R, Habon T (2021) The effect of resveratrol on the cardiovascular system from molecular mechanisms to clinical results. Int J Mol Sci 22(18):10152

Gayathri K, Bhaskaran M, Selvam C, Thilagavathi R (2023) Nano formulation approaches for curcumin delivery review. J Drug Delivery Sci Technol 82:104326

Ghasemi-Gojani E, Kovalchuk I, Kovalchuk O (2022) Cannabinoids and terpenes for diabetes mellitus and its complications: from mechanisms to new therapies. Trends Endocrinol Metabol 33(12):828–849

Gholami L, Momtazi-Borojeni AA, Malaekeh-Nikouei B, Nikfar B, Amanolahi F, Mohammadi A, Kazemi Oskuee R (2023) Selective cellular uptake and cytotoxicity of curcumin-encapsulated SPC and HSPC liposome nanoparticles on human bladder cancer cells. Curr Pharm Des 29(13):1046–1058

Hafez Ghoran S, Calcaterra A, Abbasi M, Taktaz F, Nieselt K, Babaei E (2022) Curcumin-based nanoformulations: a promising adjuvant towards cancer treatment. Molecules 27(16):5236. https://doi.org/10.3390/molecules271652

Halamoda-Kenzaoui B, Vandebriel RJ, Howarth A, Siccardi M, David CAW, Liptrott NJ, Santin M, Borgos SE, Bremer-Hoffmann S, Caputo F (2021) Methodological needs in the quality and safety characterisation of nanotechnology-based health products: priorities for method development and standardisation. J Control Release 336:192–206

Hammami I, Alabdallah NM (2021) Gold nanoparticles: synthesis, properties, and applications. J King Saud Univ-Sci 33(7):101560

Han HS, Koo SY, Choi KY (2022) Emerging nanoformulation strategies for phytocompounds and applications from drug delivery to phototherapy to imaging. Bioact Mater 14:182–205

Hano C, Tungmunnithum D (2020) Plant polyphenols, more than just simple natural antioxidants: oxidative stress, aging and age-related diseases. Medicine 7(5):26

Harish V, Tewari D, Gaur M, Yadav AB, Swaroop S, Bechelany M, Barhoum A (2022) Review on nanoparticles and nanostructured materials: bioimaging, biosensing, drug delivery, tissue engineering, antimicrobial, and agro-food applications. Nanomaterials 12(3):457

Hasan AA, Tatarskiy V, Kalinina E (2022) Synthetic pathways and the therapeutic potential of quercetin and curcumin. Int J Mol Sci 23(22):14413

Henein MY, Vancheri S, Longo G, Vancheri F (2022) The role of inflammation in cardiovascular disease. Int J Mol Sci 23(21):12906

Hu F, Sun DS, Wang KL, Shang DY (2022) Nanomedicine of plant origin for the treatment of metabolic disorders. Front Bioeng Biotechnol 9:811917

Ielciu I, Sevastre B, Olah N-K, Turdean A, Chişe E, Marica R, Oniga I, Uifălean A, Sevastre-Berghian AC, Niculae M, Benedec D, Hanganu D (2021) Evaluation of Hepatoprotective activity and oxidative stress reduction of *Rosmarinus officinalis L.* Shoots tincture in rats with experimentally induced hepatotoxicity. Molecules (Basel, Switzerland) 26(6):1737. https://doi.org/10.3390/molecules26061737

Iqbal I, Wilairatana P, Saqib F, Nasir B, Wahid M, Latif MF, Iqbal A, Naz R, Mubarak MS (2023) Plant polyphenols and their potential benefits on cardiovascular health: a review. Molecules 28:6403

Ismail AMA (2022) Metabolic syndrome components' response to the conducted 16-week randomised-controlled training trial on an elliptical trainer. Eur J Physiother:1–7. https://doi.org/10.1080/21679169.2021.2022756

Jaiswal SG, Naik SN (2021) Turmeric oil: composition, extraction, potential health benefits, and other useful applications. Avicenna J Med Biochem 9(2):93–106

Jeyasri R, Muthuramalingam P, Priya A, Alexpandi R, Shanmugam NS, Nivetha S, Shin H, Pandian SK, Ravi AV, Ramesh M (2024) Comprehensive in vitro and in vivo evaluation of therapeutic potential of Bacopa-derived asiatic acid against a human oral pathogen Streptococcus mutans. Front Microbiol 15:140401

Ju J, Santana de Oliveira M, Qiao Y (2023) Antioxidant activity and mechanism of cinnamon. In: Cinnamon: a medicinal plant and a functional food systems. Springer International Publishing, Cham, pp 69–83

Jurczyk M, Kasperczyk J, Wrześniok D, Beberok A, Jelonek K (2022) Nanoparticles loaded with docetaxel and resveratrol as an advanced tool for cancer therapy. Biomedicine 10(5):1187

Kardani SL (2024) Nanocarrier-based formulations: regulatory challenges, ethical and safety considerations in pharmaceuticals. Asian J Pharma (AJP) 18(02)

Kaur P (2021) Biosynthesis of nanoparticles using eco-friendly factories and their role in plant pathogenicity: a review. Biotechnol Res Innov J 2(1):63–73

Kaur A, Tiwari R, Tiwari G, Ramachandran V (2022) Resveratrol: a vital therapeutic agent with multiple health benefits. Drug research 72(01):5–17

Kim T, Cho AY, Lee SW, Lee HJ (2024) Controlled quercetin release by fluorescent mesoporous nanocarriers for effective anti-adipogenesis. Int J Nanomedicine 19:5441–5458

Kumar M, Barbhai MD, Hasan M, Punia S, Dhumal S, Rais N, Chandran D, Pandiselvam R, Kothakota A, Tomar M, Satankar V (2022a) Onion (Allium cepa L.) peels: a review on bioactive compounds and biomedical activities. Biomed Pharmacother 146:112498

Kumar N, Chamoli P, Misra M, Manoj MK, Sharma A (2022b) Advanced metal and carbon nanostructures for medical, drug delivery, and bio-imaging applications. Nanoscale 14(11):3987–4017

Kumar A, Kurmi BD, Singh A, Singh D (2022c) Potential role of resveratrol and its nano-formulation as anti-cancer agent. Explor Targeted Antitumor Ther 3(5):643

Kumari S, Goyal A, Sönmez Gürer E, Algın Yapar E, Garg M, Sood M, Sindhu RK (2022) Bioactive-loaded novel nano-formulations for targeted drug delivery and their therapeutic potential. Pharmaceutics 14(5):1091

Li P, Pu S, Lin C, He L, Zhao H, Yang C, Guo Z, Xu S, Zhou Z (2022) Curcumin selectively induces colon cancer cell apoptosis and S cell cycle arrest by regulating Rb/E2F/p53 pathway. J Mol Struct 1263:133180

Lillich FF, Imig JD, Proschak E (2021) Multi-target approaches in metabolic syndrome. Front Pharmacol 11:554961. https://doi.org/10.3389/fphar.2020.554961. https://www.frontiersin.org

Lim CL, Raju CS, Mahboob T, Kayesth S, Gupta KK, Jain GK, Dhobi M, Nawaz M, Wilairatana P, de Lourdes Pereira M, Patra JK (2022) Precision and advanced nano-phytopharmaceuticals for therapeutic applications. Nanomaterials 12(2):238

Lippman D, Stump M, Veazey E, Guimarães ST, Rosenfeld R, Kelly JH, Ornish D, Katz DL (2024) Foundations of lifestyle medicine and its evolution. Mayo Clin Proc: Innov, Qual Outcomes 8(1):97–111

Liu Y, Bravo KM, Liu J (2021) Targeted liposomal drug delivery: a nanoscience and biophysical perspective. Nanoscale Horiz 6(2):78–94

Lord MS, Berret JF, Singh S, Vinu A, Karakoti AS (2021) Redox active cerium oxide nanoparticles: current status and burning issues. Small 17(51):2102342

Luo H, Li Z, Yao M, McClements DJ, Xiao H (2022) Impact of excipient emulsions made from different types of oils on the bioavailability and metabolism of curcumin in the gastrointestinal tract. Food Chem 370:130980

Mabrouk M, Das DB, Salem ZA, Beherei HH (2021) Nanomaterials for biomedical applications: production, characterisation, recent trends and difficulties. Molecules 26(4):1077

Malode LL, Manwar JV, Panchale WA, Bartere SA, Bakal RL (2021) Potential of medicinal plants in management of diabetes: an update. GSC Adv Res Rev 8(1):149–159
Man AW, Zhou Y, Xia N, Li H (2023) Perivascular adipose tissue oxidative stress in obesity. Antioxidants 12(8):1595
Manzoor MF, Hussain A, Sameen A, Sahar A, Khan S, Siddique R, Aadil RM, Xu B (2021) Novel extraction, rapid assessment and bioavailability improvement of quercetin: a review. Ultrason Sonochem 78:105686
Masenga SK, Kabwe LS, Chakulya M, Kirabo A (2023) Mechanisms of oxidative stress in metabolic syndrome. Int J Mol Sci 24(9):7898
Mashaqbeh H, Obaidat R, Al-Shar'i N (2021) Evaluation and characterization of curcumin-β-cyclodextrin and cyclodextrin-based nanosponge inclusion complexation. Polymers 13(23):4073
Mazzoni L, Ariza Fernández MT, Capocasa F (2021) Potential health benefits of fruits and vegetables. Appl Sci 11(19):8951
Mitchell MJ, Billingsley MM, Haley RM, Wechsler ME, Peppas NA, Langer R (2021) Engineering precision nanoparticles for drug delivery. Nat Rev Drug Discov 20(2):101–124
Mohamed SM, Shalaby MA, El-Shiekh RA, El-Banna HA, Emam SR, Bakr AF (2023) Metabolic syndrome: risk factors, diagnosis, pathogenesis, and management with natural approaches. Food Chem Adv 3:100335
Naghipour M, Joukar F, Nikbakht HA, Hassanipour S, Asgharnezhad M, Arab-Zozani M, Mansour-Ghanaei F (2021) High prevalence of metabolic syndrome and its related demographic factors in North of Iran: results from the Persian Guilan cohort study. Int J Endocrinol 2021(1):8862456
Nardini M (2022) Phenolic compounds in food: characterization and health benefits. Molecules 27(3):783
Nazir F, Tabish TA, Tariq F, Iftikhar S, Wasim R, Shahnaz G (2022) Stimuli-sensitive drug delivery systems for site-specific antibiotic release. Drug Discov Today 27(6):1698–1705
Nguyen LA, Simons CW, Thomas R (2025) Nootropic foods in neurodegenerative diseases: mechanisms, challenges, and future. Transl Neurodegener 14(1):17
Obeid MA, Alsaadi M, Aljabali AA (2023) Recent updates in curcumin delivery. J Liposome Res 33(1):53–64
Orsini F, D'Ambrosio F, Scardigno A, Ricciardi R, Calabrò GE (2023) Epidemiological impact of metabolic syndrome in overweight and obese European children and adolescents: a systematic literature review. Nutrients 15(18):3895
Pandey P, Verma M, Lakhanpal S, Pandey S, Kumar MR, Bhat M, Sharma S, Alam MW, Khan F (2025) An updated review summarizing the anticancer potential of poly (lactic-co-glycolic acid) (PLGA) based curcumin, epigallocatechin gallate, and resveratrol nanocarriers. Biopolymers 116(1):e23637
Panigrahy SK, Bhatt R, Kumar A (2021) Targeting type II diabetes with plant terpenes: the new and promising antidiabetic therapeutics. Biologia 76(1):241–254
Patel JK, Patel A, Bhatia D (2021) Introduction to nanomaterials and nanotechnology. In: Emerging technologies for nanoparticle manufacturing. Springer International Publishing, Cham, pp 3–23
Pedroso RD, Andrade G, Pires RH (2021) Medicinal plants: an approach to rational and safe use. Physis Rev Saude Coletiva 31:e310218
Petrovic S, Bita B, Barbinta-Patrascu ME (2024) Nanoformulations in pharmaceutical and biomedical applications: green perspectives. Int J Mol Sci 25(11):5842
Rajnath YK, Verma VK, Tiwari S (2024) Engineering approaches in pharmaceutical research. In: Converging pharmacy science and engineering in computational drug discovery. IGI Global, pp 190–211
Rani M, Jindal S, Anand R, Sharma N, Ranjan KR, Mukherjee MD, Mishra V (2023a) Plant essential oils and their constituents for therapeutic benefits. In: Essential oils: extraction methods and applications. Wiley, pp 977–1008

Rani N, Singh P, Kumar S, Kumar P, Bhankar V, Kumar K (2023b) Plant-mediated synthesis of nanoparticles and their applications: a review. Mater Res Bull 163:112233

Riaz M, Khalid R, Afzal M, Anjum F, Fatima H, Zia S, Rasool G, Chukwuebuka Egbuna, Mtewa AG, Uche CZ, Aslam MA (2023a) Phytobioactive compounds as therapeutic agents for human diseases: a review. Food Sci Nutr 11(6):2500–2529. https://doi.org/10.1002/fsn3.3308

Riaz M, Khalid R, Afzal M, Anjum F, Fatima H, Zia S, Rasool G, Egbuna C, Mtewa AG, Uche CZ, Aslam MA (2023b) Phytobioactive compounds as therapeutic agents for human diseases: a review. Food Sci Nutr 11(6):2500–2529

Rodríguez-Gómez FD, Penon O, Monferrer D, Rivera-Gil P (2023) Classification system for nanotechnology-enabled health products with both scientific and regulatory applications. Front Med 10:1212949

Rodríguez-Gómez FD, Monferrer D, Penon O, Rivera-Gil P (2025) Regulatory pathways and guidelines for nanotechnology-enabled health products: a comparative review of EU and US frameworks. Front Med 12:1544393

Rodríguez-Negrete EV, Morales-González Á, Madrigal-Santillán EO, Sánchez-Reyes K, Álvarez-González I, Madrigal-Bujaidar E, Valadez-Vega C, Chamorro-Cevallos G, Garcia-Melo LF, Morales-González JA (2024) Phytochemicals and their usefulness in the maintenance of health. Plants 13(4):523. https://doi.org/10.3390/plants13040523

Rudrapal M, Khairnar SJ, Khan J, Dukhyil AB, Ansari MA, Alomary MN, Alshabrmi FM, Palai S, Deb PK, Devi R (2022) Dietary polyphenols and their role in oxidative stress-induced human diseases: insights into protective effects, antioxidant potentials and mechanism (s) of action. Front Pharmacol 13:806470

Sakhale BK, Giri NA, Borse BB (2023) Role of phytochemicals in human health. In: Innovel processing methods for plant-based health foods. Apple Academic Press, pp 187–205

Salehirozveh M, Dehghani P, Mijakovic I (2024) Synthesis, functionalization, and biomedical applications of iron oxide nanoparticles (IONPs). J Funct Biomater 15(11):340

Salla M, Karaki N, El Kaderi B, Ayoub AJ, Younes S, Abou Chahla MN, Baksh S, El Khatib S (2024) Enhancing the bioavailability of resveratrol: combine it, derivatize it, or encapsulate it? Pharmaceutics 16(4):569. https://doi.org/10.3390/pharmaceutics16040569

Sazdova I, Keremidarska-Markova M, Dimitrova D, Mitrokhin V, Kamkin A, Hadzi-Petrushev N, Bogdanov J, Schubert R, Gago H, Dimiter Avtanski et al (2023) Anticarcinogenic potency of EF24: an overview of its pharmacokinetics, efficacy, mechanism of action, and nanoformulation for drug delivery. Cancer 15(22):5478. https://doi.org/10.3390/cancers15225478

Sharifi-Rad J, Quispe C, Mukazhanova Z, Knut E, Turgumbayeva A, Kipchakbayeva A, Seitimova G, Mahomoodally MF, Lobine D, Koay A, Wang J (2021a) Resveratrol-based nanoformulations as an emerging therapeutic strategy for cancer. Front Mol Biosci 8:649395

Sharifi-Rad J, Quispe C, Butnariu M, Rotariu LS, Sytar O, Sestito S, Rapposelli S, Akram M, Iqbal M, Krishna A, Kumar NV (2021b) Chitosan nanoparticles as a promising tool in nanomedicine with particular emphasis on oncological treatment. Cancer Cell Int 21(1):318

Sharma N, Saini D, Kesharwani RK, Gupta PC, Keservani RK (eds) (2024) Advances in flavonoids for human health and prevention of diseases. CRC Press

Shivani, Singh G, Narwal S, Chopra B, Dhingra AK (2023) Quercetin-based nanoformulation: a potential approach for cancer treatment. Anti-Cancer Agents Med Chem 23(18):1983–2007

Silva TJ, Okuro PK, da Silva MG, Ribeiro AP, da Cunha RL (2023a) Lecithin. In: Fat mimetics for food applications, pp 57–87

Silva PM, Gonçalves C, Pastrana LM, Coimbra MA, Vicente AA, Cerqueira MA (2023b) Recent advances in oral delivery systems of resveratrol: foreseeing their use in functional foods. Food Funct 14(23):10286–10313

Silveira Rossi JL, Barbalho SM, Reverete de Araujo R, Bechara MD, Sloan KP, Sloan LA (2022) Metabolic syndrome and cardiovascular diseases: going beyond traditional risk factors. Diabetes Metab Res Rev 38(3):e3502

Sruthi D, Dhanalakshmi M, Rao HY, Parthasarathy R, Jayabaskaran C (2023) Extraction, isolation, and characterization of phytochemicals, the bioactive compounds of plants. In: Recent frontiers of phytochemicals. Elsevier, pp 1–8

Stielow M, Witczyńska A, Kubryń N, Fijałkowski Ł, Nowaczyk J, Nowaczyk A (2023) The bioavailability of drugs—the current state of knowledge. Molecules 28(24):8038

Su M, Zhao W, Xu S, Weng J (2022) Resveratrol in treating diabetes and its cardiovascular complications: a review of its mechanisms of action. Antioxidants 11(6):1085

Sultana A, Zare M, Thomas V, Kumar TS, Ramakrishna S (2022) Nano-based drug delivery systems: conventional drug delivery routes, recent developments and prospects. Med Drug Discov 15:100134

Tagde P, Tagde P, Islam F, Tagde S, Shah M, Hussain ZD, Rahman MH, Najda A, Alanazi IS, Germoush MO, Mohamed HR (2021) The multifaceted role of curcumin in advanced nanocurcumin form in treating and managing chronic disorders. Molecules 26(23):7109

Theodoridis S, Drakou EG, Hickler T, Thines M, Nogues-Bravo D (2023) Evaluating natural medicinal resources and their exposure to global change. Lancet Planet Health 7(2):e155–e163

Urošević M, Nikolić L, Gajić I, Nikolić V, Dinić A, Miljković V (2022) Curcumin: biological activities and modern pharmaceutical forms. Antibiotics 11(2):135

Vesa CM, Zaha DC, Bungău SG (2024) Molecular mechanisms of metabolic syndrome. Int J Mol Sci 25(10):5452

Vollmannová A, Bojňanská T, Musilová J, Lidiková J, Ňorbová M (2024) Quercetin as one of the most abundant represented valuable biological plant components with remarkable chemoprotective effects review. Heliyon 10(12):e33342

Wahi A, Bishnoi M, Raina N, Singh MA, Verma P, Gupta PK, Kaur G, Tuli HS, Gupta M (2023) Recent updates on nano-phyto-formulations-based therapeutic intervention for cancer treatment. Oncol Res 32(1):19–47. https://doi.org/10.32604/or.2023.042228. PMID: 38188681; PMCID: PMC10767243

Wang N, An J, Zhang Z, Liu Y, Fang J, Yang Z (2022) The antimicrobial activity and characterization of bioactive compounds in Peganum harmala L. based on HPLC and HS-SPME-GC-MS. Front Microbiol 13:916371

Willis SA, Bawden SJ, Malaikah S, Sargeant JA, Stensel DJ, Aithal GP, King JA (2021) The role of hepatic lipid composition in obesity-related metabolic disease. Liver Int 41(12):2819–2835

Xia D, Mo Q, Yang L, Wang W (2022) Crosstalk between mycotoxins and intestinal microbiota and the alleviation approach via microorganisms. Toxins 14(12):859

Xiong Y, Xu Y, Zhou F, Hu Y, Zhao J, Liu Z, Zhai Q, Qi S, Zhang Z, Chen L (2023) Bio-functional hydrogel with antibacterial and anti-inflammatory dual properties to combat burn wound infection. Bioeng Transl Med 8(1):e10373

Yang L, Yin J, Wu J, Qiao L, Zhao EM, Cai F, Ye H (2021) Engineering genetic devices for in vivo control of therapeutic T cell activity triggered by the dietary molecule resveratrol. Proc Natl Acad Sci 118(34):e2106612118

Zagoskina NV, Zubova MY, Nechaeva TL, Kazantseva VV, Goncharuk EA, Katanskaya VM, Baranova EN, Aksenova MA (2023) Polyphenols in plants: structure, biosynthesis, abiotic stress regulation, and practical applications. Int J Mol Sci 24(18):13874

Zamanian MY, Alsaab HO, Golmohammadi M, Yumashev A, Jabba AM, Abid MK, Joshi A, Alawadi AH, Jafer NS, Kianifar F, Obakiro SB (2024) Nf-κb pathway as a molecular target for curcumin in diabetes mellitus treatment: focusing on oxidative stress and inflammation. Cell Biochem Funct 42(4):e4030

Zeng J, Fan J-G, Francque SM (2023) Therapeutic management of metabolic dysfunction associated with steatotic liver disease. United European Gastroenterol J 12(2):177–186. https://doi.org/10.1002/ueg2.12525

Zhao W, Zeng M, Li K, Pi C, Liu Z, Zhan C, Yuan J, Su Z, Wei Y, Wen J, Pi F (2022) Solid lipid nanoparticle as an effective drug delivery system of a novel curcumin derivative: formulation, release in vitro and pharmacokinetics in vivo. Pharm Biol 60(1):2300–2307

Zhu Y, Lao F, Pan X, Wu J (2022) Food protein-derived antioxidant peptides: molecular mechanism, stability, and bioavailability. Biomolecules 12(11):1622

Zinatullina KM, Kasaikina OT, Khrameeva NP, Indeykina MI, Kononikhin AS (2021) Interaction between glutathione and resveratrol in the presence of hydrogen peroxide: a kinetic model. Kinet Catal 62:255–263

Zoi V, Galani V, Lianos GD, Voulgaris S, Kyritsis AP, Alexiou GA (2021) The role of curcumin in cancer treatment. Biomedicine 9(9):1086

Chapter 18
Environmental Implications of Plant-Based Nanomaterials in Drug Formulation and Delivery

Abhishek Singh and V. K. Tripathi

18.1 Introduction

Nanotechnology is identified as a pivotal generation of the twenty first century, producing widespread hobby globally. Nevertheless, its development has been hindered because of a loss of complete information concerning its capacity dangers and the insufficiency of rules to mitigate rising risks (Verma et al. 2019). Despite these challenges, researchers are diligently striving to address various issues associated with nanotechnology, including its management, production, funding, regulation, and technical obstacles.

Within this context, green nanotechnology emerges as a segment of green technology that integrates principles from Green chemistry, Green engineering. The term green subject to the use of materials sourced from plants, which aims to enhnce energy efficiency and conserve resources by minimizing material consumption and utilizing renewable resources whenever possible. Innovations driven by nanotechnology are anticipated to make substantial contributions to environmental protection and climate change mitigation, facilitating the conservation of raw materials, energy, and water, while also mitigating greenhouse gas emissions and waste generation. Key advantages of green nanotechnology comprise enhanced energy efficiency, reduced waste and emissions, and diminished reliance on non renewable resources (Verma et al. 2019; Cuong et al. 2022).

Copper oxide nanoparticles rank among the most prevalent nanomaterials utilized today. They find extensive application across various domains, including biology, medicine, energy systems, environmental renovation, and industrial processes. As energy crises intensify and concerns regarding the limitations of conventional

A. Singh (✉) · V. K. Tripathi
Fish Biology Lab, Department of Zoology, T.D.P.G. College, Jaunpur, Uttar Pradesh, India

V. Ghosh et al. (eds.), *Applications of Plant Bioactive Compounds-Based Nanoformulations*, Nanotechnology in the Life Sciences,
https://doi.org/10.1007/978-3-032-13086-0_18

chemical and physical methods for nanoparticle synthesis grow, there is a burgeoning interest in novel fabrication techniques. Biologically-based methods, which generate minimal harmful waste and don't necessitate extensive purification processes, extensively investigated. Plants offer promising avenues for the biosynthesis of metal nanoparticles (Cuong et al. 2022).

Furthermore, research has proliferated around the diverse applications of CuO-NPs synthesized from various plant extracts, environmentally related repair, detection applications, catalytic reduction, photocatalysts, photocatalysts, various biological functions, energy storage, energy storage, and many chemical processes such as reduction, Coupl reactions (Cuong et al. 2022).

Copper nanoparticles are one of the most frequently employed nanomaterials in many sectors with copper-based nanostructured materials that can be used in biotechnology, traditional medicine, electronics and industrial environments. Given this escalating energy problem and the drawbacks of traditional methods of chemical and physical synthesis of metal nanoparticles, the focus is on alternative approaches that affect traditional chemistry. This has increased the green synthesis method for copper nanoparticles containing CuO, Cu_2O, CuS and other copper types. This is typically frequently imobilized with various support such as Soda, Bentonite, ZnO, MgO, MnO_2 and with various support such as ASS (RGO) and Fe_3O_4 (Nasrollahzadeh et al. 2019a).

Nanotechnology is a young and fast developing field dealing with the processing and use of materials with dimensions of 1–100 nm. The cross-section of current synthesis protocols of nanoparticles (NPs) is diverse; however, the use of natural resources as a way of catalysis reduction in metallic precursor to specific NPs forms an ecologically benign, cost-effective, and safe source. These bottom-up approaches are similar to chemical ones but differ: biologically-derived extracts replace synthetic reagents to carry out nanoparticle synthesis. These extracts stabilize concurrently, control the particle's growth and provide protective cover to the resultant NPs (Hussain et al. 2016).

Silver nanoparticles (Ag NPs) in particular occupy an especially prominent niche in the sphere of materials, with a decisive importance of silver nanoparticles in biomedicine, biosensors, and catalysis, pharmacology and other nanotechnological subdivisions, in particular nanomedicine. Traditionally synthesised Ag NPs, however, require vigorous reductants and produce toxic by-products and significant volumes of toxic effluents (Nasrollahzadeh et al. 2019b).

The popularity of synthetic procedures to NP synthesis is thus, requiring novel, environment-friendly approaches, with immediate need. The paradigm of the so-called green synthesis has thus become increasingly popular, being characterized by the relative ease, low cost, minimal environmental impact and renewable reagents. In this paradigm, plant extracts that come from roots, leaves and flowers are possible not only for the reproduction formation of NPs, but also for further stabilization, hence meet the core requirements of sustainability that prioritise waste minimisation, pollution prevention, benign solvents usage and work on renewable resources (Shiraz et al. 2024).

Presently large volumes of fruits are produced in the form of waste each day world over. The production of metal nanoparticles is sought in an innovative, cheap and eco-friendly approach which entails the utilization of fruit peels. Current study shows how copper nanoparticles (CuNPs) are made using a green method. In this process, pineapple peel extract (PLX) is combined with copper sulfate. Formation of CuNPs was seen visually and confirmed using UV-Visible spectroscopy. To understand the properties of the nanoparticles, several techniques were used, including Fourier-transform infrared (FTIR) spectroscopy, particle size analysis, scanning electron microscopy (SEM), and X-ray diffraction (XRD). To find out the antioxidant and reducing abilities of the CuNPs, DPPH scavenging assays and ferric reducing antioxidant power (FRAP) tests were carried out (Mitra et al. 2024).

Extremely important to hear the development of copper nanoparticles according to environmentally friendly lines. This is a viable direction since a plant component is a promising path since compounds isolated as plant components are both capping and reducing agents. This article tried to make CuNO using the plant called *Cissus arnotiana*, and then they checked how well these nanoparticles worked against both gram-positive and gram-negative bacteria (Rajeshkumar et al. 2019).

18.2 Synthesis and Characterization

18.2.1 Utilizing Plant Extracts as a Mechanism for Green Synthesis

Nanoparticle synthesis can be performed both chemically and physically, but this is not ideal in a chemical environment. Since nanomaterials are also synthesized utilizing plant extracts, animal proteins, agricultural waste, pigments, bacteria, fungi, and small viruses are materials that are hard to make with traditional methods. For these, green synthesis methods are more effective (Dos Santos et al. 2014).

Certainly, plant substances, especially plant species, contain chemicals like aldehydes, ketones, terpenoids, polyphenols, tannins, polysaccharides, flavonoids, proteins, amines, and alkaloids that can help keep nanoparticles stable., offering protection against aggregation, and metal ion reduction to metal nanoparticles. The characteristics such as high catalytic activity, enhanced surface activity and regulated surface morphology, which are provided by these syntheses, are well reported. An example is Known as the soapberry family, *Acer oblongifolium* is also known as Himalayan maple, Kashmir maple or evergreen maple. The tree is of medium size and can grow up to 15–22 m tall, and is evergreen throughout the winter, made of a semi-deciduous or an evergreen woody shrub. The variety is distributed throughout eastern, central, and southeastern Asia including northeast India, Tibet, Japan, northern Indochina and southern China (Naveed et al. 2022).

Based on this study, silver nanoparticles (AgNPs) were prepared using a natural reducing agent, medicinal plant Asplenium dalhousiae. The absorption maxima at

Table 18.1 List of some nanomaterials produced by plants species

Nanoparticles	Species	Application	References
Ag	*Abelmoschus esculentus (Extract)*	Sensors	Huston et al. (2021) and Gnanaprakasam and Selvaraju (2014)
	Camellia sinensis (Extract)	Antibacterial	Huston et al. (2021) and Flores-González et al. (2018)
	Mangifera indica (Extract)	Medical (antibacterial)	Huston et al. (2021) and Horta-Piñeres et al. (2019)
	Syzygium Aromaticum (Extract)	Electrical	Huston et al. (2021) and Jeevika and Ravi Shankaran (2015)
Au	*Beta vulgaris (Extract)*	Sensors	Huston et al. (2021) and Castro et al. (2011)
	Phormidium valderianum	Medical	Huston et al. (2021) and Parial et al. (2012)
	Rhizopus oryzae	Medical	Huston et al. (2021) and Das et al. (2010)
Cu	*Eucalyptus globulus (Extract)*	Electronic and Optical	Huston et al. (2021) and Pinto et al. (2019)
Se	*Zooglea ramigera (Bacteria, Live)*	Medical	Huston et al. (2021) and Srivastava and Mukhopadhyay (2013)

approximately 420 nm, 443 mm, and 439 ppm were observed in UV-Vis spectrophotometry, while the Fourier-transform infrared (FTIR) spectra identified the functional groups that impact the reduction phenomena. Through the use of SEM and X-ray diffraction, they were able to determine more clearly the physical characteristics and composition of the AgNPs (Parveen et al. 2025).

Astragalus fasciculifolius is a plant of the Fabaceae family, often called Anzaroot, and has therapeutic properties associated with saponins, flavonoids, and polysaccharides. The plant has been demonstrated to contain efficacy in managing cardiovascular disorders, in inhibiting cancerous cell growth, and alleviating the side effects caused by chemotherapy. Recently, therefore, effectiveness of Anzaroot extract as a cytotoxic agent against invasive human breast cancer cell line MCF-7 has been investigated through the green synthesis of AgNPs (Nosrati et al. 2025).

Ethanolic extract of *Nepeta leucophylla* (NLe) has also been used to synthesize (AgNPs) and (AuNPs) nanoparticles. NLe has been observed to contain a lot of natural polyphenols and antioxidants that possess high reducing potential. Characterization techniques used included: FTIR, XRD, and UV are all types of ultraviolet-visible spectrophotometers, and did show a UV absorption peak of 434 nm of NLe-AgNPs and 535 nm of NLe-AuNPs (Verma et al. 2025) (Table 18.1).

18.2.2 Characterization Techniques and Properties

Creation of a sustainable approach to nanoparticle (NP) production has been supported by the use of green chemistry. "The green synthetic methods utilize multiple biological substances, such as microbial, higher plant and other biowaste derived substances to prepare the nanoscale particulate systems. Active biomolecules from natural sources the substances can reduce and stabilize. Enzymes produced by microorganisms help in reducing metal ions can transform into nanoparticle. Plant-based modalities hold a unique attraction over comparable chemical approaches to nanoparticle production, due to high precursor availability, the simplicity of the methodology, and its low economic requirement. Silver nanoparticle (AgNPs) is an example of a system that has attracted a lot of academic literature consideration, considering their many capabilities in chemistry, food industry, microbiology or biomedicine (Thomas et al. 2024).

Green synthesis of AgNPs methods have several advantages over chemical and physical methods, such as cost-effective synthesis, zero toxic byproducts environmentally friendly production additionally, and easy regeneration. Plants derived materials have been shown to have potential in reducing particles (Arshad et al. 2024).

The silver nanoparticles made through nano and chemical methods were tested using the techniques used in this area include UV-Vis spectroscopy, Nanotrac, DLS (Dynamic Light Scattering), ZP (Zeta Potential), and Transmission Electron Microscopy. According to the Nanotrac and TEM results, the AgNPs produced by green synthesizers are cylindrical in shape with an average size of 94 nm, while those produced chemically are roughly 104 м. Zeta potential was −1202 mV for green AgNPs and −10.4 mV in chemical AgNV (Kummara et al. 2016).

The current study promotes a biological approach to the creation of silver nanoparticles, known as AgNPs: sodium para-hydroxybenzoate tetrahydrate (SPHT) extracted from leaves of *Vitex negundo*. The characterization of the synthesized SPHT-AgNPs was achieved through the use of UV-Vis spectroscopy, HRTEM with SAED patterns, FESEM imaging, EDX, and FT-IR analysis. In order to determine the stability of the SPHT-AgNPs synthesis peak at approximately 430 nm, we examined different pH and temperature ranges (Durai et al. 2014).

18.3 Pharmaceuticals Application and Drug Delivery

18.3.1 Pharmaceuticals Application

Nanoparticles (AuNPs) possess tunable properties that make them valuable tools in biomedical applications, particularly in the identification and treatment of malignancies, neurodegenerative diseases (Muddapur et al. 2022). The use of green-synchronized AUNPs (antibody uv-light-activated nanoparticles) in treating cancer is based on the following sequential biochemical events: mitochondrial damage and the activation of ROS lead to caspase activity followed by apoptosis. Understanding their therapeutic mechanisms can aid in the development of personalized medicines and treatments, making them a promising cancer therapy (Bharadwaj et al. 2021).

AuNPs undergo green synthesis of *Rumex obtusifolius* extract, which contains bioactive compounds, antioxidant and anticancer properties. The extract process of synthesizing AuNPs resulted in their effect was evaluated against lung cancer in in vitro experiments (Yayintas et al. 2024).

AuNPs were found to significantly inhibit the growth of A549 cells in a dose-dependent manner by the MTT assay. The findings highlight that AuNPs synthesized from *M. tenasishima* extract functions as an effective toxic active ingredient and is good stabilization against lung cancer cell lines (A549) (Sun et al. 2019).

18.3.2 Drug Delivery of Nanoparticles

Plant-based nanomaterials have become a big topic of interest because they show promise in fighting harmful microbes and dealing with antimicrobial resistance. These materials are small, they have a vast surface area and can navigate through biological barriers, which makes them special (Okeke et al. 2023). Using nanomaterials in drug delivery systems is a new way to get medicines to work better. These systems can help improve how well drugs dissolve, stay stable, and reach the right place in the body. They can also reduce side effects, but there are still some concerns to address (Obeid et al. 2025).

Nanomaterials can include metal nanoparticles, carbon fiber materials, and other variations, and structures based on polymers, can destroy bacteria in various ways. They can produce harmful substances known as reactive oxygen species, damage the membranes of bacteria, and make antibiotics work better. Research shows that these nanomaterials can also overcome typical methods bacteria use to become resistant, such as forming biofilms and using pumps to remove drugs. A new treatment called nanomaterial-mediated sonodynamic therapy uses sound waves, the production of reactive oxygen species can eliminate bacteria without risk of antibiotic resistance (Parvin et al. 2025; Zheng et al. 2025).

Another treatment, called antibacterial photodynamic therapy, uses special substances called reactivity of light and the photosensitizer is used to produce reactive oxygen species. These extremely harmful to microorganisms and can kill them without resistance. A newer version of this treatment, antimicrobial sonodynamic therapy, uses sound instead of light to activate the sonosensitizers, making the treatment more versatile (Javanmard et al. 2025).

18.4 Impact of Nanomaterial on the Environmental Sustainability

18.4.1 In Use of Waste Water Treatment

The use of green nanomaterials can help to reduce the environmental impact caused by toxic substances. Biosynthesis has become a subject of interest due to the lack of chemical and physical methods for nanoparticle production (Bhushan et al. 2024).

The inadequate supply is exacerbated by the increasing population, new industries, expanding cities and changing weather patterns. However, the situation is already dire. Only this approach can provide a solution to the problem. Although there are nanomaterials based on FeS, including nanocomposites of the same structure and shape, that can be used to treat water as suitable for remove heavy metals and contaminants from wastewater through adsorptive and photocatalytic processes, they are particularly problematic due to their persistence and recalcitrance in the environment. Bioadsorbents, biosorption, and photocatalysis are the most suitable techniques for treating contaminated wastewater (Nawaz et al. 2025; Zhao et al. 2016).

Extraction, chemical coagulation, membrane separation, biological precipitation (stomach activation), adsorption, and ion exchange are all methods used to remove heavy metals from water solutions. While adsorption, which is generally considered the most effective way to remove heavy metals from wastewater, can lead to waste product from other sources that create environmental problems. The Pb^{2+}, Cd^{2+1}, and Ni^2 are all made from metal oxides. By collecting and processing the + ions from agricultural waste, carbon nanomaterials can be produced. It is possible to investigate the reusability of heavy metal spent adsorbents in BFP (Nthwane et al. 2025).

Carbohydrate-based biopolymers in the form of chitosan, carboxymethylcellulose, alginate and lignin have drawn much interest in environmental applications because of their biodegradability, biocompatibility, non-toxicity and low-cost. These polymers have also been used in recent years to develop various nanocomposified adsorbents in treatment of waste water. The efficacy of these adsorbents in removing inorganic pollutants from eutrophic solutions has been extensively researched (Bayuo et al. 2024; Fouda-Mbanga et al. 2021).

A new method for cleaning water uses materials known as adsorbents, which can be made by humans or found in nature, to get rid of heavy metals in the water. These materials can be used again after their first use, which saves money and helps the environment (Mata et al. 2010). The use of sugar-beet pectin xerogels has been considered to remove heavy metal contaminants like cadmium, lead and copper. Xerogels are able to sequentially bind many metals in a multistep process without need of substantial cleaning between steps, and the sorption efficiency of the xerogel remains at a high level over several successive cycles. A sequence of tests indicates that the xerogel bead sorbent is characterized by a high level of mechanical stability: the structural and electrostatic characteristics did not change significantly during the nine repetitive adsorption/desorption cycles with minimal attenuation (Mata et al. 2010; Anuta et al. 2025).

18.4.2 Enhanced Agricultural Sustainability

Nanoparticles can enhance plant growth, stress tolerance, and productivity. Nanoparticles can cause physiological and biochemical modifications in plants, which enhance plant growth in favorable and stressful growth conditions (Anuta et al. 2025).Cerium dioxide nanoparticles (CeO_2-NPs) are applied in many industrial processes, which results in their inevitable release into the environment, including the soil ecosystem (Ukwattage and Zhiyong 2025).

Carbon quantum dots (CQDs) possess widespread applications as non-toxic and multi-functional nanomaterials, particularly in soil fertility enhancement, plant stress resistance, and green agriculture. Nanoparticles and nanomaterials have a beneficial influence on plant growth and development, seed germination, plant stress regulation, and postharvest management of vegetables and fruits (Andleeb et al. 2025; Sharma et al. 2022).

18.5 Adverse Effects of Nanomaterials on Environment and Human Health

Toxic character of nanomaterials is quite different from the manner they appear when they are in raw and natural forms. It is because they possess especial physical and chemical structures. These nanomaterials are toxic to the environment and to people. When nanomaterials are emitted in the environment, they settle over time and turn into huge issues to the natural environment and living organisms, leading to health problems. As we are introducing more nanoparticles into our daily life, more people are exposed and, therefore, more susceptible to the ill effects of nanomaterials. Due to their highly tiny size, nanomaterials can penetrate biological membranes and travel to cells, tissues, and organs. Hence, the impact of nanomaterials on the environment as well as the health of human beings is a matter of serious concern (Qamar et al. 2024). To learn about the actual or perceived risk of nanomaterials, one should learn about their toxic personality and how they move their bodies, and how to expose people and their environment by chance and intentionally for medical reasons (Oberdörster 2010).

In the field of nanomedicine, when nanoparticles are administered intravenously or subcutaneously, they enter the body directly and bypass the normal digestive process. In addition to potentially interacting with vital bodily molecules, these nanoparticles themselves may be dangerous. Furthermore, some nanoparticles that are difficult to disintegrate can remain in organs or tissues. As such, it is crucial to consider the potential hazards to nanomedicine involves the use of nanomaterials, which can have implications for health and safety (Zhao and Castranova 2011).

Recent study demonstrated that mice's lungs had an uncommon inflammatory response upon inhaling single-walled carbon nanotubes. Collagen and elastin

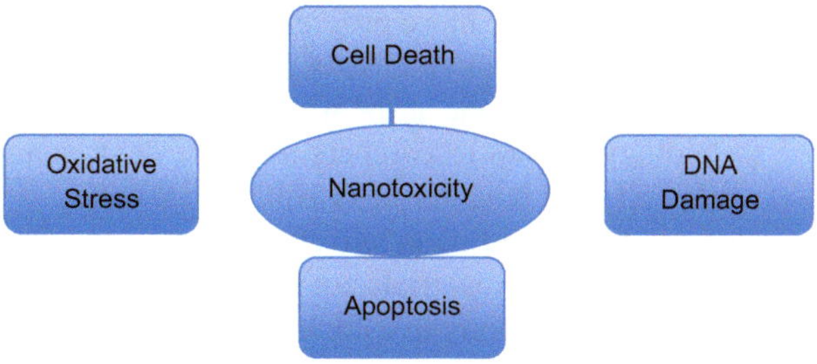

Fig. 18.1 Showing adverse effect of nanotoxiciy (Kumah et al. 2023)

accumulated in the lungs as a result of this response, which swiftly changed from an acute inflammation to tissue damage (Kagan et al. 2005; Yu et al. 2008).

Because engineered nanomaterials are the primary products of nanotechnology, researchers and policymakers need to know how much of these materials are present in the environment and in people. This information is critical to understanding how these materials interact with the environment at many scales, from inside living things to entire ecosystems, and to guiding the industry towards more sustainable development. Effects of nanomaterials on individual creatures have the potential to spread through populations and alter food chain ecosystems composed of a wide variety of species, including bacteria, fish, and plants. These community shifts can subsequently affect ecosystem services, which are the natural processes that regulate the Earth's water and air and transport carbon, nutrients, and energy (Holden et al. 2013) (Fig. 18.1).

18.6 Future Perspectives of Nanomaterials

- Dental applications are increasingly incorporating smart nanomaterials and nanotechnology, such as nanoclays, nanofibers, or nanocomposites, which can be modified to form solid-lipid nanoparticles (including dental implants), while nanocapsules also exist. When compared to other nanomaterials, smart nanoparticles have numerous advantages, which makes them beneficial for a variety of dental applications, including gum disease treatment, fillings, root canal therapy, and prevention (Vasiliu et al. 2021).
- People with irregular heartbeats may benefit from safer, more significant, and more effective treatments made possible by nanotechnology and nanomaterials. Nano biomaterials are excellent for applications such as creating tissues, delivering medications to the heart, and diagnosing and treating illnesses because of their enormous surface area, regulated physical and chemical properties, and superior compatibility with the body (Lu and Fan 2024).
- To help deliver drugs better, nanomaterials are changing drug delivery systems, leading to more effective, personalized, and easier treatments for many diseases (Al Ayidh et al. 2025).

- There is a need for more progress in nanomaterials to improve drug delivery for eye diseases like conjunctivitis, keratitis, uveitis, and scleritis (Wei et al. 2023).
- The structure of bone at the nano level, along with recent advances in nanobiotechnology, allows researchers to create devices and drug delivery systems that work well with bone tissue (J Hill et al. 2019).
- Nanomedicine and nanodelivery systems are a new and quickly growing area where very small materials are used to help diagnose or deliver treatment to specific parts of the body in a controlled way (Patra et al. 2018).
- The potential for medical research is growing as a result of nanotechnology's ongoing development, particularly in the area of drug delivery for cancer.It's critical to create improved techniques for evaluating the efficacy and safety of nanomedicine as it develops. The application of nanotechnology in drug delivery can enhance therapeutic efficacy by providing greater control over the location of pharmaceuticals throughout the body (Swierczewska et al. 2018).
- Treatment options for breast cancer that show promise include medication delivery systems based on nanoparticles. Through direct drug delivery to tumors, increased drug levels at the tumor site, gradual drug release, stabilization, and decreased adverse effects, they can support chemotherapy (Du et al. 2019).

18.7 Conclusion

A design of Nanomaterials derived from plants for medication delivery and formulation spells out a major transformation in pharmaceutical sciences to make things sustainable, efficient, and innovative. During the discussion, it has become apparent that such naturally derived nanostructures can specifically combat several of the loopholes presented by their synthetic counter nanomaterials, while also being strongly aligned with the global push for green technologies and reduced impact on the environment. Hence, while perpetually stabbing the enhancement of drug efficacy, their strong respect for the environment also recognizes the principles of circular economy frameworks.

Another interesting thing to look in toward the shortcomings of plant-derived nanomaterials is that they do some dual responsibilities-some scientific and some ecological. In pharmaceutical terms, these substances are highly biocompatible, good efficiency for drug loading, targeted in drug delivery, and an acceptable hazard profile overall. Conversely, unlike synthetics, they can often be obtained from renewable plant biomass, from agricultural waste, or from other naturally occurring sources, dramatically lowering the energy-imbalance burden of their production. This renewable nature causes a reduction in reliance on petroleum-based precursors and lessens the associated life-cycle carbon footprint of pharmaceutical nanocarriers.

Also, the deployment of plant-based nanomaterials in drug delivery systems can solve problems in various therapeutic areas. Their morphological versatility allows for the adjustment of particle size, surface chemistry, and kinetics of release into specialized drug delivery vehicles. Therefore, these characteristics ideally fit the precision medicine approaches where dosing and action require individualization. Also, plant nanocarriers tend to biodegrade better and pose less risk of toxic buildup in biological systems, the two major concerns implicated in safety issues and long-term use of a patient.

From an environmental standpoint, their advantages become very evident. Generally, synthetic nanomaterials, although performant, remain in ecosystems long after their use has passed, and therefore have been of concern regarding nano-pollution and bioaccumulation in different aquatic and terrestrial organisms. Conversely, plant-based nanomaterials tend to degrade faster, thereby reducing environmental persistence and associated toxicity. This biodegradability prevents post-consumption residues from posing serious ecological problems, hence bearing a constructive promise towards environmental conservation.

However, there exist some bottlenecks on the other side of widespread use of plant-based nanomaterials in drug formulation. More efficient for use have been demonstrated in laboratory-scale research; however, scaling up procedures for commercial production is very complicated. Hence, batch-to-batch consistency, mastering physicochemical properties of the product, and regulations related to pharmaceuticals must be accounted for. The stability of these materials during manufacturing, storage, and transportation is also a factor that requires due consideration, especially in regions where the climate is high in humidity or temperature tends to fluctuate.

In spite of these handicaps, extraction techniques, nanoparticle synthesis, and functionalization methods continue to advance rapidly. In fact, green synthesis methodologies using water-based solvents, enzymatic catalysts, and energy-efficient processing techniques are being increasingly fine-tuned so that they become economically viable and environment compatible on a large scale. Such processes correspond well with the burgeoning aspirations of the pharmaceutical community trying to adopt environment-friendly production lines while generating minimal hazardous waste and complying with environmental regulations evolving through time.

An equally important concern would be applications of plant-based nanomaterials for enabling circular economy entrances. If raw materials are being generated from agricultural by-products or plant waste streams, then pharmaceutical manufacturers are just actively valorizing waste. This simultaneously reduces landfill pressure and offers some income streams to the farming community. Thus, combining waste use with manufacturing of high-value products creates a win–win solution fostering environmental conservation and socio-economic development.

The potential plant-based nanomaterials have been most beautifully attributed in drug delivery to enable therapeutic improvements through drug solubility for poorly water-soluble drugs, release profiles of sustained type, and delivery that is composed of specific tissues and cells. Functionalization of these nanoparticles further adds to their therapeutic avenues by allowing specific targeting using ligands, antibodies, peptides, etc., in cancer, neurodegenerative, and infectious diseases. This sort of specificity reduces systemic side effects, largely improving patient compliance, which is an essential factor in the management of any chronic disorder.

Environmental concerns do not affect materials alone. Plant-based nanomaterials could bring about systemic changes in pharmaceutical supply chains. Moving from resource-intensive synthetic nanomaterials to plant-based forms of nanomaterials could provide one means to reduce industrial greenhouse gas emissions, dependency on non-renewable feedstocks, and generation of hazardous chemical waste. In addition, these kinds of products embrace the solution to more environmentally friendly end-of-life management of pharmaceutical products, such as biodegradable packaging and ecologically friendly disposal practices.

The alignment of plant nanomaterials with global health and sustainability priorities should also be taken into account. One example of this is the Sustainable Development Goals (SDGs) of the United Nations, which emphasize health and well-being, climate action, and responsible consumption and production. Incorporating plant-based nanotechnologies into pharma pipelines is directly aligned with these agendas, therefore making them an attractive option for future health care innovation.

Future policy support, green nanotechnology research funding, and academia-industry-government collaboration will be required to speed up the application of plant-based nanomaterials. The regulatory agencies will have to revise guidelines to include the special properties and the environmental benefits of these materials while ensuring that safety assessment is rigorous. Public awareness and acceptance will also enter the scene, especially in stimulating demand for sustainable healthcare options.

Advances in technology like sophisticated nanoencapsulation methods, microfluidics for highly controlled particle preparation, and hybrid nanomaterials that integrate plant components with other biocompatible entities will continue to add to the list of applications. Integration with future technologies like biopharmaceuticals, gene therapy, and personalized medicine will place plant-derived nanomaterials at the forefront of next-generation drug delivery platforms. The materials' versatility allows them to be applicable for both generic low-cost drugs and cutting-edge therapeutic interventions.

In short, plant-based nanomaterials in pharmaceutical delivery and formulation are a significant step towards matching pharmaceutical innovation with sustainability. They offer a promising path towards maximizing curative effectiveness, reducing ecological footprints, and encouraging resource optimization. While technical and regulatory challenges are there, the vigor of research and development activities in this field suggests that these challenges can be overcome. Capitalizing on the advancements in green chemistry, material science, and biotechnology, the pharmaceutical industry can drive itself towards a future where medical innovation and care for the environment go hand-in-hand. The long-term vision is clear: a healthcare system in which life-saving medicines are not only safe and effective but also produced, distributed, and disposed of in a way that respects and preserves the ecological balance of the earth.

Switching to plant-based nanomaterials is more than just a technological advancement; it's a transformation. It alters our views of the relationship between human health and the environment, and it shows that it is possible to innovate without harming nature. In making this transition, we can open the door to a new era in pharmaceutical science—one where green leaf is as powerful a symbol of healing as the medicines they deliver.

References

Al Ayidh A, Abbas M, Parayangat M, Ijyas T (2025) Advances in nanomaterials for targeted drug delivery: emerging trends and future prospects in nanodrug development. Curr Cancer Drug Targets. https://doi.org/10.2174/0115680096362452250301054711. Advance online publication

Andleeb N, Zafar S, Rahim Z, Iqbal MM, Lal R, Farooq MA (2025) Carbon quantum dots as versatile nanomaterials for improving soil health and plant stress tolerance: a comprehensive review. Planta 262(2):44. https://doi.org/10.1007/s00425-025-04758-2

Anuta V, Blidaru A, Dinu-Pîrvu CE, Fierascu RC, Fierascu I, Toma Sărdărescu DI, Popa L, Ghica MV, Prisada RM (2025) Metal-based nanoparticles with biostimulatory effects: harnessing nanotechnology for enhanced agricultural sustainability. Materials (Basel) 18(13):3142. https://doi.org/10.3390/ma18133142

Arshad F, Naikoo GA, Hassan IU, Chava SR, El-Tanani M, Aljabali AA, Tambuwala MM (2024) Bioinspired and green synthesis of silver nanoparticles for medical applications: a green perspective. Appl Biochem Biotechnol 196(6):3636–3669. https://doi.org/10.1007/s12010-023-04719-z

Bayuo J, Rwiza MJ, Choi JW, Mtei KM, Hosseini-Bandegharaei A, Sillanpää M (2024) Adsorption and desorption processes of toxic heavy metals, regeneration and reusability of spent adsorbents: economic and environmental sustainability approach. Adv Colloid Interf Sci 329:103196. https://doi.org/10.1016/j.cis.2024.103196

Bharadwaj KK, Rabha B, Pati S, Sarkar T, Choudhury BK, Barman A, Bhattacharjya D, Srivastava A, Baishya D, Edinur HA, Abdul Kari Z, Mohd Noor NH (2021) Green synthesis of gold nanoparticles using plant extracts as beneficial prospect for cancer theranostics. Molecules 26(21):6389. https://doi.org/10.3390/molecules26216389

Bhushan D, Shoran S, Kumar R, Gupta R (2024) Plant biomass-based nanoparticles for remediation of contaminants from water ecosystems: recent trends, challenges, and future perspectives. Chemosphere 365:143340. https://doi.org/10.1016/j.chemosphere.2024.143340

Castro L, Blázquez ML, Muñoz JA, González F, García-Balboa C, Ballester A (2011) Biosynthesis of gold nanowires using sugar beet pulp. Process Biochem 46:1076–1082. https://doi.org/10.1016/j.procbio.2011.01.025

Cuong HN, Pansambal S, Ghotekar S, Oza R, Thanh Hai NT, Viet NM, Nguyen VH (2022) New frontiers in the plant extract mediated biosynthesis of copper oxide (CuO) nanoparticles and their potential applications: a review. Environ Res 203:111858. https://doi.org/10.1016/j.envres.2021.111858

Das SK, Das AR, Guha AK (2010) Microbial synthesis of multishaped gold nanostructures. Small 6(9):1012–1021. https://doi.org/10.1002/smll.200902011

Dos Santos CA, Seckler MM, Ingle AP, Gupta I, Galdiero S, Galdiero M, Gade A, Rai M (2014) Silver nanoparticles: therapeutical uses, toxicity, and safety issues. J Pharm Sci 103:1931–1944

Du M, Ouyang Y, Meng F, Ma Q, Liu H, Zhuang Y, Pang M, Cai T, Cai Y (2019) Nanotargeted agents: an emerging therapeutic strategy for breast cancer. Nanomedicine (Lond) 14(13):1771–1786. https://doi.org/10.2217/nnm-2018-0481

Durai P, Chinnasamy A, Gajendran B, Ramar M, Pappu S, Kasivelu G, Thirunavukkarasu A (2014) Synthesis and characterization of silver nanoparticles using crystal compound of sodium para-hydroxybenzoate tetrahydrate isolated from Vitex negundo. L leaves and its apoptotic effect on human colon cancer cell lines. Eur J Med Chem 84:90–99. https://doi.org/10.1016/j.ejmech.2014.07.012

Flores-González M, Talavera-Rojas M, Soriano-Vargas E, Rodríguez-González V (2018) Practical mediated-assembly synthesis of silver nanowires using commercial Camellia sinensis extracts and their antibacterial properties. New J Chem 42:2133–2139. https://doi.org/10.1039/C7NJ03812G

Fouda-Mbanga, B. G., Prabakaran, E., & Pillay, K. (2021). Carbohydrate biopolymers, lignin based adsorbents for removal of heavy metals (Cd2+, Pb2+, Zn2+) from wastewater, regeneration and reuse for spent adsorbents including latent fingerprint detection: A review. Biotechnology reports (Amsterdam, Netherlands), 30, e00609. https://doi.org/10.1016/j.btre.2021.e00609

Gnanaprakasam P, Selvaraju T (2014) Green synthesis of self assembled silver nanowire decorated reduced graphene oxide for efficient nitroarene reduction. RSC Adv 4:24518–24525. https://doi.org/10.1039/C4RA01798F

Holden PA, Nisbet RM, Lenihan HS, Miller RJ, Cherr GN, Schimel JP, Gardea-Torresdey JL (2013) Ecological nanotoxicology: integrating nanomaterial hazard considerations across the subcellular, population, community, and ecosystems levels. Acc Chem Res 46(3):813–822. https://doi.org/10.1021/ar300069t

Horta-Piñeres S, Britto Hurtado R, Avila-Padilla D, Cortez-Valadez M, Flores-López NS, Flores-Acosta M (2019) Silver nanoparticle-decorated silver nanowires: a nanocomposite via green synthesis. Appl Phys A 126:15. https://doi.org/10.1007/s00339-019-3178-4

Hussain I, Singh NB, Singh A, Singh H, Singh SC (2016) Green synthesis of nanoparticles and its potential application. Biotechnol Lett 38(4):545–560. https://doi.org/10.1007/s10529-015-2026-7

Huston M, DeBella M, DiBella M, Gupta A (2021) Green synthesis of nanomaterials. Nanomaterials 11(8):2130. https://doi.org/10.3390/nano11082130

J Hill M, Qi B, Bayaniahangar R, Araban V, Bakhtiary Z, Doschak MR, Goh BC, Shokouhimehr M, Vali H, Presley JF, Zadpoor AA, Harris MB, Abadi PP, Mahmoudi M (2019) Nanomaterials for bone tissue regeneration: updates and future perspectives. Nanomedicine (Lond) 14(22):2987–3006. https://doi.org/10.2217/nnm-2018-0445

Javanmard Z, Pourhajibagher M, Bahador A (2025) New strategies to enhance antimicrobial photo-sonodynamic therapy based on nanosensitizers against bacterial infections. Folia Microbiol 70(1):55–70. https://doi.org/10.1007/s12223-024-01206-8

Jeevika A, Ravi Shankaran D (2015) Seed-free synthesis of 1D silver nanowires ink using clove oil (Syzygium Aromaticum) at room temperature. J Colloid Interface Sci 458:155–159. https://doi.org/10.1016/j.jcis.2015.07.045

Kagan VE, Bayir H, Shvedova AA (2005) Nanomedicine and nanotoxicology: two sides of the same coin. Nanomedicine 1(4):313–316. https://doi.org/10.1016/j.nano.2005.10.003

Kumah EA, Fopa RD, Harati S et al (2023) Human and environmental impacts of nanoparticles: a scoping review of the current literature. BMC Public Health 23:1059. https://doi.org/10.1186/s12889-023-15958-4

Kummara S, Patil MB, Uriah T (2016) Synthesis, characterization, biocompatible and anticancer activity of green and chemically synthesized silver nanoparticles – a comparative study. Biomed Pharmacother 84:10–21. https://doi.org/10.1016/j.biopha.2016.09.003

Lu D, Fan X (2024) Insights into the prospects of nanobiomaterials in the treatment of cardiac arrhythmia. J Nanobiotechnol 22(1):523. https://doi.org/10.1186/s12951-024-02805-w

Mata YN, Blázquez ML, Ballester A, González F, Muñoz JA (2010) Studies on sorption, desorption, regeneration and reuse of sugar-beet pectin gels for heavy metal removal. J Hazard Mater 178(1–3):243–248. https://doi.org/10.1016/j.jhazmat.2010.01.069

Mitra S, Dua TK, Easmin S, Sarkar S, Roy AP, Sahu R, Nandi G, Haydar MS, Roy S, Paul P (2024) Green synthesis of copper nanoparticles by using pineapple peel waste: in vitro characterizations and antibacterial potential. Bioprocess Biosyst Eng 47(8):1151–1161. https://doi.org/10.1007/s00449-024-02982-w

Muddapur UM, Alshehri S, Ghoneim MM, Mahnashi MH, Alshahrani MA, Khan AA, Iqubal SMS, Bahafi A, More SS, Shaikh IA, Mannasaheb BA, Othman N, Maqbul MS, Ahmad MZ (2022) Plant-based synthesis of gold nanoparticles and theranostic applications: a review. Molecules 27(4):1391. https://doi.org/10.3390/molecules27041391

Nasrollahzadeh M, Ghorbannezhad F, Issaabadi Z, Sajadi SM (2019a) Recent developments in the biosynthesis of Cu-based recyclable nanocatalysts using plant extracts and their application in the chemical reactions. Chem Rec 19(2–3):601–643. https://doi.org/10.1002/tcr.201800069

Nasrollahzadeh M, Mahmoudi-Gom Yek S, Motahharifar N, Ghafori Gorab M (2019b) Recent developments in the plant-mediated green synthesis of Ag-based nanoparticles for environmental and catalytic applications. Chem Rec 19(12):2436–2479. https://doi.org/10.1002/tcr.201800202

Naveed M, Bukhari B, Aziz T, Zaib S, Mansoor MA, Khan AA, Shahzad M, Dablool AS, Alruways MW, Almalki AA, Alamri AS, Alhomrani M (2022) Green synthesis of silver nanoparticles using the plant extract of Acer oblongifolium and study of its antibacterial and antiproliferative activity via mathematical approaches. Molecules 27(13):4226. https://doi.org/10.3390/molecules27134226

Nawaz A, Khalid A, Qayyum W, Bibi R, Qamar MA, Zahid M, Farhan A, Rayaroth MP, Cichocki Ł, Boczkaj G (2025) FeS-based nanocomposites: A promising approach for sustainable environmental remediation – focus on adsorption and photocatalysis – A review. J Environ Manag 373:123530. https://doi.org/10.1016/j.jenvman.2024.123530

Nosrati F, Fakheri B, Ghaznavi H, Mahdinezhad N, Sheervalilou R, Fazeli-Nasab B (2025) Green synthesis of silver nanoparticles from plant Astragalus fasciculifolius Bioss and evaluating cytotoxic effects on MCF7 human breast cancer cells. Sci Rep 15(1):25474. https://doi.org/10.1038/s41598-025-05224-5

Nthwane YB, Fouda-Mbanga BG, Thwala M, Pillay K (2025) A comprehensive review of heavy metals (Pb^{2+}, Cd^{2+}, Ni^{2+}) removal from wastewater using low-cost adsorbents and possible revalorisation of spent adsorbents in blood fingerprint application. Environ Technol 46(3):414–430. https://doi.org/10.1080/09593330.2024.2358450

Obeid MA, Alyamani H, Alenaizat A, Tunç T, Aljabali AAA, Alsaadi MM (2025) Nanomaterial-based drug delivery systems in overcoming bacterial resistance: current review. Microb Pathog 203:107455. https://doi.org/10.1016/j.micpath.2025.107455

Oberdörster G (2010) Safety assessment for nanotechnology and nanomedicine: concepts of nanotoxicology. J Intern Med 267(1):89–105. https://doi.org/10.1111/j.1365-2796.2009.02187.x

Okeke ES, Nweze EJ, Anaduaka EG, Okoye CO, Anosike CA, Joshua PE, Ezeorba TPC (2023) Plant-derived nanomaterials (PDNM): a review on pharmacological potentials against pathogenic microbes, antimicrobial resistance (AMR) and some metabolic diseases. 3 Biotech 13(9):291. https://doi.org/10.1007/s13205-023-03713-w

Parial D, Patra HK, Dasgupta AKR, Pal R (2012) Screening of different algae for green synthesis of gold nanoparticles. Eur J Phycol 47:22–29. https://doi.org/10.1080/09670262.2011.653406

Parveen S, Iqbal S, Maher S, Wilson P, Nasir H, Soomro S, Nisar S, Imran M, Riaz M, Urooj R, Bakhshi MG, Essa M, Mukhtar F, Bhatti A, Janjua HA, Faisal A, Saleem A (2025) Biosynthesis and characterization of silver nanoparticles from Asplenium dalhousiae and their potential biological properties. PLoS One 20(6):e0325533. https://doi.org/10.1371/journal.pone.0325533

Parvin N, Joo SW, Mandal TK (2025) Nanomaterial-based strategies to combat antibiotic resistance: mechanisms and applications. Antibiotics 14(2):207. https://doi.org/10.3390/antibiotics14020207

Patra JK, Das G, Fraceto LF, Campos EVR, Rodriguez-Torres MDP, Acosta-Torres LS, Diaz-Torres LA, Grillo R, Swamy MK, Sharma S, Habtemariam S, Shin HS (2018) Nano based drug delivery systems: recent developments and future prospects. J Nanobiotechnol 16(1):71. https://doi.org/10.1186/s12951-018-0392-8

Pinto RJB, Lucas JMF, Silva FM, Girão AV, Oliveira FJ, Marques PAAP, Freire CSR (2019) Bio-based synthesis of oxidation resistant copper nanowires using an aqueous plant extract. J Clean Prod 221:122–131. https://doi.org/10.1016/j.jclepro.2019.02.189

Qamar W, Gulia S, Athar M, Ahmad R, Imam MT, Chandra P, Singh BP, Haque R, Hassan MI, Rahman S (2024) An insight into impact of nanomaterials toxicity on human health. PeerJ 12:e17807. https://doi.org/10.7717/peerj.17807

Rajeshkumar S, Menon S, Venkat Kumar S, Tambuwala MM, Bakshi HA, Mehta M, Satija S, Gupta G, Chellappan DK, Thangavelu L, Dua K (2019) Antibacterial and antioxidant potential of biosynthesized copper nanoparticles mediated through Cissus arnotiana plant extract. J Photochem Photobiol B 197:111531. https://doi.org/10.1016/j.jphotobiol.2019.111531

Sharma S, Shree B, Aditika, Sharma A, Irfan M, Kumar P (2022) Nanoparticle-based toxicity in perishable vegetable crops: molecular insights, impact on human health and mitigation strategies for sustainable cultivation. Environ Res 212(Pt A):113168. https://doi.org/10.1016/j.envres.2022.113168

Shiraz M, Imtiaz H, Azam A, Hayat S (2024) Phytogenic nanoparticles: synthesis, characterization, and their roles in physiology and biochemistry of plants. Biometals 37(1):23–70. https://doi.org/10.1007/s10534-023-00542-5

Srivastava N, Mukhopadhyay M (2013) Biosynthesis and structural characterization of selenium nanoparticles mediated by Zooglea ramigera. Powder Technol 244:26–29. https://doi.org/10.1016/j.powtec.2013.03.050

Sun B, Hu N, Han L, Pi Y, Gao Y, Chen K (2019) Anticancer activity of green synthesised gold nanoparticles from Marsdenia tenacissima inhibits A549 cell proliferation through the apoptotic pathway. Artif Cells Nanomed Biotechnol 47(1):4012–4019. https://doi.org/10.1080/21691401.2019.1575844

Swierczewska M, Crist RM, McNeil SE (2018) Evaluating nanomedicines: obstacles and advancements. Methods Mol Biol 1682:3–16. https://doi.org/10.1007/978-1-4939-7352-1_1

Thomas S, Gonsalves RA, Jose J, Zyoud SH, Prasad AR, Garvasis J (2024) Plant-based synthesis, characterization approaches, applications and toxicity of silver nanoparticles: a comprehensive review. J Biotechnol 394:135–149. https://doi.org/10.1016/j.jbiotec.2024.08.009

Ukwattage NL, Zhiyong Z (2025) Impacts of cerium dioxide nanoparticles on the soil-plant system and their potential agricultural applications. Nanomaterials 15(12):950. https://doi.org/10.3390/nano15120950

Vasiliu S, Racovita S, Gugoasa IA, Lungan MA, Popa M, Desbrieres J (2021) The benefits of smart nanoparticles in dental applications. Int J Mol Sci 22(5):2585. https://doi.org/10.3390/ijms22052585

Verma A, Gautam SP, Bansal KK, Prabhakar N, Rosenholm JM (2019) Green nanotechnology: advancement in phytoformulation research. Medicines (Basel) 6(1):39. https://doi.org/10.3390/medicines6010039

Verma K, Kathuria D, Ram A, Verma K, Sharma S, Tohra SK, Sharma A (2025) Evaluation of cytotoxic and antioxidant potential of green-synthesized silver and gold nanoparticles from Nepeta leucophylla Benth. Chem Biodivers 22(7):e202402679. https://doi.org/10.1002/cbdv.202402679

Wei J, Mu J, Tang Y, Qin D, Duan J, Wu A (2023) Next-generation nanomaterials: advancing ocular anti-inflammatory drug therapy. J Nanobiotechnol 21(1):282. https://doi.org/10.1186/s12951-023-01974-4

Yayintas OT, Demir N, Canbolat F, Ayna TK, Pehlivan M (2024) Characterization, biological activity, and anticancer effect of green-synthesized gold nanoparticles using Nasturtium officinale L. BMC Complement Med Ther 24(1):346. https://doi.org/10.1186/s12906-024-04635-7

Yu Y, Zhang Q, Mu Q, Zhang B, Yan B (2008) Exploring the immunotoxicity of carbon nanotubes. Nanoscale Res Lett 3(8):271–277. https://doi.org/10.1007/s11671-008-9153-1

Zhao J, Castranova V (2011) Toxicology of nanomaterials used in nanomedicine. J Toxicol Environ Health B Crit Rev 14(8):593–632. https://doi.org/10.1080/10937404.2011.615113

Zhao M, Xu Y, Zhang C, Rong H, Zeng G (2016) New trends in removing heavy metals from wastewater. Appl Microbiol Biotechnol 100(15):6509–6518. https://doi.org/10.1007/s00253-016-7646-x

Zheng G, Tang Z, Peng F (2025) Ultrasound-activated inorganic nanomaterials to generate ROS for antibacterial applications. Biomater Sci 13(10):2628–2641. https://doi.org/10.1039/d5bm00121h

Chapter 19
Economic Feasibility and Potential Market for Plant-Based Nanomaterials

Swetha Panneerselvam, P. Mani, Ponnulakshmi Rajagopal, Vishnu Priya Veeraraghavan, Selvaraj Jayaraman, Janani Muthukumar, and A. P. B. Balaji

19.1 Introduction

Nanotechnology has been a major area of scientific research in recent years thanks to developments, control, and use of nanoscale materials. More specifically, nanotechnology includes one or more dimensions less than 100 nm (1×10^{-9} m) including material engineering at the nanoscale and device fabrication. Owing to their superior chemical, electrical, optical, and physical characteristics, nanoparticles, be they organic or inorganic particles with diameters less than 100 nm, are of distinct importance. Through an extraordinary atomic structure that makes them functionalizable and useable as molecules' carriers, liquid droplets' carriers, or gas carriers, the nanoparticles are either amorphous or crystalline (Khan et al. 2019; Sharma et al. 2019a, b; Kumari et al. 2020). Nanoparticles are increasingly being desired in

S. Panneerselvam · V. P. Veeraraghavan · S. Jayaraman (✉)
Centre of Molecular Medicine and Diagnostics (COMManD), Department of Biochemistry, Saveetha Dental College & Hospitals, Saveetha Institute of Medical & Technical Sciences, Saveetha University, Chennai, India
e-mail: selvarajj.sdc@saveetha.com

P. Mani
PG and Research Department of Biotechnology, Annai college Of Arts and Science, Kumbakonam, Tamil Nadu, India

P. Rajagopal
Central Research Laboratory, Meenakshi Ammal Dental College and Hospital, Meenakshi Academy of Higher Education and Research (Deemed to be University), Chennai, India

J. Muthukumar
Department of Biotechnology, Easwari Engineering College, Chennai, India

A. P. B. Balaji
ESSEM Compliance Solutions, SEZ MIHAN, Nagpur, Maharashtra, India

V. Ghosh et al. (eds.), *Applications of Plant Bioactive Compounds-Based Nanoformulations*, Nanotechnology in the Life Sciences,
https://doi.org/10.1007/978-3-032-13086-0_19

nearly every interdisciplinary field of research due to their unique properties. Nanoparticles find many applications in the biomedical sector, for instance, diagnosis of diseases (McNamara and Tofail 2017), targeted drug and gene delivery, biosensors, bioimaging (Gayda et al. 2019), and other therapeutic applications different from the conventional practice, e.g., demonstration of antimicrobial activity (Razavi et al. 2020). The life sciences have taken advantage of wide use of nanotechnology and, in the process, given birth to the intriguing interdisciplinary science of nanobiotechnology.

The various classes of nanoparticles consist of ceramic, polymeric, metallic, non-metallic, carbon-based, and lipid-based materials (Elegbede et al. 2020). Production of nanoparticles using various physicochemical methods is usually followed by the utilization of toxic materials, the utilization of a significant amount of energy, strict conditions of reactions, and an outlay that is very costly. Additionally, the traditional synthetic processes yield harmful byproducts that could potentially damage ecosystems and living organisms (Gour and Jain 2019). Biological or "green" synthesis has emerged as a feasible method of producing nanoparticles as a sustainable alternative. This green process efficiently synthesizes functional nanoparticles with fungi, bacteria, and plant extracts. Green synthesis is a strong, economically viable, and time-saving process that further greatly reduces the generation of toxic wastes (Kharissova et al. 2019). Bacteria, yeast, and fungi and plants are few examples of the biogenic precursors applied in the synthesis of nanoparticles. Microorganisms have, in recent work, been reported to serve as effective biofactories for nanoparticles synthesis, alongside their many other uses in the field of biotechnology (Sharma et al. 2019a, b).

Biological synthesis and reduction of metal nanoparticles by bacteria, fungi, algae, and yeast are tremendous. Two are the primary methods by which microbes synthesize nanoparticles: extracellularly and intracellularly (Shukla and Iravani 2017). Metal ions are reduced to nanoparticles outside the cell by enzymes that exist on the cell walls of bacteria or get secreted in the culture medium in the extracellular pathway (Slavin et al. 2017; Ovais et al. 2018). Cations of the medium move inside the cell by the intracellular route, reverse conjunction, where they experience an interaction with negatively charged ions of the bacterial cell wall. Intracellular enzymes enable reduction of these metal ions, thus causing metal nanoparticle formation and accumulation within the cell (Subbaiya et al. 2017). Owing to their ease of handling and cultivation as well as their high yield, bacteria are responsible for playing an important role in the synthesis of nanoparticles and other biotechnological uses (Dauthal and Mukhopadhyay 2016). However, there are also some disadvantages that are linked to bacterial systems, including a reduced rate of synthesis, limited control over particle shape and size, and need for a sterile environment (Marooufpour et al. 2019). Therefore, plant-mediated synthesis of nanoparticles is a new technique that is gaining prominence.

Because of their benign properties, natural capacity to stabilize and lower metal ions, and propensity to absorb heavy metals in their cells, scientists all over find plants appealing. Thanks in great part to the explosive expansion of nanotechnology, numerous fields including environmental science, health, and agriculture have seen notable progress. Plant-based nanomaterials (PBNMs) have become a good option among several kinds of nanomaterials due of its possible uses and environmentally acceptable synthesis techniques. In reaction to the move toward sustainability, green chemistry which emphasizes the use of renewable feedstocks in the manufacturing of materials has been in more and more demand (Fariq et al. 2017). Plant extracts that act as reducing and stabilizing agents are used to create PBNMs, which provide a safer and biocompatible alternative over conventional chemical synthesis methods (Singh et al. 2016).

Plant extracts or derivatives are generally used to create nanoparticles that are characterized by size and shape, non-toxic, biocompatible, and with increased penetration ability along with increased functional activity. For reducing, coating, and stabilizing metal ions and also the formation of nanoparticles, a number of phytochemicals like alkaloids, terpenoids, flavonoids, polyphenols, organic acids, polysaccharides, amino acids, and vitamins play important roles (Kim et al. 2016; Singh et al. 2018). Further, plant-assisted biosynthesis is one efficient method for the production of nanoparticles on an industrial scale. These benefits rendered green synthesis techniques an appropriate alternative in numerous industries, further advancing environmentally friendly technology. Science and the search for plant-mediated methods of preparing nanomaterials have hence been the topic of extensive study (Osman et al. 2024). The ecological impact of traditional nanoparticle synthesis is mitigated by following this biogenic pathway, while it also upgrades the biological function of the derived products, suitable for diverse use in environmental decontamination, agriculture, and nanomedicine. The PBNMs also have economic competitiveness as a determining factor that needs to be given attention.

As more industries seek to implement sustainable alternatives to conventional materials, it is crucial to examine the cost factors involved in the manufacture of PBNMs. Availability of raw materials, production scale, and demand significantly influence the economic feasibility of the materials. Current studies indicate that green synthesis procedures hold the promise to reduce energy consumption by up to 30% and lower production costs by up to 40% from traditional procedures (Hano and Abbasi 2021). The market for nanomaterials is set to transform immensely in the years to come due to their multistranded applications in different sectors such as healthcare, agriculture, and electronics. In particular, PBNMs are being increasingly popularized for their antimicrobial nature and medicinal application potential. The agroindustry also explores the use of PBNMs for the enhancement of crop yield and resistance to pests using eco-friendly formulations (Kulkarni et al. 2023). Although such prospective advancements are present, it is challenging to upscale manufacturing processes from laboratory to industrial scales. Problems of standardization, conformity to regulation, and consumer acceptability need to be addressed in order to facilitate broader commercial adoption of PBNMs (Okeke et al. 2023).

Economic sustainability and market potential study for plant-based nanomaterials not only facilitate the goal of universal sustainability but also provide opportunities for new business across industries. This chapter will discuss in more detail the various aspects of PBNMs like their synthesis routes, cost analysis, market trends, applications, and commercialization challenges.

19.2 Types of Nanomaterials

Nanomaterials can be classified based on their dimensionality & composition. Zero-dimensional (0D) nanomaterials have nanoscale dimensions in every direction, and examples of such nanomaterials are nanoparticles like gold nanoparticles and quantum dots. These materials find extensive applications in biomedical fields because of their special optical properties. One-dimensional (1D) nanomaterials have one dimension greater than the nanoscale; examples are nanowires and nanotubes. Carbon nanotubes (CNTs), tubular forms composed of rolled graphene sheets, are especially famous for high damage resistance and electrical conductivity. Two-dimensional (2D) nanomaterials comprise graphene and other materials with very large lateral dimensions but nanoscale thickness. Graphene is particularly famous for high electrical, thermal, and mechanical qualities. Lastly, three-dimensional (3D) nanomaterials are made up of bulk nanocomposites that integrate nanoparticles with macro materials to bring about such attributes as strength and heat resistance (Sajid 2022). Nanomaterials fall under four general categories based on composition.

Fullerenes, carbon nanotubes, and graphene together make carbon nanomaterials. Their low weight and great strength help them to be employed in electronics and nanocomposites. Widely used metal nanomaterials include silver, gold, and iron oxide nanoparticles; their catalytic and antibacterial effects drive their application. Furthermore, in this group are metal oxides such as titanium dioxide since they have photocatalytic activity (Bhardwaj et al. 2020). Because they may enclose other molecules, dendrimers branched macromolecules with great surface functionality are used as catalysts and also in drug delivery systems. Composites combine nanoparticles with larger materials that are made to be better at mechanical or thermal properties to enable application across industries from car accessories to packaging materials. Nanomaterials possess unique characteristics that enable them to be highly relevant to many applications. One of the prominent characteristics is the greater surface area, which enhances reactivity and interaction with other compounds.

Moreover, quantum effects become significant at the nanoscale to determine the behavior of the materials, impacting electronic and optical features. Many nanomaterials also exhibit improved mechanical strength compared to bulk materials, which makes them suitable for light structures applications (Saleh 2020). Furthermore, some nanomaterials show remarkable thermal conductivity, which is quite helpful in uses involving effective heat dissipation. Medical diagnosis, medicine delivery devices, renewable energy technologies, and environmental cleaning technologies are only a few of the novel uses for nanomaterials made possible by their special qualities (Thatyana et al. 2023).

19.3 Economic Feasibility Analysis

Plant-based nanomaterials (PBNMs) are economically feasible only depending on several important factors like cost of manufacturing, market demand, and probably profitability. Among the factors primarily influencing PBNMs' economic viability is raw material cost. Plant extracts serve as reducing agents not just to minimize costs but also to possess a renewable supply of raw materials. For instance, it can cost anything from $380/kg to $2255/kg depending on the functionality as well as nature of the synthesized nanoparticles (Augusto et al. 2020). Such variation brings with it the fact that the option of an easily affordable but widely available plant source must be properly selected. Moreover, commercial nanoparticle market price varies extensively based on their utility and nature; hence, market trends knowledge is essential in deciding pricing strategies.

Emerging advances in synthesis technologies have significantly improved the scalability and efficiency of producing plant-based nanomaterials. Techniques such as microwave-assisted synthesis and ultrasonic methods have been discovered to lower production time without compromising high-quality nanoparticle characteristics (Osman et al. 2024). These advancements not only facilitate mass production but also lower energy consumption compared to traditional techniques. Moreover, ongoing research into new plant sources for nanoparticle synthesis continuously extends the range of materials available for particular applications. For instance, studies on employing agricultural waste as a precursor for synthesizing nanoparticles not only enhance sustainability but also reduce raw material acquisition costs. As these technologies continue to mature, they will be instrumental in reducing costs and boosting accessibility for PBNMs in various industries. Labor costs are also important in the economic evaluation. Labor costs are estimated to average between $10,000 per employee per year in nanoparticle manufacturing plants. This number has to be weighed against the productivity levels realized through effective manufacturing operations. Additionally, energy usage is another essential aspect; conventional nanoparticle synthesis methods technically require high energy usage, whereas green synthesis methods using plant extracts generally require less energy (Mpongwana and Rathilal 2022). This decrease in energy usage can greatly improve the economic feasibility of PBNMs.

Lifecycle analysis is necessary to understand the environmental impacts of plant-based nanomaterials versus traditional materials. LCAs examine resource use, emissions, and waste generation during the production process from raw material extraction to disposal. New studies demonstrate that the green synthesis methods significantly reduce the toxic by-products inherent in traditional nanoparticle synthesis (Kharissova et al. 2019). Furthermore, metrics such as carbon footprint and water usage can be quantified to provide a clearer view of environmental effects. By pointing out these environmental benefits through LCA results, manufacturers can better communicate the advantages of using PBNMs over traditional materials to consumers and stakeholders. This not only aids in global sustainability goals but also assists consumers in making informed choices when looking for environmentally friendly alternatives. There is increasingly steady demand in the marketplace for PBNMs from several sectors like medicine, agriculture, and remediation.

For instance, the international market for nanoparticles in wastewater treatment is expected to expand dramatically as industries pursue efficient solutions for heavy metal elimination. The possible revenue from PBNMs can be considerable; research shows that well-designed business models could bring annual revenues ranging from €40 million for particular applications (Su et al. 2022). This future profitability is further complemented by the emerging trend towards more sustainable manufacturing practices and product design. In spite of these favorable considerations, obstacles continue to hinder production scale-up from laboratory conditions to industrial applications.

Considerations like compliance with regulatory requirements, standardization of manufacturing processes, and competition with entrenched synthetic nanoparticle producers need to be handled with caution. For example, although the initial investment in setting up a manufacturing plant can be estimated at about $1.425 million for some applicationsthe eventual operating cost savings and market prospects are warranted by this investment (Kulkarni et al. 2023). Plant nanomaterials are economically viable based on reduced production costs, rising market demand, and likelihood of profitability on diverse applications. Yet, adequate attention to the sourcing of raw materials, work efficiency, power usage, as well as to market forces is critical to capture their full potential in commerce.

19.4 Market Trends for Plant-Based Nanomaterials

The regulatory environment for plant-based nanomaterials is complex and differs significantly by location. In the United States, the FDA demands thorough safety assessments of nanomaterials used in food and medicine, which can be overwhelming for companies entering the market. In Europe, systems such as REACH require full information on the properties and potential risks of nanomaterials prior to sale (Subhan and Subhan 2022). These regulatory hurdles obligate companies to not only adhere to safety standards but also to maintain effective dialogue with regulators so that their products are compatible with evolving guidelines. Moreover, standardization work is also essential in making uniform testing practices for PBNMs, which can enhance consumer trust and facilitate market acceptance. Addressing these legislative matters is vital in terms of enabling the development of PBNMs for various uses but with consideration for consumer protection and environmental protection. The market for plant-based nanomaterials (PBNMs) has been growing enormously, driven by increasing needs for various application sectors like biomedicine, agriculture, and environment cleanup.

This is mostly a consequence of increased focus on sustainability and use of green chemistry protocols in nanotechnology. Plant extract-based green synthesis of nanoparticles not only reduces environmental risk but also offers a cost-effective alternative to conventional chemical synthesis. Existing research indicates that green synthesis methods are able to reduce energy consumption by up to 30% and production costs by approximately 40% in comparison to traditional processes, hence enhancing the economic value of PBNMs for use in industry (Osman et al. 2024). In the biomedical field, PBNMs are emerging as a central component of drug delivery agents, antimicrobial medications, and imaging diagnostics.

The physicochemical properties of plant-derived nanoparticles, such as biocompatibility, biodegradability, and the ability to penetrate biological barriers, make them highly suitable for healthcare applications. Silver nanoparticles synthesized by various plant extracts have shown considerable antimicrobial activity against varied pathogens in view of the high demand for new antibiotics to overcome antibiotic resistance (Afonso et al. 2024). The escalating frequency of multidrug-resistant infections has heighted the demand for new therapeutic methods, making PBNMs a leading figure in the pharmaceutical sector. On the agricultural front, applying PBNMs is the most popular trend because they are able to promote the yield of plants and provide effective pest control.

Nanoparticles have the potential to act as delivery agents for fertilizers and pesticides, which will allow for targeted delivery and improved nutrient uptake efficiency with decreased environmental footprint. The nanofertilizer market worldwide is expected to expand at a large scale as agricultural stakeholders seek more sustainable and efficient agricultural practices (Kandav and Sharma 2023). Besides, PBNMs are also under consideration for water treatment purposes, driven by the pressure to offer efficient remediation technologies to combat water pollution and supply clean drinking water. Changing lifestyles with consumers opting for sustainable and green products are largely compelling market demand for plant-based nanomaterials. Recent surveys have determined that consumers are willing to pay more for products that include natural ingredients and green processes.

This trend is stronger in industries like cosmetics & personal care, where companies are adding PBNMs to formulations to maximize performance while creating eco-friendly products. The growing trend towards naturally sourced alternatives offers lucrative opportunities for PBNM manufacturers to expand market share by positioning their products around consumer values around sustainability. The application of plant-based nanoparticles in cosmetics is also increasing due to their beneficial impacts. The nanoparticles find broad applications for their antimicrobial properties and product and shelf life stabilizing property. Demand for green and natural products by consumers also contributes to the trend in the cosmetics industry. Despite all these promising trends, several challenges remain to scale up production processes and pass the regulatory tests. Standardization of production processes and quality control procedures are essential preconditions for facilitating broader application of PBNMs in commercial markets (Khan and Khan 2023). Furthermore, consumer knowledge of the benefits of PBNMs such as their reduced toxicity and enhanced performance can also promote acceptance and demand in various markets.

19.5 Applications of Plant-Mediated Nanoparticles

Nanoparticles (NPs) are coming into play with substantial business value since they possess multiple applications in multiple industries such as biomedicine, agriculture, environmental remediation, and energy production. Particularly sought for are plant-based or green-synthesized nanoparticles since they are shown to have fewer toxicity and negative impacts than chemically produced particles. In many disciplines, including nanomedicine and health protection, these nanoparticles show antimicrobial, antiparasitic, antiproliferative, pro-apoptotic, antioxidant and pro-oxidant activity depending on the biological environment as well as anti-inflammatory activities (Saleem et al. 2019). In the biomedical field, PBNMs potential in drug delivery systems is being explored, cancer treatment, and diagnostic imaging.

These most appropriate options for targeted drug administration are these nanoparticles as they can cross biological barriers without endangering the biological system. Against many infections, silver nanoparticles made from plant extracts have shown great antibacterial action, therefore addressing the increasing risk of antimicrobial resistance (Nadeem et al. 2020). Moreover, use of plant nanomaterials as cancer treatments shows promise; studies demonstrate that nanoparticles made from plants have the ability to improve chemotherapy medications and minimize side effects (Andleeb et al. 2021). Moreover, PBNMs find applications in diagnostic imaging owing to their unique optical properties, including enhanced contrast and visualization in medical imaging techniques. PBNMs in agriculture has the capacity to increase crop yield and shield plants from diseases.

By allowing particular delivery of pesticides and fertilizers, nanoparticle carriers help to improve the absorption of nutrients and lower environmental effect. Made from plant extracts, for instance, zinc oxide nanoparticles have shown to improve plant growth and disease resistance to pathogens (Anjum et al. 2019). Moreover, PBNMs can be used in the treatment of phytopathogens and soil remediation so enabling the introduction of important agricultural practices by means of the minimizing of the use of harmful pesticides (Shafiq et al. 2020). In food science and technology, they are fundamental in food processing, storage, and packaging as well as in bioengineering uses where they function as biosensors, biocatalysts, and photocatalysts. Another important area of use for PBNMs is environmental remedial action. Their great help in pollution control projects comes from their capacity to remove contaminants from ground resources and aquatic bodies.

Effective heavy metal and organic pollution removal from wastewater treatment systems has been achieved with green-synthesised nanoparticles (Anjum et al. 2021). Reflecting their usage in treating industrial effluencies (Pineda-Vásquez et al. 2023), plant extract-synthesized manganese oxide nanoparticles have shown for effective absorption against chromium (VI). In addition, PBNMs can also serve as photocatalysts in solar power technology, making the efficiency of solar cells higher and contributing to alternative energy solutions. Beyond these primary applications, plant nanomaterials are becoming increasingly popular in the cosmetic industry. Their antimicrobial properties and ability to stabilize formulas render them attractive additives to skincare products. The growing consumer preference for natural and eco-friendly products also drives the use of PBNMs in cosmetic products further (Fytianos et al. 2020).

19.6 Regulatory and Safety Considerations

With growing uses of nanotechnology in several spheres like agricultural, environmental remediation & medicinal, the regulatory environment for plant-based nanomaterials (PBNMs) is also changing. Safety, efficiency, and environmental sustainability of these items depend on regulatory mechanisms in great part. PBNMs fall under generally applicable rules in many countries that apply to nanoparticles in general and can incorporate instructions from agencies such as the Food and Drug Administration (FDA) and the U.S. Environmental Protection Agency (EPA). These companies track possible hazards connected to the use of nanomaterials by means of pre-market testing and post-market monitoring, therefore assessing their safety (Rasmussen et al. 2016). The OECD's "Guidance Manual for the Testing of Manufactured Nanomaterials," which offers best practices to evaluate the safety of nanomaterials, such PBNMs, is among the most well-known regulating rules. This guidance emphasizes the integrated risk assessments including special nanoscale characteristics including improved chemical activity and modified biological accessibility (Rasmussen et al. 2019). Furthermore, the European Union's REACH (Registration, Evaluation, Authorisation and Restriction of Chemicals) rule mandates manufacturers of nanomaterials register them and submit details on their environmental impact and safety before commercialisation. This legislative approach is meant to give guarantee that any potential hazards connected with PBNMs are found and properly controlled (Domb et al. 2021). Safety assessment of PBNMs usually consists in toxicity studies evaluating their effects on the surroundings and human health.

To ascertain cytotoxicity, genotoxicity, and other potential ecological impacts, these studies usually combine in vivo (animal) tests with in vitro (cell culture). Studies have shown, for example, that some plant-based silver nanoparticles have great antibacterial action (Tirumala et al. 2021) with extremely little toxicity against human cells, hence fit for use in biomedicals. To fully understand the safety profile of these compounds, though, thorough toxicity testing across several routes of contact—respiratory, oral, and cutaneous exposure—is crucial. Furthermore, crucial for evaluating the possible impact of PBNMs on the ecology is environmental toxicity testing. Studies of aquatic life have revealed that only few nanoparticles cause bioaccumulation and probable trophic transfer (Fabrega et al. 2011).

Therefore, the long-term ecological impact of PBNMs depends on an awareness of their behavior and persistence in different ecosystems matrix. Although regulatory frameworks for plant-based nanomaterials (PBNMs) are still under development, current rules underline the need of comprehensive safety assessments considering environmental issues as well as human health. Development of strong safety procedures for the responsible commercialization of PBNMs depends on ongoing cooperation among researchers, regulatory authorities, and industry players.

19.7 Challenges in the Market Adoption of Plant-Based Nanomaterials

T Plant-based nanomaterials (PBNMs) have a range of obstacles that hinder their large-scale implementation over the several sectors. These difficulties include customer knowledge and acceptance problems, competition with the current synthetic nanomaterials, and technological constraints in the upscaling of manufacturing. The main technological obstacle to the acceptance of PBNMs is the inability to upscale the manufacturing techniques from laboratory to commercial scale. Often combined with intricate biological processes that, at the moment, remain difficult to replicate in mass quantities, the green synthesis method utilized in making PBNMs is also linked with consistent manufacture can be particularly difficult depending on variations in the plant's extract and possible implications thereof upon the size, shape, and stability of the nanoparticle produced (Besley 2010). Furthermore, the lack of consistent methods for synthesizing and analyzing PBNMs can lead to product variation, which could deter possible industrial consumers from using these materials. According to the research, overcoming such challenges and reaching the dependability of PBNMs in industrial applications depends on the development of efficient and repeatable manufacturing techniques (Lee et al. 2013). Another main obstacle for PBNMs is competition from traditional synthetic nanomaterials.

Established products that have been well-characterized and supported by significant research and development expenditure control the synthetic nanoparticle industry. Years of research have helped synthetic materials to have improved performance qualities. Companies could thus be reluctant to move to PBNMs, which are seen as less effective or dependable because they are somewhat new to the market (Aithal and Aithal 2016). Furthermore, existing competitors in the nanomaterials market benefit from established distribution networks and economies of scale; so, it is challenging for new market players specialized in PBNMs to compete on price and availability. Producers of PBNMs must be able to demonstrate their unique advantages including ecofriendliness and biocompatibility as well as competitive costs and performance qualities if they are to overcome this obstacle. Key obstacles to market penetration of PBNMs include consumer acceptance and awareness. Demand may be limited if many possible consumers lack knowledge of the benefits or uses for plant-based nanomaterials.

Stakeholders should be informed about the advantages of PBNMs over accepted synthetic substitutes by educational initiatives (Shapira and Youtie 2015). New technology skepticism will further hinder adoption; customers will be worried that PBNMs are not safe and/or effective when compared to established synthetic rivals. Open conversation regarding safety testing, regulatory approval, and performance statistics will only help to remove such mistrust. Emphasizing successful case studies and testimonials, targeted marketing initiatives can also help to build confidence among potential users (Maksimović and Omanović-Mikličanin 2017). Further market acceptance of plant-based nanomaterials (PBNMs) depends on resolution of these issues for development of the market. Giving technology developments top priority, competitive positioning, and consumer education top importance will help stakeholders to improve PBNM adoption and visibility in several sectors.

19.8 Successful Implementation of PBNMs in Various Industries

One noteworthy example study is the antibacterial use of positively charged silver nanoparticles generated from Matricaria chamomilla (chamomile). Against species of Mycobacterium including Escherichia coli and Staphylococcus aureus, the nanoparticles displayed decent antibacterial action. Apart from lowering manufacturing costs, the green synthesis technique enhanced the biocompatibility of the nanoparticles so enabling their application on medical devices and wound dressings (Chung et al. 2016).

Furthermore, another effective use of zinc oxide nanoparticle derived from Cymbopogon citratus is seen in increasing crop development and disease resistance. Field experiments showed that the nanoparticles improved crop fertilizer usage efficiency and stopped fungal infections including those of tomatoes and peppers. Combining the cost-effective alternative to conventional chemical pesticides and fertilizers provided by the economically feasible synthesis technique with the effectiveness of the nanoparticles gave especially in water treatment, manganese oxide nanoparticles produced from leaf extract of Abutilon indicum show promise for environmental use. Potential agents for large-scale environmental remediation systems, nanoparticles have been shown to efficiently absorb heavy metals including chromium (VI) from hazardous water sources. Apart from lowering manufacturing costs, the green synthesis approach also lessened ecological effects when compared to more traditional approaches (Khan et al. 2020; Ahmad et al. 2022).

19.9 Comparative Analysis of Plant-Based Nanomaterials (PBNMs) in Various Industries

Application of plant-based nanomaterials (PBNMs) in diversified industries signifies the versatility of the technology and scope for making production more efficient and sustainable. In this comparative assessment, important applications in biomedicine, agriculture, environmental cleaning, and cosmetics are noted to identify their costs of production, market size, regulatory hurdles, and economic viability as a whole. PBNMs in the biomedical field find most applications as drug delivery devices, antimicrobial agents & tissue repair processes. Silver nanoparticles produced from Azadirachta indica (neem) have proved to exhibit strong antibacterial activities against microbes such as Escherichia coli and Staphylococcus aureus. The green approach used not only lowers the cost of production but also enhances the biocompatibility of the nanoparticles with biological systems, making them appropriate for applications in medicine. The terpenoids and flavonoids in neem extract serve as reducing and stabilizing agents, which are responsible for the antimicrobial activity of the nanoparticles (Dutt et al. 2023). The nanoparticles have also shown promise in catalyzing reactions for wastewater treatment, indicating their multifunctional applications.

In agriculture, PBNMs are used as nanofertilizers and nanopesticides for promoting the growth of crops and disease resistance. Zinc oxide nanoparticles from Eucalyptus lanceolata has proven effective in improving the nutrient absorption potential and preventing fungal disease occurrence on crops such as tomatoes and peppers. Economical & ecofriendly green synthesis technique is an economical and eco-friendly method compared to traditional agrochemicals that provides a sustainable solution. Current research has shown that these nanoparticles have the ability to improve photosynthetic performance by enhancing the content of chlorophyll and supporting root growth (Sharma et al. 2022; Raj et al. 2021). Additionally, nanotechnology in agriculture has also been associated with enhanced soil health through increased microbial activity, bringing other advantages to agriculture aside from direct plant applications.

Abutilon indicum leaf extract-derived manganese oxide nanoparticles have been found to have potential in environmental remediation, especially in wastewater treatment (Table 19.1).

These efficiently absorb heavy elements such as chromium (VI), therefore demonstrating their relevance for large-scale environmental treatment. The green synthesis technique lowers manufacturing costs and environmental consequences. High photocatalytic activity of these manganese oxide nanoparticles has been demonstrated by studies, which helps them to effectively breakdown certain organic contaminants (Khan et al. 2020). Furthermore, applications of nano bioremediation techniques have proven a successful method to treat organic contamination in soils and proof of the adaptability of PBNMs for environmental reasons (Ahmad et al. 2022). PBNMs are increasingly found in skincare products in the cosmetics industry since they have antioxidant action and natural preservation properties. The anti-senescence action of gallic acid-coated gold nanoparticles was investigated by means of matrix metalloproteinase-1 (MMP-1), a component influencing skin aging.

Environmentally friendly synthesis of these nanoparticles accentuates their appeal in a market that values natural components more and more (Wu et al. 2021; Burketová et al. 2022). Moreover, studies show that because of their nanoscale dimensions, gold nanoparticles are fit for usage in several cosmetic preparations meant to improve skin health and can effectively pass the stratum corneum (Noah 2019).

Table 19.1 Comparative analysis of economic viability in different applications

Industry	Key applications	Production methods	Economic viability	Market demand	Regulatory challenges	References
Biomedical	Drug delivery systems, antimicrobial agents, wound healing	Green synthesis using plant extracts (e.g., silver nanoparticles from Azadirachta indica)	High	High	Moderate; requires extensive safety assessments and clinical trials	Dutt et al. (2023)
Agriculture	Fertilizers, pest control, growth enhancers	Green synthesis using plant materials (e.g., zinc oxide from *Eucalyptus lanceolata leaf*)	Moderate to high	Growing	Low; fewer regulatory hurdles compared to pharmaceuticals	Sharma et al. (2022)
Environmental	Water treatment, soil remediation	Green synthesis from plant extracts (e.g., manganese oxide from *Abutilon indicum*)	High	Increasing	Moderate; environmental impact assessments required	Khan et al. (2020)
Cosmetics	Skincare products, natural preservatives	Green synthesis using plant extracts (e.g., gallic acid mediated gold nanoparticles)	Moderate	High	Low; less stringent compared to pharmaceuticals	Wu et al. (2021)

19.10 Conclusion

Plant-based nanomaterials (PBNMs) offer tremendous potential in a range of industries like biomedicine, agriculture, environmental clean-up, and cosmetics. These biocompatible and biodegradable materials resulting from plant extract-based eco-friendly synthesis find precise solutions to crucial issues like antimicrobial resistance in medicine, increased crop production in agroindustry, and green measures for detoxification of the environment. Yet, the issues of technological limitations to scaling production, competition from entrenched synthetic counterparts, and the necessity for heightened consumer awareness need to be overcome for wider market take-up. Future research must focus on the establishment of standardized production procedures & performing in-depth safety evaluations to ensure the long-term viability of PBNMs. Further, identifying novel synthesis techniques that maximize efficiency with reduced environmental footprint is important. Cooperation between scientists, industry, and regulators is crucial for handling commercialization issues and optimizing the potential of PBNMs as durable substitutes in different applications.

References

Afonso IS et al (2024) Green synthesis of nanoparticles from olive oil waste for environmental and health applications: a review. J Environ Chem Eng 12(5):114022. https://doi.org/10.1016/j.jece.2024.114022

Ahmad KS, Yaqoob S, Gul MM (2022) Dynamic green synthesis of iron oxide and manganese oxide nanoparticles and their cogent antimicrobial, environmental and electrical applications. Rev Inorg Chem 42(3):239–263. https://doi.org/10.1515/revic-2021-0033

Aithal PS, Aithal S (2016) Nanotechnology innovations and commercialization-opportunities, challenges & reasons for delay. Int J Eng Manuf 6(6):15–25. https://doi.org/10.5815/ijem.2016.06.02

Andleeb A et al (2021) A systematic review of biosynthesized metallic nanoparticles as a promising anti-cancer-strategy. Cancer 13(11):2818. https://doi.org/10.3390/cancers13112818

Anjum S et al (2019) Advances in nanomaterials as novel elicitors of pharmacologically active plant specialized metabolites: current status and future outlooks. RSC Adv 9(69):40404–40423. https://doi.org/10.1039/c9ra08457f

Anjum S et al (2021) Emerging applications of nanotechnology in healthcare systems: grand challenges and perspectives. Pharmaceuticals 14(8):707. https://doi.org/10.3390/ph14080707

Augusto PA et al (2020) Upscale design, process development, and economic analysis of industrial plants for nanomagnetic particle production for environmental and biomedical use. Materials 13(11):2477. https://doi.org/10.3390/ma13112477

Besley J (2010) Current research on public perceptions of nanotechnology. Emerg Health Threats J 3(1):7098. https://doi.org/10.3402/ehtj.v3i0.7098

Bhardwaj B et al (2020) Eco-friendly greener synthesis of nanoparticles. Adv Pharm Bull 10(4):566–576. https://doi.org/10.34172/apb.2020.067

Burketová L et al (2022) Noble metal nanoparticles in agriculture: impacts on plants, associated microorganisms, and biotechnological practices. Biotechnol Adv 58:107929. https://doi.org/10.1016/j.biotechadv.2022.107929

Chung I-M et al (2016) Plant-mediated synthesis of silver nanoparticles: their characteristic properties and therapeutic applications. Nanoscale Res Lett 11(1):40. https://doi.org/10.1186/s11671-016-1257-4
Dauthal P, Mukhopadhyay M (2016) Noble metal nanoparticles: plant-mediated synthesis, mechanistic aspects of synthesis, and applications. Ind Eng Chem Res 55(36):9557–9577. https://doi.org/10.1021/acs.iecr.6b00861
Domb AJ et al (2021) Safety evaluation of nanotechnology products. Pharmaceutics 13(10):1615. https://doi.org/10.3390/pharmaceutics13101615
Dutt Y et al (2023) Silver nanoparticles phytofabricated through azadirachta indica: anticancer, apoptotic, and wound-healing properties. Antibiotics 12(1):121. https://doi.org/10.3390/antibiotics12010121
Elegbede JA et al (2020) Biofabrication of gold nanoparticles using xylanases through valorization of corncob by Aspergillus niger and trichodermalongibrachiatum: antimicrobial, antioxidant, anticoagulant and thrombolytic activities. Waste Biomass Valor 11(3):781–791. https://doi.org/10.1007/s12649-018-0540-2
Fabrega J et al (2011) Silver nanoparticles: behaviour and effects in the aquatic environment. Environ Int 37(2):517–531. https://doi.org/10.1016/j.envint.2010.10.012
Fariq A, Khan T, Yasmin A (2017) Microbial synthesis of nanoparticles and their potential applications in biomedicine. J Appl Biomed 15(4):241–248. https://doi.org/10.1016/j.jab.2017.03.004
Fytianos G, Rahdar A, Kyzas GZ (2020) Nanomaterials in cosmetics: recent updates. Nanomaterials 10(5):979. https://doi.org/10.3390/nano10050979
Gayda GZ et al (2019) Metallic nanoparticles obtained via "green" synthesis as a platform for biosensor construction. Appl Sci 9(4):720. https://doi.org/10.3390/app9040720
Gour A, Jain NK (2019) Advances in green synthesis of nanoparticles. Artif Cells Nanomed Biotechnol 47(1):844–851. https://doi.org/10.1080/21691401.2019.1577878
Hano C, Abbasi BH (2021) Plant-based green synthesis of nanoparticles: production, characterization and applications. Biomolecules 12(1):31. https://doi.org/10.3390/biom12010031
Kandav G, Sharma T (2023) Green synthesis: an eco friendly approach for metallic nanoparticles synthesis. Part Sci Technol 42(5):874–894. https://doi.org/10.1080/02726351.2023.2281452
Khan MF, Khan MA (2023) Plant-derived metal nanoparticles (pdmnps): synthesis, characterization, and oxidative stress-mediated therapeutic actions. Futur Pharmacol 3(1):252–295. https://doi.org/10.3390/futurepharmacol3010018
Khan I et al (2019) Nanoparticles: properties, applications and toxicities. Arab J Chem 12(7):908–931. https://doi.org/10.1016/j.arabjc.2017.05.011
Khan SA et al (2020) Green synthesis of mno nanoparticles using abutilon indicum leaf extract for biological, photocatalytic, and adsorption activities. Biomolecules 10(5):785. https://doi.org/10.3390/biom10050785
Kharissova OV et al (2019) Greener synthesis of chemical compounds and materials. R Soc Open Sci 6(11):191378. https://doi.org/10.1098/rsos.191378
Kim Y-J et al (2016) Green synthesis of multifunctional silver and gold nanoparticles from the oriental herbal adaptogen: Siberian ginseng. Int J Nanomedicine 11:3131–3143. https://doi.org/10.2147/ijn.s108549
Kulkarni D et al (2023) Biofabrication of nanoparticles: sources, synthesis, and biomedical applications. Front Bioeng Biotechnol 11:1159193. https://doi.org/10.3389/fbioe.2023.1159193
Kumari RM et al (2020) Application of nanotechnology in diagnosis and therapeutics. In: Nanotechnology for energy and environmental engineering. Springer, Cham, pp 413–440. https://doi.org/10.1007/978-3-030-33774-2_18
Lee C-J et al (2013) Factors influencing nanotechnology commercialization: an empirical analysis of nanotechnology firms in South Korea. J Nanopart Res 15(2):1444. https://doi.org/10.1007/s11051-013-1444-5
Maksimović M, Omanović-Mikličanin E (2017) Towards green nanotechnology: maximizing benefits and minimizing harm. In: IFMBE proceedings. Springer Singapore, Singapore, pp 164–170. https://doi.org/10.1007/978-981-10-4166-2_26

Marooufpour N et al (2019) Biological synthesis of nanoparticles by different groups of bacteria. In: Nanotechnology in the life sciences. Springer, Cham, pp 63–85. https://doi.org/10.1007/978-3-030-16383-9_3
McNamara K, Tofail SA (2017) Nanoparticles in biomedical applications. Adv Phys X 2(1):54–88. https://doi.org/10.1080/23746149.2016.1254570
Mpongwana N, Rathilal S (2022) A review of the techno-economic feasibility of nanoparticle application for wastewater treatment. Water 14(10):1550. https://doi.org/10.3390/w14101550
Nadeem M et al (2020) Green synthesis of cerium oxide nanoparticles (CEONPS) and their antimicrobial applications: a review. Int J Nanomedicine 15:5951–5961. https://doi.org/10.2147/ijn.s255784
Noah N (2019) Green synthesis: characterization and application of silver and gold nanoparticles. In: Green synthesis, characterization and applications of nanoparticles. Elsevier, pp 111–135. https://doi.org/10.1016/b978-0-08-102579-6.00006-x
Okeke ES et al (2023) Plant-derived nanomaterials (PDNM): a review on pharmacological potentials against pathogenic microbes, antimicrobial resistance (AMR) and some metabolic diseases. 3 Biotech 13(9):291. https://doi.org/10.1007/s13205-023-03713-w
Osman AI et al (2024) Synthesis of green nanoparticles for energy, biomedical, environmental, agricultural, and food applications: a review. Environ Chem Lett 22(2):841–887. https://doi.org/10.1007/s10311-023-01682-3
Ovais M et al (2018) Biosynthesis of metal nanoparticles via microbial enzymes: a mechanistic approach. Int J Mol Sci 19(12):4100. https://doi.org/10.3390/ijms19124100
Pineda-Vásquez T et al (2023) From e-waste to high-value materials: sustainable synthesis of metal, metal oxide, and MOF nanoparticles from waste printed circuit boards. Nanomaterials 14(1):69. https://doi.org/10.3390/nano14010069
Raj S, Trivedi R, Soni V (2021) Biogenic synthesis of silver nanoparticles, characterization and their applications – a review. Surfaces 5(1):67–90. https://doi.org/10.3390/surfaces5010003
Rasmussen K et al (2016) Review of achievements of the OECD working party on manufactured nanomaterials' testing and assessment programme. From exploratory testing to test guidelines. Regul Toxicol Pharmacol 74:147–160. https://doi.org/10.1016/j.yrtph.2015.11.004
Rasmussen K et al (2019) Developing OECD TEST guidelines for regulatory testing of nanomaterials to ensure mutual acceptance of test data. Regul Toxicol Pharmacol 104:74–83. https://doi.org/10.1016/j.yrtph.2019.02.008
Razavi R et al (2020) Biosynthesis of metallic nanoparticles using mulberry fruit (morus Alba L.) extract for the preparation of antimicrobial nanocellulose film. Appl Nanosci 10(2):465–476. https://doi.org/10.1007/s13204-019-01137-8
Sajid M (2022) Nanomaterials: types, properties, recent advances, and toxicity concerns. Curr Opin Environ Sci Health 25:100319. https://doi.org/10.1016/j.coesh.2021.100319
Saleem K et al (2019) Applications of nanomaterials in leishmaniasis: a focus on recent advances and challenges. Nanomaterials 9(12):1749. https://doi.org/10.3390/nano9121749
Saleh TA (2020) Nanomaterials: classification, properties, and environmental toxicities. Environ Technol Innovation 20:101067. https://doi.org/10.1016/j.eti.2020.101067
Shafiq M et al (2020) An overview of the applications of nanomaterials and nanodevices in the food industry. Foods 9(2):148. https://doi.org/10.3390/foods9020148
Shapira P, Youtie J (2015) The economic contributions of nanotechnology to green and sustainable growth. In: Green processes for nanotechnology. Springer, Cham, pp 409–434. https://doi.org/10.1007/978-3-319-15461-9_15
Sharma D, Kanchi S, Bisetty K (2019a) Biogenic synthesis of nanoparticles: a review. Arab J Chem 12(8):3576–3600. https://doi.org/10.1016/j.arabjc.2015.11.002
Sharma G et al (2019b) Novel development of nanoparticles to bimetallic nanoparticles and their composites: a review. J King Saud Univ Sci 31(2):257–269. https://doi.org/10.1016/j.jksus.2017.06.012

Sharma P et al (2022) Green synthesis of zinc oxide nanoparticles using eucalyptus lanceolata leaf litter: characterization, antimicrobial and agricultural efficacy in maize. Physiol Mol Biol Plants 28(2):363–381. https://doi.org/10.1007/s12298-022-01136-0
Shukla AK, Iravani S (2017) Metallic nanoparticles: green synthesis and spectroscopic characterization. Environ Chem Lett 15(2):223–231. https://doi.org/10.1007/s10311-017-0618-2
Singh P et al (2016) Biological synthesis of nanoparticles from plants and microorganisms. Trends Biotechnol 34(7):588–599. https://doi.org/10.1016/j.tibtech.2016.02.006
Singh J et al (2018) Green synthesis of metals and their oxide nanoparticles: applications for environmental remediation. J Nanobiotechnol 16(1):84. https://doi.org/10.1186/s12951-018-0408-4
Slavin YN et al (2017) Metal nanoparticles: understanding the mechanisms behind antibacterial activity. J Nanobiotechnol 15(1):65. https://doi.org/10.1186/s12951-017-0308-z
Su Y et al (2022) Cost–benefit analysis of nanofertilizers and nanopesticides emphasizes the need to improve the efficiency of nanoformulations for widescale adoption. Nat Food 3(12):1020–1030. https://doi.org/10.1038/s43016-022-00647-z
Subbaiya R et al (2017) Biomimetic synthesis of silver nanoparticles from streptomyces atrovirens and their potential anticancer activity against human breast cancer cells. IET Nanobiotechnol 11(8):965–972. https://doi.org/10.1049/iet-nbt.2016.0222
Subhan MA, Subhan T (2022) Safety and global regulations for application of nanomaterials. In: Nanomaterials recycling. Elsevier, pp 83–107. https://doi.org/10.1016/b978-0-323-90982-2.00005-6
Thatyana M et al (2023) Advances in phytonanotechnology: a plant-mediated green synthesis of metal nanoparticles using Phyllanthus plant extracts and their antimicrobial and anticancer applications. Nanomaterials 13(19):2616. https://doi.org/10.3390/nano13192616
Tirumala MG et al (2021) Novel methods and approaches for safety evaluation of nanoparticle formulations: a focus towards in vitro models and adverse outcome pathways. Front Pharmacol 12:612659. https://doi.org/10.3389/fphar.2021.612659
Wu Y-Z et al (2021) Evaluation of gallic acid-coated gold nanoparticles as an anti-aging ingredient. Pharmaceuticals 14(11):1071. https://doi.org/10.3390/ph14111071

Chapter 20
Summary of Recent Trends and Future Prospects of Nano Formulations and Applications

Monadeepa Sengupta and Hemant Kumar

20.1 Introduction

The potential of medicinal plants exists in past, present and future also. It helps naturally in every field of living steps from birth to death. It has power to save our generations from futuristic disasters. The combination of medicinal plants and Nano formulations scope is vast and rapidly evolving. Medicinal plants potentials are Discovery of new bioactive compounds, Development of novel drugs, Alternative to synthetic pharmaceuticals, Traditional medicine revival and Holistic approach to healthcare (Heya et al. 2022). They contain so vast scope of plant-based Nano formulations are Enhanced bioavailability and solubility, Targeted drug delivery, Controlled release and sustained action, Improved stability and shelf-life, Reduced toxicity and side effects.

20.2 Nanoformulation

It is the process of creating tiniest particles of nanoscale (1–100 nm) by using nanotechnology (Hano and Abbasi 2021). It helps to enhance the efficiency in medical field, pharmaceutical field, agriculture field, material science field, industry field, cosmetic field, energy storage field, fuel science field, aerospace and defense material field, etc.

M. Sengupta (✉)
Department of Physics, OP Jindal University, Raigarh, Chhattisgarh, India

H. Kumar
Department of Biotechnology, OP Jindal University, Raigarh, Chhattisgarh, India

V. Ghosh et al. (eds.), *Applications of Plant Bioactive Compounds-Based Nanoformulations*, Nanotechnology in the Life Sciences,
https://doi.org/10.1007/978-3-032-13086-0_20

Nanoparticles are tiniest particle colloidal structures made up of natural or synthetic or semi-synthetic polymers. The drug is dissolved, encapsulated, attached or entrapped to a nanoparticle matrix.

Due to its high efficiency less amount of dose required therefore increase its therapeutic efficacy, reduces side effects and bioavailability. Without injection also we can provide it through various routes like oral, nasal, skin absorption property, parenteral, intra-ocular etc.

It improved stability of Nano formulation Renewable resources, cost-effectiveness, biodegradability. Plants contain various biological biochemical used in medicinal and clinical fields. Every part of the plants such as leaves, fruits, stems, flowers, seeds, roots, barks, and fruit peels contain useful biochemical which is investigated to use to synthesise plant-based nanoparticles. The product which is made from plants called as Biomarkers which are having wide range of applications in bioengineering sciences after a long period of Research and Development to check its availability and toxic level. Formation of nanoparticles from plant-based are silver nitrate, gold from gold chloride, zine oxide from Zinc nitrate and zinc acetate, cadmium sulphide and zinc sulphide from cadmium sulphate and zinc sulphate etc. (Siddiqui et al. 2015). The therapeutic development of metal nanoparticles works against many viral, fungal pathogens and bacterial.

To understand the different aspects of the plants and nanoparticles interaction is necessary to achieve our higher potential futuristic goals. For better result it is Important to understand the plant metabolism for synthesis the plant-based Nano formulations (Fang et al. 2019).

In plant body there is hundreds of reactions simultaneously taking place in a living cell, in a well-organized manner and also follows integrated manners. The entire Chemical reactions, occurring in the living system of a plant, is collectively called as Plant metabolism.

The map of or pathway of metabolism consist a series of enzymatic reactions to produce final product. So that we can say that Plant Kingdom produce a vast range of metabolites diversity which give biological and pharmacological properties.

There are two types of metabolites in Plants are Primary and secondary metabolites Elshafie et al. (2023) and Dewanjee et al. (2020), (Tables 20.1, 20.2).

Functions of secondary metabolites is Providing flavour, colour and aroma, Defence against fungus pathogens and pests, attracting pollinators, Regulating plant

Table 20.1 Primary metabolites in plants

S.N.	Primary metabolites	Examples	Reference
1	Carbohydrates	Sugars, starches, cellulose	Elshafie et al. (2023)
2	Amino acids	Building blocks of proteins	
3	Proteins	Enzymes, structural proteins, storage proteins	
4	Nucleotides	Building blocks of DNA and RNA	
5	Fatty acids	Components of lipids and membranes	
6	Glycerol	Components of lipids and membranes	

Table 20.2 Secondary metabolites in plants

S.N.	Secondary metabolites	Compound	Examples	Reference
1	Alkaloids	Nitrogen-containing compounds	Caffeine, nicotine	Dewanjee et al. (2020)
2	Phenolics	Aromatic	Flavonoids, lignin	
3	Terpenoids	Hydrocarbon	Essential oils, steroids	
4	Glycosides	Sugar-containing	Saponins, anthocyanins	
5	Saponins	Soap-like	Glycosides of steroids	
6	Tannins	Polyphenolic	Gallic acid, ellagic acid	

growth and development, Protection against environmental stresses. Secondary metabolites used as pesticides and fertilizers. Food and beverage Industry: Secondary metabolites used as flavourings, colouring, and preservatives. Cosmetics and Personal Care: Secondary metabolites are used in skincare and hair care products.

Plant-Based materials are phytochemicals (e.g., phenolic acids, terpenes, saponins, flavonoids, alkaloids) (Shogren et al. 2019; Dada and Popoola 2024; Velavan 2015); plant proteins (e.g., albumin, legumin, vicilin, zein, gelatin) (Rausch and Greiner 2004); plant oils (e.g., neem, sunflower, soybean, jojoba, castor, coconut, olive (Antunes Filho et al. 2023); plant extracts, a concentrated preparations derived from plants, containing bioactive compounds. (e.g., algae, tea) (Ikram 2015; Mittal et al. 2013). Cellulose is derived from plant cell walls (Somerville 2006). Chitosan is derived from crustacean shells. (Bodnar et al. 2005; Yu et al. 2019; Ahmed and Aljaeid 2016). Pectin is a natural, soluble fiber and structural polymer found in the cell walls of plants. (Thankappan et al. 2020; Einhorn-Stoll et al. 2021). Starch is derived from plant roots, tubers, or seeds (Pfister and Zeeman 2016). Alginate is derived from brown algae (Paques et al. 2014; Einhorn-Stoll et al. 2021; Severino et al. 2019). Gum Arabic is derived from sap of acacia tree (Hassani et al. 2020a; Iravani 2020). Carrageenan: is derived from various species of red algae (Grenha et al. 2010; Gün Gök et al. 2021).

Some examples of Plant-based Nano formulations are Curcumin-loaded nanoparticles from turmeric, Quercetin-loaded liposomes from onion extract, Nanoparticle-based delivery of plant-derived antioxidants, Green tea catechin-loaded nanoparticles, Ginger-based nanoemulsion for anti-inflammatory applications and etc. (Wickramasinghe et al. 2022; Kokkinis et al. 2024).

Table 20.3 Factors affecting plant-based metabolism

Environmental factors				
S.N.	Source	Function	Example	Reference
1	Light	Light intensity and quality affect photosynthesis and plant growth.	Plants grown in low-light conditions may have slower growth rates and altered metabolic pathway.	Qaderi et al. (2023)
2	Temperature	Temperature affects enzyme activity, photosynthesis, and plant growth.	Plants grown in high-temperature conditions may have increased respiration rates and altered metabolic pathways.	
3	Water	Water availability affects plant growth, photosynthesis, and plant growth.	Plants grown in drought conditions may have altered metabolic pathways to conserve water.	
Nutritional factors				
S.N.	Source	Function	Example	Qaderi et al. (2023)
1	Nutrient availability	Nutrient availability affects plant growth, photosynthesis, and metabolic pathways.	Plants grown in nutrient-poor soil may have altered metabolic pathways to optimize nutrient use.	
2	Soil pH	Soil pH affects nutrient availability and plant growth.	Plants grown in acidic soil may have altered metabolic pathways to optimize nutrient use.	
Hormonal factors				
S.N.	Source	Function	Example	Sosnowski et al. (2023)
1	Auxins	Auxins regulate cell elongation, cell division, and plant growth.	Plants treated with auxins may have increased cell elongation and altered metabolic pathways.	
2	Cytokinins	Cytokinins regulate cell division, plants growth, and senescence.	Plants treated with cytokinins may have increased cell division and altered metabolic pathways.	
Genetic factors				
S.N.	Source	Function	Example	Kaur et al. (2022)
1	Genetic variation	Genetic variation affects plant metabolism, growth, and development.	Plants with genetic variations in photosynthetic genes may have altered metabolic pathways.	
2	Epigenetic modifications	Epigenetic modifications affect gene expression and plant metabolism.	Plants with epigenetic modifications in stress-response genes may have altered metabolic pathways.	

(continued)

Table 20.3 (continued)

Environmental factors				
S.N.	Source	Function	Example	Reference
Stress factors				
S.N.	Source	Function	Example	Qaderi et al. (2023)
1	Drought stress	Drought stress affects plant growth, photosynthesis, and metabolic pathways.	Plants grown in drought conditions may have in drought conditions may have altered metabolic pathways to conserve water.	
2	Temperature stress	Temperature stress affects plant growth, photosynthesis, and metabolic pathways.	Plants grown in high-temperature conditions may have increased respiration rates and altered metabolic pathways.	

20.3 Applications

20.3.1 In Medical Field

Nanoformulation medicament is also used for more specific role to kill cancer cell only in living body does not damage any healthy cell work like as a doctor work inside the body. This process called targeted drug delivery system (Dewanjee et al. 2020).

Targeted drug delivery system forms a special drug delivery system where the drug is injected to body and selectively delivered or targeted only those sites of cells or tissues or organs. As a targeted drug carrier nanoparticle if reduce drug toxicity such as in cancer treatment due to chemotherapy hair lose, nausea & vomiting, fatigue, skin changes, loss of appetite, difficulty eating, bladder and kidney problems.

Nano-emulsions, Nanocapsules, Polymeric nanoparticles, Solid-lipid nanoparticles, Quantum dots, Nanostructured lipid carriers (NLCs), Colloidal gel, Nano diamonds, Dendrimers, Nanocrystals, Liposomes, Carbon nanotubes and etc. (Mills and Needham 1999; Li et al. 2009; Kumari et al. 2022).

Targeted drug delivery, Cancer therapy: Plant-based nanoparticles for targeted delivery of anticancer drugs. Some examples of Nano capsules are: Doxil (liposomal doxorubicin for cancer treatment), Abraxane (nanoparticle albumin-bound paclitaxel for cancer treatment), Nanocrystal (nanoparticle—based pharmaceuticals for improved bioavailability) and etc. (Khan and Gurav 2018), Infectious disease treatment (Rathod et al. 2022), Neurological disorders: Plant-based nanoparticles for brain—targeted delivery (Moradi et al. 2020; Kanojia et al. 2023), Gene therapy: Plant-based Nano formulations for gene delivery (Kokkinis et al. 2024), Infectious diseases: Nano formulations for antibiotic delivery (Murthy et al. 2021), Wound healing: Nano formulations for topical delivery of growth factors. (Garg et al. 2023), Vaccine delivery (Naseer et al. 2023), Tissue engineering: Plant-based Nano formulations for scaffold development (Menaa et al. 2014; Yadav et al. 2022), Diagnostic imaging: Plant-based nanoparticles for imaging agents. (Han et al. 2022), Improved bioavailability and biocompatibility, Enhanced efficiency through solubility, Reduced toxicity (Thoms et al. 2024), Controlled or sustained released.

20.3.2 Agriculture

It is the most important sector or a field to surviving large population in past also it applies and in present also and in future it becomes the major factor to protect our civilization from vanish. When we utilize the technology in plant-based Nano formulations to enhance the scale production of food, improve plant health as well as soil health, and reduce the environmental impacts.

Pesticide delivery: Plant-based Nano formulations for targeted pesticide delivery and controlled manner, reducing the amount of chemical used and minimizing environmental impact.

Crop production: Plant-based nanoparticles for disease resistance, reducing the need for chemical pesticides and maintaining crop yields.

Fertilizers: Nano formulations for controlled release of nutrients, deliver fertilizers in a slow-release manner, reducing the amount of fertilizer used and minimizing environmental impact. (Palanisamy et al. 2021; Rai et al. 2018; Chen 2018; Chhipa 2019).

Table 20.4 Some examples of common plants used in plant-based Nano formulations

S.N.	Plants used in plant-based NPs	Used	Reference
1	Turmeric	Used for its anti-inflammatory and antioxidant properties.	Wickramasinghe et al. (2022) and Kokkinis et al. (2024)
2	Ginger	Used for its anti-inflammatory and antioxidant properties.	
3	Garlic	Used for its antibacterial and antiviral properties.	
4	Onion	Used for its antibacterial and antiviral properties.	
5	Cinnamon	Used for its antibacterial and antiviral properties.	
6	Green tea	Used for its antioxidant and anti-aging properties.	
7	Aloe vera	Used for its anti-inflammatory and antioxidant properties.	
8	Ginkgo biloba	Used for its antioxidant and anti-aging properties.	
9	St. John's Wort	Used for its antidepressant and anti-anxiety properties.	
10	Echinacea	Used for its immune-boosting properties.	

Types of Plant-based nano-fertilizers, Nanoparticle-based fertilizers (e.g., zinc, copper, iron), Nano-emulsion-based fertilizers (e.g., plant oil-based), Nano-encapsulated fertilizers (e.g., slow-release nutrients), Bio-Nano fertilizers (e.g., microbial-based) and etc.

Soil remediation: Plant-based Nano formulations can be used to remediate contaminated soil, improving soil health and fertility. Precision agriculture: Plant-based Nano formulations can be used in precision agriculture to deliver targeted and controlled amounts of pesticides, fertilizers, and other agricultural inputs.

Targeted pesticide delivery by P-B nano formulation fertilizers enhanced nutrient uptake, improved crop yield quality as well as quantity it increased, by providing targeted and controlled delivery of agricultural inputs, Disease resistance, Precision farming, Improved soil health, plant-based Nano formulations can improve soil health by providing targeted and controlled delivery of nutrients and other agriculture inputs, Reduced environmental impacts of agriculture by using plant-based Nano formulations. It helps to minimizing the use of chemical pesticides and fertilizers, increased food safety, plant-based Nano formulations can improve food safety by reducing the use of chemical pesticides and fertilizers.

Alginate-based Nano formulations for soil remediation, Bio-Nano fertilizer from microbial-based formulation, Nano-Zn fertilizer from neem oil, Nano-Cu fertilizer from algae extract, Nano-Fe fertilizer from soy protein, Nano-NPK fertilizer from plant oil-based emulsion (Singh et al. 2015; Boutchuen et al. 2019; Mittal et al. 2020; Semenova et al. 2024).

20.3.3 Food Industry

To increase the shelf lives of food like vegetable and fruits, then there is need of packing to preserve it for long time for surviving for this we need a biodegradable and sustainable food industry. Plant-based Nano formulations in food industry helps to survive in harsh disaster. For example, in food packaging, food preservation, food safety, food texture and flavor and food fortification. (Tsekhmistrenko et al. 2021; Zhu et al. 2021) Functional foods: Nanoencapsulation of bioactive compounds (e.g.—probiotics, omega-3 fatty acids), Food preservation: Improved food shelf-life, Enhanced nutritional value, Personalized nutrition, Neutraceuticals: Plant-based Nano formulations for enhanced bioavailability.

Nano-sensors for food safety: It is used to develop nano-sensors to detect the foodborne pathogens and contaminants, Nano capsules for food preservation: It can be used to develop Nano capsules that release antimicrobial agents, such as essential oils, to preserve food. Plant-based material used is starch, Nano-encapsulated omega-3 fatty acids: It is used to encapsulate omega-3 fatty acids from flaxseed or chia seeds, enhancing their bioavailability and stability. Plant-based material used is starch, alginate, and chitosan. Plant-based nano-packaging: It can be used to develop biodegradable and sustainable food packing materials. Plant-based material used is cellulose, Nano-emulsions for food fortification: It can be used to develop nano-emulsions that encapsulate nutrients, such as vitamins and minerals, for food fortification. Plant-based material used is cellulose, alginate and starch and etc. Benefits include Enhanced nutrition; Improved food safety; Increased shelf life; Improved food texture and flavor; Sustainable and biodegradable.

20.3.4 Environmental Application

Plant-based Nano formulations are innovative products to save our environment from type of pollution, it combines the benefits of nanotechnology with the sustainability of plant-based materials to protect nature for future generation. It can play a significant role in environmental protection by providing sustainable solutions for various environmental challenges like: Phytoremediation; Nano fertilizers; Nano pesticides; Soil conservation by remediation: It can be used to clean up contaminated soil by removing pollutants and toxins; Water purification: It can be used to remove pollutants and contaminants from water, making it safe for drinking and other uses (Rajwade et al. 2022); Air purification: It can be used to remove pollutants and contaminants from the air, improving indoor and outdoor air quality; Waste management: It can be used to develop biodegradable packing materials and to improve waste management practices; Climate change mitigation: It can be used to develop sustainable energy solutions, such as biofuels and solar cells (He et al. 2023).

Some examples are listed here: Nano-cellulose for oil spill cleanup, water purification and soil remediation, Phyto-nanoparticles for air purification, Chitosan-based nanoparticles for heavy metal removal in water purification and air purification, Plant-based nano-enzymes for plastic degradation, Starch is used in water purification and air purification, Plant proteins are used in water purification and soil remediation, Alginate is used in water purification and soil remediation and etc. Benefits include Biodegradable; Targeted delivery; Enhanced bioavailability; Sustainable; It doesn't harm animal, birds and all living creatures; Cost-effective; Reduced environmental impacts.

20.3.5 Cosmetics Industry

The cosmetic industry is a significant sector contain great potential that utilizes plant-based Nano formulations to develop innovative and effective products. Nowadays its demand increases rapidly.

Skin care: It can be used to develop skincare products that provide improved hydration, anti-aging benefits, Nano emulsions for skincare, natural serum, sunscreens and skin protection. Haircare: It can be used to develop haircare products that provide improved moisturization, strengthening, and color protection. Makeups: It can be used to develop makeup products that provide improved pigmentation, texture, and wear. Sunscreen: It can be used to develop sunscreens that provide improved UV protection, moisturization, and skin comfort. Anti-aging: It is used to develop anti-aging products that provide improved skin elasticity, firmness, and wrinkle reduction (Paiva-Santos et al. 2021; Setapar et al. 2022).

Plant-Based Materials are—Phytochemicals is derived from plants, used in skincare and hair care products for antioxidant, anti-inflammatory, and antimicrobial benefits. Green tea extract is derived from green tea leaves, used in skincare products for antioxidant and anti-inflammatory benefits. Vitamin C is derived from plants, used in skincare and hair care products for moisturization and nourishment benefits. Chitosan is derived from crustacean shells, used in skincare products for skin repair and regeneration benefits. Plant oils is derived from plants used in skincare and hair care products for moisturization and nourishment benefits and etc.

Benefits include Enhanced skin penetration; Improved efficacy; Reduced irritation; Sustainable and eco-friendly; Cost-effective (Nhani et al. 2024).

Some examples of Plant-based Nano formulations in cosmetics: Plant-based Nano capsules for hair care, Nano-encapsulated green tea, Plant-based Nano emulsion sunscreen, Chitosan-based Nano gel for skin, Nano-encapsulated vitamin C for skin and etc. (Setapar et al. 2022).

There are some companies which are interested in developing Plant-based Nano formulations in cosmetics industry are: L'Oreal, Unilever, Procter & Gamble, BASF, Crode, Evonik, Gattefosse', Indena, Lipotec, Sederma (Raj et al. 2012).

20.3.6 Wastewater Treatment

Wastewater treatment using plant-based Nano formulation having a great potential for future sustainability. It is the combination of both nanotechnology and plant-based materials. Today's era is full of pollution from various sources such as increasing mismanagement of population and industrial processes direct release in water system. Untreated wastewater can harm the living organism in aquatic as well as terrestrial bodies.

Metal oxides like Titanium dioxide (TiO_2), Magnesium oxide (MgO), Calcium oxide (CaO), and Zinc Oxide (ZnO). In some research shows that ZnO having optical property, magnetic activity, high isoelectric point, biocompatibility, quick electron transfer, piezoelectricity, excellent thermal and chemical stability, and gas sensing properties, adsorption ability, greater catalytic efficiency, thus employed as sensors, fungicide, in rubber processing, paints, wastewater treatment, catalysis, solar energy conversion, cosmetics, fibers, drug delivery, and as an antibacterial.

The best part is that we can extract this metal oxide from natural process which is less toxic, more stable and environment friendly called as plant-based Nano formulation.

Some examples of plants are *Cassiafistula, Amomum longiligulare, Cannabis Jatropha Curcasa Alovera* and *Tinosporacordifolia,* waste Pineapple peels, *Melia azedarach*, Neem, *typha latifolia*, Zinc oxide nanoparticles, phytogenic magnetic nanoparticles, biopolymer-coated metal nanoparticles, silver-impregnated cyclodextrin nanocomposites, silver nanoparticles in Aloe vera plant extract and Photo catalytic titanium nanoparticle (Singh et al. 2020).

Applications in Wastewater Treatment: Municipal wastewater treatment; Industrial wastewater treatment; Textile wastewater treatment; Heavy metal removal; Organic pollutant removal; Nutrient removal; Pathogen removal (Eid et al. 2023).

Some examples of plant-based Nano formulation: Phyto-nanoparticles for municipal wastewater treatment. Plant-based Nano-fibers for textile wastewater treatment. Alginate-based Nano capsules for pathogen removal. Chitosan-based nanoparticles for heavy metal removal. Cellulose-based Nano emulsions for organic pollutant removal.

20.4 Recent Trends

20.4.1 Nano Technology-Based Precision Agriculture

Emerging technology such as Plant-based Nano formulation in agriculture giving it again a nanoparticle revolution after green revolution because of its high impact. It is an advanced solution of agricultural issues. Researchers are more interested in their affordability and eco-friendly points when they compared it with other chemical synthesis available in market. Plant-based Nano formulation shows high potential to overcome the present and future challenges in an agriculture. High demand of food is completed by large scale production of food market and in sustainable way. In this modern world the demand is fulfilled by nanotechnology precisely in agricultural field.

20.4.2 Plant-Based Nano Vaccines

Development of Plant-based Nano formulation for SARS-CoV-2 Vaccine and a Tabacco Mosaic Virus.

Stopping coronavirus in COVID-19 pandemic the Researchers construct a vaccine called CoV-RBD 121-NP from plant which is comprised of the SARS-CoV-2 receptor-binding domain (RBD) of the spike glycoprotein (S) fused to a human IgG1 F_c domain (COV-RBD121) and conjugated to a modified tobacco mosaic virus (TMV) nanoparticle. This vaccine contains inactivated (through UV inactivation) TMV viral RNA. Similar TMV NP have been tested in clinical trials with Children for malaria. In this vaccine there is no harsh conditions applied for storage (−20 to −80 °C) like Pfizer, BioTech and Moderna. It is stable under Standard refrigeration or even at room temperature in urgent need (Setapar et al. 2022; Singh et al. 2020).

20.4.3 Phyto-Nanomedicine

Recent our lifestyle mainly based on the sugar consumption food items. Which produce spike rise of the glucose levels in blood and this harm our bodies caused Diabetes. In research it is shown that many Plant-derived secondary metabolites contain antidiabetic components, like glucose absorption suppression, restoring β cells, improve insulin expression, insulin resistance reversal, enhance of utilization of glucose and regulating metabolism of carbohydrate and lipid.

In diabetes there is need of long-term treatment is required. Therefore, for more research and development pharmacotherapeutic agents needed patient feedback in developing process for managing diabetes.

From ancient time till now we use Herbal remedies in managing type 2 diabetes mellitus (type 2 DM) but there are some limitations such as low stability, limited absorption & not much development of therapeutic agents, this is all limitations overcome by advancing it with nanoscience and nanotechnology term referred for this is Plant-based Nano formulation. (Dewanjee et al. 2020; Wickramasinghe et al. 2022).

20.4.4 Green Nanotechnology

Green science helps to sustainably decrease the amount of hazardous waste released to the environment. In this direction, green science incorporates many research fields, such as the study of green solvents, bio-based transformations and materials, alternative energy science, molecular self-assembly, next-generation catalyst design, and molecular designs.

20.4.5 Tissue Engineering Nanomedicines Based on Natural Sources

Nanomedicines ready from Tissue engineering is very advance level and it is a emerging field nowadays for research point of view. It has been pursued to regenerate or repair damaged tissues that supports adhesion, cell recruitment, differentiation and proliferation which is required in this field. Plant-based nanomedicines are non-toxic, biocompatible, three—dimensional substrates and biodegradable for this tissue engineering.

20.4.6 Industry Predictions

Plant-based Nano formulations market to reach $ 10 Billion by 2050 (Eid et al. 2023). Compound Annual Growth Rate (CAGR) of 20–25% (2025–2030). Expansion into emerging markets (e.g., Asia, Latin America). Increased partnerships between industry and academia. Growing demand for sustainable and eco-friendly products.

20.5 Future Directions

Plant-nanoparticle interactions: Plant-based Nano formulations is used to promote sustainable agriculture practices and reduce environmental impacts, it also improves the crop yielding and helps to reduce the use of harmful chemical pesticides and fertilizers. Standardization of plant-based materials. Development of scalable production methods. Investigation of p-b nf for various apps. Clinical trials and regulatory approvals. Regulatory agencies: Such as the FDA, EPA, and EU's ECHA, regulate the use of plant-based Nano formulations in various products. Labelling and disclosure: Manufacturers must label and disclose the use of plant-based Nano formulations in their products. Nano-biotechnology for plant improvement. Plant based Nano formulations for human health and its Environmental impact assessments. Toxicity and safety assessments. Standardized testing protocols: Needed to ensure consistent and accurate safety assessments.

More research on long-term effects: Necessary to fully understand the potential risks and benefits of plant-based Nano formulations. Development of safer alternatives: Encouraged to minimize potential risks associated with plant-based Nano formulations. In vitro and in vivo studies Conducted to assess toxicity and safety of plant-based Nano formulations. Development of sustainable Nano formulations methods. Personalized medicine, Stimuli-responsive Nano formulations, Development of Hybrid Nano formulations, Nanotechnology-based therapeutics, Point-of-care diagnostics. Use of machine learning and artificial intelligence for design and optimization.

20.6 Challenges

Scalability; Regulatory frameworks; Scalability and cost: Trends of recent increasing number of population and limit sources we have, so it is important to scale up production and reducing costs are essential for commercial applications. Toxicity concerns and Toxicological aspects of Plant-based Nano formulations synthesized (Noga et al. 2023).

According to data, the factors depend on the effect of dose and time of exposure to the toxic substance including particles size, shape and the functional surface effect.

Toxicology of Plant-Based Nano formulations synthesized:

According to data, the factors depend on the effect of dose and time of exposure to the Types of Toxicity: Acute toxicity: Adverse effects occurring within a short period of exposure. Chronic toxicity: Adverse effects occurring after prolonged or repeated exposure. Genotoxicity: Damage to genetic material, potentially leading to mutations or cancer. Cytotoxicity: Damage to cells, potentially leading to cell death.

Factors Influencing Toxicity involves Particle size: Smaller particles may be more toxic due to increased surface area. Surface chemistry: The surface properties of nanoparticles can affect their interactions with cells and biological molecules. Dose and exposure duration: Higher doses and longer exposure times may increase toxicity. Route of administration: Different routes of administration (e.g., oral, topical, inhalation) may affect toxicity.

Plant-Based Nano formulations with Potential Toxicity some examples are: Silver nanoparticles: May cause cytotoxicity, genotoxicity, and oxidative stress. Zinc oxide nanoparticles: May cause cytotoxicity, genotoxicity, and skin irritation. Titanium dioxide nanoparticles: May cause cytotoxicity, genotoxicity, and oxidative stress. Copper oxide nanoparticles: May cause cytotoxicity, genotoxicity, and oxidative stress.

Nano formulations and Their Potential Toxicity—Silver nanoparticles from tea tree oil: May cause cytotoxicity, genotoxicity, and oxidative stress. Zinc oxide nanoparticles from aloe vera: May cause cytotoxicity, genotoxicity, and skin irritation. Curcumin nanoparticles from turmeric: Generally considered safe, but high doses may cause gastrointestinal upset (Hassani et al. 2020b). Quercetin nanoparticles from apples: Generally considered safe, but high doses may cause kidney damage. Green tea catechin nanoparticles: Generally considered safe, but high doses may cause liver damage (Radeva-Ilieva et al. 2025).

Nano formulations in Commercial Products—L'Oréal's Nano-R: A skincare product containing silver nanoparticles from tea tree oil. Neutrogena's Hydrating Facial Moisturizer: A skincare product containing zinc oxide nanoparticles from aloe vera. Jarrow Formulas' Curcumin: A dietary supplement containing curcumin nanoparticles from turmeric. Thorne Research's Quercetin: A dietary supplement containing quercetin nanoparticles from apples. The Body Shop's Green Tea Serum: A skincare product containing green tea catechin nanoparticles (Radeva-Ilieva et al. 2025). Some issues regarding it, such as: Stability issues, High production costs, Standardization, Public acceptance, Intellectual property, Protection and etc. (Royal et al. 2021; Singh et al. 2025).

Research institutions and universities; Cosmetic companies (e.g., L'Oréal, Unilever); Biotech companies (e.g., Nano cell Therapeutics); Pharmaceutical companies (e.g., Pfizer, Novartis); Agricultural companies (e.g., Monsanto, Syngenta); Government agencies (Setapar et al. 2022; Royal et al. 2021; Singh et al. 2025).

In recent development of technologies such as Machine learning and artificial intelligence using to enhance plant-based Nano formulations. In September 2022,

International Conference on Machine Learning, Optimization, and Data Science (LOD) present there so many papers and presentations regarding various fields including machine learning as well. It creates the framework that integrates molecular dynamics simulations, molecular docking, and experiment data to predict the efficacy of Nano formulations. It helps to improve the designs and optimization problems of Nano formulations (Fantatto et al. 2025).

If we see the 17 sustainable goals it fulfills every goal to protect nature and mother earth.

Plant-based Nanoformulations have promising future across various fields:

Short-Term Prospects (2025–2030) Increased adoption in pharmaceuticals and cosmetics. Expanded applications in food and nutrition. Growing demand for sustainable agriculture solutions. Development of new plant-based materials. Improvements in scalability and manufacturing process (Khan et al. 2021).

Mid-Term Prospects (2030–2040) Integration with emerging technologies (e.g. Al, loT); Personalized medicine and nutrition using plant-based Nanoformulations; Increased focus on environmental applications; Enhanced understanding of plant-based Nanoformulations interactions; Regulatory framework development (Khan et al. 2021).

Long-Term Prospects (2040–2050) Potential for plant-based Nanoformulations to address global challenges (e.g., climate change, pandemics); Widespread adoption in consumer goods; Sustainable agriculture and food production; Revolutionary changes in healthcare and medicine; Development of plant-based Nano formulations-based therapeutics (Khan et al. 2021).

References

Ahmed TA, Aljaeid BM (2016) Preparation, characterization, and potential application of chitosan, chitosan derivatives, and chitosan metal nanoparticles in pharmaceutical drug delivery. Drug Des Devel Ther 10:483–507

Antunes Filho S, Dos Santos MS, Dos Santos OAL, Backx BP, Soran ML, Opriş O, Lung I, Stegarescu A, Bououdina M (2023) Biosynthesis of nanoparticles using plant extracts and essential oils. Molecules 28(7):3060

Bodnar M, Hartmann JF, Borbely J (2005) Preparation and characterization of chitosan-based nanoparticles. Biomacromolecules 6(5):2521–2527

Boutchuen A, Zimmerman D, Aich N, Masud AM, Arabshahi A, Palchoudhury S (2019) Increased plant growth with hematite nanoparticle fertilizer drop and determining nanoparticle uptake in plants using multimodal approach. J Nanomater 2019(1):6890572

Chen H (2018) Metal based nanoparticles in agricultural system: behavior, transport, and interaction with plants. Chem Spec Bioavailab 30(1):123–134

Chhipa H (2019) Applications of nanotechnology in agriculture. In: Methods in microbiology, vol 46. Academic, pp 115–142

Dada M, Popoola P (2024) Aloe vera hydrogel for supercooling applications: a review. Discov Mater 4(1):p.10

Dewanjee S, Chakraborty P, Mukherjee B, De Feo V (2020) Plant-based antidiabetic nanoformulations: the emerging paradigm for effective therapy. Int J Mol Sci 21(6):2217

Eid AM, Fouda A, Hassan SED, Hamza MF, Alharbi NK, Elkelish A, Alharthi A, Salem WM (2023) Plant-based copper oxide nanoparticles; biosynthesis, characterization, antibacterial activity, tanning wastewater treatment, and heavy metals sorption. Catalysts 13(2):348

Einhorn-Stoll U, Archut A, Eichhorn M, Kastner H (2021) Pectin-plant protein systems and their application. Food Hydrocoll 118:106783

Elshafie HS, Camele I, Mohamed AA (2023) A comprehensive review on the biological, agricultural and pharmaceutical properties of secondary metabolites based-plant origin. Int J Mol Sci 24(4):3266

Fang C, Fernie AR, Luo J (2019) Exploring the diversity of plant metabolism. Trends Plant Sci 24(1):83–98

Fantatto RR, Constantini JV, Politi FA, Sorrechia R, Medeiros CC, Luiz MT, Bechara GH, de Souza Chagas AC, Chorilli M, Pietro RC (2025) Current tick control strategies and prospects for using nanotechnology as an efficient alternative—a review. Vet Sci 12(2):163

Garg SS, Dubey R, Sharma S, Vyas A, Gupta J (2023) Biological macromolecules-based nanoformulation in improving wound healing and bacterial biofilm-associated infection: a review. Int J Biol Macromol 247:125636

Grenha A, Gomes ME, Rodrigues M, Santo VE, Mano JF, Neves NM, Reis RL (2010) Development of new chitosan/carrageenan nanoparticles for drug delivery applications. J Biomed Mater Res A 92(4):1265–1272

Gün Gök Z, Karayel M, Yiğitoğlu M (2021) Synthesis of carrageenan coated silver nanoparticles by an easy green method and their characterization and antimicrobial activities. Res Chem Intermed 47:1843–1864

Han HS, Koo SY, Choi KY (2022) Emerging nanoformulation strategies for phytocompounds and applications from drug delivery to phototherapy to imaging. Bioact Mater 14:182–205

Hano C, Abbasi BH (2021) Plant-based green synthesis of nanoparticles: production, characterization and applications. Biomolecules 12(1):31

Hassani A, Azarian MMS, Ibrahim WN, Hussain SA (2020a) Preparation, characterization and therapeutic properties of gum arabic-stabilized gallic acid nanoparticles. Sci Rep 10(1):17808

Hassani A, Mahmood S, Enezei HH, Hussain SA, Hamad HA, Aldoghachi AF, Hagar A, Doolaanea AA, Ibrahim WN (2020b) Formulation, characterization and biological activity screening of sodium alginate-gum arabic nanoparticles loaded with curcumin. Molecules 25(9):2244

He J, Li J, Gao Y, He X, Hao G (2023) Nano-based smart formulations: a potential solution to the hazardous effects of pesticide on the environment. J Hazard Mater 456:131599

Heya MS, García-Coronado PL, Cordero-Díaz A, Torres-Hernández AD, Pérez-Narváez OA (2022) Biomedical application of nanoformulated plant. Int J Pharm Sci Res 1:1–12

Ikram S (2015) Synthesis of gold nanoparticles using plant extract: an overview. Nano Res 1(1):5

Iravani S (2020) Plant gums for sustainable and eco-friendly synthesis of nanoparticles: recent advances. Inorg Nano-Metal Chem 50(6):469–488

Kanojia N, Thapa K, Kaur G, Sharma A, Puri V, Verma N (2023) Update on therapeutic potential of emerging nanoformulations of phytocompounds in Alzheimer's and Parkinson's disease. J Drug Delivery Sci Technol 79:104074

Kaur S, Samota MK, Choudhary M, Choudhary M, Pandey AK, Sharma A, Thakur J (2022) How do plants defend themselves against pathogens-biochemical mechanisms and genetic interventions. Physiol Mol Biol Plants 28(2):485–504

Khan T, Gurav P (2018) PhytoNanotechnology: enhancing delivery of plant based anti-cancer drugs. Front Pharmacol 8:1002

Khan N, Ray RL, Sargani GR, Ihtisham M, Khayyam M, Ismail S (2021) Current progress and future prospects of agriculture technology: gateway to sustainable agriculture. Sustainability 13(9):4883

Kokkinis S, Singh M, Paudel KR, De Rubis G, Saeid AB, Jessamine V, Datsyuk J, Singh SK, Vishwas S, Adams J, Hansbro PM (2024) Plant-based therapeutics for chronic obstructive pulmonary diseases: nanoformulation strategies to overcome delivery challenges. Food Biosci 58:103761

Kumari S, Goyal A, Sönmez Gürer E, Algın Yapar E, Garg M, Sood M, Sindhu RK (2022) Bioactive loaded novel nano-formulations for targeted drug delivery and their therapeutic potential. Pharmaceutics 14(5):1091

Li DC, Zhong XK, Zeng ZP, Jiang JG, Li L, Zhao MM, Yang XQ, Chen J, Zhang BS, Zhao QZ, Xie MY (2009) Application of targeted drug delivery system in Chinese medicine. J Control Release 138(2):103–112

Menaa F, Menaa A, Menaa B (2014) Polyphenols nano-formulations for topical delivery and skin tissue engineering. In: Polyphenols in human health and disease. Academic, pp 839–848

Mills JK, Needham D (1999) Targeted drug delivery. Expert Opin Ther Pat 9(11):1499–1513

Mittal AK, Chisti Y, Banerjee UC (2013) Synthesis of metallic nanoparticles using plant extracts. Biotechnol Adv 31(2):346–356

Mittal D, Kaur G, Singh P, Yadav K, Ali SA (2020) Nanoparticle-based sustainable agriculture and food science: recent advances and future outlook. Front Nanotechnol 2:579954

Moradi SZ, Momtaz S, Bayrami Z, Farzaei MH, Abdollahi M (2020) Nanoformulations of herbal extracts in treatment of neurodegenerative disorders. Front Bioeng Biotechnol 8:p.238

Murthy NTV, Agrahari V, Chauhan H (2021) Polyphenols against infectious diseases: controlled release nano-formulations. Eur J Pharm Biopharm 161:66–79

Naseer F, Ahmad T, Kousar K, Kakar S, Gul R, Anjum S, Shareef U (2023) Formulation for the targeted delivery of a vaccine strain of oncolytic measles virus (OMV) in hyaluronic acid coated thiolated chitosan as a green nanoformulation for the treatment of prostate cancer: a viro-immunotherapeutic approach. Int J Nanomedicine 18:185

Nhani GBB, Di Filippo LD, de Paula GA, Mantovanelli VR, da Fonseca PP, Tashiro FM, Monteiro DC, Fonseca-Santos B, Duarte JL, Chorilli M (2024) High-tech sustainable beauty: exploring nanotechnology for the development of cosmetics using plant and animal by-products. Cosmet 11(4):112

Noga M, Milan J, Frydrych A, Jurowski K (2023) Toxicological aspects, safety assessment, and green toxicology of silver nanoparticles (AgNPs)—critical review: state of the art. Int J Mol Sci 24(6):5133

Paiva-Santos AC, Herdade AM, Guerra C, Peixoto D, Pereira-Silva M, Zeinali M, Mascarenhas-Melo F, Paranhos A, Veiga F (2021) Plant-mediated green synthesis of metal-based nanoparticles for dermopharmaceutical and cosmetic applications. Int J Pharm 597:120311

Palanisamy S, Subramaniam BS, Thangamuthu S, Nallusamy S, Rengasamy P (2021) Review on agro-based nanotechnology through plant-derived green nanoparticles: synthesis, application and challenges. J Environ Sci Public Health 5(1):77–98

Paques JP, van der Linden E, van Rijn CJ, Sagis LM (2014) Preparation methods of alginate nanoparticles. Adv Colloid Interf Sci 209:163–171

Pfister B, Zeeman SC (2016) Formation of starch in plant cells. Cell Mol Life Sci 73:2781–2807

Qaderi MM, Martel AB, Strugnell CA (2023) Environmental factors regulate plant secondary metabolites. Plants 12(3):447

Radeva-Ilieva M, Stoeva S, Hvarchanova N, Georgiev KD (2025) Green tea: current knowledge and issues. Foods 14(5):745

Rai PK, Kumar V, Lee S, Raza N, Kim KH, Ok YS, Tsang DC (2018) Nanoparticle-plant interaction: implications in energy, environment, and agriculture. Environ Int 119:1–19

Raj S, Jose S, Sumod US, Sabitha M (2012) Nanotechnology in cosmetics: opportunities and challenges. J Pharm Bioallied Sci 4(3):186–193

Rajwade D, Jaiswal S, Baghel VS, Kumar H (2022) Isolation, identification and characterization of novel azo dye degrading bacteria from the industrial effluents of Raipur city, Chhattisgarh. Nat Environ Pollut Technol 21(3):1175–1182

Rathod S, Arya S, Kanike S, Shah SA, Bahadur P, Tiwari S (2022) Advances on nanoformulation approaches for delivering plant-derived antioxidants: a case of quercetin. Int J Pharm 625:122093

Rausch T, Greiner S (2004) Plant protein inhibitors of invertases. Biochim Biophys Acta 1696(2):253–261

Royal JM, Simpson CA, McCormick AA, Phillips A, Hume S, Morton J, Shepherd J, Oh Y, Swope K, DeBeauchamp JL, Webby RJ (2021) Development of a SARS-CoV-2 vaccine candidate using plant-based manufacturing and a tobacco mosaic virus-like nano-particle. Vaccine 9(11):1347

Semenova NA, Burmistrov DE, Shumeyko SA, Gudkov SV (2024) Fertilizers based on nanoparticles as sources of macro-and microelements for plant crop growth: a review. Agronomy 14(8):1646

Setapar SHM, Ahmad A, Jawaid M (eds) (2022) Nanotechnology for the preparation of cosmetics using plant-based extracts. Elsevier

Severino P, da Silva CF, Andrade LN, de Lima Oliveira D, Campos J, Souto EB (2019) Alginate nanoparticles for drug delivery and targeting. Curr Pharm Des 25(11):1312–1334

Shogren R, Wood D, Orts W, Glenn G (2019) Plant-based materials and transitioning to a circular economy. Sustainable Prod Consumption 19:194–215

Siddiqui MH, Al-Whaibi MH, Mohammad F (2015) Nanotechnology and plant sciences. Springer, Cham, pp 978–973. https://doi.org/10.1007/978-3-319-14502-0_1

Singh A, Singh NB, Hussain I, Singh H, Singh SC (2015) Plant-nanoparticle interaction: an approach to improve agricultural practices and plant productivity. Int J Pharm Sci Invent 4(8):25–40

Singh P, Yadav SK, Kuddus M (2020) Green nanomaterials for wastewater treatment. In: Green nanomaterials: processing, properties, and applications. Springer Singapore, Singapore, pp 227–242

Singh GB, Sethi A, Kumari M, Panigrahi P, Chaudhary P, Sharma A (2025) Biorefining agro-industrial waste into green nanomaterials for sustainable agriculture. In: Nano-microbiology for sustainable development. Springer, Cham, pp 105–137

Somerville C (2006) Cellulose synthesis in higher plants. Annu Rev Cell Dev Biol 22(1):53–78

Sosnowski J, Truba M, Vasileva V (2023) The impact of auxin and cytokinin on the growth and development of selected crops. Agriculture 13(3):724

Thankappan DA, Raman HK, Jose J, Sudhakaran S (2020) Plant-mediated biosynthesis of zein–pectin nanoparticle: preparation, characterization and in vitro drug release study. J King Saud Univ Sci 32(2):1785–1791

Thoms S, Gonsalves RA, Jose J, Zyoud SH, Prasad AR, Garvasis J (2024) Plant-based synthesis, characterization approaches, applications and toxicity of silver nanoparticles: a comprehensive review. J Biotechnol 394:135–149

Tsekhmistrenko SC, Bityutskyy V, Tsekhmistrenko O, Merzlo S, Tymoshok N, Melnichenko A, Polishcuk S, Demchenko A, Yakymenko I (2021) Bionanotechnologies: synthesis of metals'nanoparticles with using plants and their applications in the food industry: a review. J Microbiol Biotechnol Food Sci 10(6):e1513

Velavan S (2015) Phytochemical techniques – a review. World J Sci Res 1(2):80–91

Wickramasinghe ASD, Kalansuriya P, Attanayake AP (2022) Nanoformulation of plant-based natural products for type 2 diabetes mellitus: from formulation design to therapeutic applications. Curr Ther Res 96:100672

Yadav N, Mudgal D, Anand R, Jindal S, Mishra V (2022) Recent development in nanoencapsulation and delivery of natural bioactives through chitosan scaffolds for various biological applications. Int J Biol Macromol 220:537–572

Yu S, Xu X, Feng J, Liu M, Hu K (2019) Chitosan and chitosan coating nanoparticles for the treatment of brain disease. Int J Pharm 560:282–293

Zhu M, Huan S, Liu S, Li Z, He M, Yang G, Liu S, McClements DJ, Rojas OJ, Bai L (2021) Recent development in food emulsion stabilized by plant-based cellulose nanoparticles. Curr Opin Colloid Interface Sci 56:101512

Index

V. Ghosh et al. (eds.), *Applications of Plant Bioactive Compounds-Based Nanoformulations*, Nanotechnology in the Life Sciences,
https://doi.org/10.1007/978-3-032-13086-0

MIX
Papier aus verantwortungsvollen Quellen
Paper from responsible sources
FSC® C105338

If you have any concerns about our products,
you can contact us on
ProductSafety@springernature.com

In case Publisher is established outside the EU,
the EU authorized representative is:
Springer Nature Customer Service Center GmbH
Europaplatz 3, 69115 Heidelberg, Germany

Printed by Libri Plureos GmbH
in Hamburg, Germany